Building Secure And Reliable
Network Applications

Building Secure and Reliable Network Applications

Kenneth P. Birman

MANNING

Greenwich
(74° w. long.)

For electronic browsing or ordering of this book, see:

    http://www.browsebooks.com

The publisher offers discounts on this book when ordered in quantity.

    For more information please contact:

    Special Sales Department

    Manning Publications Co.

    3 Lewis Street

    Greenwich, CT 06830

    or

    ORDERS@MANNING.COM

    Fax: (203) 661-9018

Copyright © 1996 by Manning Publications Co.

All rights reserved.

Design and typesetting: Syd Brown

Cover design: Leslie Haimes

Recognizing the importance of preserving what has been written, it is the policy of Manning Publications to have the books they publish printed on acid-free paper and we exert our best efforts to that end.

Library of Congress Cataloging-in-Publication Data

Birman, Kenneth P.

    Building secure and reliable network applications

        p. cm.

    Includes index.

    ISBN 1-884777-29-5 (alk. paper)

    1. Computer networks—Security measures 2. Computer networks—Reliability

I. Title.

TK5105.59.B57   1996

004'.36—dc20   96-30875

                          CIP

Printed in the United States of America

1 2 3 4 5 6 7 8 9 - CR - 98 97 96

TABLE OF CONTENTS

TABLE OF CONTENTS

Trademarks

UNIX is a trademark of Santa Cruz Operations, Inc. CORBA (Common Object Request Broker Architecture) and OMG IDL are trademarks of the Object Management Group. ONC (Open Network Computing), NFS (Network File System), Solaris, Solaris MC, XDR (External Data Representation), and Java are trademarks of Sun Microsystems, Inc. DCE is a trademark of the Open Software Foundation. XTP (Xpress Transfer Protocol) is a trademark of the XTP Forum. RADIO is a trademark of Stratus Computer Corporation. Isis Reliable Software Developer's Kit, Isis Reliable Network File System, Isis Reliable Message Bus, and Isis for Databases are trademarks of Isis Distributed Computing Systems, Inc. Orbix is a trademark of Iona Technologies Ltd. Orbix+Isis is a joint trademark of Iona and Isis Distributed Computing Systems, Inc. TIB (Teknekron Information Bus) and Subject-Based Addressing are trademarks of Teknekron Software Systems (although we use subject-based addressing in a more general sense in this text). Chorus is a trademark of Chorus Systemes, Inc. Power Objects is a trademark of Oracle Corporation. Netscape is a trademark of Netscape Communications. OLE, Windows, Windows New Technology (Windows NT), and Windows 95 are trademarks of Microsoft Corporation. Lotus Notes is a trademark of Lotus Computing Corporation. Purify is a trademark of Highland Software, Inc. Proliant is a trademark of Compaq Computers, Inc. VAXClusters, DEC MessageQ, and DECsafe Available Server Environment are trademarks of Digital Equipment Corporation. MQSeries and SP2 are trademarks of International Business Machines. PowerBuilder is a trademark of PowerSoft Corporation. Visual BASIC is a trademark of Microsoft Corporation. Ethernet is a trademark of Xerox Corporation.

Other products and services mentioned in this document are covered by the trademarks, service marks, or product names as designated by the companies that market those products. The author respectfully acknowledges any that may not have been included.

Preface

This book is dedicated to my family, for their support and tolerance over the two-year period during which it was written. The author is grateful to so many individuals for their technical assistance with aspects of the book's development, that to try to list them one by one would certainly omit someone whose role was vital. Instead, let me just thank my colleagues at Cornell, Isis Distributed Systems, and worldwide for their help in this undertaking. I am also grateful to Paul Jones of Isis Distributed Systems and to Francois Barrault and Yves Eychenne of Stratus France and Isis Distributed Systems, France, for providing me with resources needed to work on this book during a sabbatical that I spent in Paris (fall 1995 and spring 1996). Cindy Williams and Werner Vogels provided invaluable help in overcoming some of the details of working at such a distance from home.

A number of reviewers provided feedback on early copies of this text. Thanks are due to Marjan Bace, David Bakken, Robert Cooper, Yves Eychenne, Dalia Malkhi, Raghu Hudli, David Page, David Plainfosse, Henrijk Paszt, John Warne, and Werner Vogels. Raj Alur, Ian Service, and Mark Wood provided help in clarifying some thorny technical questions and are also gratefully acknowledged. Bruce Donald's e-mails on idiosyncrasies of the Web were extremely useful and had a surprisingly large impact on treatment of that topic in this text.

Much of the work reported here was made possible by grants from the U.S. Department of Defense through its Advanced Research Projects Agency, DARPA (administered by the Office of Naval Research, Rome Laboratories, and NASA), and by infrastructure grants from the National Science Foundation. Grants from a number of corporations have also supported this work, including IBM Corporation, Isis Distributed Systems, Inc., Siemens Corporate Research (Munich and New Jersey), and GTE Corporation. I wish to express my thanks to all of these agencies and corporations for their generosity.

The techniques, approaches, and opinions expressed here are my own; they may not represent positions of the organizations and corporations that have supported this research.

Introduction

Despite nearly 20 years of progress toward ubiquitous computer connectivity, distributed computing systems have only recently emerged to play a serious role in industry and society. Perhaps this explains why so few distributed systems are reliable in the sense of tolerating failures automatically, guaranteeing properties such as performance or response time, or offering security against intentional threats. In many ways the engineering discipline of reliable distributed computing is still in its infancy.

One might be tempted by a form of circular reasoning, concluding that reliability must not be all that important in distributed systems (otherwise, the pressure to make such systems reliable would long since have become overwhelming). Yet, it seems more likely that we have only recently begun to see the types of distributed computing systems in which reliability is critical. To the extent that existing mission- and even life-critical applications rely upon distributed software, the importance of reliability has perhaps been viewed as a narrow, domain-specific issue. On the other hand, as distributed software is placed into more and more critical applications, where safety or financial stability of large organizations depends upon the reliable operation of complex distributed applications, the inevitable result will be a growing demand for technology developers to demonstrate the reliability of their distributed architectures and solutions. It is time to tackle distributed system reliability in a serious manner. To fail to do so today is to invite catastrophic computer-system failures tomorrow.

At the time of this writing, the sudden emergence of the World Wide Web (variously called the Web, the Information Superhighway, the Global Information Infrastructure, the Internet, or just the Net) is bringing this issue to the forefront. In many respects, reliability in distributed systems is today tied to the future of the Web and the technology base that has been used to develop it. It is unlikely that any reader of this text is not familiar with the Web technology base, which has penetrated the computing industry in record time. A basic premise of our study is that the Web will be a driver for distributed computing, by creating a mass market around distributed computing. However, the term "Web" is often used loosely: Much of the public sees the Web as a single entity that encompasses all the Internet technologies that exist today and that may be introduced in the future. Thus, when we talk about the Web, we are inevitably faced with a much broader family of communication technologies.

It is clear that some form of critical mass has recently been reached: Distributed computing is emerging from its specialized and very limited niche to become a mass-market commodity—something that literally everyone depends on, such as a telephone or an automobile. The Web paradigm brings together the key attributes of this new market in a single package: easily understandable graphical displays, substantial content, unlimited information to draw upon, and virtual worlds in which to wander and work. But the Web is also stimulating growth in other types of distributed applications. In some intangible way, the experience of the Web has caused modern society to suddenly notice the potential of distributed computing.

Consider the implications of a societal transition whereby distributed computing has suddenly become a mass-market commodity. In the past, a mass-market item was something everyone "owned." With the Web, one suddenly sees a type of commodity that everyone "does." For the most part, the computers and networks were already in place. What has changed is the way people see them and use them. The paradigm of the Web is to connect useful things (and many useless things) to the network. Communication and connectivity suddenly seem to be mandatory: No company can possibly risk arriving late for the Information Revolution. Increasingly, it makes sense to believe that if an application *can* be put on the network, someone is thinking about doing so, and soon.

Whereas reliability and indeed distributed computing were slow to emerge prior to the introduction of the Web, reliable distributed computing will be necessary if networked solutions are to be used safely for many of the applications that are envisioned. In the past, researchers in the field wondered why the uptake of distributed computing had been so slow. Overnight, the question has become one of understanding how the types of computing systems that run on the Internet and the Web, or that will be accessed through them, can be made reliable enough for emerging critical uses.

If Web-like interfaces present medical status information and records to a doctor in a hospital, or are used to control a power plant from a remote console, or to guide the decision making of major corporations, reliability of those interfaces and applications will be absolutely critical to the users. Some may have life-or-death implications: If a physician bases a split-second decision on invalid data, the patient might die. Other interfaces may be critical to the efficient function of the organization that uses them: If a bank mismanages risk because of an inaccurate picture of how its investments are allocated, the bank could incur huge losses or even fail. In still other settings, reliability may emerge as a key determinant in the marketplace: The more-reliable product, at a comparable price, may simply displace the less-reliable one. Reliable distributed computing suddenly has broad relevance.

Throughout this book, the term "distributed computing" is used to describe a type of computer system that differs from what could be called a "network computing" system. The distinction illuminates the basic issues with which we will be concerned.

As we use the term here, a *computer network* is a communication technology supporting the exchange of messages among computer programs executing on computational nodes. Computer networks are *data movers*, providing capabilities for sending data from one location to another, dealing with mobility and changing topology, and automating the division of available bandwidth among contending users. Computer networks have evolved over a 20-year period, and during the mid-1990s network connectivity between computer systems became pervasive. Network bandwidth has also increased enormously, rising from hundreds of bytes per second in the early 1980s to millions of bytes per second in the mid-1990s, with gigabyte rates anticipated in the late 1990s and beyond.

Network functionality evolved steadily during this period. Early use of networks was entirely for file transfer, remote log in, and electronic mail or news. Over time, however, the expectations of users and the tools available have changed. The network user in 1996 is likely to be familiar with interactive network browsing tools, such as Netscape's browsing tool, which permit the user to wander within a huge and interconnected network of multimedia information and documents. Tools such as these permit the user to conceive a computer workstation as a window into an immense world of information, accessible using a great variety of search tools, easy to display and print, and linked to other relevant material that may be physically stored halfway around the world and yet accessible at the click of a mouse.

Meanwhile, new types of networking hardware have emerged. The first generation of networks was built using point-to-point connections: To present the illusion of full connectivity to users, the network included a software layer for routing and connection management. Over time, these initial technologies were largely replaced by high-speed, long-distance lines that route through various hubs, coupled to local area networks implemented using multiple access technologies such as Ethernet and FDDI: hardware in which a single "wire" has a large number of computers attached to it, supporting the abstraction of a shared message bus. At the time of this writing, a third generation of technologies is reaching the market: ATM hardware capable of supporting gigabyte communication rates over virtual circuits, mobile connection technologies for the office that will allow computers to be moved without rewiring, and more ambitious mobile computing devices that exploit the nationwide cellular telephone grid for communications support.

As recently as the early 1990s, computer bandwidth over wide area links was limited for most users. The average workstation had high-speed access to a local network, and perhaps the local e-mail system was connected to the Internet, but individual users (especially those working from PCs) rarely had better than 1,600-baud connections available for personal use of the Internet. This picture is changing rapidly today: More and more users have high-speed modem connections to an Internet service provider that offers megabyte-per-second connectivity to remote servers. With the emergence of ISDN services to the home, the last link of the chain will suddenly catch up with the rest. Individual connectivity has thus jumped from 1,600 baud to perhaps 28,800 baud at the time of this writing, and may jump to 1 Mbaud or more in the not-too-distant future. Moreover, this bandwidth has finally reached the PC community, which enormously outnumbers the workstation community.

It has been suggested that technology revolutions are often spurred by discontinuous, as opposed to evolutionary, improvement in a key aspect of a technology. The bandwidth improvements we are now experiencing are so disproportionate with respect to other performance changes (memory sizes, processor speeds) as to fall squarely into the discontinuous end of the spectrum. The sudden connectivity available to PC users is similarly disproportionate to anything in prior experience. The Web is perhaps just the first of a new generation of communication-oriented technologies enabled by these sudden developments.

In particular, the key enablers for the Web were precisely the availability of adequate long-distance communication bandwidth to sustain its programming model, coupled to the evolution of computing systems supporting high-performance graphical displays and sophisticated local applications dedicated to the user. It is only recently that these pieces fell into place. Indeed, the Web emerged as early as it could possibly have done, considering the state of the art in the various technologies on which it depends. Thus, while the Web is clearly a breakthrough—the "killer application" of the Internet—it is also the most visible manifestation of a variety of underlying developments that are also enabling other kinds of distributed applications. It makes sense to see the Web as the tip of an iceberg: a paradigm for something much broader that is sweeping the entire computing community.

As the trend toward better communication performance and lower latencies continues, it is certain to fuel continued growth in distributed computing. In contrast to a computer network, a *distributed computing system* refers to computing systems and applications that cooperate to coordinate actions at multiple locations in a network. Rather than adopting a perspective in which conventional (nondistributed) application programs access data remotely over a network, a distributed system includes multiple application programs that communicate over the network, but that take action at the multiple locations where the applications run. Despite the widespread availability of networking since early 1980, distributed computing has only become common in the 1990s. This lag reflects a fundamental issue: Distributed computing turns out to be much harder than nondistributed or network computing applications, especially if reliability is a critical requirement.

Our treatment explores the technology of distributed computing with a particular bias: to understand why the emerging generation of critical Internet and Web technologies is likely to require very high levels of reliability, and to explore the implications of this for distributed computing technologies. A key issue is to gain some insight into the factors that make it so hard to develop distributed computing systems that can be relied upon in critical settings, and to understand what can be done to simplify the task. In other disciplines, such as civil engineering or electrical engineering, a substantial body of practical development rules exist upon which the designer of a complex system can draw to simplify his or her task. It is rarely necessary for the company that builds a bridge to engage in theoretical analyses of stress or basic properties of the materials used, because the theory in these areas has already been reduced to collections of practical rules and formulas that the practitioner can treat as tools for solving practical problems.

This observation motivated the choice of the cover of the book. The Golden Gate Bridge is a marvel of civil engineering that reflects a very sophisticated understanding of the science of bridge building. Although located in a seismically active area, the bridge is believed capable of withstanding even an extremely severe earthquake. It is routinely exposed to violent winter storms: It may sway but is never seriously threatened. And yet the bridge is also esthetically pleasing: It is one of the truly beautiful constructions of its era. Watching the sun set over the bridge from Berkeley, where I attended graduate school, remains among the most memorable experiences of my life. The bridge illustrates that beauty can also be resilient: a fortunate development, since, otherwise, the failure of the Tacoma Narrows Bridge might have ushered in a generation of bulky and overengineered bridges. The achievement of the Golden Gate Bridge illustrates that even when engineers are confronted with extremely demanding standards, it is possible to achieve solutions that are elegant and lovely at the same time as they are resilient. This is only possible, however, to the degree that there exists an engineering science of robust bridge building.

We can build distributed computing systems that are reliable in this sense, too. Such systems would be secure, trustworthy, and would guarantee availability and consistency even when limited numbers of failures occur. Hopefully, these limits can be selected to provide adequate reliability without excessive cost. In this manner, just as the science of bridge building has yielded elegant and robust bridges, reliability need not compromise elegance and performance in distributed computing.

One could argue that in distributed computing, we are today building the software bridges of the Information Superhighway. Yet, in contrast to the disciplined engineering that enabled the Golden Gate Bridge, as one explores the underlying technology of the Internet and the Web, one discovers a disturbing and pervasive inattention to issues of reliability. It is common to read that the Internet (developed originally by the Defense Department's Advanced Research Projects Agency, DARPA) was built to withstand a nuclear war. Today, we need to adopt a similar mindset as we extend these networks into systems that must support tens or hundreds of millions of Web users, as well as a growing number of hackers whose objectives vary from the annoying to the criminal. We will see that many of the fundamental technologies of the Internet and the Web, although completely reasonable in the early days of the Internet's development, have now started to limit scalability and reliability, and the infrastructure is consequently exhibiting troubling signs of stress.

One of the major challenges, of course, is that use of the Internet has begun to expand so rapidly that the researchers most actively involved in extending its protocols and enhancing its capabilities are forced to work incrementally: Only limited changes to the technology base can be contemplated, and even small upgrades can have very complex implications. Moreover, upgrading the technologies used in the Internet is somewhat like changing the engines on an airplane while it is flying. Jointly, these issues limit the ability of the Internet community to move to a more reliable, secure, and scalable architecture. They create a background against which the goals of this book will not easily be achieved.

In early 1995, I was invited by DARPA to participate in an unclassified study concerning the survivability of distributed systems. Participants included academic experts and experts familiar with the state of the art in such areas as telecommunications, power system management, and banking. This study was undertaken against a backdrop colored by the recent difficulties of the Federal Aviation Agency, which launched a project in the late 1980s and early 1990s to develop a new generation of highly reliable distributed air traffic control software. Late in 1994, after losing a huge sum of money and essentially eliminating all distributed aspects of an architecture that was originally innovative precisely for its distributed reliability features, a prototype of the proposed new system was finally delivered, but with such limited functionality that planning of yet another new generation of software had to begin immediately. Meanwhile, article after article in the national press reported on failures of air traffic control systems, many stemming from software problems and several exposing airplanes and passengers to extremely dangerous conditions. Such a situation can only inspire the utmost concern in regard to the practical state of the art.

Although our study did not focus on the FAA's specific experience, the areas we did study are in many ways equally critical. What we learned is that situations encountered by the FAA's highly visible project are occurring, to a greater or lesser degree, within all of these domains. The pattern is one in which pressure to innovate and introduce new forms of products leads to the increasingly ambitious use of distributed computing systems. These new systems rapidly become critical to the enterprise that developed them: Too many interlocked decisions must be made to permit such steps to be reversed. Responding to the pressures of timetables and the need to demonstrate new functionality, engineers inevitably postpone considerations of availability, security, consistency, system management, and fault tolerance—what we call "reliability" in this text—until late in the game, only to find that it is then very hard to retrofit the necessary technologies into what has become an enormously complex system. Yet, when pressed on these issues, many engineers respond that they are merely following common practice: that their systems use the best generally accepted engineering practice and are neither more nor less robust than the other technologies used in the same settings.

Our group was very knowledgeable about the state of the art in research on reliability. So, we often asked our experts whether the development teams in their area were aware of one result or another in the field. What we learned was that research on reliability has often stopped too early to impact the intended consumers of the technologies we developed. It is common for work on reliability to stop after a paper or two and perhaps a splashy demonstration of how a technology can work. But such a proof of concept often leaves open the question of how the reliability technology can interoperate with the software development tools and environments that have become common in industry. This represents a serious obstacle to the ultimate use of the technique, because commercial software developers necessarily work with commercial development products and seek to conform to industry standards.

This creates a quandary: One cannot expect a researcher to build a better version of a modern operating system or communications architecture—such tasks are enormous and even very large companies have difficulty successfully concluding them. So it is hardly surprising that research results are demonstrated on a small scale. Thus, if industry is not eager to exploit the best ideas in an area such as reliability, there is no organization capable of accomplishing the necessary technology transition.

For example, we will look at an object-oriented technology called the Common Object Request Broker Architecture, or CORBA, which has become extremely popular. CORBA is a structural methodology: a set of rules for designing and building distributed systems so that they will be explicitly described and easily managed, and so that components can be interconnected as easily as possible. One would expect that researchers on security, fault tolerance, consistency, and other properties would embrace such architectures, because they are highly regular and designed to be extensible: Adding a reliability property to a CORBA application should be a very natural step. However, relatively few researchers have looked at the specific issues that arise in adapting their results to a CORBA setting (we'll hear about some of the ones that have). Meanwhile, the CORBA community has placed early emphasis on performance and interoperability, while reliability issues have been dealt with primarily by individual vendors (although, again, we'll hear about some products that represent exceptions to the rule). What is troubling is the sense of disconnection between the reliability community and its most likely users, and the implication that reliability is not accorded a very high value by the vendors of distributed system products today.

Our study contributed toward a decision by the Department of Defense (DoD) to expand its investment in research on technologies for building practical, survivable, distributed systems. This DoD effort will focus both on developing new technologies for implementing survivable systems, and on developing new approaches to hardening systems built using conventional distributed programming methodologies, and it could make a big difference. But one can also use the perspective gained through a study such as this one to look back over the existing state of the art, asking to what degree the technologies we already have in hand can, in fact, be applied to the critical computing systems that are already being developed.

As it happened, I started work on this book during the period when this DoD study was underway, and the presentation that follows is strongly colored by the perspective that emerged from it. Indeed, the study has considerably impacted my own research project. I've come to the personal conclusion that the situation could be much better if developers were simply to begin to think hard about reliability and had greater familiarity with the techniques at their disposal today. There may not be any magic formulas that will effortlessly confer reliability upon a distributed system, but, at the same time, the technologies available to us are in many cases very powerful and are frequently much more relevant to even off-the-shelf solutions than is generally recognized. We need more research on the issue, but we also need to try harder to incorporate what we already know how to do into the software development tools and environments on which the majority of distributed computing applications are now based. This said, it is also clear that researchers will need to start paying more attention to the issues that arise in moving their ideas from the laboratory to the field.

Lest these comments seem to suggest that the solution is in hand, it must be understood that there are intangible obstacles to reliability that seem very subtle and yet rather pervasive. Earlier, it was mentioned that the Internet and the Web are in some ways fundamentally unreliable and that industry routinely treats reliability as a secondary consideration, to be addressed only in mature products and primarily in a fire-fighting mode—for example, after a popular technology is somehow compromised by hackers in a visible way. Neither of these problems will be easy to fix, and they combine to have far-reaching implications. Major standards have repeatedly deferred consideration of reliability issues and security until future releases of the standards documents or prototype platforms. The message sent to developers is clear: Should they wish to build a reliable distributed system, they will need to overcome tremendous obstacles, both internal to their companies and in the search for enabling technologies, and they will find relatively little support from the vendors that sell standard computing platforms.

The picture is not uniformly grim, of course. The company I founded in 1988, Isis Distributed Systems, is one of a handful of small technology sources that do offer reliability solutions, often capable of being introduced very transparently into existing applications. (Isis now operates as a division of Stratus Computers, Inc., and my own role is limited to occasional consulting.) However, the big story is that reliability has yet to make much of a dent on the distributed computing market.

The approach of this book is to treat distributed computing technology in a uniform way, looking at the technologies used in developing Internet and Web applications, at emerging standards such as CORBA, and at the technologies available to us for building reliable solutions within these settings. Many books that set this goal would do so primarily through a treatment of the underlying theory, but our approach here is much more pragmatic. By and large, we treat the theory as a source of background information that one should be aware of, but not as the major objective. Our focus, rather, is to understand how and why practical software tools for reliable distributed programming work, and to understand how they can be brought to bear on the broad area of technology currently identified with the Internet and the Web. By building up models of how distributed systems execute, and by using these models to prove properties of distributed communication protocols, we will show how computing systems of this sort can be formalized and reasoned about; however, the treatment is consistently driven by the *practical* implications of our results.

One of the most serious concerns about building reliable distributed systems stems from more basic issues that underlie any form of software reliability. Through decades of experience, it has become clear that software reliability is a *process,* not a *property.* One can talk about design practices that reduce errors, protocols that reconfigure systems to exclude faulty components, testing and quality-assurance methods that lead to increased confidence in the correctness of software, and basic design techniques that tend to limit the impact of failures and prevent them from propagating. All of these improve the reliability of a software system, and presumably would also increase the reliability of a distributed software system. Unfortunately, however, no degree of process ever leads to more than empirical confidence in the reliability of a software system. Thus, even in the case of a nondistributed system, it is hard to say "system X guarantees reliability property Y" in a rigorous manner. This same limitation extends to distributed settings, but is made even worse by the lack of a process comparable to the one used in conventional systems. Significant advances are needed in the process of developing reliable distributed computing systems, in the metrics by which we characterize reliability, the models we use to predict their behavior in new configurations reflecting changing loads or failures, and in the formal methods used to establish that a system satisfies its reliability goals.

For certain types of applications, this creates a profound quandary. Consider the design of an air traffic control software system, which (among other services) provides air traffic controllers with information about the status of air traffic sectors (Figure 1). Web sophisticates may want to think of this system as one that provides a Web-like interface to a database of routing information maintained on a server. Thus, the controller would be presented with a depiction of the air traffic situation, with pushbutton-style interfaces or other case-specific interfaces providing access to additional information about flights, projected trajectories, possible options for rerouting a flight, and so forth. To the air traffic controller these are the commands supported by the system; the Web user might think of them as active hyperlinks. Indeed, even if air traffic control systems were not typical of what the Web is likely to support, other equally critical applications are already moving to the Web, using very much the same programming model.

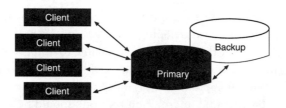

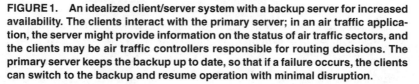

FIGURE 1. An idealized client/server system with a backup server for increased availability. The clients interact with the primary server; in an air traffic application, the server might provide information on the status of air traffic sectors, and the clients may be air traffic controllers responsible for routing decisions. The primary server keeps the backup up to date, so that if a failure occurs, the clients can switch to the backup and resume operation with minimal disruption.

A controller who depends upon a system such as this needs an absolute assurance that if the service reports that a sector is available and a plane can be routed into it, this information is correct and no other controller has been given the same information in regard to routing some other plane. An optimization criterion for such a service would be that it minimizes the frequency with which it reports a sector as being occupied when it is actually free. A fault-tolerance goal would be that the service remains operational despite limited numbers of failures of component programs, and perhaps that it performs self-checking operations so as to take a component off-line if it somehow falls out of synchronization with regard to the states of other components. Such goals would avoid scenarios such as the one illustrated in Figure 2, where the system state has become dangerously inconsistent as a result of a network failure that fools some clients into thinking the primary server has failed, and similarly fools the primary and backup servers into mutually believing one another to have crashed.

Now, suppose that the techniques of this book were used to construct such a service, using the best available technological solutions, combined with rigorous formal specifications of the software components involved and the best possible quality process. Theoretical results assure us that inconsistencies such as the one in Figure 2 cannot occur. Years of testing might yield a very high degree of confidence in the system, yet the service remains a large, complex software artifact. Even minor changes to the system, such as adding a feature, correcting a very simple bug, or upgrading the operating system version or hardware, could introduce serious problems long after the system was put into production. The question then becomes: Can complex software systems ever be used in critical settings? If so, are distributed systems somehow worse, or are the issues similar?

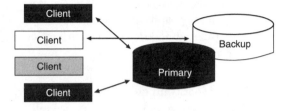

FIGURE 2.   **This figure represents a scenario that will occur in Chapter 4, when we consider the use of a standard remote procedure call methodology to build a client/server architecture for a critical setting. In the case illustrated, some of the client programs have become disconnected from the primary server, perhaps because of a transient network failure (one that corrects itself after a brief period during which message loss rates are very high). In the resulting system configuration, the primary and backup servers each consider themselves to be in charge of the system as a whole. There are two clients still connected to the primary server (black), one to the backup server (white), and one client is completely disconnected (gray). Such a configuration exposes the application user to serious threats. In an air traffic control situation, it is easy to imagine that accidents could occur if such a situation were encountered. The goal of this book is twofold: to assist the reader in understanding why such situations are a genuine threat in modern computing systems and to study the technical options for building better systems that can prevent such situations from occurring. The techniques presented will sometimes have limitations, which we will attempt to quantify and to understand any reliability implications. While many modern distributed systems have overlooked reliability issues, our working hypothesis will be that this situation is changing rapidly and that the developer of a distributed system has no choice but to confront these issues and begin to use technologies that respond to them.**

At the core of the material in this book is the consideration seen in this question. There may not be a single answer: Distributed systems are suitable for some critical applications and are not suited for others. In effect, although one can build reliable distributed software, reliability has its limits, and there are problems that distributed software should probably not be used to solve. Even given an appropriate technology, it is easy to build inappropriate solutions—and, conversely, with an inadequate technology, one can sometimes build critical services that are still useful in limited ways. The air traffic example, described previously, might or might not fall into the feasible category, depending on the detailed specification of the system, the techniques used to implement the solution, and the overall process by which the result is used and maintained.

Through the material in this book, the developer will be guided to appropriate design decisions, appropriate development methodologies, and to an understanding of the reliability limits on the solutions that result from this process. No book can expect to instill the sense of responsibility that the reader may need to draw upon in order to make such decisions wisely, but one hopes that computer system engineers, like bridge builders and designers of aircraft, are highly motivated to build the best and most reliable systems possible. Given such a motivation, an appropriate development methodology, and appropriate software tools, extremely reliable distributed software can be implemented and deployed even into critical settings. We will see precisely how this can be done in the following chapters.

Perhaps this book can serve a second purpose while accomplishing its primary one. Many highly placed industry leaders have commented to me that until reliability is forced upon them, their companies will *never* take the issues involved seriously. The investment needed is simply viewed as very large and likely to slow the frantic rate of progress on which computing as an industry has come to depend. I believe that the tide is now turning in a way that will, in fact, force change, and that this book can contribute to what will, over time, become an overwhelming priority for the industry.

Reliability is viewed as complex and costly, much as the phrase "robust bridge" conjures up a vision of a massive, expensive, and ugly artifact. Yet, the Golden Gate Bridge is robust and is anything but massive or ugly. To overcome this instinctive reaction, it will be necessary for the industry to come to understand reliability as being compatible with performance, elegance, and market success. At the same time, it will be important for pressure favoring reliability to grow, through demand by consumers for more reliable products. Together, such trends would create an incentive for reliable distributed software engineering.

As the general level of demonstrated knowledge concerning how to make systems reliable rises, the expectation of society and government that vendors will employ such technologies is also likely to rise. It will become harder and harder for corporations to cut corners by bringing an unreliable product to market and yet advertise it as "fault tolerant," "secure," or otherwise "reliable." Today, these terms are often used in advertising for products that are not reliable in any meaningful sense at all. One might similarly claim that a building or a bridge was constructed "above code" in a setting where the building code is completely ad hoc. The situation changes considerably when the building code is made more explicit and demanding, and bridges and buildings that satisfy the standard have actually been built successfully (and, perhaps, elegantly and without excessive added cost). In the first instance, a company can easily cut corners; in the second, the risks of doing so are greatly increased.

Moreover, at the time of this writing, vendors often seek to avoid software product liability by using complex contracts that stipulate the unsuitability of their products for critical uses, the near certainty that their products will fail even if used correctly, and in which it is stressed that the customer accepts full responsibility for the eventual use of the technology. It seems likely that as such contracts are put to the test, many of them will be recognized as analogous to those used by a landlord who rents a dangerously deteriorated apartment to a tenant, using a contract that warns of the possibility that the kitchen floor could collapse without warning and that the building is a firetrap lacking adequate escape routes. A landlord could certainly draft such a contract and a tenant might well sign it. But if the landlord fails to maintain the building according to the general standards for a safe and secure dwelling, the courts would still find the landlord liable if the floor indeed collapses. One cannot easily escape the generally accepted standards for one's domain of commercial activity.

By way of analogy, we may see growing pressure on vendors to recognize their fundamental responsibilities to provide a technology base adequate to the actual uses of their technologies, like it or not. Meanwhile, today a company that takes steps to provide reliability worries that in so doing, it may have raised expectations impossibly high and hence exposed itself to litigation if its products fail. As reliability becomes more and more common, such a company will be protected by having used the best available engineering practices to build the most reliable product it was capable of producing. If such a technology does fail, one at least knows that it was not the consequence of some outrageous form of negligence. Viewed in these terms, many of the products on the market today are seriously deficient. Rather than believing it safer to confront a reliability issue using the best practices available, many companies feel that they run a lower risk by ignoring the issue and drafting evasive contracts that hold themselves harmless in the event of accidents.

The challenge of reliability in distributed computing is perhaps the unavoidable challenge of the coming decade, just as performance was the challenge of the past decade. By accepting this challenge, we also gain new opportunities, new commercial markets, and help create a future in which technology is used responsibly for the broad benefit of society. There will inevitably be real limits on the reliability of the distributed systems we can build, and consequently there will be types of distributed computing systems that should not be built because we cannot expect to make them adequately reliable. However, we are far from those limits; in many circumstances we are deploying technologies known to be fragile in ways that actively encourage their use in critical settings. Ignoring this issue, as occurs too often today, is irresponsible and dangerous and increasingly unacceptable. Reliability challenges us as a community: It now falls upon us to respond.

A User's Guide to This Book

This book was written with several types of readers in mind, and consequently it weaves together material that may be of greater interest to one type of reader than that aimed at another type of reader.

Practitioners will find that the book has been constructed to be readable more or less sequentially from start to finish. The first part of the book may well be familiar material to many practitioners, but we try to approach this as a perspective of understanding reliability and consistency issues that arise even when using the standard distributed system technologies. We also look at the important roles of performance and modularity in building distributed software that can be relied upon. The second part of the book, which focuses on the Web, is of a similar character. Even if experts in this area may be surprised by some of the subtle reliability and consistency issues associated with the Web, they may find the suggested solutions useful in their work.

The third part of the book looks squarely at reliability technologies. Here, a pragmatically oriented reader may want to skim through Chapters 13 through 16, which cover the details of some fairly complex protocols and programming models. This material is included for thoroughness, and I don't think it is exceptionally hard to understand. However, the developer of a reliable system doesn't necessarily need to know every detail of how the underlying protocols work, or how they are positioned relative to some of the theoretical arguments of the decade. The remainder of the book can be read without having read through these chapters in any great detail. Chapters 17 and 18 look at the use of tools through an approach based on wrappers, and Chapters 19 through 24 look at some related issues concerning topics such as real-time systems, security, persistent data, and system management. The content is practical and the material is intended to be of a hands-on nature. Thus, the book is designed to be read more or less in order by system developers, with the exception of those parts of Chapters 13 through 16 where the going gets a bit heavy.

Where possible, the book includes general background material: There is a section on ATM networks, for example, that could be read independently of the rest of the book, one on CORBA, one on message-oriented middleware, and so forth. As much as practical, I have tried to make these sections freestanding and to index them properly, so that if one were worried about security exposures of the NFS file system, for example, it would be easy to read about that specific topic without reading the entire book as well. Hopefully, practitioners will find this book useful as a general reference for the technologies covered, and not purely for its recommendations in the area of security and reliability.

Next, here are some comments directed toward other researchers and instructors who may read or choose to teach from this book. I based the original outline of this book on a course that I have taught several times at Cornell, to a mixture of fourth-year undergraduates, professional master's degree students, and first-year Ph.D. students. To facilitate the development of course materials, I have placed my slides (created using the Microsoft PowerPoint utility) on Cornell University's public file server, where they can be retrieved using FTP. (Copy the files from ftp.cs.cornell.edu/pub/ken/slides.) The book also includes a set of problems that can be viewed either as thought-provoking exercises for the professional who wishes to test his or her own understanding of the material, or as the basis for possible homework and course projects in a classroom setting.

Any course based on this book should adopt the same practical perspective as the book itself. I suspect that some of my research colleagues will consider the treatment broad but somewhat superficial; this reflects a decision to focus primarily on system issues, rather than on theory or exhaustive detail on any particular topic. In making this decision, compromises had to be accepted: When teaching from this book, it may be necessary to ask the students to read some of the more theoretically rigorous books, which are cited in subsections of interest to the instructor, and to look in greater detail at some of the systems that are mentioned only briefly here. On the positive side, however, there are few, if any, introductory distributed system books that try to provide a genuinely broad perspective on issues in reliability. In my experience, many students are interested in this kind of material today, and, having gained a general exposure to it, would then be motivated to attend a much more theoretical course focused on fundamental issues in distributed systems theory. Thus, while this book may not be sufficient in and of itself for launching a research effort in distributed computing, it could well serve as a foundation for such an activity.

It should also be noted that, in my own experience, the book is too long for a typical 12-week semester. Instructors who elect to teach from it should be selective about the material that will be covered, particularly if they intend to treat Chapters 13 through 17 in any detail. If one has the option of teaching over two semesters, it might make sense to split the course into two parts and to include supplemental material on the Web. I suspect that such a sequence would be very popular given the current interest in network technology. At Cornell, for example, I tend to split this material into a more practical course that I teach in the fall, aiming at our professional master's degree students, followed by a more probing advanced graduate course that I or one of my colleagues teaches in the spring, drawing primarily on the original research papers associated with the topics we cover. This works well for us at Cornell, and the organization and focus of the book match with such a sequence.

A final comment regarding references. To avoid encumbering the discussion with a high density of references, the book cites relevant work the first time a reference to it occurs in the text, or where the discussion needs to point to a specific reference, but may not do so in subsequent references to the same work. These can be found in the Bibliography. References are also collected at the end of each chapter into a short section on related reading. It is hard to do adequate justice to such a large and dynamic area of research with a limited number of citations, but every effort has been made to be fair and complete.

Part I

BASIC DISTRIBUTED
COMPUTING TECHNOLOGIES

Although our treatment is motivated by the emergence of the Global
Information Superhighway and the World Wide Web, this first part of
the book focuses on the general technologies on which any distributed computing system relies. We review basic communication options and the basic
software tools that have emerged for utilizing them and for simplifying the
development of distributed applications. In the interests of generality, we
cover more than just the specific technologies embodied in the Web as it exists at the time of this writing, and, in fact, terminology and concepts specific
to the Web are not introduced until Part II of the book. However, even in
this first part, we do discuss some of the most basis issues that arise in building reliable distributed systems, and we begin to establish the context within
which reliability can be treated in a systematic manner.

CHAPTER 1 ✧ ✧ ✧ ✧

Fundamentals

CONTENTS

1.1 Introduction

Reduced to the simplest terms, *a distributed computing system* is a set of computer programs, executing on one or more computers, and coordinating actions by exchanging *messages*. A *computer network* is a collection of computers interconnected by hardware that directly supports message passing. Most distributed computing systems operate over computer networks, although this is not always the case: One can build a distributed computing system in which the components execute on a single multitasking computer, and one can also build distributed computing systems in which information flows between the components by means other than message passing. Moreover, as we will see in Chapter 24, there are new kinds of parallel computers, called clustered servers, that have many attributes of distributed systems despite appearing to the user as a single machine built using rack-mounted components.

We will use the term "protocol" in reference to an algorithm governing the exchange of messages, by which a collection of processes coordinate their actions and communicate information among themselves. Much as a *program* is the set of instructions, and a *process* denotes the execution of those instructions, a *protocol* is a set of instructions governing the communication in a distributed program, and a distributed computing system is the result of executing some collection of such protocols to coordinate the actions of a collection of processes in a network.

This book is concerned with *reliability* in distributed computing systems. Reliability is a very broad term and can have many meanings, including:

▶ *Fault tolerance:* The ability of a distributed computing system to recover from component failures without performing incorrect actions.

▶ *High availability:* In the context of a fault-tolerant distributed computing system, the ability of the system to restore correct operation, permitting it to resume providing services during periods when some components have failed. A highly available system may provide reduced service for short periods of time while reconfiguring itself.

▶ *Continuous availability:* A highly available system with a very small recovery time, capable of providing uninterrupted service to its users. The reliability properties of a continuously available system are unaffected or only minimally affected by failures.

▶ *Recoverability:* Also in the context of a fault-tolerant distributed computing system, the ability of failed components to restart themselves and rejoin the system, after the cause of failure has been repaired.

▶ *Consistency:* The ability of the system to coordinate related actions by multiple components, often in the presence of concurrency and failures. Consistency underlies the ability of a distributed system to emulate a nondistributed system.

▶ *Security:* The ability of the system to protect data, services, and resources against misuse by unauthorized users.

▶ *Privacy:* The ability of the system to protect the identity and locations of its users from unauthorized disclosure.

- *Correct specification:* The assurance that the system solves the intended problem.

- *Correct implementation:* The assurance that the system correctly implements its specification.

- *Predictable performance:* The guarantee that a distributed system achieves desired levels of performance—for example, data throughput from source to destination, latencies measured for critical paths, requests processed per second, and so forth.

- *Timeliness:* In systems subject to real-time constraints, the assurance that actions are taken within the specified time bounds, or are performed with a desired degree of temporal synchronization between the components.

Underlying many of these issues are questions of tolerating failures. Failure, too, can have many meanings:

- *Halting failures:* In this model, a process or computer either works correctly, or simply stops executing and crashes without taking incorrect actions, as a result of failure. As the model is normally specified, there is no way to detect that the process has halted except by timeout: It stops sending "keep alive" messages or responding to "pinging" messages and hence other processes can deduce that it has failed.

- *Fail-stop failures:* These are accurately detectable halting failures. In this model, processes fail by halting. However, other processes that may be interacting with the faulty process also have a completely accurate way to detect such failures—for example, a fail-stop environment might be one in which timeouts can be used to monitor the status of processes, and *no timeout occurs unless the process being monitored has actually crashed.* Obviously, such a model may be unrealistically optimistic, representing an idealized world in which the handling of failures is reduced to a pure problem of how the system should react when a failure is sensed. If we solve problems with this model, we then need to ask how to relate the solutions to the real world.

- *Send-omission failures:* These are failures to send a message that, according to the logic of the distributed computing systems, should have been sent. Send-omission failures are commonly caused by a lack of buffering space in the operating system or network interface, which can cause a message to be discarded after the application program has sent it but before it leaves the sender's machine. Perhaps surprisingly, few operating systems report such events to the application.

- *Receive-omission failures:* These are similar to send-omission failures, but they occur when a message is lost near the destination process, often because of a lack of memory in which to buffer it or because evidence of data corruption has been discovered.

- *Network failures:* These occur when the network loses messages sent between certain pairs of processes.

- *Network partitioning failures:* These are a more severe form of network failure, in which the network fragments into disconnected subnetworks, within which messages can be transmitted, but between which messages are lost. When a failure of this sort is repaired, one talks about *merging* the network partitions. Network partitioning failures are a common problem in modern distributed systems; hence, we will discuss them in detail in Part III of this book.

▶ *Timing failures:* These occur when a temporal property of the system is violated—for example, when a clock on a computer exhibits a value that is unacceptably far from the values of other clocks, or when an action is taken too soon or too late, or when a message is delayed by longer than the maximum tolerable delay for a network connection.

▶ *Byzantine failures:* This is a term that captures a wide variety of other faulty behaviors, including data corruption, programs that fail to follow the correct protocol, and even malicious or adversarial behaviors by programs that actively seek to force a system to violate its reliability properties.

An even more basic issue underlies all of these: the meaning of computation, and the model one assumes for communication and coordination in a distributed system. Some examples of models include these:

▶ *Real-world networks:* These are composed of workstations, personal computers, and other computing devices interconnected by hardware. Properties of the hardware and software components will often be known to the designer, such as speed, delay, and error frequencies for communication devices; latencies for critical software and scheduling paths; throughput for data generated by the system and data distribution patterns; speed of the computer hardware; accuracy of clocks; and so forth. This information can be of tremendous value in designing solutions to problems that might be very hard—or impossible—in a completely general sense.

   A specific issue that will emerge as being particularly important when we consider guarantees of behavior in Part III concerns the availability, or lack, of accurate temporal information. Until the late 1980s, the clocks built into workstations were notoriously inaccurate, exhibiting high drift rates that had to be overcome with software protocols for clock resynchronization. There are limits on the quality of synchronization possible in software, and this created a substantial body of research and led to a number of competing solutions. In the early 1990s, however, the advent of satellite time sources as part of the global positioning system (GPS) changed the picture: For the price of an inexpensive radio receiver, any computer could obtain accurate temporal data, with resolution in the submillisecond range. The degree to which GPS receivers actually replace quartz-based time sources remains to be seen, however. Thus, real-world systems are notable (or notorious) in part for having temporal information, but of potentially low quality.

▶ *Asynchronous computing systems:* This is a very simple theoretical model used to approximate one extreme sort of computer network. In this model, no assumptions can be made about the relative speed of the communication system, processors, and processes in the network. One message from a process p to a process q may be delivered in zero time, while the next is delayed by a million years. The asynchronous model reflects an assumption about time, but not failures: Given an asynchronous model, one can talk about protocols that tolerate message loss, protocols that overcome fail-stop failures in asynchronous networks, and so forth.

The main reason for using the model is to prove properties about protocols for which one makes as few assumptions as possible. The model is very clean and simple, and it lets us focus on fundamental properties of systems without cluttering up the analysis by including a great number of practical considerations. If a problem can be solved in this model, it can be solved at least as well in a more realistic one. On the other hand, the converse may not be true: We may be able to do things in realistic systems by making use of features not available in the asynchronous model, and in this way we may be able to solve problems in real systems that are impossible in ones that use the asynchronous model.

▸ *Synchronous computing systems:* Like the asynchronous systems, these represent an extreme end of the spectrum. In the synchronous systems, there is a very strong concept of time that all processes in the system share. One common formulation of the model can be thought of as having a systemwide gong that sounds periodically; when the processes in the system hear the gong, they run one round of a protocol, reading messages from one another, sending messages that will be delivered in the next round, and so forth. And these messages *always* are delivered to the application by the start of the next round, or not at all.

Normally, the synchronous model also assumes bounds on communication latency between processes, clock skew and precision, and other properties of the environment. As in the case of an asynchronous model, the synchronous one takes an extreme point of view because this simplifies reasoning about certain types of protocols. Real-world systems are not synchronous—it is impossible to build a system in which actions are perfectly coordinated as this model assumes. However, if one proves the impossibility of solving some problem in the synchronous model, or proves that some problem requires at least a certain number of messages in this model, one has established a sort of lower bound. In a real-world system, things can only get worse, because we are limited to weaker assumptions. This makes the synchronous model a valuable tool for understanding how hard it will be to solve certain problems.

▸ *Parallel shared memory systems:* An important family of systems are based on multiple processors that share memory. Communication is by reading and writing shared memory locations. Clearly, the shared memory model can be emulated using message passing, and it can be used to implement message communication. Nonetheless, because there are important examples of real computers that implement this model, there is considerable theoretical interest in the model per se. Unfortunately, although this model is very rich and a great deal is known about it, it would be beyond the scope of this book to attempt to treat the model in any detail.

1.2 Components of a Reliable Distributed Computing System

Reliable distributed computing systems are assembled from basic building blocks. In the simplest terms, these are just processes and messages, and if our interest were purely theoretical, it might be reasonable to stop at that. On the other hand, if we wish to apply theoretical results in practical systems, we will need to work from a fairly detailed real understanding of how practical systems actually work. In some ways, this is unfortunate, because real systems often include mechanisms that are deficient in ways that seem simple to fix, or inconsistent with one another, but have such a long history (or are so deeply embedded into standards) that there may be no way to improve on the behavior in question. Yet, if we want to actually build reliable distributed systems, it is unrealistic to insist that we will only do so in idealized environments that support some form of theoretically motivated structure. The real world is heavily committed to standards, and the task of translating our theoretical insights into practical tools that can interplay with these standards is probably the most important challenge faced by the computer systems engineer.

It is common to think of a distributed system as operating over a layered set of network services (see Table 1.1). Each layer corresponds to a software abstraction or hardware feature, and may be implemented in the application program itself, in a library of procedures to which the program is linked, in the operating system, or even in the hardware of the communications device. As an example, here is the layering of the International Organization for Standardization (ISO) Open Systems Interconnection (OSI) protocol model (see Comer, Comer and Stevens [1991, 1993], Coulouris et al., Tanenbaum):

- *Application:* This is the application program itself, up to the points at which it performs communication operations.

- *Presentation:* This is the software associated with placing data into messages in a format that can be interpreted by the destination process(es) to which the message will be sent and for extracting data from messages in the destination process.

- *Session:* This is the software associated with maintaining connections between pairs or sets of processes. A session may have reliability properties and may require some form of initialization or setup, depending on the specific setting with which the user is working. In the OSI model, any reliability properties are implemented by the session software, and lower layers of the hierarchy are permitted to be unreliable—for example, by losing messages.

- *Transport:* The transport layer is responsible for breaking large messages into smaller packets that respect size limits imposed by the network communication hardware. On the incoming side, the transport layer reassembles these packets into messages, discarding packets that are identified as duplicates, or messages for which some constituent packets were lost in transmission.

- *Network:* This is the layer of software concerned with routing and low-level flow control on networks composed of multiple physical segments interconnected by what are called bridges and gateways.

- *Data Link:* The data-link layer is normally part of the hardware that implements a communication device. This layer is responsible for sending and receiving packets, recognizing packets destined for the local machine and copying them, discarding corrupted packets, and other interface-level aspects of communication.

- *Physical:* The physical layer is concerned with representation of packets on the wire—for example, the hardware technology for transmitting individual bits and the protocol for gaining access to the wire if it is shared by multiple computers.

TABLE 1.1. OSI Protocol Layers	
Application	The program using the communication connection
Presentation	Software to encode application data into messages and to decode on reception
Session	The logic associated with guaranteeing end-to-end properties such as reliability
Transport	Software concerned with fragmenting big messages into small packets
Network	Routing functionality, usually limited to small- or fixed-size packets
Data Link	The protocol used to send and receive packets
Physical	The protocol used to represent packets on the wire

It is useful to distinguish the types of guarantees provided by the various layers: *end-to-end* guarantees in the case of the session, presentation, and application layers and *point-to-point* guarantees for layers below these. The distinction is important in complex networks where a message may need to traverse many links to reach its destination. In such settings, a point-to-point property is one that holds only on a per-hop basis—for example, the data-link protocol is concerned with a single hop taken by the message, but not with its overall route or the guarantees that the application may expect from the communication link itself. The session, presentation, and application layers, in contrast, impose a more complex logical abstraction on the underlying network, with properties that hold between the end points of a communication link that may physically extend over a complex substructure. In Part III of this book we will discuss increasingly elaborate end-to-end properties, until we finally extend these properties into a completely encompassing distributed communication abstraction that embraces the distributed system as a whole and provides consistent behavior and guarantees throughout. And, just as the OSI layering builds its end-to-end abstractions over point-to-point ones, so will we need to build these more sophisticated abstractions over what are ultimately point-to-point properties.

As seen in Figure 1.1, each layer is logically composed of transmission logic and the corresponding reception logic. In practice, this often corresponds closely to the implementation of the architecture—for example, most session protocols operate by imposing a multiple session abstraction over a shared (or multiplexed) link-level connection. The packets generated by the various higher-level session protocols can be thought of as merging into a single stream of packets that are treated by the IP link level as a single customer for its services. Nonetheless, one should not necessarily assume that the implementation of a layered protocol architecture involves some sort of separate module for each layer. To maximize performance, the functionality of a layered architecture is often compressed into a single piece of software, and in some cases layers may be completely bypassed for types of messages where the layer would take no action—

for example, if a message is very small, the OSI transport layer wouldn't need to fragment it into multiple packets, and one could imagine an implementation of the OSI stack, specialized for small messages, that omits the transport layer. Indeed, the pros and cons of layered protocol architecture have become a major topic of debate in recent years (see Abbott and Peterson, Braun and Diot, Clark and Tennenhouse, Karamcheti and Chien, Kay and Pasquale).

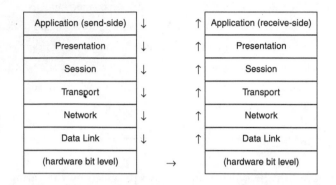

FIGURE 1.1.   Data flow in an OSI protocol stack. Each sending layer is invoked by the layer above it and passes data to the layer below it, and conversely on the receive-side. In a logical sense, however, each layer interacts with its peer on the remote side of the connection—for example, the send-side session layer may add a header to a message that the receive-side session layer strips off.

Although the OSI layering is probably the best known, the concept of layering communication software is pervasive, and there are many other examples of layered architectures and layered software systems. Later in this book we will see ways in which the OSI layering is outdated, because it doesn't directly address multiparticipant communication sessions and doesn't match very well with some new types of communication hardware, such as asynchronous transfer mode (ATM) switching systems. In discussing this point we will see that more appropriate layered architectures can be constructed, although they don't match the OSI layering very closely. Thus, one can think of layering as a methodology matched to the particular layers of the OSI hierarchy. The former perspective is a popular one that is only gaining importance with the introduction of object-oriented distributed computing environments, which have a natural form of layering associated with object classes and subclasses. The latter form of layering has probably become hopelessly incompatible with standard practice by the time of this writing, although many companies and governments continue to require that products comply with it.

It can be argued that a layered communication architecture is primarily valuable as a *descriptive abstraction*—a model that captures the essential functionality of a real communication system but doesn't need to accurately reflect its implementation. The idea of abstracting the behavior of a distributed system in order to concisely describe it or to reason about it is a very important one. However, if the abstraction doesn't accurately correspond to the implementation, this also creates a number of problems for the system designer, who now has the obligation to develop a specification and correctness proof for the abstraction; to implement, verify, and test the corresponding software; and to undertake an additional analysis that confirms that the abstraction accurately models the implementation.

It is easy to see how this process can break down—for example, it is nearly inevitable that changes to the implementation will have to be made long after a system has been deployed. If the development process is really this complex, it is likely that the analysis of overall correctness will not be repeated for every such change. Thus, from the perspective of a user, abstractions can be a two-edged sword. They offer appealing and often simplified ways to deal with a complex system, but they can also be simplistic or even incorrect. And this bears strongly on the overall theme of reliability. To some degree, the very process of cleaning up a component of a system in order to describe it concisely can compromise the reliability of a more complex system in which that component is used.

Throughout the remainder of this book, we will often have recourse to models and abstractions, in much more complex situations than the OSI layering. This will assist us in reasoning about and comparing protocols, and in proving properties of complex distributed systems. At the same time, however, we need to keep in mind that this whole approach demands a sort of meta-approach, namely a higher level of abstraction at which we can question the methodology itself, asking if the techniques by which we create reliable systems are themselves a possible source of unreliability. When this proves to be the case, we need to take the next step as well, asking what sorts of systematic remedies can be used to fight these types of reliability problems.

Can well-structured distributed computing systems be built that can tolerate the failures of their own components? In layerings such as OSI, this issue is not really addressed, which is one of the reasons that the OSI layering won't work well for our purposes. However, the question is among the most important ones that will need to be resolved if we want to claim that we have arrived at a workable methodology for engineering reliable distributed computing systems. A methodology, then, must address descriptive and structural issues, as well as practical ones such as the protocols used to overcome a specific type of failure or to coordinate a specific type of interaction.

1.2.1 Communication Technology

The most basic communication technology in any distributed system is the hardware support for message passing. Although there are some types of networks that offer special properties, most modern networks are designed to transmit data in *packets* with some fixed, but small, maximum size. Each packet consists of a *header*, which is a data structure containing information about the packet—its destination, route, and so forth. It contains a *body*, which is the bytes that make up the content of the packet. And it may contain a *trailer*, which is a second data structure that is physically transmitted after the header and body and would normally consist of a checksum for the packet that the hardware computes and appends to it as part of the process of transmitting the packet.

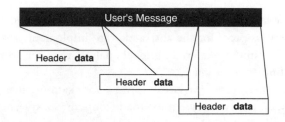

FIGURE 1.2. Large messages are fragmented for transmission.

When a user's message is transmitted over a network, the packets actually sent on the wire include headers and trailers, and may have a fixed maximum size. Large messages are sent as multiple packets. Figure 1.2 illustrates a message that has been fragmented into three packets, each containing a header and some part of the data from the original message. Not all fragmentation schemes include trailers, and in the figure no trailer is shown.

Modern communication hardware often permits large numbers of computers to share a single communication fabric. For this reason, it is necessary to specify the address to which a message should be transmitted. The hardware used for communication will therefore normally support some form of *addressing capability,* by which the destination of a message can be identified. More important to most software developers, however, are addresses supported by the transport services available on most operating systems. These *logical addresses* are a representation of location within the network, and are used to route packets to their destinations. Each time a packet makes a "hop" over a communications link, the sending computer is expected to copy the hardware address of the next machine in the path into the outgoing packet. Within this book, we assume that each computer has a logical address, but will have little to say about hardware addresses.

On the other hand, there are two hardware addressing features that have important implications for higher-level communication software. These are the ability of the hardware to *broadcast* and *multicast* messages.

A broadcast is a way of sending a message so that it will be delivered to all computers that it reaches. This may not be all the computers in a network, because of the various factors that can cause a receive omission failure to occur, but, for many purposes, absolute reliability is not required. To send a hardware broadcast, an application program generally places a special logical address in an outgoing message that the operating system maps to the appropriate hardware address. The message will only reach those machines connected to the hardware communication device on which the transmission occurs, so the use of this feature requires some knowledge of network communication topology.

A multicast is a form of broadcast that communicates to a subset of the computers that are attached to a communication network. To use a multicast, one normally starts by creating a new multicast group address and installing it into the hardware interfaces associated with a communication device. Multicast messages are then sent much as a broadcast would be, but are only accepted, at the hardware level, at those interfaces that have been instructed to install the group

FUNDAMENTALS   Chapter 1

address to which the message is destined. Many network routing devices and protocols watch for multicast packets and will forward them automatically, but this is rarely attempted for broadcast packets.

Chapter 2 discusses some of the most common forms of communication hardware in detail.

1.2.2 Basic Transport and Network Services

The layer of software that runs over the communications layer is the one most distributed system programmers deal with. This layer hides the properties of the communication hardware from the programmer (see Figure 1.3). It provides the ability to send and receive messages that may be much larger than the ones supported by the underlying hardware (although there is normally still a limit, so that the amount of operating system buffering space needed for transport can be estimated and controlled). The transport layer also implements logical addressing capabilities by which every computer in a complex network can be assigned a unique address, and can send and receive messages from every other computer.

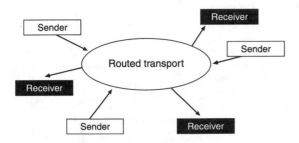

FIGURE 1.3. The routing functionality of a modern transport protocol conceals the network topology from the application designer.

Although many transport layers have been proposed, one set of standards has been adopted by almost all vendors. This standard defines the so-called "Internet Protocol" or IP protocol suite, and it originated in a research network called the ARPANET, which was developed by the U.S. government in the late 1970s (see Comer, Coulouris et al., Tanenbaum). A competing standard was introduced by the ISO organization in association with the OSI layering cited earlier, but it has not gained the sort of ubiquitous acceptance as the IP protocol suite. There are also additional proprietary standards that are widely used by individual vendors or industry groups, but rarely seen outside their community—for example, most PC networks support a protocol called NetBIOS, but this protocol is not common in any other type of computing environment.

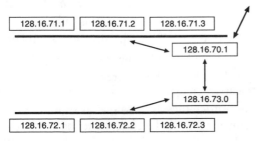

FIGURE 1.4. A typical network may have several interconnected subnetworks and a link to the Internet.

Transport services generally offer at least the features of the underlying communication hardware (see Figure 1.4). Thus, the most widely used communication services include a way to send a message to a destination, to broadcast a message, and to multicast a message. Unlike the communication hardware versions of these services, however, transport-layer interfaces tend to work with logical addresses and to automatically route messages within complex environments that may mix multiple forms of communication hardware or include multiple communication subnetworks bridged by routing devices or computers.

All of this is controlled using *routing tables,* as shown in Table 1.2. A routing table is a data structure local to each computer in a network—each computer has one, but the contents will generally not be identical from machine to machine. The table is indexed by the logical address of a destination computer, and entries contain the hardware device on which messages should be transmitted (the next hop to take). Distributed protocols for dynamically maintaining routing tables have been studied for many years and seek to minimize the number of hops a message needs to take to reach its destination, while at the same time attempting to spread the load evenly and route around failures or congested nodes. In practice, however, static routing tables are probably more common: These are maintained by a system administrator for the network and generally offer a single route from a source to each destination. Chapter 3 discusses some of the most common transport services in more detail.

TABLE 1.2. A sample routing table,
such as might be used by computer 128.16.73.0 in Figure 1.4

Destination	Route via	Forwarded by	Estimated distance
128.16.72.*	Outgoing link 1	(direct)	1 hop
128.16.71.*	Outgoing link 2	128.16.70.1	2 hops
128.16.70.1	Outgoing link 2	(direct)	1 hop
..*.*	Outgoing link 2	128.16.70.1	(infinite)

1.2.3 Reliable Transport Software and Communication Support

A limitation of the basic message passing services discussed in Section 1.2.2 is that they operate at the level of individual messages and provide no guarantees of reliability. Messages can be lost for many reasons, including link failures, failures of intermediate machines on a complex multihop route, noise that causes corruption of the data in a packet, lack of buffering space (the most common cause), and so forth. For this reason, it is common to layer a reliability protocol over the message-passing layer of a distributed communication architecture. The result is called a *reliable communication channel.* This layer of software is the one that the OSI stack calls the session layer, and it corresponds to the TCP protocol of the Internet. UNIX programmers may be more familiar with this concept from their use of pipes and streams (see Ritchie).

The protocol implementing a reliable communication channel will typically guarantee that lost messages will be retransmitted and that out-of-order messages will be resequenced and delivered in the order sent. Flow control and mechanisms that choke back the sender when data volume becomes excessive are also common in protocols for reliable transport (see Jacobson [1988]). Just as the lower layers can support one-to-one, broadcast, and multicast communication, these forms of destination addressing are also potentially interesting in reliable transport layers. Moreover, some systems go further and introduce additional reliability properties at this level, such as authentication (a trusted mechanism for verifying the identity of the processes at the ends of a communication connection) or security (trusted mechanisms for concealing the data transmitted over a channel from processes other than the intended destinations). In Chapter 3, we will begin to discuss these options, as well as some very subtle issues concerned with how and when connections report failure.

1.2.4 Middleware: Software Tools, Utilities, and Programming Languages

The most interesting issues that we will consider in this book are those relating to programming environments and tools that live in the middle, between the application program and the communication infrastructure for basic message passing and support for reliable channels.

Examples of important middleware services include the naming service, the file system, the time service, and the security key services used for authentication in distributed systems. We will be looking at all of these in more detail later, but we review them briefly here for clarity.

A naming service is a collection of user-accessible directories that map from application names (or other selection criteria) to network addresses of computers or programs. Name services can play many roles in a distributed system, and they represent an area of intense research interest and rapid evolution. When we discuss naming, we'll see that the whole question of what a name represents is itself subject to considerable debate, and raises important questions about concepts of abstraction and services in distributed computing environments. Reliability in a name service involves issues such as trust—can one trust the name service to truthfully map a name to the correct network address? How can one know that the object at the end of an address is the same one that the name service was talking about? These are fascinating issues, and we will discuss them in detail later in the book (see, for example, Section 7.2).

From the outset, though, the reader may want to consider that if an intruder breaks into a system and is able to manipulate the mapping of names to network addresses, it will be possible to interpose all sorts of snooping software components in the path of communication from an application to the services it is using over the network. Such attacks are now common on the Internet and reflect a fundamental issue, which is that most network reliability technologies tend to trust the lowest-level mechanisms that map from names to addresses and that route messages to the correct host when given a destination address.

A time service is a mechanism for keeping the clocks on a set of computers closely synchronized and close to real time. Time services work to overcome the inaccuracy of inexpensive clocks used on many types of computers, and they are important in applications that either coordinate actions using real time or that make use of time for other purposes, such as to limit the lifetime of a cryptographic key or to timestamp files when they are updated. Much can be said about time in a distributed system, and we will spend a considerable portion of this book discussing issues that revolve around the whole concept of before and after and its relation to intuitive concepts of time in the real world. Clearly, the reliability of a time service will have important implications for the reliability of applications that make use of time, so time services and associated reliability properties will prove to be important in many parts of this book.

Authentication services are, perhaps surprisingly, a new technology that is lacking in most distributed computing environments. These services provide trustworthy mechanisms for determining who sent a message, for making sure that the message can only be read by the intended destination, and for restricting access to private data so that only authorized access can occur. Most modern computing systems evolved from a period when access control was informal and based on a core principle of trust among users. One of the really serious implications is that distributed systems that want to superimpose a security or protection architecture on a heterogeneous environment must overcome a pervasive tendency to accept requests without questioning them, to believe the user-id information included in messages without validating it, and to route messages wherever they may wish to go.

If banks worked this way, one could walk up to a teller in a bank and pass that person a piece of paper requesting a list of individuals who have accounts in the branch. Upon studying the response and learning that W. Gates is listed, one could then fill out an account balance request in the name of W. Gates, asking how much money is in that account. And, after this, one could withdraw some of that money, up to the bank's policy limits. At no stage would one be challenged: The identification on the various slips of paper would be trusted for each operation. Such a world model may seem strangely trusting, but it is the model from which modern distributed computing systems emerged.

1.2.5 Distributed Computing Environments

An important topic around which much of this book is oriented concerns the development of general-purpose tools from which specialized distributed systems can be constructed. Such tools can take many forms and can be purely conceptual—for example, a methodology or theory that offers useful insight into the best way to solve a problem or that can help the developer confirm that a proposed solution will have a desired property. A tool can offer practical help at a very low level—for example, by eliminating the relatively mechanical steps required to encode the arguments for a remote procedure call into a message to the server that will perform the action. A tool can embody complex higher-level behavior, such as a protocol for performing some action or overcoming some class of errors. Tools can even go beyond this, taking the next step by offering mechanisms to control and manage software built using other tools.

It has become popular to talk about distributed systems that support *distributed operating environments*—well-integrated collections of tools that can be used in conjunction with one another to carry out potentially complex distributed programming tasks. Examples of distributed programming environments are the Open Network Computing (ONC) environment of Sun Microsystems, the Distributed Computing Environment (DCE) of the Open Software Foundation, the various CORBA-compliant programming tools that have become popular among C++ programmers who work in distributed settings, and the Isis Toolkit and the Horus system—these last two systems were developed by my colleagues and me and will be discussed in Chapter 18.

Distributed system architectures undertake to step even beyond the concept of a distributed computing environment. An architecture is a general set of design principles and implementation standards by which a collection of compliant systems can be developed. In principle, multiple systems that implement the same architecture will interoperate, so that if vendors implement competing solutions, the resulting software can still be combined into a single system with components that might even be able to communicate and cooperate with one another. The Common Object Request Broker Architecture, or CORBA, is probably the best-known distributed computing architecture; it is useful for building systems using an object-oriented approach in which the systems are developed as modules that cooperate. Thus, CORBA is an architecture, and the various CORBA-based products that comply with the architecture are distributed computing environments.

1.2.6 End-User Applications

One might expect that the end of the line for a layered distributed system architecture would be the application level, but this is not necessarily the case. A distributed application might also be some sort of operating system service built over the communication tools we have been discussing—for example, the distributed file system is an application in the sense of the OSI layering, but the user of a computing system might think of the file system as an operating system service over which applications can be defined and executed. Within the OSI layering, then, an application is any freestanding solution to a well-defined problem that presents something other than a communication abstraction to its users. The distributed file system is just one example among many. Others include message bus technologies, distributed database systems, electronic mail, network bulletin boards, and the World Wide Web. In the near future, computer-supported collaborative work systems and multimedia digital library systems are likely to emerge as further examples in this area.

A limitation of a layering such as the OSI hierarchy is that it doesn't really distinguish these sorts of applications, which provide services to higher-level distributed applications, from what might be called end-user solutions—namely, programs that operate over the communication layer to directly implement commands for a person. One would like to believe that there is much more structure to a distributed air traffic control system than to a file transfer program, yet the OSI hierarchy views both as examples of applications. We lack a good classification system for the various types of distributed applications.

In fact, even complex distributed applications may merely be components of even larger-scale distributed systems—one can easily imagine a distributed system that uses a distributed computing toolkit to integrate an application that exploits distributed files with one that stores information into a distributed database. In an air traffic control environment, availability may be so critical that one is compelled to run multiple copies of the software concurrently, with one version backing up the other. Here, the entire air traffic control system is at one level a complex distributed application in its own right, but, at a different meta level, it is just a component of an over-arching reliability structure visible on a scale of hundreds of computers located within multiple air traffic centers.

1.3 Critical Dependencies

One of the major challenges to building reliable distributed systems is that computer networks have evolved that have many dependencies on a variety of technologies. Some of the major ones are identified in Figure 1.5; however the number is growing steadily and this figure is not necessarily complete. Critical applications often introduce new servers and critical components not shown in the figure. Also, the figure does not treat dependencies on hardware components of the distributed infrastructure, such as the communication network itself, power supply, or hardware routers. Moreover, the telecommunication infrastructure underlying a typical network application is itself a complex network with many of the same dependencies, together with additional ones such as the databases used to resolve mobile telephone numbers or to correctly account for use of network communication lines.

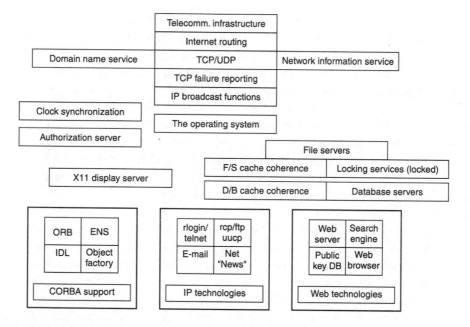

FIGURE 1.5. Technologies on which a distributed application may depend in order to provide correct, reliable behavior. The figure is organized so that dependencies are roughly from top to bottom (the lower technologies being dependent upon the upper ones), although a detailed dependency graph would be quite complex. Failures in any of these technologies can result in visible application-level errors, inconsistency, security violations, denial of service, or other problems. These technologies are also interdependent in complex and often unexpected ways—for example, some types of UNIX workstations will hang (freeze) if the NIS (network information service) server becomes unavailable, even if there are duplicate NIS servers that remain operational. Moreover, such problems can impact an application that has been running normally for an extended period and is not making any explicit new use of the server in question.

Fortunately, many of these services are fairly reliable, and one can plan around potential outages of such critical services as the network information service. The key issue is to understand the technology dependencies that can impact reliability issues for a specific application and to program solutions into the network to detect and work around potential outages. In this book we will be studying technical options for taking such steps. The emergence of integrated environments for reliable distributed computing will, however, require a substantial effort from the vendors offering the component technologies: An approach in which reliability is left to the application inevitably overlooks the problems that can be caused when such applications are forced to depend upon technologies that are themselves unreliable for reasons beyond the control of the developer.

1.4 Next Steps

While distributed systems are certainly layered, Figure 1.5 makes it clear that one should question the adequacy of any simple layering model for describing reliable distributed systems. We noted, for example, that many governments have mandated the use of the OSI layering for description of distributed software. Yet, there are important reliability technologies that require structures inexpressible in this layering, and it is unlikely that those governments intended to preclude the use of reliable technologies. More broadly, the types of complex layerings that can result when tools are used to support applications that are in turn tools for still higher-level applications are not amenable to any simple description of this nature. Does this mean that users should refuse the resulting complex software structures, because they cannot be described in terms of the standard? Should they accept the perspective that software should be used but not described, because the description methodologies seem to have lagged behind the state of the art? Or should governments insist on new standards each time a new type of system finds it useful to circumvent the standard?

Questions such as these may seem narrow and almost pointless, yet they point to a deep problem. Certainly, if we are unable to even describe complex distributed systems in a uniform way, it will be very difficult to develop a methodology within which one can reason about them and prove that they respect desired properties. On the other hand, if a standard proves unwieldy and constraining, it will eventually become difficult for systems to adhere to it.

Perhaps for these reasons, there has been little recent work on layering in the precise sense of the OSI hierarchy: Most researchers view this as an unpromising direction. Instead, the concepts of structure and hierarchy seen in the OSI protocol have reemerged in much more general and flexible ways: the object-class hierarchies supported by technologies in the CORBA framework, the layered protocol stacks supported in operating systems such as UNIX or the x-Kernel, or in systems such as Horus. We'll be reading about these uses of hierarchy later in the book, and the OSI hierarchy remains popular as a simple but widely understood framework within which to discuss protocols.

1.5 Related Reading

General discussion of network architectures and the OSI hierarchy: (see Architecture Projects Management Limited [1989, April 1991, November 1991], Comer, Comer and Stevens [1991, 1993], Coulouris et al., Cristian and Delancy, Tanenbaum, XTP Forum).

Pros and cons of layered architectures: (see Abbott and Peterson, Braun and Diot, Clark and Tennenhouse, Karamcheti and Chien, Kay and Pasquale, Ousterhout [1990], van Renesse et al. [1988, 1989]).

Reliable stream communication: (see Comer, Comer and Stevens [1991, 1993], Coulouris et al., Jacobson [1988], Ritchie, Tanenbaum).

Failure models and classification: (see Chandra and Toueg [1991], Chandra et al. [1992], Cristian [February 1991], Christian and Delancy, Fischer et al. [April 1985], Gray and Reuter, Lamport [1978, 1984], Marzullo [1990], Sabel and Marzullo, Skeen [June 1982], Srikanth and Toueg).

CHAPTER 2 ✧ ✧ ✧ ✧

Communication Technologies

CONTENTS

Historically, it has rarely been necessary to understand details of the hardware components from which a computing system was constructed if one merely wishes to develop software for it. The pressure to standardize operating systems, and the presentation of devices within them, created a situation in which it sufficed to understand the way that the operating system abstracted a device in order to use it correctly—for example, there are a great many designs for computer disk storage units and the associated device controllers. Each design offers its own advantages and disadvantages when compared with the others, and any system architect charged with selecting a data storage device would be wise to learn about the state of the art before making a decision. Yet, from a software perspective, device attributes are largely hidden. The developer normally considers a disk to be a device on which files, having various layout parameters, which can be tuned to optimize I/O performance and are characterized by a set of speed and reliability properties, can be stored. Developers of special classes of applications, such as multimedia image servers, may prefer to work with a less abstracted software interface to the hardware, exploiting otherwise hidden features at the cost of much greater software complexity. But, for the normal user, one disk is much like any other.

To a considerable extent, the same is true for computer networking hardware. There are a number of major classes of communication devices, differing in speed, average access latency, maximum capacity (packets per second, bytes of data per second), support for special addressing modes, and so forth. However, since most operating systems implement the lowest layers of the OSI hierarchy as part of the device driver or communication abstraction of a system, applications can treat these devices interchangeably. Indeed, it can be quite difficult to determine just what the communication topology of a system actually is, because many operating systems lack services that would permit the user to query for this information.

In the remainder of this chapter, we review communication hardware in very superficial terms, giving just enough detail so that the reader is familiar with technology names and properties, but without getting into the level of technical issues that would be important in designing the network topology for a demanding enterprise.

Throughout this chapter, the reader will notice that we use the term "packet" to refer to the type of messages that can be exchanged between communication devices. The distinction between a packet and a *message,* throughout this book, is that a message is a logical object generated by the application for transmission to one or more destinations. A message can be quite large, and can easily exceed the limits imposed by the operating system or the communication hardware. For transmission, messages are therefore fragmented into one or more packets, if necessary. A packet, then, is a hardware-level message, which often respects hardware-imposed size and format constraints, and may contain just a fragment of an application-level message.

2.1 Types of Communication Devices

Communication devices can be coarsely partitioned into functional classes:

▶ *Point to point:* This is a class of devices implementing packet or data passing between two computers. A good example is a pair of modems that communicate over a telephone wire.

The Internet is composed of point-to-point communication devices that form a wide area architecture, to which individual local area networks are connected through Internet gateway devices.

▸ *Multiple access:* This class of devices permits many computers to share a single communications medium—for example, using the popular Ethernet architecture, a single coaxial cable can be used to wire a floor of a building or some other moderately large area. Computers can be connected to this cable by tapping into it, which involves inserting a special type of needle through the outer cover of the coaxial conductor and into the signal-conducting core. The device interfaces implement a protocol in hardware to avoid collisions, which can occur if several machines attempt to send packets at the same time.

▸ *Mesh, tree, and crossbar architectures:* This class of devices consists of point-to-point links, which connect the individual computer to some form of switching mechanism. Messages are typically very small and are routed at hardware link speeds through the switches and to their destinations. Connections of this sort are most often used in parallel computers, but they are also being adapted for very high-speed communication in clusters of more conventional computing nodes.

▸ *ATM switches:* Asynchronous Transfer Mode, or ATM, is an emerging standard for packet switching between computers, over communication links and switches of varied speeds and properties. ATM is based on a star architecture, in which optical fibers connect individual computers to switches; these behave much like the communication buses seen in parallel computers. ATM is designed for very high-speed communications, including optical fiber, that can operate at speeds of 2.5 GB/sec or more. Even the first-generation systems are extremely fast, giving performance of 155 MB/sec for individual connections ("OC3" in the ATM terminology).

▸ *Bridges:* A bridge, or *router* (we'll use the term "bridge" to avoid confusion with the concept of routing) is a special-purpose communication computer that links multiple networking devices by forwarding the packets received on either device onto the other. Bridges introduce some latency, which is called a "hop delay," but normally they operate as fast as the devices they interconnect. Bridges tend to lose packets if a destination network is heavily loaded and data cannot be forwarded as fast as they are received from the originating network. Bridges can be programmed to forward messages selectively; this is often exploited to avoid a problem whereby the load on a network can grow without limit as the size of the network is increased—the load on a bridged network is the sum of the load local to a segment, and the load forwarded over the bridge, and can be much less than the sum of the loads on all segments.

2.2 Properties

Communication devices vary enormously in their properties, although this variability is often concealed by the layers of system software through which applications operate. In simple terms, communication devices can be rated by a series of metrics:

- *The maximum data throughput of the device:* Speed is normally measured in terms of the number of bytes of data per second that can be transmitted. Vendors often quote figures in terms of bits per second, referring to the performance seen on the wire as information is transmitted. In either case, one should be aware that speed figures often do not include overhead such as start and stop bits, headers and trailers, and mandatory dead space between packets. These factors can greatly reduce the effective performance of a device, making it difficult to obtain even as much as half of the maximum theoretical performance from an application program. Indeed, it is not uncommon for the most easily used communication primitives to offer performance that is an order of magnitude or more poorer than that of the hardware! This often forces the application designer to choose between performance and software complexity.

- *The number of packets per second that can be sent:* Many devices have a startup overhead associated with sending packets, and for some devices the sending interface must wait for access to the communication medium. These factors combine to limit the number of packets per second that a device can send and, when packets can be of variable size, can also imply that to achieve maximum data throughput, the application must send large packets.

- *The end-to-end latency of the device:* This is a measure of how much time elapses from when a packet starts to be transmitted and when it is first presented to the receiving machine, and it is generally quoted as an average figure, which will actually vary depending on the degree to which the network is loaded at the time a packet is transmitted.

- *The reliability of the device:* All commonly used communication hardware includes automatic mechanisms for detection and rejection of corrupted data. These methods operate using checksums (CRC computations) and are not infallible, but in practice it is normal to treat communication hardware as failing only by packet loss. The reliability of a communication technology is a measure of the percentage of packets that can be lost in the interface or on the wire, as an average. Average reliability is often very high—not uncommonly, hardware approaches perfect reliability. However, it should be kept in mind that an average loss rate may not apply in an exceptional situation, such as an interface that is experiencing intermittent failures, a poorly connected Ethernet tap, or a pattern of use that stresses some sort of uncommon loss problem associated with a technology. From the perspective of the builder of a reliable distributed system, these factors imply that a communication device should be considered somewhat bimodal: having one reliability level in the normal case, but perhaps having a second, much poorer reliability level, in exceptional cases, which the system may need to tolerate or reconfigure.

- *Security:* This is a measure of the extent to which the device protects the contents of packets from eavesdroppers. Some devices achieve very high levels of security in hardware; others are completely open to eavesdropping and require that measures be taken in software if data security is desired.

- *Privacy:* This is a measure of the extent to which the device conceals the source and destination of packets from eavesdroppers. Few devices offer privacy properties, and many security features are applied only to the data portion of a packet, and hence offer little help if privacy is desired. However, there are technologies for which privacy is a meaningful concept—for example, on an Ethernet, interfaces can be programmed to respond to a small set of addresses within a very large space of potential addresses. This feature potentially allows the destination of a packet to be concealed from listeners. On the other hand, the Ethernet standard never permits the address of the sender to be reprogrammed, and, consequently, will always reveal the address of the communication interface from which a packet was sent.

2.3 Ethernet

At the time of this writing, Ethernet is the most widely used communication technology for local area networks (networks within a limited physical region, such as a single floor of a building). Bridged Ethernets are the most common technology for networks within small enterprises, such as a large company at a single site.

As summarized earlier, the basic technology underlying an Ethernet is a shared coaxial cable, on which signals are transmitted using a modulation technology similar to that of a radio. Packets have a fixed maximum size of 1,400 bytes, but the size can be varied as long as this limit is not exceeded. In practice, software that runs over the Ethernet will often be limited to approximately 1,024 bytes of payload in each packet; the remaining 376 bytes are then available for headers, trailers, and data representation information. The Ethernet itself, however, treats the entire object as data. The specific encoding used to represent packets will not be important to us here, but the basic idea is that each interface is structured into a sending side and a listening side, and the latter is continuously active.

To receive a message, the listening side waits until it senses a packet header. The incoming packet is then copied to a memory buffer internal to the Ethernet interface. To be accepted, a packet must have a valid checksum and must specify a destination address that the interface has been preprogrammed to accept. Specifically, each interface has some number of programmable address registers, consisting of a pattern mask and a corresponding value mask, each 32 bits in length. The pattern mask specifies bits within the destination address that must exactly match the corresponding bits of the value mask. A pattern mask that selects for all bits of the address will require an exact match between the packet and the value mask. A pattern mask that selects no bits will match every incoming packet—an interface with such an address loaded is said to be in *promiscuous mode.*

A received packet is copied into memory in the host computer, or discarded if no memory for an incoming packet is available. The host is then interrupted. Most Ethernet interfaces permit the host to enqueue at least two memory regions for incoming messages, and some permit the host to chain a list of memory regions. Most also permit multiple (address, mask) pairs to be loaded into the interface, often as many as 64 (see Figure 2.1).

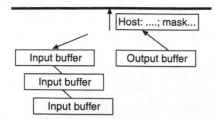

FIGURE 2.1. Ethernet interface with one queued output buffer and three available input buffers. A table of up to 64 (host, mask) pairs controls input selectivity.

To send a packet, the Ethernet interface waits for a pause between packets—a time when its listening side is idle. It then transmits the packet, but also listens to its own transmission. The idea is that if two Ethernets both attempt to send at the same time, a collision will occur and the messages will overwrite one another, causing a noise burst. The receive logic will either fail immediately, or the checksum test will fail, since anything the interfaces read back in will be damaged by the collision, and the sending logic will recognize that a problem has occurred. The hardware implements an *exponential back-off* algorithm, in which the sending side delays for a randomly selected period of time within an interval that starts at a small value but doubles with each successive collision up to a maximum upper value. Although the probability of a collision on a first attempt to send can be high when the Ethernet becomes loaded, exponential back-off has been shown to give very good average access behavior with excellent fairness properties.[1] Because collisions are often detectable within a few bits after starting to send, Ethernets lose little data to collisions even under heavy load, and the back-off algorithm can be shown to provide very uniform delays for access to the medium over very large numbers of senders and excess loads.

As a general rule, although a single interface can send multiple packets, a small amount of dead space will separate each packet in the stream. This is because some work by the operating system is normally needed before each successive packet can be transmitted, and also because the Ethernet hardware logic requires time to compare the checksum on the echo of the outgoing packet, and to trigger an interrupt to the device driver, before starting to send a new packet. In contrast, when more than one interface is used to send data, sequences of back-to-back packets can be generated, potentially forcing the interface to accept several packets in a row with almost no delay between them. Obviously, this can result in packet loss if the chain of memory for incoming messages is exhausted. However, precisely because an Ethernet is shared, the probability that any one interface will be the destination for any large number of back-to-back packets is low. File system servers and bridges, which are more likely to receive back-to-back packets, compensate for this by using long chains of buffers for incoming messages, and implementing lightweight logic for dealing with received messages as quickly as possible.

An interesting feature of the Ethernet is that it supports both broadcast and multicast in hardware. The two features are implemented in the same way. Before any communication is undertaken, the Ethernet interface is preloaded with a special address—one that is the same on all machines within some set. Now, if a packet that contains this address is transmitted, all machines in that set will receive a copy, because all of their interfaces will detect a match.

---

[1]Developers of real-time computing systems have developed deterministic back-off algorithms, which have extremely predictable behavior under heavy load, for settings such as process control. In these approaches, the loaded behavior of an Ethernet is completely characterized—provided, of course, that all interfaces use the same algorithm.

　　　　　COMMUNICATION TECHNOLOGIES　　　Chapter 2

To support broadcast, a special address is agreed upon and installed on all interfaces in the entire network, typically at the time the machine is first booted. Broadcast packets will now be received by every machine, offering a way to distribute the same data to a great many machines at very low cost. However, one should keep in mind that each interface individually computes the checksum; hence, some interfaces may discard a packet as corrupted, while others may accept it. Moreover, some machines may lack input buffers and be incapable of accepting packets that are correctly received by the interface. Thus, Ethernet broadcast can potentially send a single packet to all machines on a network; in practice, however, the technology is not a reliable one.

Multicast uses precisely the same approach, except that a subset of machines picks a group address, which is unique within the network, and installs this into the interfaces. Messages destined to a multicast address will be accepted (up to checksum failures) by just these machines, and will be ignored by others. The maximum number of multicast addresses that an interface can support varies from vendor to vendor. As in the case of broadcast, hardware multicast is reasonably reliable but not absolutely so.

Earlier, we commented that even a very reliable communications device may exhibit modal behavior whereby reliability can be much poorer for certain communication patterns. One example of this problem is termed the *broadcast,* or *multicast storm,* and occurs when broadcast or multicast is used by multiple senders concurrently. In this situation, it becomes much more likely that a typical network interface will actually need to accept a series of back-to-back packets—with a sufficient number of senders, chains of arbitrary length can be triggered. The problem is that in this situation the probability that two back-to-back messages will be destined to the same machine, or set of machines, becomes much higher than for a more typical point-to-point communication load. As a result, the interface may run out of buffers for incoming packets and begin to drop them.

In a broadcast storm situation, packet loss rises to very high levels, because network interfaces become chronically short of memory for incoming packets. The signature of a broadcast storm is that heavy use of broadcast or multicast by multiple senders causes a dramatic increase in the packet loss rate throughout the network. Notice that the problem will affect any machine that has been programmed to accept incoming broadcasts, not just the machines that make meaningful use of the packets after they occur. Fortunately, the problem is uncommon if just a single sender is initiating the broadcasts, because a single sender will not generate back-to-back packets.

2.4 Fiber Distributed Data Interface

Fiber Distributed Data Interface (FDDI) is a multiaccess communication technology based upon a ring architecture, in which interfaces are interconnected by shielded twisted-pair wiring. An interface plays a dual role:

▶ As a *repeater,* an FDDI interface receives messages from the interface to its left, accepts those messages that match an incoming address pattern (similar to Ethernet), and then (accepted or not) forwards the message to the interface on the right. Forwarding occurs bit by bit or in small blocks of bits, so the delay associated with forwarding packets can be very low.

> ◗ As a *transmitter,* an FDDI interface waits until it has permission to initiate a packet, which occurs when there is no other packet being forwarded, and then sends its packet to the interface on its right. When the packet has completed its trip around the ring, the transmitter receives it from the interface to the left and deletes it. Status information is available in the packet trailer, and can be used to immediately retransmit a packet that some intended destination was unable to accept because of a shortfall of memory or because it detected a checksum error.

Finally, FDDI has a built-in fault-tolerance feature: If a link fails, FDDI will automatically reconfigure itself to route around it, as illustrated in Figure 2.2. In contrast, a severed Ethernet will either become inoperative, or will partition into two or more segments, which are disconnected from one another.

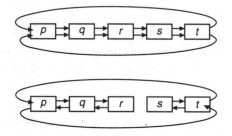

FIGURE 2.2. An FDDI ring is able to heal itself if a link breaks, as seen in the lower example.

FDDI throughput (150 MB/sec) is about 15 times greater than standard 10 MB Ethernet, although high-speed 100 MB Ethernet interfaces have recently been introduced that can approach FDDI performance. Latency for an FDDI ring, in particular, is poorer than that of an Ethernet, because of the protocol used to wait for permission to send, and because of delays associated with forwarding the packet around the ring. In complex environments, Ethernets or FDDI rings may be broken into segments, which are connected by some form of bridge or routing device; these environments have latency and throughput properties that are more difficult to quantify, because the topology of the interconnection path between a pair of computers can significantly impact the latency and throughput properties of the link.

2.5   B-ISDN and the Intelligent Network

B-ISDN is a standard introduced by telecommunication switching companies for advanced telephone services. ISDN stands for "integrated services digital network," and is based on the idea of supporting a communication system that mixes digital voice with other types of digital data, including video. The B stands for "broadband" and is a reference to the underlying data links, which operate at the extremely high speeds needed to handle these kinds of data.

Layered over B-ISDN, which is an infrastructure standard, is the emerging *intelligent network*, an elaborate software architecture for supporting next-generation telecommunication services. Such services have been slower to emerge than had originally been predicted, and many companies have become cautious in accessing intelligent network prospects for the near-term future. However, it may simply be the case that the intelligent network has been harder to deploy than was originally predicted: There are many signs that the long-predicted revolution in telecommunications is really happening today. Break-out No. 2.1 discusses the challenges of the emerging information superhighway.

The B-ISDN architecture is elaborate, and the intelligent network is even more so. Our focus on reliability of general distributed computing systems prevents us from discussing either in any detail here. However, one interesting feature of B-ISDN merits mention. Although destination addresses in ISDN are basically telephone numbers, ISDN interprets these in a very flexible manner, which permits the architecture to do far more than just create connections between telephones and other paired devices. Instead, the ISDN architecture revolves around the concept of an intelligent switching system: Each packet follows a route from the source through a series of switches to its destination. When this route is first established, each switch maps the destination telephone number to an appropriate outgoing connection (to another switch), to a local point of data delivery (if this switch serves the destination), or to a service. The decision is made by looking up the telephone number in a database, using a software procedure that can be reprogrammed to support very elaborate behaviors—for example, suppose that a telephone company wanted to offer a special communication service to computer vendors in a particular metropolitan area. This service would offer a single telephone number to each vendor and would arrange to automatically route a call to the mobile telephone of the computer repair person physically closest to the caller.

To implement this service using B-ISDN, the telephone company would make use of a database it already maintains, giving the locations of mobile telephone units. As a call is routed into a switch, the company would sense that the destination is one of the special telephone numbers assigned to this new service. It would invoke a database search algorithm programmed to look up the physical address of the caller, and then match this with the locations of service vehicles for the called organization to pick the closest one, routing the call to the computer vendor's switchboard if the lookup fails. Although timing constraints for this process are demanding (actions are generally required within a small fraction of a second), modern computers are becoming fast enough to work within these sorts of deadlines. And, one can imagine a great number of other B-ISDN services, including message centers, automatic playback of prerecorded information, telephone numbers that automatically patch into various types of public or private information bases, and so forth. The potential is huge.

B-ISDN is also illustrative of how advances in telecommunication switching technology are creating new demands for reliable distributed software services. It is common to require that telephone systems maintain extremely high levels of reliability—a typical requirement is that

Challenges of an Emerging Information Superhighway

Challenges of an emerging information superhighway include providing guarantees of privacy, availability, consistency, security, and trustworthiness in a wide range of innovative applications. Prototypes of the new services that may someday become critical often lack the sorts of strong guarantees that will eventually be required if these types of systems are to play the role that is envisioned by the public and government.

Today, the telephone system is relatively inflexible: One can telephone from more or less any telephone to any other, but the telephone numbers correspond to static locations. If mobile telephones are used, one calls the mobile device using a scheme in which requests are routed through the home location. In the future, however, a great variety of innovative telephone-based communication services will become available. These will include telecommunication services that mix various forms of media—voice, image, computer data—and abstract data types capable of performing computations or retrieving desired information. Mobility of users and services will greatly increase, as will the sophistication of the routing mechanisms used to track down the entity to which a call should be connected. Moreover, whereas contemporary telecommunication systems are very different from computer communication architectures such as the World Wide Web, these will be increasingly integrated in the future and will eventually merge into a single infrastructure with the properties of both.

In many communities, ISDN will bring high-bandwidth connections into the home at reasonable cost. This development will revolutionize use of the Web, which is currently bandwidth limited from the curb to the home and hence only usable in limited ways from home computing platforms.

High-speed communication is already available in the workplace, however, and this trend is already creating new businesses. It is widely expected that a boom in commerce associated with the communication infrastructure will follow, as it reaches the average household—for example, one can imagine small companies that offer customized services to a worldwide clientele over the network, permitting entrepreneurial activities that will tap into an immense new market.

Of course, the promise of this new world comes with challenges. As we come to rely more and more heavily on innovative communication technologies for day-to-day activities, and these new forms of information-based work and commerce grow in importance, the reliability requirements placed on the underlying technology infrastructure will also grow. Thus, there is a great potential for economic growth and increased personal freedom associated with the new communication technologies, but there is also a hidden implication that the software implementing such systems be secure, private, fault-tolerant, consistent in its behavior, and trustworthy.

These properties do not hold for many of the prototype systems that have so excited the public, and a significant change in the mindset of the developers of such applications will be needed before reliability of this sort becomes routine—for example, fraudulent use of telephone systems has increased with the introduction of new forms of flexibility in the system, and attacks on secure or critical information-based applications have risen steadily in recent years. These attacks range from actions by malicious or careless insiders, whose actions disrupt critical systems, to aggressive attacks by hackers, terrorists, and other agents, whose goal is to cause damage or to acquire confidential information.

not more than one call in 100,000 be dropped, and downtime for an entire switch may be limited to seconds per year or less—switches are increasingly used to support critical services such as 911 emergency numbers, as well as communication among air traffic controllers, emergency services vehicles, police vehicles, and so forth.

Reliability of this sort has many implications for developers of advanced switching systems. The switches themselves must be paired, and protocols for doing so have been standardized as part of an architecture called Signaling System 7 (SS7), which is gradually entering into worldwide use. The coprocessors on which intelligent services reside are often constructed using fault-tolerant computing hardware. The software that implements the switching logic must be self-managing, fault-tolerant, and capable of supporting on-line upgrades to new versions of applications and of the operating system itself. And, because many services require some form of distributed database, sets of coprocessors will often need to be organized into distributed systems that manage dynamically changing replicated data and take actions in a consistent but decentralized manner—for example, routing a call may require independent routing decisions by the service programs associated with several switches, and these decisions need to be based upon consistent data or the call will eventually be dropped or be handled incorrectly.

B-ISDN, as well as the intelligent network it is intended to support, represents a good example of a setting where the technology of reliable distributed computing is required, and it will have a major impact on society as a whole. Given solutions to reliable distributed computing problems, a vast array of useful telecommunication services will become available starting in the near future and continuing over the decades to come. One can imagine a telecommunication infrastructure that is nearly ubiquitous and elegantly integrated into the environment, providing information and services to users without the constraints of telephones, which are physically wired to the wall, and computer terminals or televisions, which weigh many pounds and are physically attached to a company's network. But the dark side of this vision is that without adequate attention to reliability and security, this exciting new world will also be erratic and prone to failure.

2.6   Asynchronous Transfer Mode

Asynchronous Transfer Mode, or *ATM,* is an emerging technology for routing small digital packets in telecommunication networks. When used at high speeds, ATM networking is the broadband layer underlying B-ISDN; thus, an article describing a B-ISDN service is quite likely to be talking about an application running on an ATM network that is designed using the B-ISDN architecture.

ATM technology is considered especially exciting both because of its extremely high bandwidth and low latencies, and because this connection to B-ISDN represents a form of direct convergence between the telecommunication infrastructure and the computer communication infrastructure. With ATM, for the first time, computers are able to communicate directly over the data transport protocols used by the telephone companies. Over time, ATM networks will be more and more integrated with the telephone system, offering the possibility of new kinds of telecommunication applications that can draw immediately upon the worldwide telephone network. Moreover, ATM opens the door for technology migration from those who develop software for computer networks and distributed systems into the telecommunication infrastructure and environment.

The packet switches and computer interfaces needed in support of ATM standards are being deployed rapidly in industry and research settings, with performance expected to scale from rates comparable to those of a fast Ethernet for first-generation switches to gigabyte rates in the late 1990s and beyond. ATM is defined as a routing protocol for very small packets, containing 48 bytes of payload data with a five-byte header. These packets traverse routes that must be prenegotiated among the sender, the destination, and the switching network. The small size of the ATM packets leads some readers to assume that ATM is not really about networking in the same sense as an Ethernet, with its 1,400-byte packets. In fact, the application programmer normally would not need to know that messages are being fragmented into such a small size, tending instead to think of ATM in terms of its speed and low latency. Indeed, at the highest speeds, ATM cells can be thought of almost as if they were fat bits, or single words of data being transferred over a backplane.

ATM typically operates over point-to-point fiber-optic cables, which route through switches. Thus, a typical ATM installation might resemble the one shown in Figure 2.3. Notice that in this figure, some devices are connected directly to the ATM network itself and not handled by any intermediary processors. The rationale for such an architecture is that ATM devices may eventually run at such high data rates[2] (today, an OC3 ATM network operates at 155 MB/sec, and future OC24 networks will run at a staggering 1.2 GB/sec) that any type of software intervention on the path between the data source and the data sink would be out of the question. In such environments, application programs will more and more be relegated to a supervisory and control role, setting up the links and turning the devices on and off, but not accessing the data flowing through the network in a direct way. Not shown in Figure 2.3 are adapters, which might be used to interface an ATM directly to an Ethernet or some other local area technology, but these are available on the market today and will play a big role in many future ATM installations. These devices allow an ATM network to be attached to an Ethernet, Token Ring, or FDDI network, with seamless communication through the various technologies. They should be common late in the 1990s.

The ATM header consists of a VCI (two bytes, giving the virtual circuit ID); a VPI (one byte, giving the virtual path ID); a flow-control data field for use in software; a packet-type bit (normally used to distinguish the first cell of a multicell transmission from the subordinate ones, for reasons that will become clear momentarily); a cell loss priority field; and a one-byte error-

---

[2]ATM data rates are typically quoted on the basis of the maximum that can be achieved through any single link. However, the links multiplex through switches and when multiple users are simultaneously active, the maximum individual performance may be less than the maximum performance for a single dedicated user. ATM bandwidth allocation policies are an active topic of research.

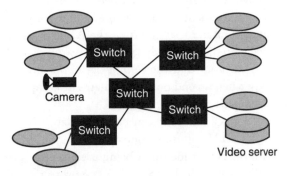

FIGURE 2.3. Client systems (gray ovals) connected to an ATM switching network. The client machines can be PCs or workstations, but they can also be devices such as ATM frame grabbers, file servers, or video servers. Indeed, the very high speed of some types of data feeds may rule out any significant processor intervention on the path from the device to the consuming application or display unit. Over time, software for ATM environments may be split more and more into a managerial and control component, which sets up circuits and operates the application, and a data-flow component, which moves the actual data without a direct program. In contrast to a standard computer network, an ATM network can be integrated directly into the networks used by the telephone companies themselves, offering a unique route toward eventual convergence of distributed computing and telecommunications.

checking field, which typically contains a checksum for the header data. Of these, the VCI and the packet-type (PTI) bit are the most heavily used, and the ones we will discuss in more detail. The VPI is intended for use when a number of virtual circuits connect the same source and destination; it permits the switch to multiplex such connections in a manner that consumes less resources than if the VCIs were used directly for this purpose. However, most current ATM networks set this field to 0, and hence we will not discuss it further here.

There are three stages to creating and using an ATM connection. First, the process initiating the connection must construct a route from its local switch to the destination. Such a route consists of a path of link addresses—for example, suppose that each ATM switch is able to accept up to eight incoming links and eight outgoing sinks. The outgoing links can be numbered 0–7, and a path from any data source to any data sink can then be expressed as a series of 3-bit numbers, indexing each successive hop that the path will take. Thus, a path written as 4.3.0.7.1.4 might describe a route through a series of six ATM switches. Having constructed this path, a virtual circuit identifier is created, and the ATM network is asked to open a circuit with that identifier and path. The ATM switches, one by one, add the identifier to a table of open identifiers and record the corresponding out-link to use for subsequent traffic. If a bidirectional link is desired, the same path can be set up to operate in both directions. The method generalizes to also include multicast and broadcast paths. The VCI, then, is the virtual circuit identifier used during the open operation.

Having described this, however, it should be stressed that many early ATM applications depend upon what are called "permanent virtual channels"—namely, virtual channels that are preconfigured by a system administrator at the time the ATM is installed, and rarely (if ever) changed. Although it is widely predicted that dynamically created channels will eventually dominate the use of ATM, it may turn out that the complexity of opening channels and of ensuring that they are closed correctly when an end point terminates its computation or fails will present an obstacle that could prevent this from happening.

In the second stage, the application program can send data over the link. Each outgoing message is fragmented by the ATM interface controller into a series of ATM packets, or cells. These cells are prefixed with the circuit identifier being used (which is checked for security purposes), and the cells then flow through the switching system to their destination. Most ATM devices will discard cells in a random manner if a switch becomes overloaded, but there is a great deal of research underway on ATM scheduling, and a variety of so-called *quality-of-service* options will become available over time. These might include guarantees of minimum bandwidth, priority for some circuits over others, or limits on the rate at which cells will be dropped. Fields such as the packet-type field and the cell loss priority field are intended for use in this process.

It should be noted, however, that just as many early ATM installations use permanent virtual circuits instead of supporting dynamically created circuits, many also treat the ATM as an Ethernet emulator and employ a fixed bandwidth allocation corresponding roughly to what an Ethernet might offer. It is possible to adopt this approach because ATM switches can be placed into an emulation mode in which they support broadcast, and early ATM software systems have taken advantage of this to layer the TCP/IP protocols over ATM much as they are built over an Ethernet. However, fixed bandwidth allocation is inefficient, and treating an ATM as if it were an Ethernet somewhat misses the point! Looking to the future, most researchers expect this emulation style of network to gradually give way to direct use of the ATM itself, which can support packet-switched multicast and other types of communication services. Over time, value-added switching is also likely to emerge as an important area of competition between vendors—for example, one can easily imagine incorporating encryption and filtering directly into ATM switches and in this way offering virtual private network services to users (Chapters 17 and 19).

The third stage of ATM connection management is concerned with closing a circuit and freeing dynamically associated resources (mainly, table entries in the switches). This occurs when the circuit is no longer needed. ATM systems that emulate IP networks or that use permanent virtual circuits are able to skip this final stage, leaving a single set of connections continuously open, and perhaps dedicating some part of the aggregate bandwidth of the switch to each such connection. As we evolve to more direct use of ATM, one of the reliability issues that may occur will be that of detecting failures so that any ATM circuits opened by a process that later crashed will be safely and automatically closed on its behalf. Protection of the switching network against applications that erroneously (or maliciously) attempt to monopolize resources by opening a great many virtual circuits will also need to be addressed in future systems.

ATM poses some challenging software issues. Communication at gigabyte rates will require substantial architectural evolution and may not be feasible over standard OSI-style protocol stacks, because of the many layers of software and protocols that messages typically traverse in these architectures. As noted above, ATM seems likely to require that video servers and disk data servers be connected directly to the wire, because the overhead and latency associated with fetching data into a processor's memory before transmitting can seem very large at the extremes of performance for which ATM is intended. These factors make it likely that although ATM will be usable in support of networks of high-performance workstations, the technology will really take off in settings that exploit novel computing devices and new types of software architectures. These issues are already stimulating reexamination of some of the most basic operating system structures, and when we look at high-speed communication in Chapter 8, many of the technologies considered turn out to have arisen as responses to this challenge.

Even layering the basic Internet protocols over ATM has turned out to be nontrivial. Although it is easy to fragment an IP packet into ATM cells, and the emulation mode mentioned above makes it straightforward to emulate IP networking over ATM networks, traditional IP software will drop an entire IP packet if any part of the data within it is corrupted. An ATM network that drops even a single cell per IP packet would thus seem to have 0 percent reliability, even though close to 99 percent of the data might be getting through reliably. This consideration has motivated ATM vendors to extend their hardware and software to understand IP and to arrange to drop *all* of an IP packet if even a single cell of that packet must be dropped, an example of a simple quality-of-service property. The result is that as the ATM network becomes loaded and starts to shed load, it does so by beginning to drop entire IP packets, hopefully with the result that other IP packets will get through unscathed. This leads us to the use of the packet-type identifier bit: The idea is that in a burst of packets, the first packet can be identified by setting this bit to 0, and subsequent subordinate packets can be identified by setting it to 1. If the ATM must drop a cell, it can then drop all subsequent cells with the same VCI until one is encountered with the PTI bit set to 0, on the theory that all of these cells will be discarded in any case upon reception, because of the prior lost cell.

It should not be long before IP drivers or special ATM firmware is developed that can buffer outgoing IP packets briefly in the controller of the sender and selectively solicit retransmission of just the missing cells if the receiving controller notices that data are missing. One can also imagine protocols whereby the sending ATM controller might compute and periodically transmit a parity cell containing the exclusive-or of all the prior cells for an IP packet; such a parity cell could then be used to reconstruct a single missing cell on the receiving side. Quality-of-service options for video data transmission using MPEG or JPEG may soon be introduced. Although these suggestions may sound complex and costly, keep in mind that the end-to-end latencies of a typical ATM network are so small (tens of microseconds) that it is entirely feasible to solicit the retransmission of a cell or two even as the data for the remainder of the packet flow through the network. With effort, such steps should eventually lead to very reliable IP networking at ATM speeds. But the nontrivial aspects of this problem also point to the general difficulty of what, at first glance, might have seemed to be a completely obvious step to take.

2.7   Cluster and Parallel Architectures

Parallel supercomputer architectures, and their inexpensive and smaller-scale cousins, the cluster computer systems, have a natural correspondence to distributed systems. Increasingly, all three classes of systems are structured as collections of processors connected by high-speed communication buses and with message passing as the basic abstraction. In the case of cluster computing systems, these communication buses are often based upon standard technologies, such as fast Ethernet or packet switching, similar to those used in ATM. However, there are significant differences, too, both in terms of scale and properties. These considerations make it necessary to treat cluster and parallel computing as a special case of distributed computing for which a number of optimizations are possible, and where special considerations are also needed in terms of the expected nature of application programs and their goals vis-à-vis the platform.

In particular, cluster and parallel computing systems often have built-in management networks that make it possible to detect failures extremely rapidly, and they may have special-purpose communication architectures with extremely regular and predictable performance and reliability properties. The ability to exploit these features in a software system creates the possibility that developers will be able to base their work on the general-purpose mechanisms used in general distributed computing systems, optimizing them in ways that might greatly enhance their reliability or performance—for example, we will see that the inability to accurately sense failures is one of the hardest problems to overcome in distributed systems; certain types of network failures can create conditions indistinguishable from processor failure, and yet may heal themselves after a brief period of disruption, leaving the processor healthy and able to communicate again as if it had never been gone. Such problems do not arise in a cluster, or parallel, architecture, where accurate failure detection can be wired to available hardware features of the communication interconnect.

In this book, we will not consider cluster or parallel systems until Chapter 24, at which time we will consider how the special properties of such systems impact the algorithmic and protocol issues we have considered in the previous chapters. Although there are some important software systems for parallel computing (PVM is the best known [see Geist et al.]; MPI may eventually displace it), these are not particularly focused on reliability issues, and hence will be viewed as being beyond the scope of the current treatment.

2.8   Next Steps

Few areas of technology development are as active as that involving basic communication technologies. The coming decade should see the introduction of powerful wireless communication technologies for the office, permitting workers to move computers and computing devices around a small space without the rewiring that contemporary devices often require. Bandwidth delivered to the end user can be expected to continue to increase, although this will also require substantial changes in the software and hardware architecture of computing devices, which currently limits the achievable bandwidth for traditional network architectures. The emergence of exotic computing devices targeted to single applications should begin to displace general computing systems from some of these very demanding settings.

As speeds increase, so too will congestion and contention for network resources. It is likely that virtual private networks, supported through a mixture of software and hardware, will soon become available to organizations able to pay for dedicated bandwidth and guaranteed latency. Such networks will need to combine strong security properties with new functionality, such as conferencing and multicast support. Over time, it can be expected that these data-oriented networks will merge into the telecommunication intelligent network architecture, which provides support for voice, video and other forms of media, and mobility. All of these features will present the distributed application developer with new options, as well as new reliability challenges.

Reliability of the telecommunication architecture is already a concern, and that concern will only grow as the public begins to insist on stronger guarantees of security and privacy. Today, the rush to deploy new services and to demonstrate new communication capabilities has somewhat overshadowed robustness issues of these sorts. One consequence, however, has been a rash of dramatic failures and attacks on distributed applications and systems. Shortly after work on this book began, a telephone "phreak" was arrested for reprogramming the telecommunication switch in his home city in ways that gave him nearly complete control over the system, from the inside. He was found to have used his control to misappropriate funds through electronic transfers, and the case is apparently not an isolated event. Meanwhile, new services such as caller ID have turned out to have unexpected side-effects, such as permitting companies to build databases of the telephone numbers of the individuals who contact them. Obviously, not all of these individuals would have agreed to divulge their numbers.

Such events, understandably, have drawn considerable public attention and protest. As a consequence, they contribute toward a mindset in which the reliability implications of technology decisions are being given greater attention. Should the trend continue, it could eventually lead to wider use of technologies that promote distributed computing reliability, security, and privacy over the coming decades.

2.9 Related Reading

Additional discussion of the topics covered in this chapter can be found in Comer, Comer and Stevens (1991, 1993), Coulouris et al., Tannenbaum.

An outstanding treatment of ATM is given in Handel et al.

CHAPTER 3 ✧ ✧ ✧ ✧

Basic Communication Services

CONTENTS

3.1 Communication Standards

A communication standard is a collection of specifications governing the types of messages that can be sent in a system, the formats of message headers and trailers, the encoding rules for placing data into messages, and the rules governing format and use of source and destination addresses. In addition to this, a standard will normally specify a number of protocols that a provider should implement.

Examples of communication standards that are used widely, although not universally, are:

▸ *The Internet protocols:* These protocols originated in work done by the Defense Department's Advanced Research Projects Agency, or DARPA, in the 1970s, and have gradually grown into a wider-scale, high-performance network interconnecting millions of computers. The protocols employed in the Internet include IP, the basic packet protocol; and UDP, TCP, and IP-multicast, each of which is a higher-level protocol layered over IP. With the emergence of the Web, the Internet has grown explosively during the mid-1990s.

▸ *The Open Systems Interconnection protocols:* These protocols are similar to the Internet protocol suite, but employ standards and conventions that originated with the ISO organization.

▸ *Proprietary standards:* Examples include the Systems Network Architecture, developed by IBM in the 1970s and widely used for mainframe networks during the 1980s; DECnet, developed at Digital Equipment Corporation but discontinued in favor of open solutions in the 1990s; NetWare, Novell's widely popular networking technology for PC-based client/server networks; and Banyan's VINES system, also intended for PCs used in client/server applications.

During the 1990s, the emergence of open systems—namely, systems in which computers from different vendors could run independently developed software—has been an important trend. Open systems favor standards, but also must support current practice, since vendors otherwise find it hard to move their customer base to the standard. At the time of this writing, the trend clearly favors the Internet protocol suite as the most widely supported communication standard, with the Novell protocols strongly represented by force of market share. However, their protocol suites were designed long before the advent of modern high-speed communication devices, and the commercial pressure to develop and deploy new kinds of distributed applications that exploit gigabyte networks could force a rethinking of these standards. Indeed, even as the Internet has become a de facto standard, it has turned out to have serious scaling problems, which may not be easy to fix in less than a few years (see Break-out 3.1).

The remainder of this chapter focuses on the Internet protocol suite, because this is the one used by the Web. Details of how the suite is implemented can be found in Comer, Comer and Stevens (1991, 1993).

3.2 Addressing

The *addressing* tools in a distributed communication system provide unique identification for the source and destination of a message, together with ways of mapping from symbolic names for resources and services to the corresponding network address, and for obtaining the best route to use for sending messages.

Addressing is normally standardized as part of the general communication specifications for formatting data in messages, defining message headers, and communicating in a distributed environment.

Within the Internet, several address formats are available, organized into classes aimed at different styles of application. Each class of address is represented as a 32-bit number. Class A Internet addresses have a 7-bit network identifier and a 24-bit host identifier and are reserved for very large networks. Class B addresses have 14 bits for the network identifier and 16 bits for the host ID, and class C has 21 bits for the network identifier and 8 bits for the host ID. These last two classes are the most commonly used. Eventually, the space of Internet addresses is likely to be exhausted, at which time a transition to an extended IP address is planned; the extended format increases the size of addresses to 64 bits, but does so in a manner that provides backward compatibility with existing 32-bit addresses. However, there are many problems raised by such a transition and industry is clearly hesitant to embark on what will be a hugely disruptive process.

Internet addresses have a standard ASCII representation, in which the bytes of the address are printed as signed decimal numbers in a standardized order—for example, this book was edited on host gunnlod.cs.cornell.edu, which has Internet address 128.84.218.58. This is a class B Internet address, with network address 42 and host ID 218.58. Network address 42 is assigned to Cornell University, as one of several class B addresses used by the university. The 218.xxx addresses designate a segment of Cornell's internal network—namely, the Ethernet to which my computer is attached. The number 58 was assigned by the Computer Science Department to identify my host on this Ethernet segment.

A class D Internet address is intended for a special use: IP multicasting. These addresses are allocated for use by applications that exploit IP multicast. Participants in the application join the multicast group, and the Internet routing protocols automatically reconfigure themselves to route messages to all group members.

The string gunnlod.cs.cornell.edu is a symbolic name for an IP address. The name consists of a machine name (gunnlod, an obscure hero of Norse mythology) and a suffix (cs.cornell.edu), designating the Computer Science Department at Cornell University, which is an educational institution in the United States. The suffix is registered with a distributed service called the domain name service, or DNS, which supports a simple protocol for mapping from string names to IP network addresses.

Here's the mechanism used by the DNS when it is asked to map my host name to the appropriate IP address for my machine. DNS has a top-level entry for edu but doesn't have an Internet address for this entry. However, DNS resolves cornell.edu to a gateway address for the Cornell domain—namely, host 132.236.56.6. Finally, DNS has an even more precise address stored for cs.cornell.edu—namely, 128.84.227.15, a mail server and gateway machine in the Computer Science Department. All messages to machines in the Computer Science Department pass through this machine, which intercepts and discards messages to all but a select set of application programs.

Internet Brownouts:
Power Failures on the Data Superhighway?

The data superhighway is experiencing serious growing pains. Growth in load has vastly exceeded the capacity of the protocols used in the Internet and the World Wide Web. Issues of consistency, reliability, and availability of technologies, such as the ones that support these applications, are at the core of this book.

Beginning in late 1995, clear signs emerged that the Internet was beginning to overload. One reason is that the root servers for the DNS architecture are experiencing exponential growth in the load of DNS queries that require action by the top levels of the DNS hierarchy. A server that saw ten queries per minute in 1993 was up to 250 queries per second in early 1995, and traffic was doubling every three months. Such problems point to fundamental aspects of the Internet that were based on assumptions of a fairly small and lightly loaded user population repeatedly performing the same sorts of operations. In this small world, it makes sense to use a single hierarchical DNS structure with caching, because cache hits are possible for most data. In a network that suddenly has millions of users, and that will eventually support billions of users, such design considerations must be reconsidered: Only a completely decentralized architecture can possibly scale to support a truly universal and worldwide service.

These problems have visible but subtle impact on the Internet user: They typically cause connections to break or alert boxes to appear on your Web browser warning you that the host possessing some resource is unavailable. There is no obvious way to recognize that the problem is not one of local overload or congestion, but in fact is an overloaded DNS server or one that has crashed at a major Internet routing point. Unfortunately, such problems have become increasingly common: The Internet is starting to experience brownouts. Indeed, the Internet became largely unavailable for many hours during one crash in September 1995 because of failures of this nature, and this was hardly an unusual event. As the data superhighway becomes increasingly critical, such brownouts represent increasingly serious threats to reliability.

Conventional wisdom has it that the Internet does not follow the laws of physics; there is no limit to how big, fast, and dense the Internet can become. As with the hardware itself, which seems outmoded almost before it reaches the market, we assume that the technology of the network is also speeding up in ways that exceed demand. But the reality of the situation is that the *software architecture* of the Internet is in some basic ways *not* scalable. Short of redesigning these protocols, the Internet won't keep up with growing demands. In some ways, it already can't.

Several problems are identified as the most serious culprits at the time of this writing. Number one in any ranking: the World Wide Web. The Web has taken over by storm, but it is inefficient in the way it fetches documents. In particular, as we will see in Chapter 10, the HyperText Transfer Protocol (HTTP) often requires that large numbers of connections be created for typical document transfers, and these connections (even for a single HyperText Markup Language [HTML] document) can involve contacting many separate servers. Potentially, each of these connection requests forces the root nodes of the DNS to

Internal Brownouts:
Power Failures on the Data Superhighway?
(continued)

respond to a query. With millions of users surfing the network, DNS load is skyrocketing.

Bandwidth requirements are also growing exponentially. Unfortunately, the communication technology of the Internet is scaling more slowly than this. So overloaded connections, particularly near hot sites, are a tremendous problem. A popular Web site may receive hundreds of requests per second, and each request must be handled *separately*. Even if identical bits are being transmitted concurrently to hundreds of users, each user is sent his or her own private copy. And this limitation means that as soon as a server becomes useful or interesting, it also becomes vastly overloaded. Yet even though identical bits are being sent to hundreds of thousands of destinations, the protocols offer no obvious way to multicast the desired data, in part because Web browsers explicitly make a separate connection for each object fetched; they only specify the object to send after the connection is in place. At the time of this writing, the best hope is that popular documents can be cached with increasing efficiency in Web proxies, but, as we will see, doing so also introduces tricky issues of reliability and consistency. Meanwhile, the bandwidth issue is with us to stay.

Internet routing is another area that hasn't scaled very well. In the early days of the Internet, routing was a major area of research, and innovative protocols were used to route around areas of congestion. But these protocols were eventually found to be consuming too much bandwidth and imposing considerable overhead: Early in the 1980s, 30 percent of Internet packets were associated with routing and load balancing. A new generation of relatively static routing protocols was proposed at that time and remains in use today. But the assumptions underlying these new protocols reflected a network that, at the time, seemed large because it contained hundreds of nodes. A network of tens of millions or billions of nodes poses problems that could never have been anticipated in 1985. Now that we have such a network, even trying to understand its behavior is a major challenge. Meanwhile, when routers fail (for reasons of hardware, software, or simply because of overload), the network is tremendously disrupted.

The Internet Engineering Task Force (IETF), a governing body for the Internet and for Web protocols, is working on these problems. This organization sets the standards for the network and has the ability to legislate solutions. A variety of proposals are being considered: They include ways of optimizing the Web protocol called HTTP, as well as other protocol optimizations.

Some service providers are urging the introduction of mechanisms that would charge users based on the amount of data they transfer and thus discourage overuse. There is considerable skepticism regarding the practicality of such measures. Bill Gates has suggested that in this new world, one can easily charge for the size of the on ramp (the bandwidth of one's connection), but not for the amount of information a user transfers, and early evidence supports his perspective. In Gates's view, this is simply a challenge of the new Internet market.

There is no clear solution to the Internet bandwidth problem. However, as we will see, there are some very powerful technologies that could begin to offer answers: Coherent replication and caching are the most obvious remedies for many of the problems cited above. The financial motivations for being first to market with the solution are staggering, and history shows that this is a strong incentive indeed.

DNS is itself structured as a hierarchical database of slowly changing information. It is hierarchical in the sense that DNS servers form a tree, with each level providing addresses of objects in the level below it, but also *caching* remote entries that are frequently used by local processes. Each DNS entry tells how to map some form of ASCII host name to the corresponding IP machine address or, in the case of commonly used services, how to find the service representative for a given host name.

Thus, DNS has an entry for the IP address of gunnlod.cs.cornell.edu (somewhere), and can track it down using its resolution protocol. If the name is used rapidly, the information may become cached locally to the typical users and will resolve quickly; otherwise, the protocol sends the request up the hierarchy to a level at which DNS knows how to resolve some part of the name, and then back down the hierarchy to a level that can fully resolve it. Similarly, DNS has a record telling how to find a mail transfer agent running the SMTP protocol for gunnlod.cs.cornell.edu—this may not be the same machine as gunnlod itself, but the resolution protocol is the same.

The Internet address specifies a machine, but the identification of the specific application program that will process the message is also important. For this purpose, Internet addresses contain a field called the port number, which at present is a 16-bit integer. A program that wants to receive messages must bind itself to a port number on the machine to which the messages will be sent. A predefined list of port numbers is used by standard system services, and has values ranging from 0 to 1,023. Symbolic names have been assigned to many of these predefined port numbers, and a table mapping from names to port numbers is generally provided—for example, messages sent to gunnlod.cs.cornell.edu that specify port 53 will be delivered to the DNS server running on machine gunnlod, or will be discarded if the server isn't running. E-mail is sent using a subsystem called Simple Mail Transfer Protocol (SMTP) on port 25. Of course, if the appropriate service program isn't running, messages to a port will be silently discarded. Small port numbers are reserved for special services and are often trusted, in the sense that it is assumed that only a legitimate SMTP agent will ever be connected to port 25 on a machine. This form of trust depends upon the operating system, which decides whether or not a program should be allowed to bind itself to a requested port.

Port numbers larger than 1,024 are available for application programs. A program can request a specific port, or allow the operating system to pick one randomly. Given a port number, a program can register itself with the local Network Information Service (NIS) program, giving a symbolic name for itself and the port number on which it is listening. Or, it can send its port number to some other program—for example, by requesting a service and specifying the Internet address and port number to which replies should be transmitted.

The randomness of port selection is, perhaps unexpectedly, an important source of security in many modern protocols. These protocols are poorly protected against intruders, who could attack the application if they were able to guess the port numbers being used. By virtue of picking port numbers randomly, the protocol assumes that the barrier against attack has been raised substantially and that it need only protect against accidental delivery of packets from other sources (presumably an infrequent event, and one that is unlikely to involve packets that could be confused with the ones legitimately used by the protocol on the port). Later, however, we will see that such assumptions may not always be safe: Modern network hackers may be able to steal port numbers out of IP packets; indeed, this has become a serious enough problem so that proposals for encrypting packet headers are being considered by the IETF.

Not all machines have identical byte orderings. For this reason, the Internet protocol suite specifies a standard byte order that must be used to represent addresses and port numbers. On a host that does not use the same byte order as the standard requires, it is important to byte-swap these values before sending a message, or after receiving one. Many programming languages include communication libraries with standard functions for this purpose.

Finally, notice that the network service information specifies a protocol to use when communicating with a service: TCP, when communicating with the UUCP service; UDP when communicating with the TFTP service (a file transfer program); and so forth. Some services support multiple options, such as the domain name service. As we discussed earlier, these names refer to protocols in the Internet protocol suite.

3.3   Internet Protocols

This section presents the three major components of the Internet protocol suite: the IP protocol, on which the others are based, and the TCP and UDP protocols, which are the ones normally employed by applications. We also discuss some recent extensions to the IP protocol layer in support of IP multicast protocols. There has been considerable discussion of security for the IP layer, but no single proposal has gained wide acceptance as of the time of this writing, and we will say very little about this ongoing work for reasons of brevity.

3.3.1   Internet Protocol: IP Layer

The lowest layer of the Internet protocol suite is a connectionless packet transmission protocol called IP. IP is responsible for unreliable transmission of variable-size packets (but with a fixed maximum size, normally 1,400 bytes), from the sender's machine to the destination machine. IP packets are required to conform to a fixed format consisting of a variable-length packet header, a variable-length body, and an optional trailer. The actual lengths of the header, body, and trailer are specified through length fields, which are located at fixed offsets into the header. An application that makes direct use of IP is expected to format its packets according to this standard. However, direct use of IP is normally restricted because of security issues raised by the prospect of applications that might exploit such a feature to mimic some standard protocol, such as Transmission Control Protocol (TCP), doing this in a nonstandard manner that could disrupt remote machines or create security loopholes.

Implementations of IP normally provide routing functionality, using either a static or dynamic routing architecture. The type of routing used will depend upon the complexity of the installation and its configuration of the Internet software, and this is a topic beyond the scope of this book.

In 1995, IP was enhanced to provide a security architecture, whereby packet payloads can be encrypted to prevent intruders from determining packet contents, and to provide options for signatures or other authentication data in the packet trailer. Encryption of the packet header is also possible within this standard, although use of this feature is possible only if the routing layers and IP software implementation on all machines in the network agree upon the encryption method to use.

3.3.2 Transmission Control Protocol: TCP

TCP is a name for the connection-oriented protocol within the Internet protocol suite. TCP users start by making a TCP connection, which is done by having one program set itself up to listen for and accept incoming connections, while the other program connects to it. A TCP connection guarantees that data will be delivered in the order sent, without loss or duplication, and will report an end of file if the process at either end exits or closes the channel. TCP connections are byte stream oriented: Although the sending program can send blocks of bytes, the underlying communication model views this communication as a continuous sequence of bytes. TCP is thus permitted to lose the boundary information between messages, so that what is logically a single message may be delivered in several smaller chunks, or delivered together with fragments of a previous or subsequent message (always preserving the byte ordering, however!). If very small messages are transmitted, TCP will delay them slightly to attempt to fill larger packets for efficient transmission; the user must disable this behavior if immediate transmission is desired.

Applications that involve concurrent use of a TCP connection must interlock against the possibility that multiple write operations will be done simultaneously on the same channel; if this occurs, then data from different writers can be interleaved when the channel becomes full.

3.3.3 User Datagram Protocol: UDP

UDP is a message- or datagram-oriented protocol. With this protocol, the application sends messages, which are preserved in the form sent and delivered intact, or not at all, to the destination. No connection is needed, and there are no guarantees that the message will get through, or that messages will be delivered in any particular order, or even that duplicates will not arise. UDP imposes a size limit of 8 KB on each message: an application needing to send a large message must fragment it into 8 KB chunks.

Internally, UDP will normally fragment a message into smaller pieces, which correspond to the maximum size of an IP packet and match closely with the maximum-size packet that an Ethernet can transmit in a single hardware packet. If a UDP packet exceeds the maximum IP packet size, the UDP packet is sent as a series of smaller IP packets. On reception, these are reassembled into a larger packet. If any fragment is lost, the UDP packet will eventually be discarded.

The reader may wonder why this sort of two-level fragmentation scheme is used—why not simply limit UDP to 1,400 bytes, too? To understand this design, it is helpful to start with a measurement of the cost associated with a communication system call. On a typical operating system, such an operation has a minimum overhead of twenty thousand to fifty thousand instructions, regardless of the size of the data object to be transmitted. The idea, then, is to avoid repeatedly traversing long code paths within the operating system. When an 8 KB UDP packet is transmitted, the code to fragment it into smaller chunks executes deep within the operating system. This can save tens of thousands of instructions.

One might also wonder why communication needs to be so expensive, in the first place. In fact, this is a very interesting and rather current topic, particularly in light of recent work that has reduced the cost of sending a message (on some platforms) to as little as six instructions. In this approach, which is called *Active Messages* (see von Eicken et al. [1992, 1995]), the operating system is kept completely off the message path, and if one is willing to pay a slightly higher price, a similar benefit is possible even in a more standard communication architecture (see Section 8.3). It is entirely plausible to believe that commercial operating systems products offering comparably low latency and high throughput will start to be available in the late 1990s. However, the average operating system will certainly not catch up with the leading-edge approaches for many years. Thus, applications may have to continue to live with huge and in fact unnecessary overheads for the time being.

3.3.4   Internet Packet Multicast Protocol: IP Multicast

IP multicast is a relatively recent addition to the Internet protocol suite (see Deering [1988, 1989], Deering and Cheriton). With IP multicast, UDP or IP messages can be transmitted to groups of destinations, as opposed to a single point-to-point destination. The approach extends the multicast capabilities of the Ethernet interface to work even in complex networks with routing and bridges between Ethernet segments.

IP multicast is a session-oriented protocol: Some work is required before communication can begin. The processes that will communicate must create an IP multicast address, which is a class D Internet address containing a multicast identifier in the lower 28 bits. These processes must also agree upon a single port number, which all will use for the communication session. As each process starts, it installs an IP address into its local system, using system calls that place the IP multicast address on the Ethernet interface(s) to which the machine is connected. The routing tables used by IP, discussed in more detail later in this chapter, are also updated to ensure that IP multicast packets will be forwarded to each destination and network on which group members are found.

Once this setup has been done, an IP multicast is initiated by simply sending a UDP packet with the IP multicast group address and port number in it. As this packet reaches a machine that is included in the destination list, a copy is made and delivered to local applications receiving on the port. If several are bound to the same port on the same machine, a copy is made for each.

Like UDP, IP multicast is an unreliable protocol: Packets can be lost, duplicated, or delivered out of order, and not all members of a group will see the same pattern of loss and delivery. Thus, although one can build reliable communication protocols over IP multicast, the protocol itself is inherently unreliable.

When used through the UDP interface, a UDP multicast facility is similar to a UDP datagram facility, in that each packet can be as long as the maximum size of UDP transmissions, which is typically 8 KB. However, when sending an IP or UDP multicast, it is important to remember that the reliability observed may vary from destination to destination. One machine may receive a packet that others drop, because of memory limitations or corruption caused by a weak signal on the communication medium, and the loss of even a single fragment of a large UDP message will cause the entire message to be dropped. Thus, one talks more commonly about IP multicast than UDP multicast, and it is uncommon for applications to send very large messages using the UDP interface. Any application that uses this transport protocol should carefully instrument loss rates, because the effective performance for small messages may actually be better than for large ones due to this limitation.

3.4 Routing

Routing is the method by which a communication system computes the path by which packets will travel from source to destination. A routed packet is said to take a series of *hops,* as it is passed from machine to machine. The algorithm used is generally as follows:

- An application program generates a packet, or a packet is read from a network interface.

- The packet destination is checked and, if it matches with any of the addresses that the machine accepts, delivered locally (one machine can have multiple addresses—a feature that is sometimes exploited in networks with dual hardware for increased fault tolerance).

- The *hop count* of the message is incremented. If the message has a maximum hop count and would exceed it, the message is discarded. The hop count is also called the *time to live,* or TTL, in some protocols.

- For messages that do not have a local destination, or class D multicast messages, the destination is used to search the routing table. Each entry specifies an address, or a pattern covering a range of addresses. An outgoing interface is computed for the message (a list of outgoing interfaces, if the message is a class D multicast). For a point-to-point message, if there are multiple possible routes, the least costly route is employed. For this purpose, each route includes an estimated cost, in hops.

- The packet is transmitted on interfaces in this list, other than the one on which the packet was received.

A number of methods have been developed for maintaining routing tables. The most common approach is to use *static routing.* In this approach, the routing table is maintained by system administrators, and it is never modified while the system is active.

Dynamic routing is a class of protocols by which machines can adjust their routing tables to benefit from load changes, route around congestion and broken links, and reconfigure to exploit links that have recovered from failures. In the most common approaches, machines periodically distribute their routing tables to nearest neighbors, or periodically broadcast their routing tables within the network as a whole. For this latter case, a special address is used, which causes the packet to be routed down every possible interface in the network; a hop-count limit prevents such a packet from bouncing endlessly.

The introduction of IP multicast has resulted in a new class of routers, which are static for most purposes but that maintain special dynamic routing policies for use when an IP multicast group spans several segments of a routed local area network. In very large settings, this *multicast routing daemon* can take advantage of the *multicast backbone,* or *mbone,* network to provide group communication or conferencing support to sets of participants working at physically remote locations. However, most use of IP multicast is limited to local area networks at the time of this writing, and wide area multicast remains a somewhat speculative research topic.

3.5   End-to-End Argument

The reader may be curious about the following issue. The architecture described above permits packets to be lost at each hop in the communication subsystem. If a packet takes many hops, the probability of loss would seem likely to grow proportionately, causing the reliability of the network to drop linearly with the diameter of the network. There is an alternative approach in which error correction would be done hop by hop. Although packets could still be lost if an intermediate machine crashes, such an approach would have loss rates that are greatly reduced, at some small but fixed background cost (when we discuss the details of reliable communication protocols, we will see that the overhead need not be very high). Why, then, do most systems favor an approach that seems likely to be much less reliable?

In a classic paper, Jerry Saltzer and others took up this issue in 1984 (see Saltzer et al.). This paper compared end-to-end reliability protocols, which operate only between the source and destination of a message, with hop-by-hop reliability protocols. They argued that even if reliability of a routed network is improved by the use of hop-by-hop reliability protocols, it will still not be high enough to completely overcome packet loss. Packets can still be corrupted by noise on the lines, machines can crash, and dynamic routing changes can bounce a packet around until it is discarded. Moreover, they argued, the measured average loss rates for lightly to moderately loaded networks are extremely low. True, routing exposes a packet to repeated threats, but the overall reliability of a routed network will still be very high on the average, with worst-case behavior dominated by events such as routing table updates and crashes that hop-by-hop error correction would not overcome. From this the authors concluded that since hop-by-hop reliability methods increase complexity and reduce performance, and must still be duplicated by end-to-end reliability mechanisms, one might as well use a simpler and faster link-level communication protocol. This is the end-to-end argument, and it has emerged as one of the defining principles governing modern network design.

Saltzer's paper revolves around a specific example, involving a file transfer protocol (FTP). The paper makes the point that the analysis used is in many ways tied to the example and the actual reliability properties of the communication lines in question. Moreover, Saltzer's interest was specifically in reliability of the packet transport mechanism: failure rates and ordering. These points are important because many authors have come to cite the end-to-end argument in a much more expansive way, claiming that it is an absolute argument against putting any form of property or guarantee within the communication subsystem. Later, we will be discussing protocols that *need* to place properties and guarantees into subsystems, as a way of providing systemwide properties that would not otherwise be achievable. Thus, those who accept the generalized end-to-end argument would tend to oppose the use of these sorts of protocols on philosophical (one is tempted to say "religious") grounds.

A more mature view is that the end-to-end argument is one of those situations where one should accept its point with a degree of skepticism. On one hand, the end-to-end argument is clearly correct in situations where an analysis comparable to Saltzer's original one is possible. However, the end-to-end argument cannot be applied blindly: There are situations in which low-level properties are beneficial and genuinely reduce complexity and cost in application software, and, for these situations, an end-to-end approach might be inappropriate, leading to more complex applications that are error prone or, in a practical sense, impossible to construct.

In a network with high link-level loss rates, or one that is at serious risk of running out of memory unless flow control is used link to link, an end-to-end approach may result in near-total packet loss, while a scheme that corrects packet loss and does flow control at the link level could yield acceptable performance. This, then, is a case in which Saltzer's analysis could be applied as he originally formulated it, but it would lead to a different conclusion. When we look at the reliability protocols presented in Part III of this book, we will see that certain forms of consistent distributed behavior (such as is needed in a fault-tolerant coherent caching scheme) depend upon systemwide agreement, which must be standardized and integrated with low-level failure-reporting mechanisms. Omitting such a mechanism from the transmission layer merely forces the application programmer to build it as part of the application; if the programming environment is intended to be general and extensible, this may mean that one makes the mechanism part of the environment or gives up on it entirely. Thus, when we look at distributed programming environments such as the CORBA architecture, discussed in Chapter 6, there is in fact a basic design choice to be made: Either such a function is made part of the architecture, or, by omitting it, no application can achieve this type of consistency in a general and interoperable way except with respect to other applications implemented by the same development team. These examples illustrate that, like many engineering arguments, the end-to-end approach is highly appropriate in certain situations, but not uniformly so.

3.6 O/S Architecture Issues: Buffering and Fragmentation

We have reviewed most stages of the communication architecture that interconnects a sending application to a receiving application. But what of the operating system software at the two ends?

The communication software of a typical operating system is modular, organized as a set of components that subdivide the tasks associated with implementing the protocol stack or stacks in use by application programs. One of these components is the *buffering* subsystem, which maintains a collection of kernel memory buffers that can be used to temporarily store incoming or outgoing messages. On most UNIX systems, these are called *mbufs,* and the total number available is a configuration parameter, which should be set when the system is built. Other operating systems allocate buffers dynamically, competing with the disk I/O subsystem and other I/O subsystems for kernel memory. All operating systems share a key property, however: The amount of buffering space available is limited.

The TCP and UDP protocols are implemented as software modules that include interfaces up to the user and down to the IP software layer. In a typical UNIX implementation, these protocols allocate some amount of kernel memory space for each open communication socket, at the time the socket is created. TCP, for example, allocates an 8 KB buffer, and UDP allocates two 8 KB buffers, one for transmission and one for reception (both can often be increased to 64 KB). The message to be transmitted is copied into this buffer (in the case of TCP, this is done in chunks if necessary). Fragments are then generated by allocating successive memory chunks for use by IP, copying the data to be sent into them, prepending an IP header, and then passing them to the IP sending routine. Some operating systems avoid one or more of these copying steps, but this can increase code complexity, and copying is sufficiently fast that many operating systems simply copy the data for each message multiple times. Finally, IP identifies the network interface to use by searching the routing table and queues the fragments for transmission. As might be expected, incoming packets trace the reverse path.

An operating system can drop packets or messages for reasons unrelated to the hardware corruption or duplication. In particular, an application that tries to send data as rapidly as possible, or a machine that is presented with a high rate of incoming data packets, can exceed the amount of kernel memory that can safely be allocated to any single application. Should this happen, it is common for packets to be discarded until memory usage drops back below threshold. This can result in unexpected patterns of message loss—for example, consider an application program that simply tests packet loss rates. One might expect that as the rate of transmission is gradually increased, from one packet per second to ten, then 100, then 1,000, the overall probability that a packet loss will occur would remain fairly constant; hence, packet loss will rise in direct proportion to the actual number of packets sent. Experiments that test this case, running over UDP, reveal quite a different pattern, illustrated in Figure 3.1; the upper graph is for a sender and receiver on the same machine (the messages are never actually put on the wire in this case), and the lower graph shows the case of a sender and receiver on identical machines connected by an Ethernet.

As can be seen from the figure, the packet loss rate is a serious problem when processes send very frequently (left side of both curves). The loss rate is evident because the receiver's incoming data rate is erratic and low, growing stable only as packets grow in size and the rate of sending drops. For high rates of communication, one sees bursty behavior, in which some groups of packets are delivered and others are completely lost. Moreover, the aggregate throughput can be quite low in these overloaded cases, and the operating system often reports no errors at all for the sender and destination—on the sending side, the loss occurs after UDP has accepted a packet,

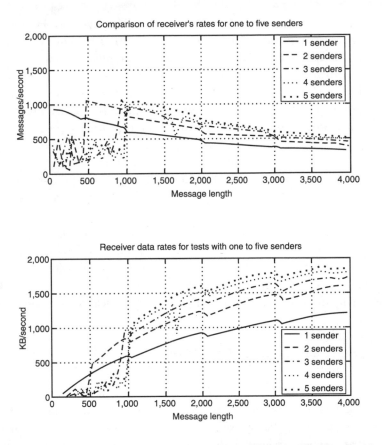

FIGURE 3.1. Amount of data received by a process compared to the amount of data that was sent, as a function of message size and the rate of sending. Both graphs are for UNIX and a 10 MB Ethernet. These data are based on a study reported as part of a doctoral dissertation by Guerney Hunt (see Hunt).

when it is unable to obtain memory for the IP fragments. On the receiving side, the loss occurs when UDP packets turn out to be missing fragments, or when the queue of incoming messages exceeds the limited capacity of the UDP input buffer.

The quantized scheduling algorithms used in multitasking operating systems such as UNIX probably account for the bursty aspect of the loss behavior. UNIX tends to schedule processes for long periods, permitting the sender to send many packets during congestion periods, without allowing the receiver to run to clear its input queue in the local case or giving the interface time to transmit an accumulated backlog in the remote case. The effect is that once a loss starts to occur, many packets can be lost before the system recovers. Interestingly, packets can also be delivered out of order when tests of this sort are done, presumably reflecting some sort of stacking mechanisms deep within the operating system. Thus, the same measurements might yield different results on other versions of UNIX or other operating systems. However, with the exception of special-purpose, communication-oriented operating systems such as QNX (a real-time system for embedded applications), one would expect a similar result for most of the common platforms used in distributed settings today.

TCP behavior is much more reasonable for the same tests, but there are other types of tests for which TCP can behave poorly—for example, if one process makes a great number of TCP connections to other processes, and then tries to transmit multicast messages on the resulting one-to-many connections, the measured throughput drops worse than linearly, as a function of the number of connections, for most operating systems. Moreover, if groups of processes are created and TCP connections are opened between them, pairwise, performance is often found to be extremely variable—latency and throughput figures can vary wildly even for simple patterns of communication.

UDP or IP multicast gives the same behavior as UDP. However, the user of multicast should also keep in mind that many sources of packet loss can result in different patterns of reliability for different receivers. Thus, one destination of a multicast transmission may experience high loss rates even if many other destinations receive all messages with no losses at all. Problems such as this are potentially difficult to detect and are very hard to deal with in software.

3.7   Xpress Transfer Protocol

Although widely available, TCP, UDP, and IP are also limited in the functionality they provide and their flexibility. This has motivated researchers to investigate new and more flexible protocol development architectures that can coexist with TCP/IP but support varying qualities of transmission service that can be matched closely to the special needs of demanding applications.

Prominent among such efforts is the Xpress Transfer Protocol (XTP), which is a toolkit of mechanisms that can be exploited by users to customize data transfer protocols operating in a point-to-point or multicast environment. All aspects of the protocol are under control of the developer, who sets option bits during individual packet exchanges to support a highly customizable suite of possible communication styles. References to this work include Dempsey et al., Strayer et al., XTP Forum.

XTP is a connection-oriented protocol, but one in which the connection setup and closing protocols can be varied depending on the needs of the application. A connection is identified by a 64-bit key; 64-bit sequence numbers are used to identify bytes in transit. XTP does not define any addressing scheme of its own, but is normally combined with IP addressing. An XTP protocol is defined as an exchange of XTP messages. Using the XTP toolkit, a variety of options can be specified for each message transmitted; the effect is to support a range of possible qualities of service for each communication session—for example, an XTP protocol can be made to emulate UDP- or TCP-style streams, to operate in an unreliable source to destination mode, with selective retransmission based on negative acknowledgments, or can even be asked to go back to a previous point in a transmission and to resume. Both rate-based and windowing-flow control mechanisms are available for each connection, although one or both can be disabled if desired. The window size is configured by the user at the start of a connection, but can later be varied while the connection is in use; a set of traffic parameters can be used to specify requirements, such as the maximum size of data segments that can be transmitted in each packet, maximum or desired burst data rates, and so forth. Such parameters permit the development of general-purpose transfer protocols that can be configured at run time to match the properties of the hardware environment.

This flexibility is exploited in developing specialized transport protocols that may look like highly optimized versions of the standard ones, but that can also provide very unusual properties—for example, one could develop a TCP-style of stream that will be reliable provided that the packets sent arrive on time, using a user-specific concept of time, but that drops packets if they timeout. Similarly, one can develop protocols with out-of-band or other forms of priority-based services.

At the time of this writing, XTP was gaining significant support from industry leaders, whose future product lines potentially require flexibility from the network. Video servers, for example, are poorly matched to the communication properties of TCP connections; hence, companies that are investing heavily in video on demand face the potential problem of having products that work well in the laboratory but not in the field, because the protocol architecture connecting customer applications to the server is inappropriate. Such companies are interested in developing proprietary data transport protocols, which would essentially extend their server products into the network itself, permitting fine-grained control over the communication properties of the environment in which their servers operate and overcoming limitations of the more traditional but less flexible transport protocols.

In Chapters 13 through 16 we will discuss special-purpose protocols designed for settings in which reliability requires data replication or specialized performance guarantees. Although we will generally present these protocols in the context of streams, UDP, or IP multicast, it is likely that the future will bring a considerably wider set of transport options that can be exploited in applications with these sorts of requirements.

There is, however, a downside associated with the use of customized protocols based on technologies such as XTP: They can create difficult management and monitoring problems, which will often go well beyond those seen in standard environments where tools can be developed to monitor a network and to display, in a well-organized manner, the status of the network and applications. Such tools benefit from being able to intercept network traffic and to associate the message sent with the applications sending them. To the degree that technologies such as XTP lead to extremely specialized patterns of communication that work well for individual applications, they may also reduce this desirable form of regularity and hence impose obstacles to system control and management.

Broadly, one finds a tension within the networking community today. On the one side are developers convinced that their special-purpose protocols are necessary if a diversity of communication products and technologies is to be feasible over networks such as the Internet. In some sense this community generalizes to also include the community that develops special-purpose reliability protocols and that may need to place properties within the network to support those protocols. On the other side are the system administrators and managers, whose lives are already difficult and who are extremely resistant to technologies that might make this problem worse. Sympathizing with them are the performance experts of the operating system communication community: This group favors an end-to-end approach because it greatly simplifies their task. They tend to oppose technologies such as XTP because these technologies result in irregular behaviors that are harder to optimize in the general case. For these researchers, knowing more about the low-level requirements (and keeping them as simple as possible) makes it more practical to optimize the corresponding code paths for extremely high performance and low latency.

From a reliability perspective, one must sympathize with both points of view: This book will look at problems for which reliability requires high performance or other guarantees, as well as problems for which reliability implies the need to monitor, control, or manage a complex environment. If there is a single factor that prevents a protocol suite such as XTP from sweeping the industry, it seems likely to be this. More likely, however, is an increasingly diverse collection of low-level protocols, creating ongoing challenges for the community that must administer and monitor the networks in which those protocols are used.

3.8 Next Steps

It is not surprising that problems such as the performance anomalies cited in the previous sections are common on modern operating systems, because the communication subsystems have rarely been designed or tuned to guarantee good performance for communication patterns such as those shown in Figure 3.1. As will be seen in the next few chapters, the most common communication patterns are very regular ones that would not trigger the sorts of pathological behaviors caused by memory resource limits and stressful communication loads.

However, given a situation in which most systems must in fact operate over protocols such as TCP and UDP, these behaviors do create a context that should concern students of distributed system reliability. They suggest that even systems that behave well most of the time may break down catastrophically because of something as simple as a slight increase in load. Software designed on the assumption that message loss rates are low may, for reasons completely beyond the control of the developer, encounter loss rates that are extremely high. All of this can lead the researcher to question the appropriateness of modern operating systems for reliable distributed applications. Alternative operating system architectures, which offer more controlled degradation in the presence of excess load, represent a potentially important direction for investigation and discussion.

3.9 Related Reading

On the Internet protocols: (see Comer, Comer and Stevens [1991, 1993], Coulouris et al., Tanenbaum).

Performance issues for TCP and UDP: (see Brakmo et al., Comer, Comer and Stevens [1991, 1993], Hunt, Kay and Pasquale, Partridge and Pink).

IP multicast: (see Deering [1988, 1989], Deering and Cheriton, Frank et al., Hunt).

Active messages: (see von Eicken et al. [1992, 1995]).

End-to-end argument: (see Saltzer et al.).

Xpress Transfer Protocol: (see Dempsey et al., Strayer et al., XTP Forum).

CHAPTER 4   ✧ ✧ ✧ ✧

Remote Procedure Calls and the Client/Server Model

CONTENTS

4.1 The Client/Server Model

The emergence of real distributed computing systems is often identified with the *client/server* paradigm and a protocol called *remote procedure call* (RPC), which is normally used in support of this paradigm. The basic idea of a client/server system architecture involves a partitioning of the software in an application into a set of *services,* which provide a set of operations to their users, and *client programs,* which implement applications and issue requests to services as needed to carry out the purposes of the application. In this model, the application processes do not cooperate directly with one another, but instead share data and coordinate actions by interacting with a common set of servers and by the order in which the application programs are executed.

There are a great number of client/server system structures in a typical distributed computing environment. Some examples of servers include the following:

▶ *File servers:* These are programs (or, increasingly, combinations of special-purpose hardware and software) that manage disk storage units on which file systems reside. The operating system on a workstation that accesses a file server acts as the client, thus creating a two-level hierarchy: The application processes talk to their local operating system. The operating system on the client workstation functions as a single client of the file server, with which it communicates over the network.

▶ *Database servers*: The client/server model operates in a similar way for database servers, except that it is rare for the operating system to function as an intermediary in the manner that it does for a file server. In a database application, there is usually a library of procedure calls with which the application accesses the database, and this library plays the role of the client in a client/server communication protocol to the database server.

▶ *Network name servers:* Name servers implement some form of map from a symbolic name or service description to a corresponding value, such as an IP address and port number for a process capable of providing a desired service.

▶ *Network time servers:* These are processes that control and adjust the clocks in a network, so that clocks on different machines give consistent time values (values with limited divergence from one another). The server for a clock is the local interface by which an application obtains the time. The clock service, in contrast, is the collection of clock servers and the protocols they use to maintain clock synchronization.

▶ *Network security servers:* Most commonly, these consist of a type of directory in which public keys are stored, as well as a key generation service for creating new secure communication channels.

▶ *Network mail and bulletin board servers:* These are programs for sending, receiving, and forwarding e-mail and messages to electronic bulletin boards. A typical client of such a server would be a program that sends an electronic mail message or that displays new messages to a user who is using a newsreader interface.

▶ *WWW servers:* As we learned in the introduction, the World Wide Web is a large-scale distributed document management system developed at CERN in the early 1990s and subsequently commercialized. The Web stores hypertext documents, images, digital movies, and other information on *Web servers,* using standardized formats that can be displayed through various browsing programs. These systems present point-and-click interfaces to hypertext documents, retrieving documents using Web document locators from Web servers, and then displaying them in a type-specific manner. A Web server is thus a type of enhanced file server on which the Web access protocols are supported.

In most distributed systems, services can be instantiated multiple times—for example, a distributed system can contain multiple file servers or multiple name servers. We normally use the term *service* to denote a set of servers. Thus, the *network file system service* consists of the network file servers for a system, and the *network information service* is a set of servers, provided on UNIX systems, that maps symbolic names to ASCII strings encoding values or addresses. An important question to ask about a distributed system concerns the binding of applications to servers.

We say that a *binding* occurs when a process that needs to talk to a distributed service becomes associated with a specific server that will perform requests on its behalf. Various binding policies exist, differing in how the server is selected. For an NFS distributed file system, binding is a function of the file path name being accessed—in this file system protocol, the servers all handle different files, so that the path name maps to a particular server that owns that file. A program using the UNIX network information server normally starts by looking for a server on its own machine. If none is found, the program broadcasts a request and binds to the first NIS that responds, the idea being that this NIS representative is probably the least loaded and will give the best response times. (On the negative side, this approach can reduce reliability: Not only will a program now be dependent on availability of its file servers, but it may be dependent on an additional process on some other machine, namely the NIS server to which it became bound.) The CICS database system is well known for its explicit load-balancing policies, which bind a client program to a server in a way that attempts to give uniform responsiveness to all clients.

Algorithms for binding, and for dynamically rebinding, represent an important topic to which we will return in Chapter 17, once we have the tools at our disposal to solve the problem in a concise way.

A distributed service may or may not employ *data replication,* whereby a service maintains more than one copy of a single data item to permit local access at multiple locations or to increase availability during periods when some server processes may have crashed—for example, most network file services can support multiple file servers, but they do not replicate any single file onto multiple servers. In this approach, each file server handles a partition of the overall file system, and the partitions are disjoint from one another. A file can be replicated, but only by giving each replica a different name, placing each replica on an appropriate file server, and implementing hand-crafted protocols for keeping the replicas coordinated. Replication, then, is an important issue in designing complex or highly available distributed servers.

Caching is a closely related issue. We say that a process has *cached* a data item if it maintains a copy of that data item locally, for quick access if the item is required again. Caching is widely used in file systems and name services, and permits these types of systems to benefit from locality of reference. A *cache hit* is said to occur when a request can be satisfied out of cache, avoiding the expenditure of resources needed to satisfy the request from the *primary store* or *primary service*. The Web uses document caching heavily, as a way to speed up access to frequently used documents.

Caching is similar to replication, except that cached copies of a data item are in some ways second-class citizens. Generally, caching mechanisms recognize the possibility that the cache contents may be stale, and they include a policy for validating a cached data item before using it. Many caching schemes go further, and include explicit mechanisms by which the primary store or service can invalidate cached data items that are being updated, or refresh them explicitly. In situations where a cache is actively refreshed, caching may be identical to replication—a special term for a particular style of replication.

However, "generally" does not imply that this is always the case. The Web, for example, has a cache validation mechanism but does not actually require that Web proxies validate cached documents before providing them to the client; the reasoning is presumably that even if the document were validated at the time of access, nothing prevents it from changing immediately afterwards and hence being stale by the time the client displays it. Thus, a periodic refreshing scheme in which cached documents are refreshed every half hour or so is in many ways equally reasonable. A caching policy is said to be *coherent* if it guarantees that cached data are indistinguishable to the user from the primary copy. The Web caching scheme is thus one that does not guarantee coherency of cached documents.

4.2 RPC Protocols and Concepts

The most common communication protocol for communication between the clients of a service and the service itself is remote procedure call. The basic idea of an RPC originated in work by Birrell and Nelson in the early 1980s (see Birrell and Nelson). Nelson worked in a group at Xerox PARC that was developing programming languages and environments to simplify distributed computing. At that time, software for supporting file transfer, remote log in, electronic mail, and electronic bulletin boards had become common. PARC researchers, however, had ambitious ideas for developing other sorts of distributed computing applications, with the consequence that many researchers found themselves working with the lowest-level-message-passing primitives in the PARC distributed operating system, which was called Cedar.

Much like a more modern operating system, message communication in Cedar supported three communication models:

- Unreliable datagram communication, in which messages could be lost with some (hopefully low) probability
- Broadcast communication, also through an unreliable datagram interface
- Stream communication, in which an initial connection was required, after which data could be transferred reliably

Programmers found these interfaces hard to work with. Any time a program, *p*, needed to communicate with another program, *s,* it was necessary for *p* to determine the network address o f s, encode its requests in a way that s would understand, send off the request, and await a reply. Programmers soon discovered that certain basic operations needed to be performed in almost any network application and that each developer was developing his or her own solutions to these standard problems. Some programs used broadcasts to find a service with which they needed to communicate; others stored the network address of services in files or hard-coded them into the application; and still others supported directory programs with which services could register themselves, supporting queries from other programs at run time. Not only was this situation confusing, it turned out to be difficult to maintain the early versions of PARC software: A small change to a service might break all sorts of applications that used it, so it became hard to introduce new versions of services and applications.

Surveying this situation, Bruce Nelson started by asking what sorts of interaction programs were really needed in distributed settings. He concluded that the problem was really no different from a function or procedure call in a nondistributed program that uses a presupplied library; that is, most distributed computing applications would prefer to treat other programs with which they interact much as they treat presupplied libraries, with well-known, documented, procedural interfaces. Talking to another program would then be as simple as invoking one of its procedures—a remote procedure call (RPC).

The idea of a remote procedure call is compelling. If distributed computing can be transparently mapped to a nondistributed computing model, all the technology of nondistributed programming could be brought to bear on the problem. In some sense, we would already know how to design and reason about distributed programs; how to prove them to be correct; how to test, maintain, and upgrade them; and all sorts of preexisting software tools and utilities would be readily applicable to the problem.

Unfortunately, the details of supporting a remote procedure call turn out to be nontrivial, and some aspects result in visible differences between remote and local procedure invocations. Although this wasn't evident in the 1980s when RPC really took hold, the subsequent ten or 15 years saw considerable theoretical activity in distributed computing, out of which ultimately emerged a deep understanding of how certain limitations on distributed computing are reflected in the *semantics,* or properties, of a remote procedure call. In some ways, this theoretical work finally led to a major breakthrough in the late 1980s and early 1990s, when researchers learned how to create distributed computing systems in which the semantics of RPC are precisely the same as for local procedure calls (LPC). In Part III of this book, we will study the results and necessary technology underlying such a solution, and we will see how to apply it to RPC. We will also see, however, that such approaches involve subtle tradeoffs between the semantics of the RPC and the performance that can be achieved; the faster solutions also weaken semantics in fundamental ways. Such considerations ultimately lead to the insight that RPC cannot be transparent, however much we might wish that this were not the case.

Making matters worse, during the same period of time a huge engineering push behind RPC elevated it to the status of a standard—and this occurred *before* it was understand how RPC could be made to accurately mimic LPC. The result of this is that the standards for building RPC-based computing environments (and, to a large extent, the standards for object-based computing that followed RPC in the early 1990s) embody a nontransparent and unreliable RPC model, and this design decision is often fundamental to the architecture in ways that the developers who formulated these architectures probably did not appreciate. In the next chapter, when we discuss stream-based communication, we will see that the same sort of premature standardization affected the standard stream technology, which as a result also suffers from serious limitations that could have been avoided had the problem simply been better understood at the time the standards were developed.

In the remainder of this chapter, we will focus on standard implementations of RPC. We will look at the basic steps by which a program RPC is coded in a program, how that program is translated at compile time, and how it becomes bound to a service when it is executed. Then, we will study the encoding of data into messages and the protocols used for service invocation and for collecting replies. Finally, we will try to pin down a semantic for RPC: a set of statements that can be made about the guarantees of this protocol and that can be compared with the guarantees of LPC.

We do not, however, give detailed examples of the major RPC programming environments: the Distributed Computing Environment (DCE) and Open Network Computing (ONC). These technologies, which emerged in the mid 1980s, represented proposals to standardize distributed computing by introducing architectures within which the major components of a distributed computing system would have well-specified interfaces and behaviors and within which application programs could interoperate using RPC by virtue of employing standard RPC interfaces. DCE, in particular, has become relatively standard, and it is available on many platforms today (see Open Software Foundation). However, in the mid-1990s, a new generation of RPC-oriented technology emerged through the Object Management Group (OMG), which set out to standardize object-oriented computing. In a short period of time, the CORBA (see Object Management Group and X/Open) technologies defined by OMG swept past the RPC technologies, and it now makes more sense to focus on CORBA, which we discuss in Chapter 6. CORBA has not so much changed the basic issues, as it has broadened the subject of discourse by covering more kinds of system services than did previous RPC systems. Moreover, many CORBA systems are implemented as a layer over DCE or ONC. Thus, although RPC environments are important, they are more and more hidden from typical programmers and hence there is limited value in seeing examples of how one would program applications using them directly.

Many industry analysts talk about CORBA implemented over DCE, meaning that they like the service definitions and object orientation of CORBA, and they feel it makes sense to assume that these were built using the service implementations standardized in DCE. In practice, however, CORBA makes as much sense on a DCE platform as on a non-DCE platform; hence, it would be an exaggeration to claim that CORBA on DCE is a de facto standard today, as one sometimes reads in the popular press.

The use of RPC leads to interesting problems of reliability and fault handling. As we will see, it is not hard to make RPC work if most of the system is working well. When a system malfunctions, however, RPC can fail in ways that leave the user with no information at all about what has occurred and with no apparent strategy for recovering from the situation. There is nothing new about the situations we will be studying—indeed, for many years, it was simply assumed that RPC was subject to intrinsic limitations, and since there was no obvious way to improve on the situation, there was no reason that RPC shouldn't reflect these limitations in its semantic model. As we advance through the book, however, and it becomes clear that there *are* realistic alternatives that might be considered, this point of view becomes increasingly open to question.

Indeed, it may now be time to develop a new set of standards for distributed computing. The existing standards are flawed, and the failure of the standards community to repair these flaws has erected an enormous barrier to the development of reliable distributed computing systems. In a technical sense, these flaws are not tremendously hard to overcome—although the solutions would require some reengineering of communication support for RPC in modern operating systems. In a practical sense, however, one wonders if it will take a Tacoma Narrows event to create real industry interest in taking such steps.

One could build an RPC environment that would have few, if any, user-visible incompatibilities from a more fundamentally rigorous approach. The issue then is one of education—the communities that control the standards need to understand the issue better, and they need to understand the reasons that this particular issue represents such a huge barrier to progress in distributed computing. They also need to recognize that the opportunity vastly outweighs the reengineering costs that would be required to seize it. With this goal in mind, let's take a close look at RPC.

4.3 Writing an RPC-Based Client or Server Program

The programmer of an RPC-based application employs what is called a *stub-generation* tool. Such a tool is somewhat like a macro preprocessor: It transforms the user's original program into a modified version, which can be linked to an RPC run-time library.

From the point of view of the programmer, the server or client program looks much like any other program. Normally, the program will *import* or *export* a set of interface definitions, covering the remote procedures that will be obtained from remote servers or offered to remote clients, respectively. A server program will also have a name and a version, which are used to connect the client to the server. Once coded, the program is compiled in two stages: First the stub generator is used to map the original program into a standard program with added code to carry out the RPC, and then the standard program is linked to the RPC run-time library for execution.

RPC-based application or server programs are coded in a programming style very similar to a nondistributed program written in C for UNIX: There is no explicit use of message passing. However, there is an important aspect of RPC programming that differs from programming with local procedure calls: the separation of the service interface definition, or IDL,[1] from the code that implements it. In an RPC application, a service is considered to have two parts. The

---

[1]It is common to call the interface to a program its IDL, although IDL is actually shorthand for Interface Definition Language, which is the language used to write down the description of such an interface. Historically, this seems to represent a small degree of resistance to the overuse of acronyms by the distributed system standardization

interface definition specifies the way that the service can be located (its name), the data types used in issuing requests to it, and the procedure calls that it supports. A *version number* is included to provide for evolution of the service over time—the idea being that if a client is developed to use version 1.1 of a service, there should be a way to check for compatibility if it turns out that version 1.0 or 2.3 is running when the client actually gets executed.

The basic actions of the RPC library were described earlier. In the case of a server program, the library is responsible for registering the program with the RPC directory service program, which is normally provided as part of the RPC run-time environment. An RPC client program will automatically perform the tasks needed to query the directory to find this server and to connect to it, creating a client/server binding. For each of the server operations it invokes, code will be executed to marshal a representation of the invocation into a message—that is, information about the way that the procedure was called and values of the parameters that were passed. Code is included to send this message to the service and to collect a reply; on the server side, the stub generator creates code to read such a message, invoke the appropriate procedure with the arguments used by the remote caller, and marshal the results for transmission back to the caller. Issues such as user-ID handling, security and privacy, and handling of exceptions are often packaged as part of a solution. Finally, back on the caller side, the returning message will be demarshaled and the result made to look like the result of a local procedure.

Although much of this mechanism is automatic and hidden from the programmer, RPC programming differs from LPC programming in many ways. Most noticeable is that most RPC packages limit the types of arguments that can be passed to a remote server, and some also limit the size (in bytes) of the argument information—for example, suppose that a local procedure is written to search a list, and an LPC is performed to invoke this procedure, passing a pointer to the head of the list as its argument. One can ask whether this should work in an RPC environment—and, if so, how it can be supported. If a pointer to the head of the list is actually delivered to a remote program, that pointer will not make sense in the remote address space where the operation will execute. So, it would be natural to propose that the pointer be dereferenced, by copying the head of the list into the message. Remotely, a pointer to the copy can be provided to the procedure. Clearly, however, this will only work if one chases *all* the pointers in question—a problem because many programs that use pointers have some representation for an uninitialized pointer, and the RPC stub generator may not know about this.

In building a balanced tree, it is common to allocate nodes dynamically as items are inserted. A node that has no descendents would still have left and right pointer fields, but these would be initialized to *nil* and the procedure to search nodes would check for the nil case before dereferencing these pointers. If an RPC marshaling procedure were to automatically make a copy of a structure to send to the remote server (see Figure 4.1), it would need to realize that for this particular structure, a pointer value of nil has a special meaning and should not be chased.

---

community. Unfortunately, the resistance seems to have been short-lived: CORBA introduces at least a dozen new three-letter acronyms, ATM has swept the networking community and four- and five-letter acronyms (as the available three-letter combinations are used up) seem certain to follow!

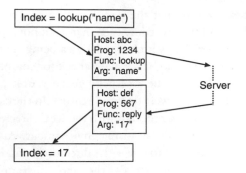

FIGURE 4.1. Remote procedure call involves creating a message that can be sent to the remote server, which unpacks it, performs the operation, and sends back a message encoding the result.

The RPC programmer sees issues such as these as a set of restrictions. Depending on the RPC package used, different approaches may be used to attack them. In many packages, pointers are simply not legal as arguments to remote procedures. In others, the user can control a copying mechanism to some degree, and in still fancier systems, the user must provide general-purpose structure traversal procedures, which will be used by the RPC package to marshal arguments. Further complications can occur if a remote procedure can modify some of its arguments. Again, the degree to which this is supported, and the degree to which the programmer must get involved, varies from package to package.

Perhaps ironically, RPC programmers tend to complain about this aspect of RPC no matter how it is handled. If a system is highly restrictive, the programmer finds that remote procedure invocation is annoying, because one is constantly forced to work around the limitations of the invocation package—for example, if an RPC package imposes a size limit on the arguments to a procedure, an application that works perfectly well in most situations may suddenly fail because some dynamically defined object has grown too large to be accepted as an RPC parameter. Suddenly, what was a single RPC becomes a multi-RPC protocol for passing the large object in chunks, and a perfectly satisfied programmer has developed distinct second thoughts about the transparency of RPC. At the other extreme are programming languages and RPC packages in which RPC is extremely transparent. These, however, often incur high overheads to copy information in and out, and the programmer is likely to be very aware of these because of their cost implications—for example, a loop that repeatedly invokes a procedure having one changing parameter as well as others (including a pointer to some large object) may be quite inexpensive to invoke in the local case. But if the large object will be copied to a remote program on every invocation, the same loop may cost a fortune when coded as part of a distributed client/server application, forcing the program to be redesigned to somehow pass the object to the remote server prior to the computational loop. These sorts of issues, then, make programming with RPC quite different from programming with LPC.

RPC also introduces error cases that are not seen in LPC, and the programmer needs to deal with these. An LPC would never fail with a binding error, a version mismatch, or a timeout. In the case of RPC, all of these are possibilities—a binding error would occur if the server were not running when the client was started. A version mismatch might occur if a client were compiled against version 1 of a server, but the server has now been upgraded to version 2. A timeout could result from a server crash, a network problem, or even a problem on the client's computer. Many RPC applications would view these sorts of problems as unrecoverable errors, but fault-tolerant systems will often have alternative sources for critical services and will need to fail-over from a primary server to a backup. The code to do this is potentially complex, and in most RPC environments, it must be implemented by the application developer on a case-by-case basis.

4.4   The RPC Binding Problem

The *binding* problem occurs when an RPC client program needs to determine the network address of a server capable of providing some service it requires. Binding can be approached from many perspectives, but the issue is simplified if issues associated with the name service used are treated separately, as we do here.

Disregarding its interactions with the name service, a binding service is primarily a protocol by which the RPC system verifies compatibility between the client and server and establishes any connections needed for communication.

The compatibility problem is important in systems that will operate over long periods of time, during which maintenance and the development of new versions of system components will inevitably occur. Suppose that a client program, c, was developed and tested using server s, but that we now wish to install a new version of s, c, or both. Upgrades such as these create a substantial risk that some old copy of c will find itself talking to a new copy of s, or vice versa—for example, in a network of workstations it may be necessary to reload c onto the workstations one by one, and if some machines are down when the reload occurs, an old copy of c could remain on its disk. Unless c is upgraded as soon as the machine is rebooted—and this may or may not occur, depending on how the system is administered—one would find an old c talking to an upgraded s. It is easy to identify other situations in which problems such as this could occur.

It would be desirable to be able to assume that all possible versions of s and c could somehow communicate with all other versions, but this is not often the case. Indeed, it is not necessarily even desirable. Accordingly, most RPC environments support a concept of *version number,* which is associated with the server IDL. When a client program is compiled, the server IDL version is noted in software. This permits the inclusion of the client's version of the server interface directly in the call to the server. When the match is not exact, the server could reject the request as being incompatible, perform some operation to map the old-format request to a new-format request, or even preserve multiple copies of its functionality, running the version matched to the caller.

Connection establishment is a relatively mechanical stage of binding. Depending on the type of client/server communication protocol that will be used, messages may be transmitted using unreliable datagrams or over reliable communication streams such as X.25 or TCP. Unreliable datagram connections normally do not require any initial setup, but stream connections typically involve some form of open or initialization operation. Having identified the server to which a request will be issued, the binding mechanism would normally perform this open operation.

The binding mechanism is sometimes used to solve two additional problems. The first of these is called the factory problem and involves starting a server when a service has no currently operational server. In this approach, the first phase of binding looks up the address of the server and learns that the server is not currently operational (or, in the connection phase, a connection error is detected and from this the binder deduces that the server has failed). The binder then issues a request to a factory in which the system designer has stored instructions for starting up a server when needed. After a suitable pause, the binder cycles back through its first phase, which presumably succeeds.

The second problem occurs in the converse situation, when the binder discovers multiple servers that could potentially handle this client. The best policy to use in such situations depends very much on the application. For some systems, a binder should always pick a server on the same machine as the client, if possible, and should otherwise pick randomly. Other systems require some form of load-balancing, while still others may implement an affinity policy under which a certain server might be especially well suited to handling a particular client for reasons such as the data it has cached in memory or the type of requests the client is expected to issue once binding has been completed.

Binding is a relatively expensive operation—for example, in the DCE RPC environment, binding can be more than ten times as costly as RPC. However, since binding only occurs once for each client/server pair, this high cost is not viewed as a major problem in typical distributed computing systems.

4.5 Marshaling and Data Types

The purpose of a data marshaling mechanism is to represent the caller's arguments in a way that can be efficiently interpreted by a server program. In the most general cases, this mechanism deals with the possibility that the computer on which the client is running uses a data representation different from the computer on which the server is running.

Marshaling has been treated at varying levels of generality, and there exists a standard, ASN.1, for *self-describing data objects* in which a specific representation is recommended. In addition to ASN.1, several major vendors have adopted data representations of their own, such as Sun Microsystem's External Data Representation (XDR) format, which is used in the widely popular Network File System (NFS) protocol.

The basic issues that occur in a data marshaling mechanism, then, are these. First, integer representations vary for the most common CPU chips. On some chips the most significant byte of an integer is also the low byte of the first word in memory, while on others the most significant byte is stored in the high byte of the last word of the integer. These are called little-endian and big-endian representations. At one point in the 1980s, computers with other representations—other byte permutations—were on the market, but at the time of this writing I am not aware of any other surviving formats.

A second representation issue concerns data alignment. Some computers require that data be aligned on 32-bit or even 64-bit boundaries, while others may have weaker alignment rules—for example, by supporting data alignment on 16-bit boundaries. Unfortunately, such issues are extremely common. Compilers know about these rules, but the programmer is typically unaware of them. However, when a message arrives from a remote machine that may be using some other alignment rule, the issue becomes an important one. An attempt to fetch data directly from a message without attention to this issue could result in some form of machine fault, or it could result in retrieval of garbage. Thus, the data representation used in messages must encode sufficient information to permit the destination computer to find the start of object in the message, or the sender and destination must agree in advance on a packed representation that will be used for messages on the wire even if the sender and destination themselves share the same rules and differ from the standard. Needless to say, this is a topic capable of generating endless and fascinating debate among computer vendors whose machines use different alignment or data representations.

A third issue arises from the existence of multiple floating-point representations. Although there is an IEEE standard floating-point representation, which has become widely accepted, some computer vendors use nonstandard representations for which conversion would be required, and even within computers using the standard, byte-ordering issues can still occur.

A fourth issue concerns pointers. When transmitting a complex structure in which there are pointers, the marshaling mechanism needs to either signal that the user has requested something illegal, or somehow represent these pointers in a way that will permit the receiving computer to fix them upon reception of the request. This is especially tricky in languages such as LISP, which requires pointers and hence cannot easily legislate against them in RPC situations. On the other hand, passing pointers raises additional problems: Should the pointed-to object be included in the message, transferred only upon use (a "lazy" scheme), or handled in some other way?

Finally, a marshaling mechanism may need to deal with incompatibilities in the basic data types available on computers (see Figure 4.2)—for example, a pair of computers supporting 64-bit integers in hardware may need to exchange messages containing 64-bit integer data. The marshaling scheme should therefore be able to represent such integers. On the other hand, when this type of message is sent to a computer that uses 32-bit integers, the need arises to truncate the 64-bit quantities so that they will fit in the space available, with an exception being generated if data would be lost by such a truncation. Yet, if the message is merely being passed through some sort of intermediary, one would prefer that data not be truncated, since precision would be lost. In the reverse direction, sign extension or padding may need to be performed to convert a 32-bit quantity into an equivalent 64-bit quantity, but only if the data sent are a signed integer. Thus, a completely general RPC package needs to put a considerable amount of information into each packet, and it may need to do quite a bit of work to represent data in a universal manner. Such an approach may be much more costly than one that supports only a very limited set of possible representations or that compiles the data marshaling and demarshaling operations directly into in-line code.

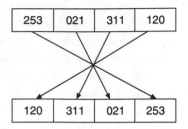

FIGURE 4.2. **The same number (here, a 32-bit integer) may be represented very differently on different computer architectures. One role of the marshaling and demarshaling process is to modify data representations (here, by permuting the bytes) so that values can be interpreted correctly upon reception.**

The approach taken to marshaling varies from RPC package to package. Sun's XDR system is extremely general, but requires the user to code marshaling procedures for data types other than the standard base types of the system. With XDR, one can represent any desired data structure, even dealing with pointers and complex padding rules. At the other end of the spectrum are marshaling procedures that transmit data in the binary format used by the sender, are limited to only simple data types, and perhaps do little more than compatibility checking on the receive-side. Finally, schemes such as ISDN.1 are often used with RPC stub generators, which automatically marshal and demarshal data, but impose some restrictions on the types of objects that can be transmitted.

As a general rule of thumb, users will want to be aware that the more general solutions to these problems are also more costly. If the goal is extreme speed, it may make sense to design the application itself to produce data in a form that is inexpensive to marshal and demarshal. The cost implications of failing to do so can be surprising, and, in many cases, it is not difficult to redesign an interface so that RPCs to it will be inexpensive.

4.6 Associated Services

No RPC system lives in isolation. As we will see, RPC is often integrated with a security mechanism and security keys, using timestamps with a clock synchronization mechanism. For this reason, one often talks about distributed computing environments that include tools for implementing client/server applications such as RPC mechanisms, security services, and time services. Elaborate environments may go well beyond this, including system instrumentation, management interfaces and tools, fault-tolerant tools, and so-called Fourth-Generation Language (4GL) tools for building applications using graphical user interfaces (GUIs). Such approaches can empower even unskilled users to develop sophisticated distributed solutions. In this section we briefly review the most important of these services.

4.6.1 Naming Services

A naming service maintains one or more *mappings* from some form of name (normally symbolic) to some form of value (normally a network address). Naming services can operate in a very narrow, focused way—for example, the Domain Name Service of the TCP/IP protocol suite maps short service names, in ASCII, to IP addresses and port numbers, requiring exact matches. At the other extreme, one can talk about general naming services, which are used for many sorts of data, allow complex pattern matching on the name, and may return other types of data in addition to, or instead of, an address. One can even go beyond this to talk about secure naming services, which can be trusted to give out only validated addresses for services, and dynamic naming services, which deal with applications such as mobile computing systems in which hosts have addresses that change constantly.

In standard computer systems at the time of this writing, three naming services are widely supported and used. As previously mentioned, the Domain Name Service (DNS) is the least functional but most widely used. It responds to requests on a standard network port address, and for the domain in which it is running can map short (eight-character) strings to Internet port numbers. DNS is normally used for static services, which are always running when the system is operational and do not change port numbers at all—for example, the e-mail protocol uses DNS to find the remote mail daemon capable of accepting incoming e-mail to a user on a remote system.

The Network Information Service (NIS), previously called Yellow Pages (YP), is considerably more elaborate. NIS maintains a collection of maps, each of which has a symbolic name (e.g., hosts, services, etc.) and maps ASCII keywords to an ASCII value string. NIS is used on UNIX systems to map host names to Internet addresses, service names to port numbers, and so forth. Although NIS does not support pattern matching, there are ways for an application to fetch the entire NIS database, one line at a time, and it is common to include multiple entries in an NIS database for a single host that is known by a set of aliases. NIS is a distributed service that supports replication: The same data are normally available from any set of servers, and a protocol is used to update the full set of servers if an entry changes. However, NIS is not designed to support rapid updates: The assumption is that NIS data consist of mappings, such as the map from host name to Internet address, which change very rarely. A 12-hour delay before NIS information is updated is not unreasonable given this model—hence, the update problem is solved by periodically refreshing the state of each NIS server by having it read the contents of a set of files in which the mapping data are actually stored. As an example, NIS is often used to store password information on UNIX systems.

X.500 is an international standard that many expect will eventually replace NIS. This service, which is designed for use by applications running the ISO standard remote procedure call interface and ISDN.1 data encoding, operates much like an NIS server. No provision has been made in the standard for replication or high-performance update, but the interface does support some limited degree of pattern matching. As might be expected from a standard of this sort, X.500 addresses a wide variety of issues, including security and recommended interfaces. However, reliability issues associated with availability and consistency of the X.500 service (i.e., when data are replicated) have not yet been tackled by the standards organization.

There is considerable interest in using X.500 to implement general-purpose White Pages (WP) servers, which would be explicitly developed to support sophisticated pattern matching on very elaborate databases with detailed information about abstract entities. Rapid update rates, fault-tolerant features, and security are all being considered in these proposals. At the time of this writing, it appears that the Web will require such services and that work on universal resource naming for use in the Web will be a major driving force for evolution in this overall area.

4.6.2 Time Services

With the launch of the so-called Global Positioning System (GPS) satellites, microsecond accuracy became possible in workstations equipped with inexpensive radio receivers. Unfortunately, however, accurate clocks remain a major problem in the most widely used computer workstations and network technologies. We will discuss this in more detail in Chapter 20, but some background may still be useful here.

At the time of this writing, the usual clock for a PC or workstation consists of a quartz-based chip much like the one in a common wristwatch, accurate to within a few seconds per year. The initial value of such a clock is either set by the vendor or by the user, when the computer is booted. As a result, in any network of workstations, clocks can give widely divergent readings and can drift with respect to one another at significant rates. For these reasons, there has been considerable study of algorithms for clock synchronization, whereby the clocks on individual machines can be adjusted to give behavior approximating that of a shared global clock. In Chapter 20, we will discuss some of the algorithms that have been proposed for this purpose, their ability to tolerate failures, and the analyses used to arrive at theoretical limits on clock accuracy.

However, much of this work has a limited lifetime. GPS receivers can give extremely accurate time, and GPS signals are transmitted frequently enough so that even inexpensive hardware can potentially maintain time accurate to microseconds. By broadcasting GPS time values, this information can be propagated within a network of computers, and although some accuracy is necessarily lost when doing so, the resulting clocks are still accurate and comparable to within tens of microseconds. This development can be expected to have a major impact on the way that distributed software is designed—from a world of asynchronous communication and clocks that can be inaccurate by many times the average message latency in the network, GPS-based time could catapult us into a domain in which clock resolutions considerably exceed the average latency between sending a message and when it is received. Such developments make it very reasonable to talk about synchronous (time-based) styles of software design and the use of time in algorithms of all sorts.

Even coarsely synchronized clocks can be of value in distributed software—for example, when comparing versions of files, microsecond accuracy is not needed to decide if one version is more current than another: Accuracy of seconds or even tens of seconds may be adequate. Security systems often have a concept of expiration associated with keys, but for these to be at risk of attacks an intruder would need a way to set a clock back by days, not fractions of a second. And, although we will see that RPC protocols use time to detect and ignore very old, stale messages, as in the case of a security mechanism a clock would need to be extremely inaccurate for such a system to malfunction.

4.6.3 Security Services

In the context of an RPC environment, security is usually concerned with the *authentication* problem. Briefly stated, this is the problem of providing applications with accurate information about the user-ID on behalf of which a request is being performed. Obviously, one would hope that the user-ID is related in some way to the user, although this is frequently the weak link in a security architecture. Given an accurate source of user identifications, however, the basic idea is to avoid intrusions that can compromise user-ID security through break-ins on individual computers and even replacements of system components on some machines with versions that have been compromised and hence could malfunction. As in the case of clock services, we will look more closely at security later in the book (Chapter 19) and hence limit ourselves to a brief review here.

To accomplish authentication, a typical security mechanism (e.g., the Kerberos security architecture for DCE [see Schiller, Steiner et al.]) will request some form of password or one-time key from the user at log-in time, and periodically thereafter, as keys expire on the basis of elapsed time. This information is used to compute a form of secure user identification, which can be employed during connection establishment. When a client binds to a server, the security mechanism authenticates both ends and (at the option of the programmer) arranges for data to be encrypted on the wire, so that intruders who witness messages being exchanged between the client and server have no way to decode the data contained within them. (Unfortunately, however, this step is so costly that many applications disable encryption and simply rely upon the security available from the initial connection setup.) Notice that for such a system to work correctly, there must be a way to trust the authentication server itself: The user needs a way to confirm that it is actually talking to the authentication server and to legitimate representatives of the services it wishes to use. Given the anonymity of network communication, these are potentially difficult problems.

In Chapter 19, we will look closely at distributed security issues (e.g., we will discuss Kerberos in much more detail) and also at the relationship between security and other aspects of reliability and availability—problems that are often viewed as mutually exclusive, since one replicates information to make it more available, and the other tends to restrict and protect the information to make it more secure. We will also look at emerging techniques for protecting privacy, namely the true user-ID's of programs active in a network. Although the state of the art does not

Threads: A Personal Perspective

Rather than choosing between threads and event dispatch, an approach that supports threads as an option over event dispatch offers more flexibility to the developer. Speaking from personal experience, I have mixed feelings on the issue of threads (versus event dispatch). Early in my career I worked with protocols implemented directly over a UDP datagram model. This turned out to be very difficult: Such a system needs to keep track of protocol state in some form of table, matching replies with requests, and is consequently hard to program—for example, suppose that a distributed file server is designed to be single-threaded. Such a file server may handle many applications at the same time, so it will need to send off one request, perhaps to read a file, but remain available for other requests, perhaps by some other application that wants to write a file. The information needed to keep track of the first request (the read that is pending) will have to be recorded in some sort of pending activities table and later matched with the incoming reply from the remote file sys-

tem. Having implemented such an architecture once, I would not want to do it again.

This motivated me to move to RPC-style protocols, using threads. We will be talking about the Isis Toolkit, which is a system that I implemented (with help from others!) in the mid-1980s, in which lightweight threads were employed extensively. Many Isis users commented to me that they had never used threads before working with Isis, and they were surprised at how much the approach simplified things. This is certainly the case: In a threaded system, the procedure handling the read would simply block waiting for the reply, while other procedures could be executed to handle other requests. The necessary bookkeeping is implicit: The blocked procedure has a local state consisting of its calling stack, local variables, and so forth. Thus, there is no need to constantly update a table of pending activities.

Of course, threads are also a potential source of insidious programming bugs. In Isis, the benefits of threads certainly outweighed the

yet support construction of high-performance, secure, private applications, this should be technically feasible within the not-too-distant future. Of course, technical feasibility does not imply that the technology will become widely practical and therefore useful in building reliable applications, but at least the steps needed to solve the problems are increasingly understood.

4.6.4   Threads Packages

A fourth component of a typical RPC system is the lightweight threads package, which enables a single program to handle multiple tasks at the same time. Although threads are a general concept and indeed have rather little to do with communication per se, they are often viewed as necessary in distributed computing systems because of the potential for deadlock if threads are *not* present.

To understand this point, it is helpful to contrast three ways of designing a communication system. A single-threaded, message-based approach would correspond to a conventional style of programming extended directly to message passing. The programmer would use system calls problems associated with them, but it is also clear that this model requires a degree of programming sophistication that goes somewhat beyond standard single-threaded programming. It took me at least a year to get in the habit of thinking through the potential reentrance and ordering issues associated with concurrency and to become comfortable with the various styles of locking needed to overcome these problems. Many users report the same experience. Isis, however, is perhaps an unusually challenging case because the order in which events happen is very important in this system, for reasons that we will study in Part III.

In more recent work, I have teamed up with Robbert van Renesse, who is the primary author of the Horus system (we discuss this in considerable detail in Chapter 18). Horus, like Isis, was initially designed to use threads and is extremely sensitive to event ordering. But when testing very demanding applications, van Renesse found that threads were a serious source of overhead and code bloat: overhead because a stack for a thread consumes 16 KB or more of space, which is a lot of space in a system that can handle tens of thousands of messages per second, and excess code because of the necessary synchronization. Yet, as in the case of Isis, Horus sometimes needs threads: They often make it easy to do things that would be very hard to express in a nonthreaded manner.

van Renesse eventually extended Horus to use an event dispatch model similar to the one in Windows NT, which offers threads as an option over a basic event dispatch mechanism. This step, which substantially simplified many parts of Horus, left me convinced that supporting threads over an event dispatch architecture is the right way to go. For cases in which a thread is needed, it is absolutely vital that it be available. However, threads bring a considerable amount of baggage, which may be unnecessary in many settings. An event dispatch style of system gives the developer freedom to make this choice and has a lightweight and fast default behavior. I am, however, still convinced that event dispatch systems that lack the option of forking a thread when one is needed are often unwieldy and very difficult to use; this approach should be avoided.

such as *sendto* and *recvfrom* as needed to send and receive messages. If there are several things happening at the same time in a program structured this way, however, the associated bookkeeping can be a headache (see Break-out 4.1).

Threads offer a simple way to eliminate this problem: Each thread executes concurrently with the others, and each incoming request spawns a new thread to handle it. While an RPC is pending, the thread that issues it blocks (waits) in the procedure call that invoked the RPC. To the degree that there is any bookkeeping to worry about, the associated state is represented directly in the local variables of this procedure and in the call itself: When the reply is received, the procedure returns (the thread resumes execution), and there is no need to track down information about why the call was being done—this is obvious to the calling procedure. Of course, the developer does need to implement adequate synchronization to avoid concurrency-related bugs, but in general this is not a difficult thing to do. The approach overcomes many forms of problems that are otherwise difficult to address.

Consider a situation in which an RPC server is also the client of some other server, which is in turn the client of still additional servers. It is entirely possible that a cycle could form, in which RPC a by process x on process y leads to an RPC b by y on z, and so forth, until finally some process in the chain makes a request back to the original process, x. If these calls were LPC calls, such a sequence would simply be a form of recursion. For a single-threaded RPC system, however, x will be busy performing RPC a and hence would be unresponsive, creating a deadlock. Alternatively, x would need to somehow save the information associated with sending RPC a while it is handling this new incoming request. This is the bookkeeping problem alluded to above.

Yet a third option is known as event dispatch and is typical of windowing systems, in which each action by the user (mouse motion or clicks, keyboard entries) results in delivery of an event record to a central dispatching loop. The application program typically registers a set of procedure callbacks to perform when events of interest are received: If the left mouse button is pressed, invoke *left_button()*. Arguments to these callbacks tell the program exactly what occurred: The cursor was at position 132,541 when the mouse button was pressed, this is inside such and such a window, and so forth. One can use the same approach to handle event dispatch in message-based systems: Incoming messages are treated as events and result in callbacks to handler procedures.

The approaches can also be combined: Event dispatch systems can, for example, fork a new thread for each incoming message. In the most general approach, the callback is registered with some indication of how it should be performed: by forking a thread, by direct procedure call, or perhaps even by some other method, such as enqueuing the event on an event queue. This last approach is used in the Horus system, which we will discuss in Chapter 18.

At the time of this writing, although this is not universally the case, many RPC systems are built directly over a lightweight threads package. Each incoming RPC is handled by a new thread, eliminating the risk of deadlock, but forcing the programmer to learn about lightweight threads, preemption, mutual exclusion mechanisms, and other issues associated with concurrency. In this book, we will present some protocols in which processes are assumed to be multithreaded, so that the initiator of a protocol can also be a participant in it. However, we will not explicitly discuss thread packages or make use of any special features of particular packages.

The use of threads in this manner remains debatable. UNIX programs have heavily favored this approach, and the UNIX community generally understands the issues that must be addressed and minimizes their difficulty. Indeed, with experience, threaded programming is not all that difficult. One merely needs to get in the habit of enforcing necessary synchronization using appropriate interlocks. However, the PC community tends to work with an event-based model that lacks threads, in which the application is visualized as a dispatcher for incoming events and all callbacks are by procedure invocation. Thus, the PC community has its own style of programming, and it is largely nonthreaded. Windows NT further complicates this picture: It supports threads, and yet uses an event-oriented style of dispatching throughout the operating system; if a user wants to create a thread to handle an event, this is easily done but not forced upon the programmer.

4.7 The RPC Protocol

The discussion up to this point has focused on client/server computing and RPC from the perspective of the user. A remote procedure call *protocol* is concerned with the actual mechanism by which the client process issues a request to a server and by which the reply is transmitted back from the server to the client. We now look at this protocol in more detail.

Abstractly, the remote procedure call problem, which an RPC protocol undertakes to solve, consists of emulating LPC using message passing. LPC has a number of properties—a single procedure invocation results in exactly one execution of the procedure body, the result returned is reliably delivered to the invoker, and exceptions are raised if (and only if) an error occurs.

Given a completely reliable communication environment, which never loses, duplicates, or reorders messages, and given client and server processes that never fail, RPC would be trivial to solve. The sender would merely package the invocation into one or more messages and transmit these to the server. The server would unpack the data into local variables, perform the desired operation, and send back the result (or an indication of any exception that occurred) in a reply message. The challenge, then, is created by failures.

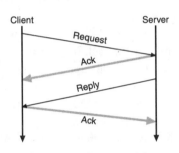

FIGURE 4.3. Simple RPC interaction, showing packets that contain data (dark) and acknowledgments (light).

Were it not for the possibility of process and machine crashes, an RPC protocol capable of overcoming limited levels of message loss, disorder, and even duplication would be easy to develop (Figure 4.3). For each process to which it issues requests, a client process maintains a message sequence number. Each message transmitted carries a unique sequence number, and (in most RPC protocols) a timestamp from a global clock—one that returns roughly the same value throughout the network, up to clock synchronization limits. This information can be used by the server to detect very old or duplicate copies of messages, which are discarded, and to identify received messages using what are called *acknowledgment protocol messages*.

The basic idea, then, is that the client process transmits its request and, until acknowledgments have been received, continues to retransmit the same messages periodically. The server collects messages and, when the full request has been received, performs the appropriate procedure invocation. When it transmits its reply, the same sort of reliable communication protocol is used. Often, the acknowledgment is delayed briefly in the hope that the reply will be sent soon and can be used in place of a separate acknowledgment.

A number of important optimizations have been proposed by developers of RPC-oriented distributed computing environments—for example, if one request will require the transmission of multiple messages, because the request is large, it is common to inhibit the sending of acknowledgments during the transmission of the burst of messages. In this case, a *negative acknowledgment* is sent if the receiver detects a missing packet; a single acknowledgment confirms reception of the entire burst when all packets have been successfully received (Figure 4.4). Similarly, it is common to delay the transmission of acknowledgment packets in the hope that the reply message itself can be transmitted instead of an acknowledgment; obviously, the receipt of

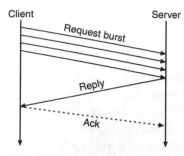

FIGURE 4.4. RPC using a burst protocol; here the reply is sent soon enough so that an acknowledgment to the burst is not needed.

a reply implies that the corresponding request was delivered and executed.

Process and machine failures, unfortunately, render this very simple approach inadequate. The essential problem is that because communication is over unreliable networking technologies, when a process is unable to communicate with some other process, there is no way to determine whether the problem is a network failure, a machine failure, or both (if a process fails but the machine remains operational, the operating system will often provide some status information, permitting this one case to be accurately sensed).

When an RPC protocol fails by timing out, but the client or server (or both) remains operational, it is impossible to know what has occurred. Perhaps the request was never received, perhaps it was received and executed but the reply was lost, or perhaps the client or server crashed while the protocol was executing. This creates a substantial challenge for the application programmer who wishes to build an application that will operate reliably despite failures of some of the services upon which it depends.

A related problem concerns the issue of what are called *exactly once semantics*. When a programmer employs LPC, the invoked procedure will be executed exactly once for each invocation. In the case of RPC, however, it is not evident that this problem can be solved. Consider a process, c, that issues an RPC to a service offered by process s. Depending upon the assumptions we make, it may be very difficult even to guarantee that s performs this request *at most* once. (Obviously, the possibility of a failure precludes a solution in which s would perform the operation exactly once.)

To understand the origin of the problem, consider the possible behaviors of an arbitrary communication network. Messages can be lost in transmission, and as we have seen this can prevent process c from accurately detecting failures of process s. But, the network might also misbehave by delivering a message after an unreasonably long delay—for example, suppose that a network router device fails by jamming up in such a manner that until the device is serviced, the software within it will simply wait for the hardware to be fixed. Obviously, there is no reason to simply assume that routers won't behave this way, and in fact it is known that some routers definitely could behave this way. Moreover, one can imagine a type of attack upon a network in which an intruder records messages for future replay.

One could thus imagine a situation in which process s performs a request from c, but then is presented with the same request after a very long delay (Figure 4.5). How can process s recognize this as a duplicate of the earlier request?

Depending upon the specific protocol used, an RPC package can use a variety of barriers to protect itself against replays of long-delayed messages—for example, the pack-

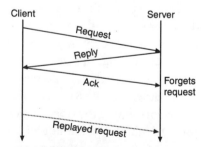

FIGURE 4.5. If an old request is replayed, perhaps because of a transient failure in the network, a server may have difficulty protecting itself against the risk of reexecuting the operation.

age might check timestamps in the incoming messages, rejecting any that are very old. Such an approach, however, presumes that clocks are synchronized to a reasonable degree and that there is no danger that a message will be replayed with a modified timestamp—an action that might well be within the capabilities of a sophisticated intruder. The server could use a connect-based binding to its clients, but this merely pushes the same problem into the software used to implement network connections—and, as we shall see shortly, the same issues occur and remain just as intractable at that level of a system. The server might maintain a list of currently valid users and could insist that each message be identified by a monotonically increasing sequence number—but a replay could, at least theoretically, reexecute the original binding protocol.

Analyses such as these lead us to two possible conclusions. One view of the matter is that an RPC protocol should take reasonable precautions against replay but not be designed to protect against extreme situations such as replay attacks. In this approach, an RPC protocol might claim to guarantee *at most once semantics,* meaning that provided that the clock synchronization protocol has not been compromised or some sort of active attack been mounted upon the system, each operation will result in either a single procedure invocation or, if a communication or process failure occurs, in no invocation. An RPC protocol can similarly guarantee at least once semantics, meaning that if the client system remains operational indefinitely, the operation will be performed at least once but perhaps more than once. Notice that both types of semantics come with caveats: conditions (hopefully very unlikely ones) under which the property would still not be guaranteed. In practice, most RPC environments guarantee a weak form of at most once semantics: Only a mixture of an extended network outage and a clock failure could cause such systems to deliver a message twice, and this is not a very likely problem.

A different approach, also reasonable, is to assume a very adversarial environment and protect the server against outright attacks that could attempt to manipulate the clock, modify messages, and otherwise interfere with the system. Security architectures for RPC applications commonly start with this sort of extreme position, although it is also common to weaken the degree of protection to obtain some performance benefits within less hostile subsets of the overall computing system. We will return to this issue and discuss it in some detail in Chapter 19.

4.8   Using RPC in Reliable Distributed Systems

The uncertainty associated with RPC failure notification and the weak RPC invocation semantics seen on some systems pose a challenge to the developer of a reliable distributed application.

A reliable application would typically need multiple sources of critical services, so that if one server is unresponsive or faulty the application can reissue its requests to another server. If the server behaves as a read-only information source, this may be an easy problem to solve. However, as soon as the server is asked to deal with dynamically changing information, even if the changes are infrequent compared to the rate of queries, a number of difficult consistency and fault-tolerant issues arise. Even questions as simple as load-balancing, so that each server in a service spanning multiple machines will do a roughly equal share of the request processing load, can be very difficult to solve.

Suppose that an application will use a primary-backup style of fault tolerance, and the requests performed by the server affect its state. The basic idea is that an application should connect itself to the primary, obtaining services from that process as long as it is operational. If the primary fails, the application will fail-over to the backup. Such a configuration of processes is illustrated in Figure 4.6. Notice that the figure includes multiple client processes, since such a service might well be used by many client applications at the same time.

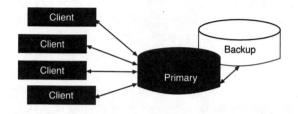

FIGURE 4.6. Idealized primary-backup server configuration. Clients interact with the primary and the primary keeps the backup current.

Consider now the design of a protocol by which the client can issue an RPC to the primary-backup pair such that if the primary performs the operation, the backup learns of the associated state change. In principle, this may seem simple: The client would issue an RPC to the server, which would compute the response and then issue an RPC to the backup, sending it the request it performed, the associated state change, and the reply being returned to the client. Then the primary would return the reply, as shown in Figure 4.7.

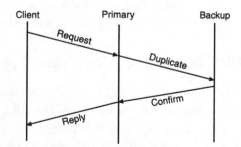

FIGURE 4.7. Simplistic RPC protocol implementing primary-backup replication.

This simple protocol is, however, easily seen to be flawed if the sorts of problems we discussed in the previous section might occur while it were running (see Birman and Glade). Take the issue of timeout (see Figure 4.8). In this solution, two RPCs occur, one nested within the other.

Either of these, or both, could fail by timeout, in which case there is no way to know with certainty in what state the system was left. If, for example, the client sees a timeout failure, there are quite a few possible explanations: The request may have been lost, the reply may have been lost, and either the primary or the primary and the backup may have crashed. Fail-over to the backup would only be appropriate if the primary were indeed faulty, but there is no accurate way to determine if this is the case, except by waiting for the primary to recover from the failure—not a very available approach.

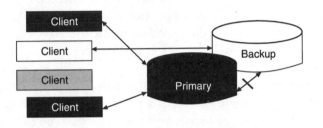

FIGURE 4.8. RPC timeouts can create inconsistent states, such as this one, in which two clients are connected to the primary, one to the backup, and one is disconnected from the service. Moreover, the primary and backup have become disconnected from one another—each considers the other faulty. In practice, such problems are easily provoked by transient network failures. They can result in serious application-level errors—for example, if the clients are air traffic controllers and the servers advise them on the safety of air traffic routing changes, this scenario could lead two controllers to route different planes into the same sector of the airspace! The matter is further complicated by the presence of more than one client. One could easily imagine that different clients could observe different and completely uncorrelated outcomes for requests issued simultaneously but during a period of transient network or computer failures. Thus, one client might see a request performed successfully by the primary, another might conclude that the primary is apparently faulty and try to communicate with the backup, and yet a third may have timed out both on the primary *and* the backup! We use the term "inconsistent" in conjunction with this sort of uncoordinated and potentially incorrect behavior. An RPC system clearly is not able to guarantee the consistency of the environment, at least when the sorts of protocols discussed above are employed, and hence reliable programming with RPC is limited to very simple applications.

The line between easily solved RPC applications and very difficult ones is not a very clear one—for example, one major type of file server accessible over the network is accessed by an RPC protocol with very weak semantics, which can be visible to users. Yet this protocol, called the Network File System protocol, is widely popular and has the status of a standard, because it is easy to implement and widely available on most vendor computing systems. NFS is discussed in more detail in Section 7.3 and so we will be very brief here.

One example of a way in which NFS behavior reflects an underlying RPC issue occurs when creating a file. NFS documentation specifies that the file creation operation should return the error code EEXISTS if a file already exists at the time the create operation is issued. However, there is also a case in which NFS can return error EEXISTS even though the file did not exist when the create was issued. This occurs when the create RPC times out, even though the request was delivered to the server and was performed successfully. NFS automatically reissues requests that fail by timing out and will retry the create operation, which now attempts to reexecute the request and fails because the file is now present. In effect, NFS is unable to ensure at most once execution of the request, and hence can give an incorrect return code. Had NFS been implemented using LPC (as in the LFS or local file system), this behavior would not be possible.

NFS illustrates one approach to dealing with inconsistent behavior in an RPC system. By weakening the semantics presented to the user or application program, NFS is able to provide acceptable behavior despite RPC semantics that create considerable uncertainty when an error is reported. In effect, the erroneous behavior is simply redefined to be a feature of the protocol.

A second broad approach that will interest us here involves the use of agreement protocols by which the components of a distributed system maintain consensus on the status (operational or failed) of one another. A rigorous derivation of the obligations upon such consensus protocols, the limitations on this approach, and the efficient implementation of solutions will be discussed later in this book (see Section 13.3). Briefly, however, the idea is that any majority of the system can be empowered to vote that a minority (often, just a single component) be excluded on the basis of apparently faulty behavior. Such a component is cut off from the majority group: If it is not really faulty, or if the failure is a transient condition that corrects itself, the component will be prevented from interacting with the majority system processes and will eventually detect that it has been dropped. It can then execute a rejoin protocol, if desired, after which it will be allowed back into the system.

With this approach, failure becomes an abstract event—true failures can trigger this type of event, but because the system membership is a self-maintained property of the system, the inability to accurately detect failures need not be reflected through inconsistent behavior. Instead, a conservative detection scheme can be used, which will always detect true failures while making errors infrequently (discussed in more detail in Section 13.9).

By connecting an RPC protocol to a group membership protocol that runs such a failure consensus algorithm, a system can resolve one important aspect of the RPC error-reporting problems discussed above. The RPC system will still be unable to accurately detect failures; hence, it will be at risk of incorrectly reporting operational components as having failed. However, the behavior will now be consistent throughout the system: If component a observes the failure of component b, than component c will also observe the failure of b, unless c is also determined to be faulty. In some sense, this approach eliminates the concept of failure entirely, replacing it with an event that might be called exclusion from membership in the system. Indeed, in the case where b is actually experiencing a transient problem, the resulting execution is much like being exiled from one's country: b is prevented from communicating with other members of the system and learns this. Conversely, the concept of a majority allows the operational part of

RPC AND THE CLIENT/SERVER MODEL   Chapter 4

the system to initiate actions on behalf of the full membership in the system. The system now becomes identified with a rigorous concept: the output of the system membership protocol, which can itself be defined formally and reasoned about using formal tools.

As we move beyond RPC to consider more complex distributed programming paradigms, we will see that this sort of consistency is often required in nontrivial distributed applications. Indeed, there appears to be a dividing line between the distributed applications that give nontrivial coordinated behavior at multiple locations, and those that operate as completely decoupled interacting components, with purely local correctness criteria. The former type of system requires the type of consistency we have encountered in this simple case of RPC error reporting. The latter type of system can manage with error detection based upon timeouts—but is potentially unsuitable for supporting any form of consistent behavior.

4.9 Related Reading

A tremendous amount has been written about client/server computing, and several pages of references could easily have been included here. Good introductions to the literature, including more detailed discussions of DCE and ASN.1, can be found in Birrell and Nelson, Comer and Stevens (1993), Coulouris et al., Tanenbaum.

On RPC performance, the classic reference is Shroeder and Burrows. Critiques of the RPC paradigm appear in Birman and van Renesse, Tanenbaum and van Renesse.

On the problem of inconsistent failure detection with RPC: (see Birman and Glade).

Other relevant publications include Bal et al. (1992), Bellovin and Merritt, Berners-Lee et al. (1994, 1995), Birrell and Nelson, Braun and Diot, Brockschmidt, Engler et al., Govindran and Anderson, Heidemann and Popek (1994), Jacobsen (1988, 1990), Mullender et al., Rashid, Shroeder and Burrows, Thekkanth and Levy, von Eicken et al. (1995).

A good reference to DCE is Open Software Foundation and to OLE-2 is Brockschmidt.

Kerberos is discussed in Bellovin and Merritt, Schiller, Steiner et al.

CHAPTER 5 ✧ ✧ ✧ ✧

Streams

CONTENTS

In Section 1.2 we introduced the idea of a reliable communication channel, or stream, that could overcome message loss and out-of-order delivery in communication sessions between a source and a destination process. In this chapter we briefly discuss the mechanisms used to implement the stream abstraction (see Ritchie), including the basic sliding window numbering scheme, flow control, and error-correction mechanisms. We describe the most common performance optimizations, which are based on dynamic adjustments of window size and transmission of short packet bursts. Finally, we look more closely at the reliability properties of a sliding window protocol, including the consistency of fault notification. Recall that the inconsistency of fault notification represented a drawback to using RPC protocols in reliable distributed applications. Here, we'll ask two questions about stream reliability: how streams themselves behave and how an RPC protocol operating over a stream can be expected to behave.

5.1 Sliding Window Protocols

The basic mechanism underlying most stream protocols is based on a data structure called a sliding window. More specifically, a typical bidirectional stream requires two sliding windows, one for each direction. Moreover, each window is duplicated on the sending and receiving sides. A window has limited size and is usually organized in a list of window slots, each having a fixed maximum size—for example, the TCP protocol normally operates over the IP protocol. It uses slots limited by the IP packet size (1,400 bytes), each divided between a TCP header and a payload portion. The size of a TCP window can vary dynamically, and the default values are often adjusted by vendors to optimize performance on their platforms.

A sliding window protocol treats the data communication from sender to destination as a continuous stream of bytes. (The same approach can be applied to messages, too, but this is less common and, in any case, doesn't change the protocol in a significant way.) The window looks into this stream and represents the portion of the stream that is actively being transmitted at a given point in time. Bytes that have already passed through the window have been received and delivered; they can be discarded by the sender. A sliding window protocol is illustrated in Figure 5.1.

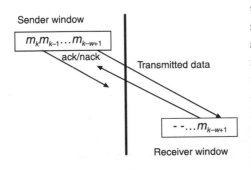

Here we see a window of size w, which is full on the sender side. Only some of the contents have been successfully received, however; in particular, m_k and m_{k-1} have yet to be received. The acknowledgment for m_{kw+1} and those bytes still in the window is in transit; the protocol will be operating on them to overcome packet loss, duplication, and out-of-order delivery. As we can see here, the sender's window can lag behind the receiver's window, so the receiver will often have holes in the window representing bytes that have not yet been received—indeed, that may not yet have been sent.

FIGURE 5.1. Sliding window protocol

In the window illustrated in Figure 5.1, the sender's window is full. If an attempt is made to transmit additional data, the sender process would be blocked, waiting until the window drains. These data are delayed on the sender side and may wait quite a while before being sent.

Each slot in the window has some form of sequence number and length. For TCP, byte numbers are used for the sequence number. Typically, these numbers can wrap (cycle), but only after a very long period of time—for example, the sequence number 0 could be reused after the TCP channel has transmitted a great deal of data. Because of this, a sliding window has the potential of becoming confused by very old duplicate packets, but the probability of this happening (except through an intruder who is intentionally attacking the system) is extremely small.

Many readers may have had the experience of using the Telnet program to connect to a remote computer over a network. Telnet operates over TCP, and you can sometimes see the effect of this queuing. Start a very large listing in your Telnet window, and then try to interrupt it. Often, many pages will scroll past before the interrupt signal breaks through! What you are seeing here is the backlog of data that was already queued up to be transmitted at the time the interrupt signal was received. These data were written by the program that creates listings to your remote terminal, though a TCP connection. The listing program is much faster than TCP—hence, it rapidly overflows the sliding window and builds a backlog of bytes. The amount of backlog, like the size of the TCP window, is variable from platform to platform, but 8 KB is typical. Finally, when this amount of data has piled up, the next attempt to write to the TCP connection will block, delaying the listing-generation program. The interrupt signal kills the program while it is in this state, but the 8 KB of queued data, plus any data that were in the window, will still be transmitted to your terminal.[1]

5.1.1 Error Correction

The most important role for a sliding window protocol is to overcome packet loss, duplication, and out-of-order delivery. This is done by using the receive-side window to reorder incoming data.

As noted earlier, each outgoing packet is labeled with a sequence number. The sender transmits the packet, then sets a timer for retransmission if an acknowledgment is not received within some period of time. Upon reception, a packet is slotted into the appropriate place in the window, or discarded as a duplicate if it repeats some packet already in the window or has a sequence number too small for the window. The sender-side queuing mechanism ensures that a packet sequence number will not be too large, since such a packet would not be transmitted.

---

[1]Some computers also clear any backlog of data to the terminal when you cause a keyboard interrupt. Thus, if you strike interrupt several times in a row, you may be able to avoid seeing all of these data print; each interrupt will clear several thousand bytes of data queued for output. This can create the impression that the first interrupt didn't actually kill the remote program and that several attempts were required before the program received the signal. In fact, the first interrupt you entered is always the one that kills the program, and the slow response time is merely a reflection of the huge amounts of data that the system has buffered in the connection from the remote program to your terminal!

The receiver now sends an acknowledgment for the packet (even if it was a duplicate), and may also send a negative acknowledgment for missing data, it if is clear that some data have been dropped.

In bidirectional communication, it may be possible to send acknowledgment and negative acknowledgment information on packets being sent in the other direction. Such piggybacking is desirable, when practical, because of the high overhead associated with sending a packet. Sending fewer packets, even if the packets would have been very small ones, is extremely beneficial for performance of a distributed application. Slightly longer packets, in fact, will often have essentially identical cost to shorter ones: The overhead of sending a packet of *any* size is considerably larger than the cost of sending a few more bytes. Thus, a piggybacked acknowledgment may be essentially free, in terms of the broader cost of communication within a system.

Streams normally implement some form of keep-alive messages, so that the sender and receiver can monitor one another's health. If a long period passes with no activity on the channel, a process can timeout and close the channel, pronouncing its remote counterpart faulty. As discussed in earlier chapters, such a timer-based failure detection mechanism can never guarantee accuracy, since communication problems will also trigger the channel shutdown protocol. However, this mechanism is part of the standard stream implementation.

5.1.2 Flow Control

A second role for the sliding window protocol is to match the speed of data generation and the speed of data consumption. With a little luck, the sender and receiver may be able to run continuously, getting the same throughput as if there were no channel between them at all. Latency, of course, may be higher than if the sender were directly connected to the receiver, but many applications are insensitive to latency—for example, a file transfer program is sensitive to latency when a request is issued to transfer a file for the first time. Once a file transfer is underway, however, throughput of the channel is the only relevant metric, and if the throughput is higher than the rate with which the receiver can consume the data, the cost of remote file access may seem very small, or negligible.

The way that this rate matching occurs is by allowing more than one packet to be in the sliding window at a time—for example, the window illustrated in Figure 5.1 is shown with w slots, all filled. In such a situation one might see w packets all on the wire at the same time. It is more likely, however, that some of these data will have already been received and that the acknowledgments are still in transit back to the sender. Thus, if the rate of data generated by the sender is close to that of the receiver, and if the window is large enough to contain the amount of data the sender can generate in a typical round-trip latency, the sender and the receiver may remain continuously active.

Of course, there are many applications in which the sender and receiver operate at different rates. If the receiver is faster, the receive window will almost always be empty—hence, the sender will never delay before sending data. If the sender is faster, some mechanism is needed to slow it down. To this end, it is not uncommon to associate what is called a *high-water/low-water* threshold with a sliding window.

The high-water threshold is the capacity of the window; when this threshold is reached, any attempt to send additional data will block. The low-water threshold is a point below which the window must drain before such a blocked sender is permitted to resume transmission. The idea of such a scheme is to avoid paying the overhead of rescheduling the sender if it would block immediately—for example, suppose that the window has a capacity of 8 KB. By setting the low-water threshold to 2 KB, the sender would sleep until there is space for 6 KB of new data before resuming communication. This would avoid a type of thrashing in which the sender might be rescheduled when the window contains some very small amount of space, which the sender would immediately overflow. The number 8 KB, incidentally, is a typical default for streaming protocols in operating systems for personal computers and workstations; higher-performance systems often use a default of 64 KB, which is also a hard limit for standard implementations of TCP.

5.1.3 Dynamic Adjustment of Window Size

Suppose that the sender and receiver operate at nearly the same rates, but that the window is too small to permit the sender and receiver rates to match. This would occur if the sender generates data fast enough to completely fill the window before a message can travel round trip to the receiver and back. In such a situation, the sender will repeatedly block, much as if the receiver were slower than it. The receiver, however, will almost always have an empty window and will be active only in brief bursts. Clearly, the system will now incur substantial idle time for both sender and receiver and will exhibit relatively low utilization of the network. The need to avoid this scenario has motivated the development of a number of mechanisms to dynamically increase the sender window.

On the other side of the coin, there are many potential problems if the window is permitted to become too large. Such a window will consume scarce kernel memory resources, reducing the memory available for other purposes, such as buffering incoming packets as they are received from the network, caching disk and virtual memory blocks, and managing windows associated with other connections. Many of these situations are manifested by increased rates of low-level packet loss, which occur when the lowest-level transport protocols are unable to find temporary buffering space for received packets.

The most widely used dynamic adjustment mechanism was proposed by Van Jacobson in a Ph.D. dissertation completed at Stanford University in 1988 (see Jacobson [1988]). Jacobson's algorithm is based on an exponential growth in the window size when it appears to be too small, that is, by doubling, and a linear backoff when the window appears to be too large. The former condition is detected when the sender's outgoing window is determined to be full, and yet only a single packet at a time is acknowledged—namely, the one with the smallest slot number. The latter condition is detected when high rates of retransmission occur. Exponential growth means that the number of window slots is increased by doubling, up to some prearranged upper limit. Linear backoff means that the window size is reduced by decrementing the number of slots on the sender's side. Under conditions where these changes are not made too frequently, they have been demonstrated to maintain a window size close to the optimal.

5.1.4 Burst Transmission Concept

When the performance of a sliding window protocol is studied closely, one finds that the window size is not the only performance-limiting factor. The frequency of acknowledgments and retransmissions is also very important and must be kept to a minimum: Both types of packets are pure overhead. As noted earlier, acknowledgments are often piggybacked on other outgoing traffic, but this optimization can only be exploited in situations where there actually are outgoing packets. Thus, the method is useful primarily in applications that exhibit a uniform, bidirectional flow of data. A separate mechanism is needed to deal with the case of messages that flow primarily in a single direction.

When the window size is small and the packets transmitted are fairly large, it is generally difficult to avoid acknowledgments of every packet. However, transmission of data in *bursts* can be a useful tactic in sliding window protocols that have relatively large numbers of slots and send relatively small packets (see Birrell and Nelson). In a burst transmission scheme, the sender attempts to schedule transmissions in bursts of several packets that are sent successively, with little or no delay between them. The sender then pauses before sending more data. The receiver uses the complementary algorithm, delaying its acknowledgments for a period of time in the hope of being able to acknowledge a burst of packets with a single message. A common variation upon this scheme uses a *burst bit* to indicate that a packet will be closely followed by some other packet; such schemes are also sometimes known as using *packet trains.* The last packet in the burst or train is recognizable as such because its burst bit is clear, and it flushes the acknowledgment back to the sender.

The biggest benefits of a burst transmission algorithm are seen in settings where very few instructions are needed to transmit a message and in which the sender and receiver are closely rate-matched. In such a setting, protocol designers strive to achieve a sort of perfect synchronization in which the transmission of data in the outgoing direction is precisely matched to a minimal flow of acknowledgments back to the sender. Tactics such as these are often needed to squeeze the maximum benefit out of a very high-performance interconnect, such as the message bus of a parallel supercomputer. However, they work well only if the communication activity is in nearly complete control of the sender and receiving machine, if message transmission latencies (and the variation in message latencies) are known and small, and if the error rate is extremely low.

5.2 Negative Acknowledgment Only

As communication bandwidths have risen, the effective bandwidth lost to acknowledgments can become a performance-limiting factor. This trend has led to the development of what are called *negative acknowledgment* protocols, in which packets are numbered in a traditional way but are not actually acknowledged upon reception. The sender uses a rate-based, flow-control method to limit the volume of outgoing data to a level the receiver is believed capable of accepting. The receiver, however, uses a traditional windowing method to reorder incoming messages and delete duplicates, and uses negative acknowledgment messages to solicit retransmission of any missing packets.

The exponential increase/linear backoff flow-control algorithm is often combined with a negative acknowledgment scheme to maximize performance. In this approach, which we first saw in conjunction with variable window-size adjustments, the sender increases the rate of transmission until the frequency of lost packets that must be retransmitted exceeds some threshold. The sender then backs off, reducing its rate steadily until the sender's rate and the receiver's rate are approximately matched, which is detected when the frequency of negative acknowledgments drops below some second (lower) threshold.

5.3 Reliability, Fault Tolerance, and Consistency in Streams

It is common for programming manuals and vendor product literature to characterize streaming protocols as reliable, since they overcome packet loss. Scrutinized closely, however, it is surprisingly difficult to identify ways in which these protocols really guarantee reliability of a sort that could be depended upon in higher-level applications (see Birman and Glade). Similar to the situation for a remote procedure call, a stream depends on timeouts and retransmissions to deal with communication errors, and reports a failure (breaking the stream) when the frequency of such events exceeds some threshold—for example, if the communication line connecting two machines is temporarily disrupted, the stream connections between them will begin to break. They will not, however, break in a coordinated manner. Quite the contrary: Each stream will break on its own, after a delay that can vary widely depending on how active the stream was immediately before the communication problem occurred. A stream that was recently active will remain open for a longer period of time, while a stream that was inactive for a period of time before communication was disrupted will be closer to its timeout and hence will break sooner. It is common to see an interval of as much as several minutes between the fastest detection of a communication failure and the slowest.

The problem that this creates is that many application programs interpret a broken stream to imply the failure of the program at the other end. Indeed, if a program fails, the streams to it will break eventually; in some situations this occurs within milliseconds (i.e., if the operating system senses the failure and closes the stream explicitly). Thus, some applications trigger failure recovery actions in situations that can also be the result of mundane communication outages that are rapidly repaired. Such applications may be left in an inconsistent state. Moreover, the long delays between when the earliest broken channel occurs and when the last one occurs can create synchronization problems, whereby one part of an application starts to reconfigure to recover from the failure while other parts are still waiting for the failed component to respond. Such skewed executions stress the application and can reveal otherwise hidden bugs. They may also have puzzling effects from the perspective of a user, who may see a screen that is partially updated and partially frozen in an old state, perhaps for an extended period of time.

As an example, consider the client/server structure shown in Figure 5.2, in which a client program maintains streams to two server programs. One server is the primary: It responds to requests in the normal mode of operation. The second is the backup, and it takes over only if the primary crashes. Streams are used to link these programs with one another. The precise replication and fault-tolerant mechanism used is not important for the point we wish to make; later in the text (Section 15.3.4) we will see how primary-backup problems such as this can be solved.

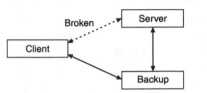

FIGURE 5.2. Inconsistently broken streams

Notice that the stream connecting the client to the server in Figure 5.2 has broken. How should the client interpret this scenario? One possibility is that the primary server has actually failed; if so, the connection between the primary server and the backup will eventually break too (although it may take some time before this occurs), in which case the backup will eventually take over the role of the primary and service to the client will be restored. A second possibility is that the client itself has failed, or is on a computer that has lost its communication connectivity to the outside world. In this case the client will eventually shut down, or will see its backup connection break too. Again, this may take some time to occur. Yet a third possibility is that the connection was lost as a consequence of a transient communication problem. In this case the client should reconnect to the server, which will have observed the situation and concluded that the client has failed.

The problem in this last situation is that the reliability properties of the stream have basically been lost. The stream was designed to overcome packet loss and communication disruptions, but in the scenario illustrated by the figure, it has done neither of those things. On the contrary, the inconsistent behavior of the separate streams present has turned out to *cause* a reliability problem. Data in the channel from the client to the server may have reached it or may have been lost when the connection was severed. Data from the server to the client may similarly have been lost. Thus, the client and the server must implement some protocol at the application level that will handle retransmission of requests that could have been dropped when a channel broke. This protocol will resolicit any data that the server may have been sending, suppressing duplicate data that the client has already seen. In effect, a client that will need to reconnect to a server in a situation such as this must implement a mechanism similar to the sliding window protocol used in the stream itself! It treats the connection much like a UDP connection: relatively reliable but not trustworthy. These types of problems should make the developer very cautious as to the reliability properties of streams.

In some ways, the inconsistency of the scenario illustrated by Figure 5.2 is even more troubling. As noted above, many applications treat the notification that a stream has broken as evidence that the end point has failed. In the illustrated setting, this creates a situation in which all three participants have differing, inconsistent views of the system membership. Imagine now that the same sort of problem occurred in a very large system, with hundreds of component programs, some of which might even have multiple connections between one another. The lack of coordination between streams means that these can break in almost arbitrary ways (e.g., one of two connections between a pair of programs can break, while the other connection remains established) and that almost any imaginable membership view can occur. Such a prospect makes it extremely difficult to use streams as a building block in applications intended to be reliable.

The success stories for such an approach are almost entirely associated with settings in which the communication hardware is very reliable, so that the failure detection used to implement the stream protocols is actually reasonably accurate, or in which one can easily give up on a transfer, as in the case of the Web (which is built over stream-style protocols, but can always abort a request just by closing its stream connections). Unfortunately, not many local area networks can claim to offer the extreme levels of reliability required for this assumption to accurately approximate the hardware, and this is almost never the case in wide area networks. And not many applications can get away with just telling the user that the "remote server is not responding or has failed." Thus, streams must be used with the utmost care in constructing reliable distributed applications.

In Chapter 13, we will discuss an approach to handling failure detection that could overcome this limitation. The idea is to introduce a protocol by which the computers in the network maintain agreement on system membership, triggering actions such as breaking a stream connection only if the agreement protocol has terminated and the full membership of the computer system agrees that the end point has failed, or at least will be treated as faulty. The approach is known to be practical and is well understood both from a theoretical perspective and in terms of the software needed to support it. Unfortunately, however, the standard stream implementations are based upon widely accepted specifications that *mandate* the use of timeout for failure detection. Thus, the stream implementations available from modern computer vendors will not become consistent in their failure reporting any time soon. Only a widespread call for consistency could possibly lead to revision of such major standards as the ones that specify how the TCP or ISO stream protocols should be implemented.

5.4 RPC over a Stream

It is increasingly common to run RPC protocols over stream protocols such as TCP, to simplify the implementation of the RPC interaction itself. In this approach, the RPC subsystem establishes a stream connection to the remote server and places it into an urgent transmission mode, whereby outgoing data are immediately transmitted to the destination. The reliability mechanisms built into the TCP protocol now subsume the need for the RPC protocol to implement any form of acknowledgment or retransmission policy of its own. In the simplest cases, this reduces RPC to a straightforward request-response protocol. When several threads multiplex the same TCP stream, sending RPCs over it concurrently, a small amount of additional code is needed to provide locking (so that data from different RPC requests are not written concurrently to the stream, which could interleave it in some undesired manner) and to demultiplex replies as they are returned from the server to the client.

It is important to appreciate that the reliability associated with a stream protocol will not normally improve (or even change) the reliability semantics of an RPC protocol superimposed upon it. As we saw above, a stream protocol can report a broken connection under the same conditions where an RPC protocol would fail by timing out, and the underlying stream-oriented acknowledgment and retransmission protocol will not affect these semantics in any useful way. The major advantage of running RPC over a stream is that by doing so, the amount of operating system software needed in support of communication is reduced: Having implemented flow control and reliability mechanisms for the stream subsystem, RPC becomes just another application-level use of the resulting operating system abstraction. Such an approach permits the operating system designer to optimize the performance of the stream in ways that might not be possible if the operating system itself were commonly confronted with outgoing packets that originate along different computational paths.

5.5 Related Reading

The best general references are Comer, Comer and Stevens (1993), Coulouris et al., Tanenbaum.

On the inconsistency of failure detection in streams: (see Birman and Glade).

There has been a considerable amount of recent work on optimizing stream protocols (particularly TCP) for high-performance network hardware. An analysis of TCP costs, somewhat along the lines of the RPC cost analysis in Shroeder and Burrows, can be found in Clark et al.

Work on performance optimization of TCP includes Jacobson (1988, 1990), Kay, Kay and Pasquale.

A summary of other relevant papers can be found in Braun and Diot, Comer, Tennenhouse.

Other pertinent publications include Brakmo et al., Clark and Tennenhouse, Comer, Comer and Stevens (1993), Coulouris et al., Drushel and Peterson, Floyd et al., Jacobson (1988, 1990), Karamcheti and Chien, Kay and Pasquale, Mullender et al., Peterson et al., Rozier et al. (Fall 1988, December 1988), Strayer et al., Tanenbaum, van Renesse (1988, 1989), von Eicken et al.

CHAPTER 6

CORBA and Object-Oriented Environments

CONTENTS

With the emergence of object-oriented programming languages, such as Modula and C++, came a recognition that object orientation could play a role similar to that of the OSI hierarchy for complex distributed systems. In this view, one would describe a computing system in terms of the set of objects from which it was assembled, together with the rules by which these objects interact with one another. Object-oriented system design became a major subject for research, with many of the key ideas pulled together for the first time by a British research effort, called the Advanced Network Systems Architecture group, or ANSA. In this chapter, we will briefly discuss ANSA, and then focus on a more recent standard, called CORBA, which draws on some of the ideas introduced by ANSA and has emerged as a widely accepted standard for objected-oriented distributed computing.

6.1   The ANSA Project

The ANSA project, headed by Andrew Herbert, was the first systematic attempt to develop technology for modeling complex distributed systems (see Architecture Projects Management Limited [1989, April 1991, November 1991]). ANSA was intended as a technology base for writing down the structure of a complex application or system and then translating the resulting description into a working version of that system in a process of stepwise refinement.

Abstractly, ANSA consists of a set of models, which deal with various aspects of distributed system design and representation problems. The enterprise model is concerned with the overall functions and roles of the organizational structure within which the problem at hand is to be solved—for example, an air traffic control system would be an application within the air traffic control organization, an enterprise. The information model represents the flow of information within the enterprise; in an air traffic application this model might describe flight control status records, radar inputs, radio communication to and from pilots, and so forth. The computation model is a framework of programming structures and program development tools that are made available to developers. The model deals with such issues as modularity of the application itself, invocation of operations, parameter passing, configuration, concurrency and synchronization, replication, and the extension of existing languages to support distributed computing. The engineering and technology models reduce these abstractions to practice, providing the implementation of the ANSA abstractions and mapping these to the underlying run-time environment and its associated technologies.

In practical terms, most users viewed ANSA as a set of rules for system design, whereby system components could be described as objects with published interfaces. An application with appropriate permissions could obtain a handle on the object and invoke its methods using the procedures and functions defined in this interface. The ANSA environment would automatically and transparently deal with such issues as fetching objects from storage, launching programs when a new instance of an object was requested, implementing the object invocation protocols, and so forth. Moreover, ANSA explicitly included features for overcoming failures of various kinds, using transactional techniques drawn from the database community, as well as process group techniques in which sets of objects are used to implement a single highly available distributed service. We will consider both types of technology in considerable detail in Part III of the book.

ANSA treated the objects that implement a system as the concrete realization of the enterprise computing model and the enterprise information model. These models captured the essence of the application as a whole, treating it as a single abstraction even if the distributed system as implemented necessarily contained many components. Thus, the enterprise computing model might support the abstraction of a collision-avoidance strategy for use by the air traffic control enterprise as a whole, and the enterprise data model might define the standard data objects used in support of this service. The actual implementation of the service would be reached by a series of refinements in which increasing levels of detail are added to this basic set of definitions. Thus, one passes from the abstraction of a collision-avoidance strategy to the more concrete concept of a collision-avoidance subsystem located at each set of primary sites and linked to one another to coordinate their actions (see Figure 6.1). This concept evolved to one with further refinements, defining the standard services composing the collision-avoidance system as used on a single air traffic control workstation, and then evolved still further to a description of how those services could be implemented.

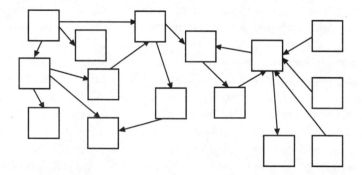

FIGURE 6.1. Distributed object abstraction. Objects are linked by object references, and the distributed nature of the environment is hidden from users. Access is uniform even if objects are implemented to have special properties or internal structure, such as replication for increased availability or transactional support for persistence. Objects can be implemented in different programming languages, but this is invisible to users.

In very concrete terms, the ANSA approach required the designer to write down the sort of knowledge of distributed system structure that, for many systems, is implicit but never encoded in a machine-readable form. The argument was that by writing down these system descriptions, a better system would emerge: one in which the rationale for the structure used was self-documenting and in which detailed information would be preserved about the design choices and objectives that the system carries out; in this manner the mechanisms for future evolution could be made a part of the system itself. Such a design promotes extensibility and interoperability, and offers a path to system management and control. Moreover, ANSA designs were expressed in terms of objects, whose locations could be anywhere in the network, with the actual issues of location developing only after the design was further elaborated, or in specific situations where location of an object might matter (Figure 6.2). This type of object-oriented, location-transparent design has proved very popular with distributed system designers.

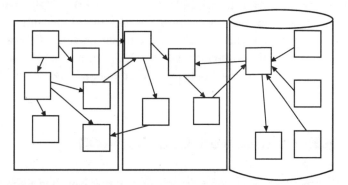

FIGURE 6.2. In practice, the objects in a distributed system execute on machines or reside in storage servers. The run-time environment works to conceal movement of objects from location to location and to activate servers when they are initially referenced after having been passively stored. This environment also deals with fault-tolerance issues, garbage collection, and other issues that span multiple objects or sites.

6.2 Beyond ANSA to CORBA

While the ANSA technology per se has not gained a wide following, these ideas have had a huge impact on the view of system design and function adopted by modern developers. In particular, as the initial stages of the ANSA project ended, a new project was started by a consortium of computer vendors. Called the Common Object Request Broker Architecture, CORBA undertakes to advance standards permitting interoperation between complex object-oriented systems built by diverse vendors (see Object Management Group and X/Open).

Although CORBA undoubtedly drew on the ANSA work, the architecture represents a consensus developed within an industry standards organization called the Object Management Group, or OMG, and differs from the ANSA architecture in many respects. The mission of OMG was to develop architecture standards promoting interoperability between systems developed using object-oriented technologies. In some ways, this represents a less-ambitious objective than the task with which ANSA was charged, since ANSA set out both to develop an all-encompassing architectural vision for building enterprise-wide distributed computing systems and to incorporate reliability technologies into its solutions. However, as CORBA has evolved, it has begun to tackle many of the same issues. Moreover, CORBA reaches well beyond ANSA by defining a very large collection of what are called services, which are CORBA-based subsystems that have responsibility for specific tasks, such as storing objects or providing reliability, and that have specific interfaces. ANSA began to take on this problem late in the project and did not go as far as CORBA.

At the time of this writing, CORBA was basically a framework for building distributed computing environments, but with the proviso that different vendors might offer their own CORBA solutions with differing properties. In principle, adherence to the CORBA guidelines should permit such solutions to interoperate—for example, a distributed system programmed using a CORBA product from Hewlett-Packard should be useful within an application developed using CORBA products from Sun Microsystems, IBM, or some other CORBA-compliant vendor. Interoperability of this sort, however, was only introduced in late 1996, and hence there has been little experience with this specific feature.

6.3   OLE-2 and Network OLE (Active/X)

As this book was bring written, Microsoft Corporation had just launched a major drive to extend its proprietary object-oriented computing standard, OLE-2, into a distributed object-oriented standard aimed squarely at the Internet. The network OLE-2 (called Active/X) specification, when it emerges, is likely to have at least as big an impact on the community of PC users as CORBA is having on the UNIX community. However, until the standard is actually released, it is impossible to comment upon it. Experts with whom I have spoken predict that network OLE will be generally similar to CORBA, but with a set of system services more closely parallel to the ones offered by Microsoft in its major network products: NT/Server and NT/Exchange. Presumably, these would include integrated messaging, e-mail, conferencing tools, systemwide security through encryption technologies, and comprehensive support for communicating using multimedia objects. One can only speculate as to more advanced features, such as group computing technologies and reliability, which will be discussed in Part III.

6.4   The CORBA Reference Model

The key to understanding the structure of a CORBA environment is the Reference Model, which consists of a set of components that a CORBA platform should typically provide. These components are fully described by the CORBA architecture, but only to the level of interfaces used by application developers and functionality. Individual vendors are responsible for deciding how to implement these interfaces and how to obtain the best possible performance; moreover, individual products may offer solutions that differ in offering optional properties such as security, high availability, or special guarantees of behavior that go beyond the basics required by the model.

At a minimum, a CORBA implementation must supply an *Object Request Broker*, or ORB, which is responsible for matching a requester with an object that will perform its request, using the object reference to locate an appropriate target object (see Figure 6.3). The implementation will also contain translation programs, responsible for mapping implementations of system components (and their IDLs) to programs which can be linked with a run-time library and executed. A set of *object services* provides the basic functionality needed to create and use objects: These include such functions as creating, deleting, copying, or moving objects; giving them names that other objects can use to bind to them; and providing security. An interesting service, which we will discuss in more detail, is the *Event Notification Service,* or ENS: This allows a program to register its interest in a class of events. All events in that class are then reported to the program. It thus represents a communication technology different from the usual RPC-style or stream-style of connection. A set of *Common Facilities* contains a collection of standardized applications that most CORBA implementations are expected to support, but that are ultimately optional: These include, for example, standards for system management and for electronic mail that may contain objects. And, finally, there are *Application Objects* developed by the CORBA user to solve a particular problem.

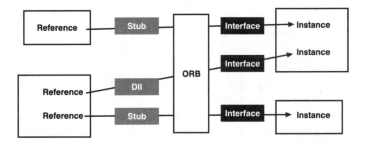

FIGURE 6.3. The conceptual architecture of CORBA uses an object request broker as an intermediary that directs object invocations to the appropriate object instances. There are two cases of invocations: the static one, which we focus on in the book, and the dynamic invocation interface (DII), which is more complex to use and is not discussed here.

In many respects the Object Request Broker is the core of a CORBA implementation. Similar to the function of a communication network or switching system, the ORB is responsible for delivering object invocations that originate in a client program to the appropriate server program and then routing the reply back to the client. The ability to invoke an object, of course, does not imply that the object that was invoked is being used correctly, has a consistent state, or is even the most appropriate object for the application to use. These broader properties fall back upon the basic technologies of distributed computing that are the general topic of this book; as we will see, CORBA is a way of *talking about* solutions, but it is not a *specific set of prebuilt solutions*. Indeed, one could say that because CORBA worries about syntax but not semantics, the technology is largely superficial: a veneer around a set of technologies. However, this particular veneer is an important and sophisticated one, and it also creates a context within which a principled and standardized approach to distributed system reliability becomes possible.

For many users, object-oriented computing means programming in C++, although SmallTalk, Java and Ada are also object-oriented languages, and one can develop object interfaces to other languages such as FORTRAN and COBOL. Nonetheless, C++ is the most widely used language, and it is the one we focus on in the examples presented in the remainder of this chapter. Our examples are drawn directly from the programmer's guide for Orbix, an extremely popular CORBA technology at the time of this writing.

An example of a CORBA object interface, coded in the Orbix interface definition language (IDL), is shown in Figure 6.4. This interface publishes the services available from a grid server, which is intended to manage two-dimensional tables such as those used in spreadsheets or relational databases. The server exports two read-only values, width and height, which can be used to query the size of a grid object. There are also two operations that can be performed upon the object: "set," which sets the value of an element, and "get," which fetches the value. Set is of type void, meaning that it does not return a result; get, on the other hand, returns a long integer.

```
//grid server example for Orbix
  //IDL -- in file grid.idl
  interface grid {
      readonly attribute short height;
      readonly attribute short width;

      void set(in short n, in short m, in long value);
      void get(in short n, in short m);
  };
```

FIGURE 6.4. IDL interface to a server for a grid object coded in Orbix, a popular CORBA-compliant technology

To build a grid server, the user would need to write a C++ program that implements this interface. To do this, the IDL compiler is first used to transform the IDL file into a standard C++ header file in which Orbix defines the information it will need to implement remote invocations on behalf of the client. The IDL compiler also produces two forms of stub files—one that implements the client side of the get and set operations; the other implements the server side. These stub files must be compiled and linked to the respective programs. (See Figure 6.5.)

If one were to look at the contents of the header file produced for the grid IDL file, one would discover that width and height have been transformed into functions; that is, when the C++ programmer references an attribute of a grid object, a function call will actually occur in the client-side stub procedures, which can perform an RPC to the grid server to obtain the current value of the attribute.

We say RPC here, but in fact a feature of CORBA is that it provides very efficient support for invocations of local objects, which are defined in the same address space as the invoking program. The significance of this is that although the CORBA IDL shown in Figure 6.4 could be used to access a remote server that handles one or more grid objects, it can also be used to communicate to a completely local instantiation of a grid object, contained entirely in the address space of the calling program. Indeed, the concept goes even further: In Orbix+Isis, a variation of Orbix, the grid server could be replicated using an object group for high availability. And in the most general case, the grid object's clients could be implemented by a server running under some other CORBA-based environment, such as IBM's DSOM product, HP's DOMF, Sun's DOE, Digital Equipment's ObjectBroker, or other object-oriented environments with which CORBA can communicate using an adapter, such as Microsoft's OLE. CORBA implementations thus have the property that object location, the technology or programming language used to build an object, and even the ORB under which they are running can be almost completely transparent to the user.

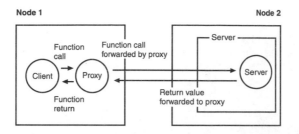

FIGURE 6.5. Orbix conceals the location of objects by converting remote operations into operations on local proxy objects, mediated by stubs. However, remote access is not completely transparent in standard CORBA applications if an application is designed for reliability—for example, error conditions differ for local and remote objects. Such issues can be concealed by integrating a reliability technology into the CORBA environment, but transparent reliability is not a standard part of CORBA, and solutions vary widely from vendor to vendor.

```
//C++ code fragment for grid implementation class
  #include "grid.hh" // generated from IDL

class grid_i: public gridBOAImpl {
        short       m_height
        short       m_width
        long        **m_a;
    public
        grid_i(short h, short w); //Constructor
        virtual ~grid_i();          //Destructor
        virtual short width(CORBA::Environment &);
        virtual short height(CORBA::Environment &);
        virtual void set(short n, short m, long value,
            CORBA::Environment &);
        virtual long get(short n, short m,
            CORBA::Environment &);
    };
```

FIGURE 6.6. Orbix example of a grid implementation class corresponding to grid IDL.

What exactly would a grid server look like? If we are working in C++, a grid would be a C++ program that includes an implementation class for grid objects. Such a class is shown in Figure 6.6, again drawing on Orbix as a source for our example. The "Environment" parameter is used for error handling with the client. The BOAImpl extension ("gridBOAImpl") designates that this is a Basic Object Adapter Implementation for the grid interface. Figure 6.7 shows the code that might be used to implement this abstract data type.

```
//Implementation of grid class
 #include "grid_i.h"

grid_i::grid_i(short h, short w) {
     m_height = h;
     m_width = w;
     m_a = new long*[h];
     for(int i = 0;i<h;i++)
            m_a[i] = new long [w];
}
grid_i::~grid_i() {
     for(int i = 0;i<m_height;i++)
            delete[]m_a[i];
     delete[]m_a;
}
short grid_i::width(CORBA::Environment &){
     return m_width;
}
short grid_i::height(CORBA::Environment &){
     return m_height;
}
void grid_i::set(short n, short m, long value, CORBA::Environment &){
     m_a[n][m]=value;
}
void grid_i::get(short n, short m, CORBA::Environment &) {
     return m_a[n][m];
}
```

FIGURE 6.7. Server code to implement the grid_i class in Orbix

Finally, our server needs an enclosing framework: the program itself that will execute this code. The code in Figure 6.8 provides this; it implements a single grid object and declares itself to be ready to accept object invocations. The grid object is not named in this example, although it could have been, and indeed the server could be designed to create and destroy grid objects dynamically at run time.

```
#include "grid_i.h"
#include<iostream.h>

void main() {
        grid_imyGrid(100,100);
        //Orbix objects can be named but this is not
        //needed for this example
        CORBA::Orbix.impl_is_ready();
        cout<<"server terminating"<<endl;
{
```

FIGURE 6.8. Enclosing program to declare a grid object and accept requests upon it

The user can now declare to Orbix that the grid server is available by giving it a name and storing the binary of the server in a file, the path name of which is also provided to Orbix (see Figure 6.9). The Orbix life-cycle service will automatically start the grid server if an attempt is made to access it when it is not running.

```
#include "grid_h.h"
  #include<iostream.h>

void main() {
        grid*p;

        p=grid::_bind(":gridSrv");
        cout<<"height is "<<p-height()<<endl;
        cout<<"width is "<<p->width()<<endl;
        p->set(2, 4, 123);
        cout<<"grid(2, 4)is "<<p->get(2, 4)<<endl:
        p->release();
}
```

FIGURE 6.9. Client program for the grid object—assumes that the grid was registered under the server name "gridSrv." This example lacks error handling; an elaborated version with error handling appears in Figure 6.10.

CORBA supports several concepts of reliability. One is concerned with recovering from failures—for example, when invoking a remote server. A second reliability mechanism is provided for purposes of reliable interactions with persistent objects and is based upon what is called a transactional architecture. We discuss transactions elsewhere in this book and will not digress onto that subject at this time. However, the basic purpose of a transactional architecture is to provide a way for applications to perform operations on complex persistent data structures, without interfering with other concurrently active but independent operations, in a manner that will leave the structure intact even if the application program or server fails while it is running. Unfortunately, as we will see in Chapter 21, transactions are primarily useful in applications that are structured as database systems on which programs operate using read and update requests. Such structures are important in distributed systems, but there are many distributed applications that match the model poorly, and, for them, transactional reliability is not a good approach.

```
#include "grid_h.h"
  #include<iostream.h>

void main() {
        grid*p;

        TRY {

                p = grid::_bind(":gridSrv");
        }
        CATCHANY {
                cerr<<"bind to object failed"<<endl;
                cerror<<"Fatal exception "<<IT_X<<endl;
                exit(1);
        }
        TRY {
                cout<<"height is "<<p->height()<<endl;
        }
        CATCHANY {
                cerr<<"call to height failed"<<endl;
                cerror<<"Fatal exception "<<IT_X<<endl;
                exit(1);
        }
        ...etc...
}
```

FIGURE 6.10. Illustration of Orbix error-handling facility. Macros are used to catch errors; if one occurs, the error can be caught and potentially worked around. Notice that each remote operation can potentially fail—hence, exception handling would normally be more standardized. A handler for a high-availability application would operate by rebinding to some other server capable of providing the same functionality. This can be concealed from the user, which is the approach used in systems such as Orbix+Isis or Electra, a CORBA technology layered over the Horus distributed system.

Outside of its transactional mechanisms, however, CORBA offers relatively little help to the programmer—for example, Orbix can be notified that a server application can be run on one of a number of machines. When a client application attempts to use the remote application, Orbix will automatically attempt to bind to each machine in turn, selecting at random the first machine that confirms that the server application is operational. However, Orbix does not provide any form of automatic mechanisms for recovering from the failure of such a server after the binding is completed. The reason for this is that a client process that is already communicating with a server may have a complex state that reflects information specific to that server, such as cached records with record identifiers that came from the server or other forms of data that differ in specific ways even among servers able to provide the same functionality. To rebind the client to a new server, one would somehow need to refresh, rebuild, or roll back this server-dependent state. And doing so is potentially very difficult; at a minimum, considerable detailed knowledge of the application will be required.

The same problems can also occur in the server itself—for example, consider a financial trading service, in which the prices of various stocks are presented, that is extremely dynamic due to rapidly changing market data. The server may need to have some form of setup that it uses to establish a client profile, and it may need to have an internal state that reflects the events that have occurred since the client first bound to it. Even if some other copy of the server is available and can provide the same services, there could be a substantial time lag when rebinding and there may be a noticeable discontinuity if the new server, lacking this state of the session, starts its financial computations from the current stream of incoming data. Such events will not be transparent to the client using the server and it is unrealistic to try to hide them.

The integration of a wider spectrum of reliability-enhancing technologies with CORBA represents an important area for research and commercial development, particularly if reliability is taken in the broad sense of security, fault tolerance, availability, and so forth. High-performance, commercially appealing products will be needed to demonstrate the effectiveness of the architectural features that result: When we discuss transactions on distributed objects, for example, we will see that merely supporting transactions through an architecture is not likely to make users happy. Even the execution of transactions on objects raises deeper issues that would need to be resolved for such a technology to be accepted as a genuinely valid reliability-enhancing tool—for example, the correct handling of a transactional request by a nontransactional service is unspecified in the architecture.

More broadly, CORBA can be viewed as the ISO hierarchy for object-oriented distributed computing: It provides us with a framework within which such systems can be described and offers ways to interconnect components without regard for the programming language or vendor technologies used in developing them. Exploiting this to achieve critical reliability in distributed settings, however, stands as a more basic technical challenge that CORBA does not directly address. CORBA tells us how to structure and present these technologies, but not how to build them.

In chapters 13 through 18 we will discuss process group computing and associated technologies. The Orbix product is unusual in supporting a reliability technology, Orbix+Isis (see Iona Ltd. and Isis Distributed Systems, Inc.), based on process groups, that overcomes these problems. Such a technology is a form of replication service, but the particular one used to implement Orbix+Isis is extremely sophisticated in comparison to the most elementary forms of replication services, and the CORBA specifications for this area remain very tentative. Thus,

Orbix+Isis represents a good response to these reliability concerns, but the response is specific to Orbix and may not correspond to a broader CORBA response to the issues that occur.

6.5 TINA

TINA-C stands for Telecommunications Information Network Architecture Consortium, and is an organization of major telecommunication service providers that set out to look at ANSA- and CORBA-like issues from a uniquely telecommunication perspective (see *Telecommunications Information Network Architecture Conference, Proceedings of*). At the time of this writing, TINA was in the process of specifying a CORBA-based architecture with extensions to deal with issues of real-time communication, reliability, and security, with standard telecommunication-oriented services that go beyond the basic list of CORBA services implemented by typical CORBA-compatible products. The TINA variant of CORBA is expected by many to have a dramatic impact on the telecommunication industry. Specific products aimed at this market are already being announced: Chorus Systems' COOL-ORB is a good example of a technology that focuses on the real-time, security, and reliability needs of the telecommunication industry by providing a set of object services and architectural features that reflects embedded applications typical of large-scale telecommunication applications, in contrast to the more general computing market to which products such as Orbix appear to be targeted.

6.6 IDL and ODL

IDL is the language used to define an object interface (in the TINA standard, there is an ODL language that goes beyond IDL in specifying other attributes of the object, and in allowing each object to export more than one interface). (See Figure 6.11.) CORBA defines an IDL for the various languages that can be supported: C++, SmallTalk, Ada 95, and so forth. The most standard of these is the IDL for C++, and the examples given are expressed in C++ for that reason. However, expanded use of IDL for other programming languages is likely in the future.

The use of C++ programs in a CORBA environment can demand a high level of sophistication in C++ programming. In particular, the operator overload functionality of C++ can conceal complex machinery behind deceptively simple interfaces. In a standard programming language one expects that an assignment statement such as $a = b$ will execute rapidly. In C++

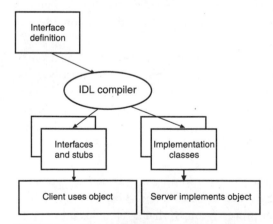

FIGURE 6.11. From the interface definition, the IDL compiler creates stub and interface files, which are used by clients that invoke the object and by servers that implement it.

such an operation may involve allocation and initialization of a new abstract object and a potentially costly copying operation. In CORBA such an assignment may involve costly remote operations on a server remote from the application program that executes the assignment statement. To the programmer, CORBA and C++ may appear as a mixed blessing: Through the CORBA IDL, operations such as assignment and value references can be transparently extended over a distributed environment, which can seem like magic. But the magic is potentially tarnished by the discovery that a single assignment might now take seconds (or hours) to complete!

Such observations point to a deficiency in the CORBA IDL language and, perhaps, the entire technology as currently conceived. IDL provides no features for specifying *behaviors* of remote objects that are desirable or undesirable consequences of distribution. There is no possibility of using IDL to indicate a performance property (or cost, in the above example) or to specify a set of fault-tolerant guarantees for an object that differs from the ones normally provided in the environment. Synchronization requirements or assumptions made by an object, or guarantees offered by the client, cannot be expressed in the language. This missing information, potentially needed for reliability purposes, can limit the ability of the programmer to fully specify a complex distributed system, while also denying the user the basic information needed to validate that a complex object is being used correctly.

One could argue that the IDL should be limited to specification of the interface to an object and that any behavioral specifications would be managed by other types of services. Indeed, in the case of the life-cycle service, one has a good example of how the CORBA community approaches this problem: The life-cycle aspects of an object specification are treated as a special type of data managed by this service and are not considered to be a part of the object interface specification. Yet I am convinced that this information often belongs in the interface specification, in the sense that these types of properties may have direct implications for the user who accesses the object and may be information of a type that is important in establishing that the object is being used correctly. I am convinced that the specification of an object involves more than the specification of its interfaces and that the interface specification involves more than just the manner in which one invokes the object. In contrast, the CORBA community considers behavior to be orthogonal to interface specification, and hence it relegates behavioral aspects of the object's specification to the special-purpose services directly concerned with that type of information. Unfortunately, it seems likely that much basic research will need to be done before this issue is addressed in a convincing manner.

6.7 ORB

An Object Request Broker, or ORB, is the component of the run-time system that binds client objects to the server objects they access and that interprets object invocations at run time, arranging for the invocation to occur on the object that was referenced. (CORBA is thus the OMG's specification of the ORB and of its associated services.) ORBs can be thought of as switching systems through which invocation messages flow. A fully compliant CORBA implementation supports interoperation of ORBs with one another over TCP connections, using what is called the Internet Inter-ORB Protocol (IIOP) protocol. In such an interoperation mode, any CORBA server can potentially be invoked from any CORBA client, even if the server and client were built and operated on different versions of the CORBA technology base.

Associated with the ORB are a number of features designed to simplify the life of the developer. An ORB can be programmed to automatically launch a server if it is not running when a client accesses it (this is called factory functionality), and it can be asked to automatically filter invocations through user-supplied code that automates the handling of error conditions or the verification of security properties. The ORB can also be programmed to make an intelligent choice of object if many objects are potentially capable of handling the same request; such a functionality would permit, for example, load-balancing within a group of servers that replicate a particular database.

6.8   Naming Service

A CORBA naming service is used to bind names to objects. Much as a file system is organized as a set of directories, the CORBA naming architecture defines a set of *naming contexts,* and each name is interpreted relative to the naming context within which that name is registered. The CORBA naming architecture is potentially a very general one, but, in practice, many applications are expected to treat it as an object-oriented generalization of a traditional naming hierarchy. Such applications would build hierarchical naming context graphs (directory trees), use ASCII-style path names to identify objects, and standardize the sets of attributes stored for each object in the naming service (size, access time, modification time, owner, permissions, etc.). The architecture, however, is sufficiently flexible to allow a much broader concept of names and naming.

A CORBA name should not be confused with an object reference. In the CORBA architecture, an object reference is essentially a pointer to the object. Although a reference need not include specific location information, it does include enough information for an ORB to find a path to the object or to an ORB that will know how to reach the object. Names, in contrast, are symbolic ways of naming these references. By analogy to a UNIX file system, a CORBA object name is similar to a path name (and, as with a path name, more than one name can refer to the same object). A CORBA object reference is similar to a UNIX *vnode* reference: a machine address and an identifier for a file *inode* stored on that machine. From the name one can look up the reference, but this is a potentially costly operation. Given the object reference one can invoke the object, and this (one hopes) will be quite a bit cheaper.

6.9   ENS—The CORBA Event Notification Service

The CORBA Event Notification Service, or ENS, provides for notifications of asynchronous events to applications that register an interest in those events by obtaining a handle to which events can be posted and on which events can be received. Reliability features are optionally supplied. The ENS is best understood in terms of what is called the *publish/subscribe* communications architecture.[1] In this approach, messages are produced by *publishers* that label each new message using a set of *subjects* or *attributes*. Separately, applications that wish to be informed when events occur on a given subject will *subscribe* to that subject or will poll for messages relating to the subject. The role of the ENS is to reliably bind the publishers to the subscribers, ensuring that even though the publishers do not know who the subscribers will be, and vice versa, messages are promptly and reliably delivered to them. (See Figure 6.12.)

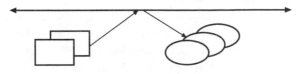

FIGURE 6.12. The CORBA ENS is a form of message bus that supports a publish/subscribe architecture. The sources of events (boxes) and consumers (ovals) need not be explicitly aware of one another, and the sets can change dynamically. A single object can produce or consume events of multiple types, and, in fact, an object can be both producer and consumer.

Two examples will make the value of such a model more clear. Suppose that one were using CORBA to implement a software architecture for a large brokerage system or a stock exchange. The ENS for such an environment could be used to broadcast stock trades as they occur. The events in this example would be named using the stock and bond names that they describe. Each broker would subscribe to the stocks of interest, again using these subject names, and the application program would then receive incoming quotes and display them to the screen. Notice that the publisher program can be developed without knowing anything about the nature of the applications that will use the ENS to monitor its outputs: It need not have compatible types or interfaces except with respect to the events that are exchanged between them. And the subscriber, for its part, does not need to be bound to a particular publisher: If a new data source of interest is developed, it can be introduced into the system without changing the existing architecture.

A second example of how the ENS can be useful would occur in system management and monitoring. Suppose that an application is being developed to automate some of the management functions occurring in a VLSI fabrication facility. As time goes by, the developers expect to add more and more sources of information and introduce more and more applications that use this information to increase the efficiency and productivity of the factory. An ENS architecture facilitates doing so, because it permits the developers to separate *the information architecture* of their application from its *implementation architecture*. In such an example, the information architecture is the structure of the ENS event space itself: the subjects under which events may be posted and the types of events that can occur in each subject. The sources and consumers of the

---

[1]It should be noted, however, that the ENS lacks the sort of subject mapping facilities that are central to many publish-subscribe message-bus architectures, and it is in this sense a more primitive facility than some of the message-bus technologies to be discussed later in this book, such as the Teknekron Information Bus (TIB).

events can be introduced later and will in general be unaware of one another. Such a design preserves tremendous flexibility and facilitates an evolutionary design for the system. After basic functionality is in place, additional functions can be introduced in a gradual way and without disrupting existing software. Here, the events would be named according to the aspect of factory function to which they relate: status of devices, completion of job steps, scheduled downtime, and so forth. Each application program would subscribe to those classes of events relevant to its task, ignoring all others by not subscribing to them.

Not all CORBA implementations include the ENS—for example, the basic Orbix product described above lacks an ENS, although the Orbix+Isis extension makes use of a technology called the Isis Message Distribution Service to implement ENS functionality in an Orbix setting. This, in turn, was implemented using the Isis Toolkit, which we will discuss in more detail in Chapter 17.

6.10   Life-Cycle Service

The Life-Cycle Service, or LCS, standardizes the facilities for creating and destroying objects and for copying them or moving them within the system. The service includes a *factory* for manufacturing new objects of a designated type. The Life-Cycle Service is also responsible for scheduling backups, periodically compressing object repositories to reclaim free space, and initiating other life-cycle activities. To some degree, the service can be used to program object-specific management and supervisory functions, which may be important to reliable control of a distributed system. However, there is at present limited experience with life-cycle issues for CORBA objects; hence, these possibilities remain an area for future development and research.

6.11   Persistent Object Service

The Persistent Object Service, or POS, is the CORBA equivalent of a file system. This service maintains collections of objects for long-term use, organizing them for efficient retrieval and working closely with its clients to give application-specific meanings to the consistency, persistency, and access-control restrictions implemented within the service. This permits the development of special-purpose POSs—for example, to maintain databases with large numbers of nearly identical objects organized into relational tables, as opposed to file system-style storage of very irregular objects.

6.12   Transaction Service

Mentioned earlier, the transaction service is an embedding of database-style transactions into CORBA architecture. If implemented, the service provides a *concurrency control* service for synchronizing the actions of concurrently active transactions; *flat* and (optionally) *nested* transactional tools and special-purpose persistent object services, which implement the transactional *commit* and *abort* mechanisms. The transaction service is often used with the *relationship service*, which tracks relationships among sets of objects—for example, if they are grouped into a database or some other shared data structure. We will discuss the transactional execution model in Section 7.4 and in Chapter 21.

6.13 Interobject Broker Protocol

The IOB, or Interobject Broker Protocol, is a protocol by which ORBs can be interconnected. The protocol is intended for use between geographically dispersed ORBs from a single vendor and to permit interoperation between ORBs developed independently by different vendors. The IOB includes definitions of a standard object reference data structure by which an ORB can recognize a foreign object reference and redirect it to the appropriate ORB, as well as definitions of the messages exchanged between ORBs for this purpose. The IOB is defined for use over a TCP channel; should the channel break or not be available at the time a reference is used, the corresponding invocation will return an exception.

6.14 Future CORBA Services

The evolution of CORBA continues to advance the coverage of the architecture, although not all vendor products will include all possible CORBA services. Future services now under discussion include archival storage for infrequently accessed objects, backup/restore services, versioning services, data interchange and internationalization services, logging and recovery services, replication services for promoting high availability, and security services. Real-time services are likely to be added to this list in a future round of CORBA enhancements, as will other sorts of reliability- and robustness-enhancing technologies.

6.15 Properties of CORBA Solutions

While the CORBA architecture is impressive in its breadth, the user should not be confused into believing that CORBA therefore embodies solutions for the sorts of problems that were raised in the first chapters of this book or the ones we consider in Chapter 15. To understand this point, it is important to again stress that CORBA is a somewhat superficial technology in specifying the way things *look* but not *how they should be implemented*. In language terminology, CORBA is concerned with syntax but not semantics. This is a position that the OMG adopted intentionally, and the key players in that organization would certainly defend it. Nonetheless, it is also a potentially troublesome aspect of CORBA, in the sense that a correctly specified CORBA application may still be underspecified (even in terms of the interface to the objects) for purposes of verifying that the objects are used correctly or for predicting the behavior of the application.

Another frequently cited concern about CORBA is that the technology can require extreme sophistication on the part of developers, who must at a minimum understand exactly how the various object classes operate and how memory management will be performed. Lacking such knowledge, which is not an explicit part of the IDL, it may be impossible to use a distributed object efficiently. Even experts complain that CORBA exception handling can be very tricky. Moreover, in very large systems there will often be substantial amounts of old code that must interoperate with new solutions. Telecommunication systems are sometimes said to involve millions or tens of millions of lines of such software, perhaps written in outmoded programming languages or incorporating technologies for which source code is not available. To gain the full benefits of CORBA, however, there is a potential need to use CORBA throughout a large distributed environment. This may mean that large amounts of old code must somehow be retrofitted with CORBA interfaces and IDLs—neither a simple nor an inexpensive proposition.

The reliability properties of a particular CORBA environment depend on a great number of implementation decisions that can vary from vendor to vendor and often will do so. Indeed, CORBA is promoted to vendors precisely because it creates a level playing field within which their products can interoperate but compete: The competition would revolve around this issue of relative performance, reliability, or functionality guarantees. Conversely, this implies that individual applications cannot necessarily count upon reliability properties of CORBA if they wish to maintain a high degree of portability: Such applications must in effect assume the least common denominator. Unfortunately, this least level of guarantees, in the CORBA architectural specification, is quite weak: Invocations and binding requests can fail, perhaps in inconsistent ways, corresponding closely to the failure conditions we identified for RPC protocols that operate over standard communication architectures. Security, being optional, must be assumed not to be present. Thus, CORBA creates a framework within which reliability technologies can be standardized, but, as currently positioned, the technology base is not necessarily one that will encourage a new wave of reliable computing systems.

On the positive side, CORBA vendors have shown early signs of using reliability as a differentiator for their products. Iona's Orbix product is offered with a high-availability technology based on process group computing (Orbix+Isis [see Iona Ltd. and Isis Distributed Systems, Inc.]) and a transactional subsystem based on a popular transactional technology (Orbix+Tuxedo [see Iona Ltd.]). Other major vendors are introducing reliability tools of their own. Thus, while reliability may not be a standard property of CORBA applications, and may not promote portability between CORBA platforms, it is at least clear that CORBA was conceived with the possibility of supporting reliable computing in mind. Most of the protocols and techniques discussed in the remainder of this book are compatible with CORBA in the sense that they could be used to implement standard CORBA reliability services, such as its replication service or the event notification service.

6.16 Related Reading

On the ANSA project and architecture: (see Architecture Projects Management Limited [1989, April 1991, November 1991]). Another early effort in the same area was Chronus: (see Gurwitz et al., Schantz et al.).

On CORBA: (see Object Management Group and X/Open). Other publications are available from the Object Management Group, a standards organization; see their Web page, http://www.omg.org.

For the CORBA products cited, such as Orbix, the reader should contact the relevant vendor.

On TINA: (see *Telecommunications Information Network Architecture Conference, Proceedings of*).

On DCE: (see Open Software Foundation).

Material discussing network OLE had not yet been made available at the time of this writing.

CHAPTER 7 ✧ ✧ ✧ ✧

Client/Server Computing

CONTENTS

7.1 Stateless and Stateful Client/Server Interactions

Chapters 4 to 6 focused on the communication protocols used to implement RPC and streams and on the semantics of these technologies when a failure occurs. Independent of the way that a communication technology is implemented, however, is the question of how the programming paradigms that employ it can be exploited in developing applications, particularly if reliability is an important objective. In this chapter, we examine client/server computing technologies, assuming that the client/server interactions are by RPC, perhaps implemented directly, and perhaps issued over streams. Our emphasis is on the interaction between architectural issues and the reliability properties of the resulting solutions. This topic will prove particularly important when we begin to look closely at the Web, which is based on what is called a stateless client/server computing paradigm, implemented over stream connections to Web servers.

7.2 Major Uses of the Client/Server Paradigm

The majority of client/server applications fall into one of two categories, which can be broadly characterized as being the file server, or *stateless,* architectures, and the database-styled transactional, or *stateful,* architectures. Although there are a great many client/server systems that neither manage files nor any other form of database, most such systems share a very similar design with one or the other of these. Moreover, although there is an important middle ground consisting of stateful distributed architectures that are not transactional, these sorts of applications have only emerged recently and continue to represent a fairly small percentage of the client/server architectures found in real systems. Accordingly, by focusing on these two very important cases, we will establish some basic facts about the broader technology areas of which each is representative and of the state of practice at the time of this writing. In Part III we will discuss stateful distributed system architectures in more general terms and in much more detail, but in doing so we will also move away from the state of practice as of the mid-1990s into technologies that may not be widely deployed until late in the decade or beyond.

A stateless client/server architecture is one in which neither the clients nor the server needs to maintain accurate information about one another's status. This is not to say that the clients cannot cache information obtained from a server; indeed, the use of caches is one of the key design features that permit client/server systems to perform well. However, such cached information is understood to be potentially stale, and any time an operation is performed on the basis of data from the cache, some sort of validation scheme must be used to ensure that the outcome will be correct even if the cached data have become invalid.

More precisely, a stateless client/server architecture has the property that servers do not need to maintain an accurate record of their current set of clients and can change state without engaging in a protocol between the server and its clients. Moreover, when such state changes occur, correct behavior of the clients is not affected. The usual example of a stateless client/server architecture is one in which a client caches records that it has copied from a name server. These

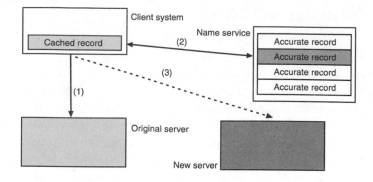

FIGURE 7.1. In this example, a client of a banking database has cached the address of the server handling a specific account. If the account is transferred, the client's cached record is said to have become stale. Correct behavior of the client is not compromised, however, because it is able to detect staleness and refresh the cached information at run time. Thus, if an attempt is made to access the account, the client will discover that it has been transferred (step 1) and will look up the new address (step 2), or it will be told the new address by the original server. The request can then be reissued to the correct server (step 3). The application program will benefit from improved performance when the cached data are correct, which is hopefully the normal case, but it never sees incorrect or inconsistent behavior if the cached data are incorrect. The key to such an architecture lies in the ability to detect that the cached data have become stale when attempting to use these data, and in the availability of a mechanism for refreshing the cache transparent to the application.

records might, for example, map from ASCII names of bank accounts to the IP address of the bank server maintaining that account. Should the IP address change (i.e., if an account is transferred to a new branch that uses a different server), a client that tries to access that account will issue a request to the wrong server. Since the transfer of the account is readily detected, this request will fail, causing the client to refresh its cache record by looking up the account's new location; the request can then be reissued and should now reach the correct server. This is illustrated in Figure 7.1. Notice that the use of cached data is transparent to (concealed from) the application program, which benefits through improved performance when the cached record is correct, but is unaffected if an entry becomes stale and must be refreshed at the time it is used.

One implication of a stateless design is that the server and client are independently responsible for ensuring the validity of their own states and actions. In particular, the server makes no promises to the client, except that the data it provides were valid at the time they were provided. The client, for its part, must carefully protect itself against the possibility that the data it obtained from the server have subsequently become stale.

Notice that a stateless architecture does not imply that there is no form of state shared between the server and its clients. On the contrary, such architectures often share state through caching, as seen in Figure 7.1. The fundamental property of the stateless paradigm is that correct function doesn't require that the server keep track of the clients currently using it and that the server can change data (reassigning the account in this example) without interacting with the clients that may have cached old copies of the data item. To compensate for this, the client side of such a system will normally include a mechanism for detecting that data have become stale and for refreshing these data when an attempt is made to use such stale data in an operation initiated by the client.

The stateless client/server paradigm is one of the most successful and widely adopted tools for building distributed systems. File servers, perhaps the single most widely used form of distributed system, are typically based on this paradigm. The Web is based on stateless servers, for example, and this is often cited as one of the reasons for its rapid success. One could conclude that this pattern repeats the earlier success of NFS—a stateless file system protocol that was tremendously successful in the early 1980s and fueled the success of Sun Microsystems, which introduced the protocol and was one of the first companies to invest heavily in it. Moreover, many of the special-purpose servers developed for individual applications employ a stateless approach. However, as we will see, stateless architectures also carry a price: Systems built this way often have limited reliability or consistency guarantees.

It should also be mentioned that stateless designs are a principle but not an absolute rule. In particular, there are many file systems (we will review some later in this chapter) that are stateless in some ways but make use of coherently shared state in other ways. In this section we will call such designs stateful, but the developers of the systems themselves might consider that they have adhered to the principle of a stateless design in making minimum use of coherently shared state. Such a philosophy recalls the end-to-end philosophy of distributed systems design (see Section 3.5), in which communication semantics are left to the application layer except when strong performance or complexity arguments can be advanced in favor of putting stronger guarantees into an underlying layer. However, to the degree that there is a design philosophy associated with stateless client/server architectures, it is probably most accurate to say that it is one that discourages the use of consistently replicated or coherently shared state except where such state is absolutely necessary.

In contrast, a stateful architecture is one in which information is shared between the client and server in such a manner that the client may take local actions under the assumption that this information is correct. In Figure 7.1, this would have meant that the client system would never need to retry a request. Clearly, to implement such a policy, the database and name mapping servers would need to track the set of clients possessing a cached copy of the server for a record that is about to be transferred. The system would need to somehow lock these records against use during the time of the transfer or invalidate them so that clients attempting to access the transferred record would first look up the new address. The resulting protocol would guarantee that if a client is permitted to access a cached record, that record will be accurate; however, it would do so at the cost of complexity (in the form of the protocol needed when a record is transferred) and delay (namely, delays visible to the client when trying to access a record that is temporarily locked and/or delays within the servers when a record being transferred is found to be cached at one or more clients).

Even from this simple example, it is clear that stateful architectures are potentially complex. If one indeed believed that the success of the NFS and Web was directly related to their statelessness, it would make sense to conclude that stateful architectures are inherently a bad idea: The theory would be that anything worth billions of dollars to its creator must be good. Unfortunately, although seductive, such a reaction also turns out to implicitly argue against reliability, coherent replication, many kinds of security, and other properties that are inherently stateful in their formulation. Thus, while there is certainly a lesson to be drawn here, the insights to be gained are not shallow ones.

Focusing on our example, one might wonder when, if ever, coherent caching is desirable. Upon close study, the issue can be seen to be a tradeoff between performance and necessary mechanism. It is clear that a client system with a coherently cached data item will obtain performance benefits by being able to perform actions correctly using local data (hence, avoiding a round-trip delay over the network) and may therefore be able to guarantee some form of real-time response to the application. For applications in which the cost of communicating with a server is very high, or where there are relatively strict real-time constraints, this could be an extremely important guarantee—for example, an air traffic controller contemplating a proposed course change for a flight would not tolerate long delays while checking with the database servers in the various air sectors that flight will traverse. Similar issues are seen in many real-time applications, such as computer-assisted conferencing systems and multimedia playback systems. In these cases one might be willing to accept additional mechanism (and hence complexity) as a way of gaining a desired property. A stateless architecture will be much simpler, but cached data may be stale and hence cannot be used in the same ways.

Another factor that favors the use of stateful architectures is the need to reduce load on a heavily used server. If a server is extremely popular, coherent caching offers the possibility of distributing the data it manages, and hence the load upon it, among a large number of clients and intermediate nodes in the network. To the degree that clients manage to cache the right set of records, they avoid access to the server and are able to perform what are abstractly thought of as distributed operations using purely local computation. A good example of a setting where this is becoming important is the handling of popular Web documents. As we will see when we discuss the Web, its architecture is basically that of a file server, and heavy access to a popular document can therefore result in enormous load on the unfortunate server on which that document is stored. By caching such documents at other servers, a large number of potential sources (shadow copies) of the document can be created, permitting the load of servicing incoming requests to be spread over many copies. But if the document in question is updated dynamically, one now faces a coherent caching question.

If records can be coherently cached while preserving the apparent behavior of a single server, the performance benefits and potential for ensuring that real-time bounds will be satisfied can be immense. In distributed systems, where reliability is important, it is therefore common to find that such coherent caching is beneficial. The key issue is to somehow package the associated mechanisms to avoid paying this complexity cost repeatedly. This is one of the topics that will be discussed in Chapter 15.

There is, however, a second way to offer clients some of the benefits of stateful architecture, but without ensuring that remotely cached data will be maintained in a coherent state. The key to this alternative approach is to use some form of abort or backout mechanism to roll back actions taken by a client on a server, under conditions where the server detects that the client's state is inconsistent with its own state, and to force the client to roll back its own state and, presumably, retry its operation with refreshed or corrected data. This underlies the transactional approach to building client/server systems. As noted above, transactional database systems are the most widely used of the stateful client/server architectures.

The basic idea in a transactional system is that the client's requests are structured into clearly delimited transactions. Each transaction begins, encompasses a series of *read* and *update* operations, and then ends by *committing* in the case where the client and server consider the outcome to be successful or *aborting* if either client or server has detected an error. An aborted transaction is backed out both by the server, which erases any effects of the transaction, and by the client, which will typically restart its request at the point of the original begin, or report an error to the user and leave it to the user to decide if the request should be retried. A transactional system is one that supports this model, guaranteeing that the results of committed transactions will be preserved and that aborted transactions will leave no trace.

The connection between transactions and statefulness is as follows. Suppose that a transaction is running, and a client has read a number of data items and issued a number of updates. Often it will have locked the data items in question for reading and writing, a topic we discuss in more detail in Chapter 21. These data items and locks can be viewed as a form of shared state between the client and the server: The client basically trusts the server to ensure that the data it has read are valid until it commits or aborts and releases the locks that it holds. Just as our cached data were copied to the client in the earlier examples, all of this information can be viewed as knowledge of the server's state that the client caches. And the relationship is mutual: The server, for its part, holds an image of the client's state in the form of updates and locks it holds on behalf of the partially completed transactions.

Now, suppose that something causes the server's state to become inconsistent with that of the client, or vice versa. Perhaps the server crashes and then recovers, and in this process some information that the client had provided to the server is lost. Or, perhaps it becomes desirable to change something in the database without waiting for the client to finish its transaction. In a stateless architecture we would not have had to worry about the state of the client. In a transactional implementation of a stateful architecture, on the other hand, the server can exploit the abort feature by arranging that the client's transaction be aborted, either immediately, or later when the client tries to commit it. This frees the server from needing to worry about the state of the client. In effect, an abort or rollback mechanism can be used as a tool by which a stateful client/server system is able to recover from a situation where the client's view of the state shared with the server has been rendered incorrect.

In the remainder of this chapter, we review examples of stateless file server architectures from the research and commercial community, stateful file server architectures (we will return to this topic in Chapter 15), and stateful transactional architectures as used in database systems. As usual, our underlying emphasis is on reliability implications of these architectural alternatives.

7.3 Distributed File Systems

We have discussed the stateless approach to file server design in general terms. In this section, we look at some specific file system architectures in more detail, to understand the precise sense in which these systems are stateless, how their statelessness may be visible to the user, and the implications of statelessness on file system reliability.

Client/server file systems normally are structured as shown in Figure 7.2. Here, we see that the client application interacts with a cache of file system blocks and file descriptor objects maintained in the client workstation. In UNIX, the file descriptor objects are called *vnodes* and are basically server-independent representations of the *inode* structures used within the server to track the file. In contrast to an inode, a vnode (virtualized inode) has a unique identifier, which will not be reused over the lifetime of the server and which omits detailed information such as block numbers. In effect, a vnode is an access key to the file, obtained by the client's file subsystem during the file open protocol, which is usable by the server to rapidly validate access and locate the corresponding inode.

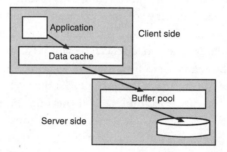

FIGURE 7.2. In a stateless file system architecture, the client may cache data from the server. Such a cache is similar in function to the server's buffer pool, but is not guaranteed to be accurate. In particular, if the server modifies the data that the client has cached, it has no record of the locations at which copies may have been cached and no protocol by which cached data can be invalidated or refreshed. The client side of the architecture will often include mechanisms that partially conceal this limitation—for example, by validating that cached file data are still valid at the time a file is opened. In effect, the cached data are treated as a set of hints that are used to improve performance but should not be trusted in an absolute sense.

On the server side, the file system is conventionally structured, but is accessed by messages containing a vnode identifier and an offset into the file, rather than block by block. The vnode identifier is termed a *file handle,* whereas the vnode itself is a representation of the contents of the inode, omitting information that is only useful on the server itself but including information that the client application programs might request using UNIX *stat* system calls.

The policies used for cache management differ widely within file systems. NFS uses a write-through policy when updating a client cache, meaning that when the client cache is changed, a write operation is immediately initiated to the NFS server. The client application is permitted to resume execution before the write operation has been completed, but because the cache has already been updated in this case, programs running on the client system will not normally observe any sort of inconsistency. NFS issues an *fsync* operation when a file is closed, causing the close operation to delay until all of these pending write operations have completed. The effect is to conceal the effects of asynchronous writes from other applications executed on the same client computer. Users can also invoke this operation, or the *close* or *fflush* operations, which also flush cached records to the server.

In NFS, the server maintains no information at all about current clients. Instead, the file handle is created in such a manner that mere possession of a file handle is considered proof that the client successfully performed an open operation. In systems where security is an issue, NFS uses digital signature technology to ensure that only legitimate clients can obtain a file handle and that they can do so only through the legitimate NFS protocol. These systems often require that file system handles be periodically reopened, so that the lifetime of encryption keys can be limited.

On the client side, cached file blocks and file handles in the cache represent the main form of state present in an NFS configuration. The approach used to ensure that this information is valid represents a compromise between performance objectives and semantics. Each time a file is opened, NFS verifies that the cached file handle is still valid. The file server, for its part, treats a file handle as invalid if the file has been written (by some other client system) since the handle was issued. Thus, by issuing a single open request, the client system is able to learn whether the data blocks cached on behalf of the file are valid or not and can discard them in the latter case.

This approach to cache validation poses a potential problem, which is that if a client workstation has cached data from an open file, changes to the file that originate at some other workstation will not invalidate these cached blocks, and no attempt to authenticate the file handle will occur. Thus, for example, if process q on client workstation a has file F open, and then process p on client workstation b opens F, writes modified data into it, and then closes it, although F will be updated on the file server, process q may continue to observe stale data for an unlimited period of time. Indeed, short of closing and reopening the file, or accessing some file block that is not cached, q might never see the updates!

One case where this pattern of behavior can occur is when a pipeline of processes is executed with each process on a different computer. If p is the first program in such a pipeline and q is the second program, p could easily send a message down the pipe to q telling it to look into the file, and q will now face the stale data problem. UNIX programmers often encounter problems such as this and work around them by modifying the programs to use *fflush* and *fsync* system calls to flush the cache at p and to empty q's cache of cached records for the shared file.

NFS vendors provide a second type of solution to this problem through an optional locking mechanism, which is accessed using the *flock* system call. If this optional interface is used, the process attempting to write the file would be unable to open it for update until the process holding it open for reads has released its read lock. Conceptually, at least, the realization that the file needs to be unlocked and then relocked would sensitize the developer of process p to the need to close and then reopen the file to avoid access anomalies, which are well documented in NFS.

At any rate, file sharing is not all that common in UNIX, as demonstrated in some studies (see Ousterhout et al. [1985]), where it was found that most file sharing is between programs executed sequentially from the same workstation.

The NFS protocol is thus stateless but there are situations in which the user can glimpse the implementation of the protocol precisely because its statelessness leads to weakened semantics compared to an idealized file system accessed through a single cache. Moreover, as noted in the previous chapter, there are also situations in which the weak error reporting of RPC protocols is reflected in unexpected behavior, such as the file *create* operation of Section 4.7, which incorrectly reported that a file couldn't be created because a reissued RPC fooled the file system into thinking the file already existed.

Similar to the basic UNIX file system, NFS is designed to prefetch records when it appears likely that they will soon be needed—for example, if the application program reads the first two blocks of a file, the NFS client-side software will typically fetch the third block of that file without waiting for a read request, placing the result in the cache. With a little luck, the application will now obtain a cache hit and be able to start processing the third block even as the NFS system fetches the fourth one. One can see that this yields performance similar to that of simply transferring the entire file at the time it was initially opened. Nonetheless, the protocol is relatively inefficient in the sense that each block must be independently requested, whereas a streaming style of transfer could avoid these requests and also handle acknowledgments more efficiently. In the following text, we will look at some file systems that explicitly perform whole-file transfers and that are able to outperform NFS when placed under heavy load.

For developers of reliable applications, the reliability of the file server is of obvious concern. One might want to know how failures would affect the behavior of operations. With NFS, as normally implemented, a failure can cause the file server to be unavailable for long periods of time, can partition a client from a server, or can result in a crash and then reboot of a client. The precise consequences depend on the file system setup. For the situations where a server becomes unreachable or crashes and later reboots, the client program may experience timeouts, which would be reported to the application layer as errors, or it may simply retry its requests periodically, for as long as necessary until the file server restarts. In the latter case, an operation could be reissued after a long delay, and there is some potential for operations to behave unexpectedly, as in the case of *create*. Client failures, on the other hand, are completely ignored by the server.

Because the NFS client-side cache uses a write-through policy, in such a situation a few updates may be lost but the files on the server will not be left in an extremely stale state. The locking protocol used by NFS, however, will not automatically break locks during a crash—hence, files locked by the client will remain locked until the application detects this condition and forces the locks to be released, using commands issued from the client system or from some other system. There is a mode in which failures automatically cause locks to be released, but this action will only occur when the client workstation is restarted, presumably to avoid confusing network partitions with failure/reboot sequences.

Thus, while the stateless design of NFS simplifies it considerably, the design also introduces serious reliability concerns. Our discussion has touched on the risk of processes seeing stale data when they access files, the potential that writes could be lost, and the possibility that a critical file server might become unavailable due to a network or computer failure. If you are building an application for which reliability is critical, any of these cases could represent a very serious failure. The enormous success of NFS should not be taken as an indication that reliable applications can in fact be built over it, but rather as a sign that failures are really not all that frequent in modern computing systems and that most applications are not particularly critical! In a world where hardware was less reliable or the applications were more critical, protocols such as the NFS protocol might be considerably less attractive.

Our discussion has concerned the case of a normal NFS server. There are versions of NFS that support replication in software for higher availability: R/NFS and Deceit (see Siegal), HA-NFS (see Bhide et al.), and Harp (see Ladin et al., Liskov et al.), as well as dual-ported NFS server units in which a backup server can take control of the file system. The former approaches employ process-group communication concepts of a sort we will discuss later, although the protocol used to communicate with client programs remains unchanged. By doing this, the possibility for load-balanced read access to the file server is created, enhancing read performance through parallelism. At the same time, these approaches allow continuous availability even when some servers are down. Each server has its own disk, permitting tolerance of media failures. And, there is a possibility of varying the level of the replication selectively, so that critical files will be replicated but noncritical files can be treated using conventional nonreplicated methods. The interest in such an approach is that any overhead associated with file replication is incurred only for files where there is also a need for high availability, and hence the multiserver configuration comes closer to also giving the capacity and performance benefits of a cluster of NFS servers. Many users like this possibility of paying only for what they use.

The dual-ported hardware approaches, in contrast, primarily reduce the time to recovery. They normally require that the servers reside in the same physical location, and are intolerant of media failure, unless a mirrored disk is employed. Moreover, these approaches do not offer benefits of parallelism: One pays for two servers, or for two servers and a mirrored disk, as a form of insurance that the entire file system will be available when needed. These sorts of file servers are, consequently, expensive. On the other hand, their performance is typically that of a normal server—there is little or no *degradation* because of the dual configuration.

Clearly, if the performance degradation associated with replication can be kept sufficiently small, the mirrored server and/or disk technologies will look expensive. Early generations of cluster server technology were slow enough to maintain mirroring as a viable alternative. However, the trend seems to be for this overhead to become smaller and smaller, in which case the greater flexibility and enhanced read performance, due to parallelism, would argue in favor of the NFS cluster technologies.

Yet another file system reliability technology has emerged recently; it involves the use of clusters or arrays of disks to implement a file system that is more reliable than any of the component disks. Such so-called RAID file systems (see Paterson et al.) normally consist of a mixture of hardware and software: the hardware for mediating access to the disks themselves and the software to handle the buffer pool, oversee file layout, and optimize data access patterns. The actual protocol used to talk to the RAID device over a network would be the same as for any other sort of remote disk: It might be the NFS protocol or some other remote file access protocol. The use of RAID in the disk subsystem itself would normally not result in protocol changes.

RAID devices typically require physical proximity of the disks to one another (needed by the hardware). The mechanism that implements the RAID is typically constructed in hardware and employs a surplus disk to maintain redundant data in the form of parity for sets of disk blocks; such an approach permits a RAID system to tolerate one or more disk failures or bad blocks, depending on the way the system is configured. A RAID is thus a set of disks that mimics a single more reliable disk unit with roughly the summed capacity of its components, minus over-head for the parity disk. However, even with special hardware, management and configuration of RAID systems can require specialized software architectures (see Wilkes et al.).

Similar to the case for a mirrored disk, the main benefits of a RAID architecture are high availability in the server itself, together with large capacity and good average seek time for information retrieval. In a large-scale distributed application, the need to locate the RAID device at a single place, and its reliance on a single source of power and software infrastructure, often means that in practice such a file server has the same distributed reliability properties any other form of file server. In effect, the risk of file server unavailability as a source of downtime is reduced, but other infrastructure-related sources of file system unavailability remain to be ad-dressed. In particular, if a RAID file system implements the NFS protocol, it would be subject to all the limitations of the NFS architecture.

7.4   Stateful File Servers

The performance of NFS is limited by its write-through caching policy, which has led developers of more advanced file systems to focus on improved caching mechanisms and, because few appli-cations actually use the optional locking interfaces, on greater attention to cache validation protocols. In this section, we briefly discuss some of the best-known stateful file systems. Break-out 7.1 discusses the Andrew File System.

Work on stateful file system architectures can be traced in part to an influential study of file access patterns in the Sprite system at Berkeley (see Baker et al.). This work sought to character-ize the file system workload along a variety of axes: read/write split, block reuse frequency, file lifetimes, and so forth. The findings, although not surprising, were at the same time eye-openers for many of the researchers in this field. In this study, it was discovered that all file access was sequential, and that there was very little sharing of files between different programs. When file sharing was observed, the prevailing pattern was the simplest one: One program tended to write the file, in its entirety, and then some other program would read the same file. Often (indeed, in most such cases), the file would be deleted shortly after it was created. In fact, most files survived for less than ten seconds or longer than 10,000 seconds. The importance of cache consistency was explored in this work (it turned out to be quite important, but relatively easy to enforce for the most common patterns of sharing), and the frequency of write/write sharing of files was shown to be so low that this could almost be treated as a special case. (Later, there was consider-

The Andrew File System (AFS)

The Andrew File System (AFS) is widely cited for its strong security architecture and consistency guarantees. It was developed at Carnegie Mellon University and subsequently used as the basis of a worldwide file system product offered by Transarc, Inc. (see Satyanarayanan et al., Spasojevic and Satyanarayanan). The basic ideas in Andrew are easily summarized.

AFS was built with the assumption that the Kerberos authentication technology would be available. We present Kerberos in Chapter 19 and therefore limit ourselves to a brief summary of the basic features of the system here. When a user logs in (and, later, periodically, if the user remains connected long enough for timers to expire), Kerberos prompts for a password. Using a secure protocol, which employs DES to encrypt sensitive data, the password is employed to authenticate the user to the Kerberos server, which will now act as a trustworthy intermediary in establishing connections between the user and the file servers he or she will access. The file servers similarly authenticate themselves to the Kerberos authentication server at startup.

File system access is by whole-file transfer, except in the case of very large files, which are treated as sets of smaller ones. Files can be cached in the AFS subsystem on a client, in which case requests are satisfied out of the cached information whenever possible (in fact, there are two caches—one of file data and one of file status information—but this distinction need not concern us here). The AFS server tracks the clients that maintain cached copies of a given file, and, if the file is opened for writing, uses callbacks to inform those clients that the cached copies are no longer valid. Additional communication from the client to the server occurs frequently enough so that if a client becomes disconnected from the server, it will soon begin to consider its cached files to be potentially stale. (Indeed, studies of AFS file server availability have noted that disconnection from the server is a more common source of denial of access to files in AFS than genuine server downtime.)

AFS provides a strong form of security guarantee, based on access control lists at the level of entire directories. Because the Kerberos authentication protocol is known to be highly secure, AFS can trust the user identification information provided to it by client systems. Short of taking over a client workstation, an unauthorized user would have no means of gaining access to cached or primary copies of a file for which access is not permitted. AFS destroys cached data when a user logs out or an authorization expires and is not refreshed (see Bellovin and Merritt, Birrell, Lampson et al., Satyanarayanan, Schiller, Steiner et al.).

In its current use as a wide area file system, AFS has expanded to include some 1,000 servers and 20,000 clients in ten countries—all united within a single file system name space (see Spasojevic and Satyanarayanan). Some 100,000 users are believed to employ the system on a regular basis. Despite this very large scale, 96 percent of file system accesses were found to be resolved through cache hits, and server inaccessibility (primarily due to communication timeouts) was as little as a few minutes per day. Moreover, this is true even when a significant fraction of file references is to remote files. AFS users are reported to have had generally positive experiences with the system, but (perhaps not surprisingly) they complain about poor performance when a file is not cached and must be copied from a remote file server. Their subjective experience presumably reflects the huge difference in performance between AFS in the case where a file is cached and when a copy must be downloaded over the network.

able speculation that on systems with significant database activity, this finding would have been challenged.) Moreover, considerable data were extracted on patterns of data transfer from server to client: rate of transfer, percentage of the typical file that was transferred, and so forth. Out of this work came a new generation of file systems that used closer cooperation between client and file system to exploit such patterns.

Examples of well-known file systems that employ a stateful approach to provide increased performance (as opposed to availability) are AFS (see Howard et al., Satyanarayanan, Satyanarayanan et al.) and Sprite (see Osterhout et al. [1988], Srinivasan and Mogul), a research file system and operating system developed at the University of California, Berkeley. On the availability side of the spectrum, the Coda project (see Kistler and Satyanarayanan, Mummert et al.), a research effort at Carnegie Mellon University, takes these ideas one step further, integrating them into a file system specifically for use on mobile computers that operate in a disconnected, or partially connected, mode. Ficus, a project at UCLA, uses a similar approach to deal with file replication in very wide area networks with nonuniform connectivity and bandwidth properties. To varying degrees, these systems can all be viewed as stateful ones in which some of the information maintained within client workstations is guaranteed to be coherent. The term stateful is used a little loosely here, particularly in comparison with the approaches we will examine in Chapter 15. Perhaps it would be preferable to say that these systems are "more stateful" than the NFS architecture, gaining performance through the additional state. Among the four, only Sprite actually provides strong cache coherence to its clients. The other systems provide other forms of guarantees, which are used either to avoid inconsistency or to resolve inconsistencies after they occur. Finally, we will briefly discuss XFS, a file system under development at the University of California, Berkeley, which exploits the file system memory of client workstations as an extended buffer pool, paging files from machine to machine over the network to avoid the more costly I/O path from a client workstation over the network to a remote disk.

Both AFS and Sprite replace the NFS write-through caching mechanism and file handle validation protocols with alternatives that reduce costs. The basic approach in AFS is to cache entire files, informing the server that a modified version of a file may exist in the client workstation. Through a combination of features, such as whole-file transfers on file open and write back to the server, and by having the file server actively inform client systems when their cached entries become invalid, considerable performance improvements are obtained with substantially stronger file access semantics than for NFS. Indeed, the workload on an AFS server can be an order of magnitude or more lower than that for an NFS server, and the performance observed by a client is comparably higher for many applications. AFS was commercialized subsequent to the initial research project at CMU, becoming the component technology for a line of enterprise file systems (worldwide file systems) marketed by Transarc, a subsidiary of IBM.

Sprite, which caches file system blocks (but uses a large 4 KB block size), takes the concept of coherent caching one step further, using a protocol in which the server actively tracks client caching, issuing callbacks to update cached file blocks if updates are received. The model is based on the caching of individual data blocks, not whole files, but the client caches are large enough to accommodate entire files. The Sprite approach leads to such high cache hit rates that the server workload is reduced to almost pure writes, an observation that triggered some extremely interesting work on file system organizations for workloads that are heavily biased toward writes. Similar to AFS, the technology greatly decreases the I/O load and CPU load on the servers that actually manage the disk.

Sprite is unusual in two ways. First, the system implements several different caching policies depending upon how the file is opened: One policy is for read-only access; a second and more expensive one is used for *sequential write access,* which occurs when a file is updated by one workstation and then accessed by a second one later (but in which the file is never written simultaneously from several systems); and a third policy is used for *concurrent write access,* which occurs when a file is written concurrently from several sources. This last policy is very rarely needed because Sprite does not cache directories and is not often used in support of database applications. Second, unlike NFS, Sprite does not use a write-through policy. Thus, a file that is opened for writing, updated, closed, and perhaps reopened by another application on the same machine, read, and then deleted, would remain entirely in the cache of the client workstation. This particular sequence is commonly seen in compilers that run in multiple passes and generate temporary results and in editors that operate on an intermediate copy of a file, which will be deleted after the file is rewritten and closed. The effect is to greatly reduce traffic between the client and the server relative to what NFS might have, but also to leave the server out of date with respect to a client system that may be writing cached files.

Sequential write sharing is handled using version numbers. When a client opens a file, the server returns the current version number, permitting the client to determine whether or not any cached records it may have are still valid. When a file is shared for concurrent writing, a more costly but simple scheme is used, whereby none of the clients are permitted to cache it. If the status of a file changes because a new open or close has occurred, Sprite issues a callback to other clients that have the file open, permitting them to dynamically adapt their caching policy in an appropriate manner. Notice that because a stateless file system such as NFS has no information as to its current client set, this policy would be impractical to implement within NFS. On the other hand, Sprite faces the problem that if the callback RPC fails, it must assume that the client has genuinely crashed; the technology is thus not tolerant of communication outages that can partition a file server from its clients. Sprite also incurs costs that NFS can sometimes avoid: Both *open* and *close* operations must be performed as RPCs, and there is at least one extra RPC required (to check consistency) in the case where a file is opened, read quickly, and then closed than is required in NFS.

The recovery of a Sprite server after a crash can be complicated, because some clients may have had files opened in a cached for writing mode. To recover, the server makes use of its knowledge of the set of clients that had cached files for writing, which are saved in a persistent storage area, and of the fact that the consistency state of a file cannot change without the explicit approval of the server. This permits the server to track down current copies of the files it manages and to bring itself back to a consistent state.

The developers of Sprite commented that most of the complexity in the recovery mechanism comes in detecting crashes and reboots, rather than in rebuilding state. This is done by tracking the passage of RPC packets, and using periodic keep-alive packets to detect when a client or server has crashed or rebooted: The same mechanism also suffices to detect network partitions. There is a cost to tracking RPC packets, but a reliable crash and reboot detection mechanism is of course useful for other purposes besides recovering file server state (see Srinivasan and Mogul). This may at first seem confusing, because we have seen that RPC mechanisms cannot reliably detect failures. However, Sprite is not subject to the restrictions we cited earlier, because it can deny access to a file while waiting to gain access to the most current version of it. Concerns about RPC arose in trying to determine the cause of an RPC failure in real time. A system that is able to wait for a server to recover is fortunate in not needing to solve this problem: If an apparent failure has occurred, it can simply wait for the problem to be repaired if doing otherwise would violate file system consistency guarantees.

Experiments have shown the Sprite cache-consistency protocols to be highly effective in reducing traffic to the file server and preserving the illusion of a single copy of each file. Performance of the system is extremely good, utilization of servers very low, and the anomalous behaviors that can occur with NFS are completely avoided. However, the technology relies on the veracity of user-ID's, and hence suffers from some of the same security concerns that we will discuss in relation to NFS in Chapter 19.

Coda is a file system for disconnected use. It can be understood as implementing a very generalized version of the whole-file caching methods first introduced in AFS: Whereas AFS caches individual files, Coda caches groups of files and directories in order to maintain a complete cached copy of the user's entire file system or application. The idea within Coda is to track updates with sufficient precision so that the actions taken by the user while operating on a cached copy of part of the file system can be merged automatically into the master file system from which the files were copied. This merge occurs when connection between the disconnected computer and the main file system server is reestablished. Much of the sophistication of Coda is concerned with tracking the appropriate sets of files to cache in this manner and with optimizing the merge mechanisms so that user intervention can be avoided when possible. (See Break-out 7.2.)

The Ficus system, developed by Jerry Popek's group at UCLA (see Reiher et al.), explores a similar set of issues but focuses on an enterprise computing environment similar to the world-wide file system problems to which AFS has been applied in recent years. (For brevity we will not discuss a previous system developed by the same group, Locus [see Walter et al.].) In Ficus, the model is one of a large-scale file system built of file servers that logically maintain replicas of a single file system image. Communication connectivity can be lost and servers can crash—hence, at any point, a server will have replicas of some parts of the file system and will be out of touch with some other replicas for the same data. This leads to an approach in which file type information is used both to limit the updates that can be performed while a portion of the file system is disconnected from other segments, and to drive a file merge process when communication is reestablished (see Heidemann and Popek [1995]). Where Coda is focused on disconnected operation, however, Ficus emphasizes support for patterns of communication seen in large organizations that experience bandwidth limits or partitioning problems that prevent servers from contacting each other for brief periods of time. The resulting protocols and algorithms are

The Challenge Faced by Coda

The challenge faced by Coda is easily appreciated when the following example is considered. Suppose that Fred and Julia are collaborating on a major report to an important customer of their company. Fred is responsible for certain sections of the report and Julia for others, but these sections are also cited in the introductory and boilerplate material used to generate the report as a whole. As many readers of this book will appreciate, there are software tools with varying degrees of ease of use for this type of collaborative work. The most primitive tools provide only for locking of some sort, so that Julia can lock Fred out of a file while she is actually editing it. More elaborate ones actually permit multiple users to concurrently edit the shared files, annotating one another's work and precisely tracking who changed what through multiple levels of revisions. Such tools typically view the document as a form of database and keep some type of log or history showing how it evolved through time.

If the files in which the report are contained can be copied onto portable computers that become disconnected from the network, however, these annotations will be introduced independently and concurrently on the various copies. Files may be split or merged while the systems are disconnected from each other, and even the time of access cannot be used to order these events, since the clocks on computers can drift or be set incorrectly for many reasons. Thus, when copies of a complex set of files are returned to the file system from which they were removed, the merge problem becomes a nontrivial one both at the level of the file system itself (which may have to worry about directories that have experienced both delete and add operations of potentially conflicting sorts by the various concurrent users of the directory) and at the level of the application and its concept of file semantics. (See Figure 7.3.)

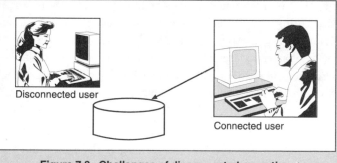

Disconnected user

Connected user

Figure 7.3. Challenges of disconnected operation.

similar to the ones used in Coda, but place greater attention on file-by-file reconciliation methods, whereas Coda is oriented toward mechanisms that deal with groups of files as an ensemble.

All of these systems are known for additional contributions beyond the ones we have discussed. Coda, for example, makes use of a recoverable virtual memory mechanism, which offers a way to back out changes made to a segment of virtual memory, using a logging facility that performs replay on behalf of the user. Ficus is also known for work on stackable file systems, in which a single file system interface is used to provide access to a variety of types of file-like abstractions. These contributions, and others not cited here, are beyond the scope of our present discussion.

Not surprisingly, systems such as Coda and Ficus incorporate special-purpose programming tools and applications that are well matched to their styles of disconnected and partially connected operation (see Mummert et al., Reiher et al.). These tools include, for example, e-mail systems that maintain logs of actions taken against mailboxes, understanding how to delete mail that has been deleted while in a disconnected mode, or merging e-mails that arrived separately in different copies of a mailbox that was split within a large-scale distributed environment. One can speculate that, over time, a small and fairly standard set of tools might emerge from such research and that developers would implement specialized disconnected applications, which rely on well-tested reconciliation methods to recorrect inconsistencies that occur during periods of disconnected interaction. At the time of this writing, however, I was not aware of any specific toolkits of this nature.

The last of the stateful file systems mentioned at the start of this section is XFS, a Berkeley project that seeks to exploit the memory of the client workstations connected to a network as a form of distributed storage region for a high-performance file server (see Anderson et al.). XFS could be called a "serverless network file system," although in practice the technology would more often be paired to a conventional file system, which would serve as a backup storage region. The basic idea of XFS, then, is to distribute the contents of a file system over a set of workstations so that when a block of data is needed, it can be obtained by a direct memory-to-memory transfer over the network rather than by means of a request to a disk server, which, having much less memory at its disposal, may then need to delay while fetching it from the disk itself.

XFS raises some very complex issues of system configuration management and fault tolerance. The applications using an XFS need to know what servers belong to it, and this set changes dynamically over time. Thus, there is a membership management problem that needs to be solved in software. Workstations are reliable, but not completely reliable—hence, there is a need to deal with failures. XFS does this by using a RAID-style storage scheme in which each set of n workstations is backed by an $n + 1$st machine, which maintains a parity block. If one of the $n + 1$ machines fails, the missing data can be regenerated from the other n. Moreover, XFS is dynamically reconfigurable, creating some challenging synchronization issues. On the positive side, all of this complexity brings with it a dramatic performance improvement when XFS is compared with more traditional server architectures. It should be noted that XFS draws heavily on the log-structured file system (see Rosenblum and Ousterhout), a technology that is beyond the scope of this book.

The reliability properties of these stateful file systems go well beyond those of NFS. For AFS and Sprite, reliability is limited by the manner in which the servers detect the failure of clients, since a failed client clears its cache upon recovery and the server needs to update its knowledge of the state of the cache accordingly. In fact, both AFS and Sprite detect failures through timeouts—hence, there can be patterns of failure that would cause a client to be sensed incorrectly as having failed, leaving its file system cache corrupted until some future attempt to validate cache contents occurs, at which point the problem would be detected and reported. In Sprite, network partition failures are considered unlikely because the physical network used at Berkeley is quite robust and, in any case, network partitions cause the client workstations to initiate a recovery protocol. Information concerning the precise handling of network partitions, or about methods for replicating AFS servers, was not available at the time of this writing. XFS is based on a failure model similar to that of AFS and Sprite, in which crash failures are anticipated and dealt with in the basic system architecture, but partitioning failures that result in the misdiagnosis of apparent crash failures is not an anticipated mode of failure.

Coda and Ficus treat partitioning as part of their normal mode of operation, dealing with partitioning failures (or client and server failures) using the model of independent concurrent operation and subsequent state merge that was presented earlier. Such approaches clearly trade higher availability for a more complex merge protocol and greater sophistications within the applications themselves. (See Break-out 7.3.)

7.5 Distributed Database Systems

Distributed database systems represent another use of client/server architectures in distributed systems. Unlike the case of distributed file systems, however, database technologies use a special programming model called the *transactional approach* and support this through a set of special protocols (see Gray [1979], Gray and Reuter). The reliability and concurrency semantics of a database are well understood through this model, and its efficient implementation is a major topic of research—and an important arena for commercial competition. For the purposes of this chapter, we will simply discuss the main issues, returning to implementation issues in Chapter 21.

Transactional systems are based upon a premise that applications can be divided into client programs and server programs, such that the client programs have minimal interactions with one another. Such an architecture can be visualized as a set of wheels, with database servers forming the hubs to which client programs are connected by communication pathways—the spokes. One client program can interact with multiple database servers, but although the issues this raises are well understood, such multidatabase configurations are relatively uncommon in commercial practice. Existing client/server database applications consist of a set of disjointed groups, each group containing a database server and its associated clients, with no interaction between client programs except through sharing a database, and with very few, if any, client programs that interact with multiple databases simultaneously. Moreover, although it is known how to replicate databases for increased availability and load-balancing (see Bernstein et al., Gray and Reuter), relatively little use is made of this option in existing systems. Thus, the hubs of distributed database systems rarely interact with one another. (We'll see why this is the case in Part III; ultimately, the issue turns out to be one of performance.)

A central premise of the approach is that each interaction by a client with the database server can be structured as a *begin* event, followed by a series of database operations (these would normally be database queries, but we can think of them as *read* and *update* operations and ignore

Lotus Notes

The Lotus Notes system is a commercial database product that uses a client/server model to manage collections of documents, which can draw upon a great variety of applications (word processing, spreadsheets, financial analysis packages, etc.). The system is widely popular because of the extremely simple sharing model it supports and its close integration with e-mail and chat facilities, supporting what has become known as a groupware collaboration model. The term "computer-supported collaborative work," or CSCW, is often used in reference to activities that are supported by technologies such as Lotus Notes.

Notes is structured as a client/server architecture. The client system is a graphical user interface, which permits the user to visualize information within the document database, create or annotate documents, "mine" the database for documents satisfying some sort of a query, and exchange e-mail or send memos containing documents as attachments. A security facility permits the database to be selectively protected by using passwords, so only designated users will have access to the documents contained in those parts of the database. If desired, portions of especially sensitive documents can be encrypted so that even a database administrator would be unable to access them without the appropriate passwords.

Lotus Notes also provides features for replication of portions of its database between the client systems and the server. Such replication permits a user to carry a self-contained copy of the desired documents (and others to which they are attached) and update them in a disconnected mode. Later, when the data-base server is back in contact with the user, updates are exchanged to bring the two sets of documents back into agreement. Replication of documents is also possible among Notes servers within an enterprise, although the Notes user must take steps to limit concurrent editing when replication is employed. (This is in contrast with Coda, which permits concurrent use of files and works to automatically merge changes.) At the time of this writing, Notes did not support replication of servers for increased availability, but treated each server as a separate security domain with its own users and passwords.

Within the terminology of this chapter, Lotus Notes is a form of partially stateful file server, although presented through a sophisticated object model and with powerful tools oriented toward cooperative use by members of workgroups. However, many of the limitations of stateless file servers are present in Notes, such as the need to restrict concurrent updates to documents that have been replicated. The Notes user environment is extremely well engineered and is largely successful in presenting such limitations and restrictions as features that the skilled Notes user learns to employ. In effect, by drawing on semantic knowledge of the application, the Lotus Notes developers were able to work around limitations associated with this style of file server. The difficulty encountered in distributed file systems is precisely that they lack this sort of semantic knowledge and are consequently forced to solve such problems in complete generality, leading to sometimes surprising or nonintuitive behavior, reflecting their distributed infrastructure.

the details), followed by a *commit* or *abort* operation. Such an interaction is called a *transaction,* and a client program will typically issue one or more transactions, perhaps interacting with a user or the outside world between the completion of one transaction and the start of the next. A transactional system should guarantee the persistence of committed transactions, although we will see that high-availability database systems sometimes weaken this guarantee to boost performance. When a transaction is aborted, on the other hand, its effects are completely rolled back, as if the transaction had never even been issued.

Transactional client/server systems are stateful: Each action by the client assumes that the database remembers various things about the previous operations done by the same client, such as locking information that comes from the database concurrency control model and updates that were previously performed by the client as part of the same transaction. The clients can be viewed as maintaining coherent caches of this same information during the period while a transaction is active (not yet committed).

The essential property of the transactional execution model, which is called the *serializability model,* is that it guarantees isolation of concurrent transactions. Thus, if transactions T_1 and T_2 are executed concurrently by client processes p and q, the effects will be as if T_1 had been executed entirely before T_2, or entirely after T_2—the database actively prevents them from interfering with one another. The reasoning underlying this approach is that it will be easier to write database application programs to assume that the database is idle at the time the program executed. Rather than force the application programmer to cope with real-world scenarios in which multiple applications simultaneously access the database, the database system is only permitted to interleave operations from multiple transactions if it is certain that the interleaving will not be noticeable to users. At the same time, the model frees the database system to schedule operations in a way that keeps the server as busy as possible on behalf of a very large number of concurrent clients. (See Figure 7.4.)

Notice that simply running transactions one at a time would achieve the serializability property.[1] However, it would also yield poor performance, because each transaction may take a long time to execute. By running multiple transactions at the same time, and interleaving their operations, a database server can give greatly improved performance, and system utilization levels will rise substantially, just as a conventional uniprocessor can benefit from multitasking. Even so, database systems sometimes need to delay one transaction until another completes, particularly when transactions are very long. To maximize performance, it is common for client/server database systems to require (or at least strongly recommend) that transactions be designed to be as short as possible. Obviously, not all applications fit these assumptions, but they match the needs of a great many computing systems.

---

[1]An important special case occurs in settings where each transaction can be represented as a single operation, performing a desired task and then committing or aborting and returning a result. Many distributed systems are said to be transactional but, in fact, operate in this much more restrictive manner. However, even if the application perceives a transaction as being initiated with a single operation, the database system itself may execute that transaction as a series of operations. These observations motivate a number of implementation decisions and optimizations, which we discuss in Chapter 21.

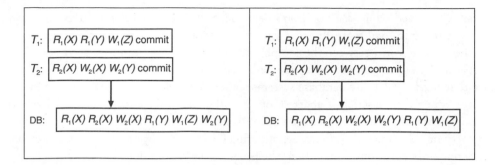

FIGURE 7.4. A nonserializable transaction interleaving (left), and one serializable in the order T_2, T_1 (right). Each transaction can be understood as a trace, which records the actions of a program that operates on the database, oblivious to other transactions that may be active concurrently. In practice, of course, the operations become known as the transaction executes, although our example shows the situation at the time these two transactions reach their commit points. The database is presented with the operations initiated by each transaction, typically one by one, and schedules them by deciding when to execute each operation. This results in an additional trace or log, showing the order in which the database actually performed the operations presented to it. A serializable execution is one that leaves the database in a state that could have been reached by executing the same transactions one by one, in some order, and with no concurrency.

There are a variety of options for implementing the serializability property. The most common is to use locking—for example, by requiring that a transaction obtain a read-lock on any data item that it will read and a write-lock on any data item it will update. Read-locks are normally nonexclusive: Multiple transactions are typically permitted to read the same objects concurrently. Write-locks, however, are mutually exclusive: Only one transaction can hold such a lock at a time. In the most standard locking protocol, called *two-phase locking,* transactions retain all of their locks until they commit or abort, and then they release them as a group. It is easy to see that this achieves serializability: If transaction Tj reads from T_i, or updates a variable after T_i does so, T_j must first acquire a lock that T_i will have held exclusively for its update operation. Transaction T_j will therefore have to wait until T_i has committed and will be serialized after T_i. Notice that the transactions can obtain read-locks on the same objects concurrently, but because read operations commute, they will not affect the serialization order (the problem gets harder if a transaction may need to upgrade some of its read-locks to write-locks).

Concurrency control (and hence locking) mechanisms can be classified as *optimistic* or *pessimistic.* The locking policy described above is a pessimistic one, because each lock is obtained before the locked data item is accessed. An optimistic policy is one in which transactions simply assume that they will be successful in acquiring locks and perform the necessary work in an opportunistic manner. At commit time, the transaction also verifies that its optimistic assumption was justified (that it got lucky, in effect) and aborts if it now turns out that some of its lock requests should in fact have delayed the computation. As one might expect, a high rate of aborts is a risk with optimistic concurrency-control mechanisms, and they can only be used in settings where the granularity of locking is small enough so that the risk of a real locking conflict between two transactions is actually very low.

The pessimistic aspect of a pessimistic concurrency-control scheme reflects the assumption that there may be frequent conflicts between concurrent transactions. This makes it necessary for a pessimistic locking scheme to operate in a more conventional manner, by delaying the transaction as each new lock request occurs until that lock has been granted; if some other transaction holds a lock on the same item, the requesting transaction will now be delayed until the lock-holding transaction has committed or aborted.

Deadlock is an important concern with pessimistic locking protocols—for example, suppose that T_i obtains a read-lock on x and then requests a write-lock on y. Simultaneously, T_j obtains a read-lock on y and then requests a write-lock on x. Neither transaction can be granted its lock, and in fact one transaction or the other (or both) must now be aborted. At a minimum, a transaction that has been waiting a very long time for a lock will normally abort; in more elaborate schemes, an algorithm can obtain locks in a way that avoids deadlock or can use an algorithm that explicitly detects deadlocks when they occur and overcomes them by aborting one of the deadlocked transactions. Deadlock-free concurrency-control policies can also be devised—for example, by arranging that transactions acquire locks in a fixed order or by using a very coarse locking granularity so that any given transaction requires only one lock. We will return to this topic, and related issues, in Chapter 21, when we discuss techniques for actually implementing a transactional system.

Locking is not the only way to implement transactional concurrency control. Other important techniques include so-called timestamped concurrency-control algorithms, in which each transaction is assigned a logical time of execution, and its operations are performed as if they had been issued at the time given by the timestamp. Timestamped concurrency control is relatively uncommon in the types of systems that we consider in this book—hence, for reasons of brevity, we omit any detailed discussion of the approach. We do note, however, that optimistic timestamped concurrency-control mechanisms have been shown to give good performance in systems where there are few true concurrent accesses to the same data items and that pessimistic locking schemes give the best performance in the converse situation, where a fairly high level of conflicting operations results from concurrent access to a small set of data items. Additionally, timestamped concurrency control is considered preferable when dealing with transactions that do a great deal of writing, while locking is considered preferable for transactions that are read-intensive. It has been demonstrated that the two styles of concurrency control cannot be mixed: One cannot use timestamps for one class of transactions and locks for another on the same database. However, a hybrid scheme, which combines features of the two approaches and works well in systems with mixtures of read-intensive and write-intensive transactions, has been proposed.

It is common to summarize the properties of a client/server database system so that the mnemonic ACID can be used to recall them:

▶ *Atomicity:* Each transaction is executed as if it were a single indivisible unit. The term *atomic* will be used throughout this book to refer to operations that have multiple suboperations but that are performed in an all-or-nothing manner.

- *Concurrency:* Transactions are executed so as to maximize concurrency, in this way maximizing the degrees of freedom available within the server to schedule execution efficiently (e.g., by doing disk I/O in an efficient order).

- *Independence:* Transactions are designed to execute independently from one another. Each client is written to execute as if the entire remainder of the system were idle, and the database server itself prevents concurrent transactions from observing one another's intermediate results.

- *Durability:* The results of committed transactions are persistent.

Notice that each of these properties could be beneficial in some settings but could represent a disadvantage in others—for example, there are applications in which one wants the client programs to cooperate explicitly. The ACID properties effectively constrain such programs to interact using the database as an intermediary. Indeed, the overall model makes sense for many classical database applications, but it is less suited to message-based distributed systems consisting of large numbers of servers and in which the programs coordinate their actions and cooperate to tolerate failures. All of this will add up to the perspective that complex distributed systems need a mixture of tools, which should include database technology but not legislate that databases be used to the exclusion of other technologies.

We turn now to the question raised earlier: the sense in which transactional systems are stateful, and the implications that this has for client/server software architectures.

A client of a transactional system maintains several forms of state during the period that the transaction executes. These include the transaction ID by which operations are identified, the intermediate results of the transactional operation (values that were read while the transaction was running or values that the transaction will write if it commits), and any locks or concurrency control information that has been acquired while the transaction was active. This state is shared with the database server, which for its part must keep original values of any data objects updated by noncommitted transactions; keep updates sorted by transactional-ID to know which values to commit if the transaction is successful; and maintain read-lock and write-lock records on behalf of the client, blocking other transactions that attempt to access the locked data items while allowing access to the client holding the locks. The server thus knows which processes are its active clients and must monitor their health in order to abort transactions associated with clients that fail before committing (otherwise, a failure could leave the database in a locked state).

The ability to use commit and abort is extremely valuable in implementing transactional systems and applications. In addition to the role of these operations in defining the scope of a transaction for purposes of serializability, they also represent a tool that can be used directly by the programmer—for example, an application can be designed to assume that a certain class of operations (such as selling a seat on an airline) will succeed, and to update database records as it runs under this assumption. Such an algorithm would be optimistic in much the same sense as a concurrency-control scheme can be optimistic. If, for whatever reason, the operation encounters an error condition (no seats available on some flight, customer credit card refused, etc.), the operation can simply abort and the intermediate actions that were taken will be erased from the database. Moreover, the serializability model ensures that applications can be written without attention to one another: Transactional serializability ensures that if a transaction would be correct when executed in isolation, it will also be correct when executed concurrently against a database server that interleaves operations for increased performance.

The transactional model is also valuable from a reliability perspective. The isolation of transactions from one another avoids inconsistencies that might occur if one transaction were to see the partial results of some other transaction—for example, suppose that transaction T_1 increments variable x by 1 and is executed concurrently, with transaction T_2, which decrements x by 1. If T_1 and T_2 read x concurrently they might base their computations on the same initial value of x. The *write* operation that completes last would then erase the other update. Many concurrent systems are prone to bugs because of this sort of mutual-exclusion problem; transactional systems avoid this issue using locking or other concurrency control mechanisms that would force T_2 to wait until T_1 has terminated, or the converse. Moreover, transactional abort offers a simple way for a server to deal with a client that fails or seems to hang: It can simply timeout and abort the transaction that the client initiated. (If the client is really alive, its attempt to commit will eventually fail: Transactional systems never guarantee that a commit will be successful.) Similarly, the client is insulated from the effects of server failures: It can modify data on the server without concern that an inopportune server crash could leave the database in an inconsistent state.

There is, however, a negative side to transactional distributed computing. As we will see in Chapter 21, transactional programming can be extremely restrictive. The model basically prevents programs from cooperating as peers in a distributed setting, and although extensions have been proposed to overcome this limitation, none seems to be fully satisfactory—that is, transactions really work best for applications in which there is a computational master process, which issues requests to a set of slave processors on which data are stored. This is, of course, a common model, but it is not the only one. Any transactional application in which several processes know about each other and execute concurrently is difficult to model in this manner.

Moreover, transactional mechanisms can be costly, particularly when a transaction is executed on data that have been replicated for high availability or distributed over multiple servers. The locking mechanisms used to ensure serializability can severely limit concurrency, and it can be very difficult to deal with transactions that run for long periods of time, since these will often leave the entire server locked and unable to accept new requests. It can also be very difficult to decide what to do if a transaction aborts unexpectedly: Should the client retry it or report to the user that it aborted? Decisions such as these are very difficult, particularly in sophisticated applications in which one is essentially forced to find a way to roll forward.

For all of these reasons, although transactional computing is a powerful and popular tool in developing reliable distributed software systems, it does not represent a complete model or a complete solution to all reliability issues that occur.

7.6 Applying Transactions to File Servers

Transactional access to data may seem extremely well matched to the issue of file server reliability. Typically, however, file servers either do not implement transactional functionality, or they do so only for the specific case of database applications. The reasons for this illustrate the sense in which a mechanism such as transactional data access may be unacceptably constraining in nontransactional settings.

General-purpose computing applications make frequent and extensive use of files. They store parameters in files, search directories for files with special names, store temporary results in files that are passed from phase to phase of a multiphase computation, implement ad hoc structures within very large files, and even use the existence or nonexistence of files and the file protection bits as persistent locking mechanisms, compensating for the lack of locking tools in operating systems such as UNIX.

As we saw earlier, file systems used in support of this model are often designed to be stateless, particularly in distributed systems—that is, each operation by a client is a complete and self-contained unit. The file system maintains no memory of actions by clients, and although the clients may cache information from the file system (such as handles pointing to open file objects), they are designed to refresh this information if it is found to be stale when referenced. Such an approach has the merit of extreme simplicity. It is certainly not the only approach: Some file systems maintain coherent caches of file system blocks within client systems, and these are necessarily stateful. Nonetheless, the great majority of distributed file systems are stateless.

The introduction of transactions on files thus brings with it stateful aspects that are otherwise avoided, potentially complicating any otherwise simple system architecture. However, transactions pose more problems than mere complexity. In particular, the locking mechanisms used by transactions are ill-matched to the pattern of file access seen in general operating system applications.

Consider the program that was used to edit this book. When started, it displayed a list of files that ended with the extension ".doc," and it waited for me to select the file on which I wished to work. Eventually, the file selected and open, an extended editing session ensued, perhaps even appearing to last overnight or over a weekend if some distraction prevented me from closing the file and exiting the program before leaving for the evening. In a standard transactional model, each of the read accesses and each of the write accesses would represent an operation associated with the transaction, and transactional serialization ordering would be achieved by delaying these operations as needed to ensure that only serializable executions are permitted—for example, with locks.

This now creates the prospect of a file system containing directories that are locked against updates (because some transaction has read the contents), files that are completely untouchable (because some transaction is updating or perhaps even deleting the contents), and long editing sessions that routinely end in failure (because locks may be broken after long delays, forcing the client program to abort its transaction and start again from scratch). It may not seem obvious that such files should pose a problem, but suppose that a transaction's behavior was slightly different as a result of seeing these transient conditions? That transaction would not be correctly serialized if the editing transaction were now aborted, resulting in some other state. No transaction should have been allowed to see the intermediate state.

Obviously, this analysis could be criticized as postulating a clumsy application of transactional serializability to the file system. In practice, one would presumably adapt the model to the semantics of the application. However, even for the specific case of transactional file systems, the system has been less than convincing—for example, at Xerox the early versions of the Clearinghouse software (a form of file system used for e-mail and other user-profile information) offered a fully transactional interface. Over time, this was greatly restricted because of the impracticality of transactional concurrency control in settings that involve large numbers of general-purpose applications.

Moreover, many file-based applications lack a practical way to assign a transaction-ID to the logical transaction. As an example, consider a version control software system. Such a system seems well matched to the transactional model: A user checks out a file, modifies it, and then checks it in; meanwhile, other users are prevented from doing updates and can only read old copies. Here, however, many individual programs may operate on the file over the period of the transaction. What is lacking is a practical way to associate an identifier with the series of operations. Clearly, the application programs themselves can do so, but one of the basic principles of reliability is to avoid placing excessive trust in the correctness of individual applications; in this example, the correctness of the applications would be a key element of the correctness of the transactional architecture, a very questionable design choice.

On the other hand, transactional file systems offer important benefits. Most often cited among these are the atomic update properties of a transaction, whereby a set of changes to files is made entirely, or not at all. This has resulted in proposals for file systems that are transactional in the limited sense of offering failure atomicity for updates, but without carrying this to the extreme of also providing transactional serializability. Hagmann's use of group commit to reimplement the Cedar file system (see Hagmann) and IBM's QuickSilver file system (see Schmuck and Wyllie) are examples of research efforts that are viewed as very successful in offering such a compromise. However, transactional atomicity remains uncommon in the mostly widely used commercial file system products because of the complexity associated with a stateful file system implementation. The appeal of stateless design, and the inherent reliability associated with an architecture in which the clients and servers take responsibility only for their own actions and place limited trust in information that they don't own directly, continues to rule the marketplace.

The most popular alternative to transactions is the atomic rename operation offered by many commercially standard file systems. For complex objects represented as a single file, or as a rooted graph of files, an application can atomically update the collection by creating a new root object containing the modifications, or by pointing to modified versions of other files, and then rename the result to obtain the equivalent effect of an atomic commit, with all the updates being installed simultaneously. If a crash occurs, it suffices to delete the partially modified copy; the original version will not be affected. Despite having some minor limitations, designers of fairly complex file system applications have achieved a considerable degree of reliability using operations such as rename, perhaps together with an *fsync* operation, which forces recent updates to an object or file out to the persistent disk storage area.

In conclusion, it is tempting to apply stateful mechanisms and even transactional techniques to file servers. Yet similar results can be obtained, for this particular application, with less costly and cumbersome solutions. Moreover, the simplicity of a stateless approach has enormous appeal in a world where there may be very little control over the software that runs on client computers, and in which trust in the client system will often be misplaced. In light of these considerations, file systems can be expected to remain predominantly stateless even in settings where reliability is paramount.

More generally, this point illustrates an insight to which we will return repeatedly in this book. Reliability is a complex goal and can require a variety of tools. While a stateless file system may be adequately reliable for one use, some other application may find its behavior hopelessly inconsistent and impossible to work around. A stateful database architecture works wonderfully for database applications, but it turns out to be difficult to adapt to general-purpose operating system applications that have less structure, or that merely have a nontransactional structure. Only a diversity of tools, integrated in an environment that encourages the user to match the tool to the need, can possibly lead to reliability in the general sense. No single approach will suffice.

7.7 Message-Oriented Middleware

An emerging area of considerable commercial importance, *message-oriented middleware* is concerned with extending the client/server paradigm so that clients and servers can be operated asynchronously. This means, for example, that a client may be able to send requests to a server that is not currently operational for batch processing later and that a server may be able to schedule requests from a request queue without fear of delaying a client application that is waiting for a reply. We discuss Message-Oriented Middleware Systems (MOMS) in Chapter 11, in conjunction with other distributed computing paradigms that fall out of the strict, synchronous-style, client/server architectures that are the focus of this chapter.

7.8 Related Topics

The discussion in this chapter has merely touched upon a very active area for both commercial products and academic research. Although NFS is probably the most widely used distributed file system technology, other major products are doing well in the field—for example, Transarc's AFS product (based on a research system developed originally at Carnegie Mellon University) is widely cited for its advanced security and scalability features. AFS is often promoted as a secure, worldwide file system technology. Later, when we discuss NFS security, it will become clear that this is potentially a very important property and represents a serious reliability exposure in distributed computing configurations that use NFS. Locus Computing Corporation's Locus product has similar capabilities, but it is designed for environments with intermittent connectivity.

On the PC side, major file system products are available from Microsoft as part of its Windows NT server technology, as well as from Banyan and Novell.

Stateful database and transactional technologies represent one of the largest existing markets for distributed computing systems. Major database products include Sybase, Informix, and ORACLE; all of these include client/server architectures. There are dozens of less well known but very powerful technologies. OnLine Transaction Processing (OLTP) technologies, which permit transaction operations on files and other special-purpose data structures, are also a major commercial market: Well-known products include Tuxedo and Encina; and there are many less well known but very successful similar technologies available in this market.

On the research side of the picture, much activity centers around the technical possibilities created by ATM communication, with its extremely high bandwidths and low latencies. File systems that page data over an ATM and that treat the client buffer pools as a large distributed buffering resource shared by all clients are being developed: Such systems gain enormous performance benefits from the resulting substantial enlargement in the file system buffer pool and because the latency incurred when fetching data over the network is orders of magnitude lower than that of fetching data from a remote disk. Examples of this style of research include the XFS project at Berkeley (see Anderson et al.), the Global Memory project at the University of Washington (see Feeley et al.), and the CRL project at MIT (see Johnson et al.). Such architectures create interesting reliability and consistency issues, which are closely related to the technologies we will be discussing in Part III.

The changing technology picture is indirectly changing the actual workload presented to the database or file server residing at the end of the line. One major area of research has concerned the creation of parallel file servers using arrays of inexpensive disks on which an error-correcting code is employed to achieve a high degree of reliability. Such RAID file servers have high capacity, because they aggregate large numbers of small disks; they have good data transfer properties, again because they can benefit from parallelism; and they have good seek time—not because the small disks are especially fast, but rather because load is shared across them and this reduces the effective length of the I/O request queue to each drive. Research on striping data across a RAID system to optimize its response time has yielded further performance improvements.

In the past, file systems and database servers saw a mixed read-write load with a bias toward read operations, and they were organized accordingly. But as the percentage of active data resident in the buffer pools of clients has risen, the percentage of read requests that actually reach the server has dropped correspondingly. A modern file system server sees a workload that is heavily biased toward update traffic. Best known of the work in this area is Rosenblum's log-structured file system (LFS) (see Rosenblum and Ousterhout), developed as part of Ousterhout's Sprite project at Berkeley. LFS implements an append-only data structure (a log), which it garbage collects and compacts using background scavenger mechanisms. Fast indexes permit rapid read access to the file system but, because most of the disk I/O is in the form of writes to the log, the system gains a tremendous performance boost. Similar issues have been studied in the context of database systems and have shown that similar benefits are possible. One can anticipate that the technology trends now seen in the broad marketplace will continue to shift basic elements of the low-level file system architecture, creating further opportunities for significant improvements in average data access latencies and in other aspects of client/server performance.

7.9 Related Reading

I am not aware of any good general references on NFS itself, although the standard is available from Sun Microsystems and is widely supported.

NFS performance and access patterns are studied in Ousterhout (1985) and extended to the Sprite file system in Baker et al.

References to NFS-like file systems supporting replication include Bhide et al., Digital Equipment Corporation, Kronenberg et al., Ladin et al., Liskov et al., Siegal.

Topics related to the CMU file system work that led to AFS are covered in Bellovin and Merritt, Birrell, Howard et al., Lampson et al., Satyanarayanan, Satyanarayanan et al., Schiller, Spector, Steiner et al.

Coda is discussed in Kistler and Satyanarayanan, Mummert et al.

RAID is discussed in Paterson et al. Sprite is discussed in Nelson et al., Ousterhout et al., Srinivasan and Mogul. Ficus is discussed in Reiher et al., Locus in Heidemann and Popek (1995), Walter et al.

XFS is discussed in Anderson et al.

Work on global memory is covered in Feeley et al., Johnson et al.

Database references for the transactional approach are studied in Bernstein et al., Gray (1979), Gray and Reuter.

Tandem's system is presented in Bartlett et al.

Nomadic transactional systems are covered in Alonso and Korth, Amir.

Transactions on file systems are discussed in Hagmann, Schmuck and Wyllie.

Related work is treated in Liskov and Scheifler, Liskov et al., Macedo et al., Moss.

CHAPTER 8

Operating System Support for High-Performance Communication

CONTENTS

The performance of a communication system is typically measured in terms of the latency and throughput for typical messages that traverse that system, starting in a source application and ending at a destination application. Accordingly, these issues have received considerable scrutiny within the operating system research community, which has developed a series of innovative proposals for improving performance in communication-oriented applications. In the following text, we review some of these proposals.

There are other aspects of communication performance that matter when building a reliable distributed application, but these have received considerably less attention. Prominent among these are the loss characteristics of the communication subsystem. In typical communication architectures, messages are generated by a source application, which passes them to the operating system. As we saw early in this book, such messages will then travel down some form of protocol stack, eventually reaching a device driver that arranges for the data to be transmitted on the wire. Remotely, the same process is repeated.

Such a path offers many opportunities for inefficiency and potential message loss. Frequently, the layer-to-layer hand-offs that occur involve copying the message from one memory region or address space to another, perhaps with a header prepended or a suffix appended. Each of these copying operations will be costly (even if other costs such as scheduling are even more costly), and if a layer is overloaded or is unable to allocate the necessary storage, a message may be lost without warning. Jointly, the consumption of CPU and memory resources by the communication subsystem can become very heavy during periods of frequent message transmission and reception, triggering overload and high rates of message loss. In Chapter 3 we saw that such losses can sometimes become significant. Thus, while we will be looking at techniques for reducing the amount of copying and the number of cross-address space control transfers needed to perform a communication operation, the reader should also keep in mind that by reducing copying, these techniques may also be reducing the rate of message loss that occurs in the protocol stack.

The *statistical properties* of communication channels represent an extremely important area for future study. Most distributed systems, and particularly the ones intended for critical settings, assume that communication channels offer identical and independent quality-of-service properties to each packet transmitted—for example, it is typically implicit in the design of a protocol that if two packets are transmitted independently, then the observed latency, data throughput, and probability of loss will be identical. Such assumptions match well with the properties of communication hardware during periods of light, uniform load, but the layers of software involved in implementing communication stacks and routing packets through a complex network can seriously distort these underlying properties.

Within the telecommunication community, bandwidth sharing and routing algorithms have been developed that are fair in the sense of dividing available bandwidth among a set of virtual circuits of known expected traffic levels. But the problem of achieving fairness in a packet-switched environment with varying loads from many sources is much harder and is not at all well understood. One way to think about this problem is to visualize the operating system layers through which packets must travel, and the switching systems used to route packets to their destinations, as a form of filter, which can distort the distribution of packets in time and superimpose errors on an initially error-free data stream. Such a perspective leads to the view that these intermediary software layers introduce noise into the distribution of intermessage latency and error rates.

This is readily confirmed by experiment. The most widely used distributed computing environments exhibit highly correlated communication properties: If one packet is delayed, the next will probably be delayed too. If one packet is dropped in transmission, the odds are that the next will be dropped as well. As one might expect, however, such problems are a direct consequence of the same memory constraints and layered architectures that also introduce the large latency and performance overheads that the following techniques are designed to combat. Thus, although the techniques discussed in this chapter were developed to provide higher performance, and were not specifically intended to improve the statistical properties of the network, they would in fact be expected to exhibit better statistical behavior than the standard distributed system architecture does simply by eliminating layers of software that introduce delays and packet loss.

8.1 Lightweight RPC

Performance of remote procedure calls has been a major topic of research since RPC programming environments first became popular. Several approaches to increasing RPC performance have had significant impact.

The study of RPC performance as a research area surged in 1989 when Shroeder and Burrows undertook to precisely measure the costs associated with RPC on the Firefly operating system (see Shroeder and Burrows). These researchers started by surveying the costs of RPC on a variety of standard platforms. Their results have subsequently become outdated because of advances in system and processor speeds, but the finding that RPC performance varies enormously even in relative terms probably remains true today. In their study, the range of performance was from 1.1 ms to do a null RPC (equivalent to 4,400 instructions) on the Cedar system, highly optimized for the Dorado multiprocessor, to 78 ms (195,000 instructions) for a very general version of RPC running on a major vendor's top-of-the-line platform (at that time). One interesting finding of this study was that the number of instructions in the RPC code path was often high (the average in the systems they looked at was approximately 6,000 for systems with many limitations and about 140,000 for the most general RPC systems). Thus, faster processors would be expected to have a big impact on RPC performance, which is one of the reasons that the situation has improved somewhat since the time of this study.

Using a bus analyzer to pin down costs to the level of individual machine cycles, this effort led to a tenfold performance improvement in the RPC technology under investigation, which was based originally on the Berkeley UNIX RPC. Among the optimizations that had the biggest impact were the elimination of copying within the application address space by marshaling data directly into the RPC packet using an in-line compilation technique, and the implementation of an RPC fast path, which eliminated all generality in favor of a hand-coded RPC protocol using the fewest instructions possible, subject to the constraint that the normal O/S protection guarantees would be respected. (It is worthwhile to note that on PC operating systems, which often lack protection mechanisms and provide applications with direct access to the I/O devices, even higher performance can often be achieved, but at the cost of substantially reduced security and hence exposure of the system as a whole to bugs and intrusion by viruses.)

Soon after this work on Firefly RPC was completed, researchers at the University of Washington became interested in other opportunities to optimize communication paths in modern operating systems. Lightweight RPC originated with the observation that as computing systems adopt RPC-based architectures, the use of RPC in *nondistributed* settings is rising as rapidly as is RPC over a network. Unlike a network, RPC in the nondistributed case can accurately sense many kinds of failures, and because the same physical memory is potentially visible to both sender and destination, the use of shared memory mechanisms represents an appealing option for enhancing performance. Bershad and others set out to optimize this common special case (see Bershad et al.).

A shared memory RPC mechanism typically requires that messages be allocated within pages, starting on page boundaries and with a limit of one message per page. In some cases, the pages used for message passing are from a special pool of memory maintained by the kernel; in others, no such restriction applies but there may be other restrictions, such as limits on passing data structures that contain pointers. When a message is sent, the kernel modifies the page table of the destination to map the page containing the message into the address space of the destination process. Depending on the operating system, the page containing the message may be mapped out of the memory of the sender, modified to point to an empty page, or marked as read-only. In this last approach (where the page is marked as read-only) some systems will trap write-faults and make a private copy if either process attempts a modification. This method is called "copy on write" and was first supported in the Mach microkernel (see Rashid).

If one studies the overheads associated with RPC in the local, shared memory case, the cost of manipulating the page tables of the sender and destination and of context switching between the sending and receiving processes emerges as a major factor. The University of Washington team focused on this problem in developing what they called a *Lightweight Remote Procedure Call* facility (LRPC). In essence, this approach reduces time for local RPC both by exploiting shared memory and by avoiding excess context switches. Specifically, the messages containing the RPC arguments are placed in shared memory, while the invocation itself is done by changing the current page table and flushing the TLB so that the destination process is essentially invoked in coroutine style, with the lowest overhead possible given that virtual memory is in use on the machine. The reply from the destination process is similarly implemented as a direct context switch back to the sender process.

Although LRPC may appear to be as costly as normal RPC in the local case, the approach actually achieves substantial savings. First, a normal RPC is implemented by having the client program perform a message send followed by a separate message receive operation, which blocks. Thus, two system calls occur, with the message itself being copied into the kernel's data space, or (if shared memory is exploited) a message descriptor being constructed in the kernel's data space. Meanwhile, the destination process will have issued a receive request and would often be in a blocked state. The arrival of the message makes the destination process runnable, and on a uniprocessor this creates a scheduling decision, since the sender process is also runnable in the first stage of the algorithm (when it has sent its request and not yet performed the subsequent receive operation). Thus, although the user might expect the sender to issue its two system calls and then block, causing the scheduler to run and activate the destination process, other sequences are possible. If the scheduler runs right after the initial send operation, it could context switch to the RPC server, leaving the client runnable. It is now possible that a context switch back to the client will occur, and then back to the server again, before the server replies. The same sequence may then occur when the reply is finally sent.

We thus see that a conventional operating system requires four system calls to implement an LRPC operation, and although a minimum of two context switches must occur, it is possible for an additional two context switches to take place. If the execution of the operating system scheduler represents a significant cost, the scheduler may run two or more times more than the minimum. All of these excess operations are potentially costly.

Accordingly, LRPC is implemented using a special system call whereby the client process combines its send and receive operations into a single request, and the server (which will normally delay waiting for a new RPC request after replying to the client) issues the reply and subsequent receive as a single request. Moreover, execution of the scheduler is completely bypassed.

As in the case of RPC, the actual performance figures for LRPC are of limited value because processor speeds and architectures have been evolving so rapidly. One can get a sense of the improvement by looking at the number of instructions required to perform an LRPC. Recall that the Shroeder and Burrows study had found that thousands of instructions were required to issue an RPC. In contrast, the LRPC team calculated that only a few hundred instructions are required to perform an LRPC—a small enough number to make such factors as TLB misses (caused when the hardware cache associated with the virtual memory mapping system is flushed) increase to have an important impact on performance. LRPC was, in any case, somewhat more expensive than the theoretical minimum: about 50 percent slower measured in terms of round-trip latency or instructions executed for a null procedure call. Nonetheless, this represents a factor of at least five, when compared to the performance of typical RPC in the local case, and ten or more when the approach is compared to the performance of a fairly heavyweight vendor-supported RPC package.

This effect is so dramatic that some operating system vendors began to support LRPC immediately after the work was first reported. Others limited themselves to fine-tuning their existing implementations or improving the hardware used to connect their processors to the network. At the time of this writing, RPC performances have improved somewhat, but faster processors are no longer bringing commensurate improvements in RPC performance. Vendors tend to point out that RPC performance, by itself, is only one of many factors that enter into overall system performance and that optimizing this one case to an excessive degree can bring diminishing returns. They also argue for generality even in the local case: that LRPC is undesirable because it requires a different RPC implementation than the remote case and thus increases the complexity of the operating system for a scenario that may not be as common in commercial computing settings as it seems to be in academic research laboratories.

To some degree, these points are undoubtedly valid ones: When an RPC arrives at a server, the program that will handle it may need to be scheduled, it may experience page faults, buffering and caching issues can severely impact its performance, and so forth. On the other hand, the performance of a null RPC or LRPC is entirely a measure of operating system overhead, and hence is wasted time by any reasonable definition. Moreover, the insights gained in LRPC are potentially applicable to other parts of the operating system: Bershad, for example, demonstrated that the same idea can be generalized using a concept of *thread activations* and *continuations,* with similar dramatic impact on other aspects of operating system performance (see Bershad et al.). This work seems not to have impacted the commercial operating system community, at least at the time of this writing.

8.2 *Fbufs* and the *x*-Kernel Project

During the same period, the University of Arizona, under Larry Peterson, developed a series of innovative operating system extensions for high-performance communication. Most relevant to the topic of this chapter are the *x*-Kernel, a stand-alone operating system for developing high-speed communication protocols, and the *fbufs* architecture (see Drushel and Peterson), which is a general-purpose technique for optimizing stack-structured protocols to achieve high performance. While these extensions were developed based on the context of a particular operating system, they are potentially applicable to most standard vendor-supported operating systems.

The *x*-Kernel (see Peterson et al.) is an operating system dedicated to the implementation of network protocols for experimental research on performance, flow control, and other issues. The assumption that *x*-Kernel applications are purely communication-oriented greatly simplified the operating system design, which confines itself to addressing those issues encountered in the implementation of protocols, while omitting support for elaborate virtual memory mechanisms, special-purpose file systems, and many of the other operating facilities that are considered mandatory in modern computing environments.

Recall from the early chapters of this book that many protocols have a layered structure, with the different layers having responsibility for different aspects of the overall communication abstraction. In *x*-Kernel, protocols having a layered structure are represented as a partially ordered graph of modules. The application process involves a protocol by issuing a procedure call to one of the root nodes in such a graph, and control then flows down the graph as the message is passed from layer to layer. *x*-Kernel includes built-in mechanisms for efficiently representing messages and managing their headers and for dynamically restructuring the protocol graph or the route that an individual message will take, depending upon the state of the protocols involved and the nature of the message. Other *x*-Kernel features include a thread-based execution model, memory management tools, and timer mechanisms.

Using the *x*-Kernel, Peterson implemented several standard RPC and stream protocols, demonstrating that his architecture was indeed powerful enough to permit a variety of such protocols to coexist and confirming its value as an experimental tool. Layered protocol architectures are often thought to be inefficient, but Peterson suggested a number of design practices that, in his experience, avoided overhead and permitted highly modular protocol implementations to perform as well as the original monolithic protocols on which his work was based. (Later, researchers such as Tennenhouse confirmed that standard implementations of layered protocols, particularly in the UNIX stream architecture, have potentially high overheads, but that appropriate design techniques can be used to greatly reduce these costs.)

Peterson's interest in layered protocols subsequently led him to look at performance issues associated with layered or pipelined architectures, in which modules of a protocol operate in protected memory regions (Figure 8.1). To a limited degree, systems such as UNIX and NT have an architecture similar to this—UNIX streams, for example, are based on a modular architecture, which is supported directly within the kernel. As an example, an incoming message is passed up a stack that starts with the device driver and then includes each of the stream modules that have been pushed onto the stream connection, terminating finally in a cross-address space transfer of control to the application program. UNIX programmers think of such a structure as a form of pipe implemented directly in the kernel. Unfortunately, like a pipe, a stream can involve significant overhead.

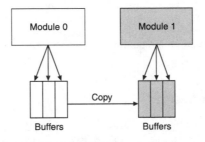

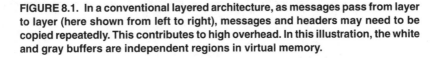

FIGURE 8.1. In a conventional layered architecture, as messages pass from layer to layer (here shown from left to right), messages and headers may need to be copied repeatedly. This contributes to high overhead. In this illustration, the white and gray buffers are independent regions in virtual memory.

Peterson's *fbufs* architecture focuses on the handling of memory in pipelined operating system contexts such as these. An *fbuf* is a memory buffer for use by a protocol; it will typically contain a message or a header for a message. The architecture concerns itself with the issue of mapping such a buffer into the successive address spaces within which it will be accessed and with the protection problems that occur if modules are to be restricted so that they can only operate on data that they own. The basic approach is to cache memory bindings, so that a protocol stack that is used repeatedly can reuse the same memory mappings for each message in a stream of messages. Ideally, the cost of moving a packet from one address space to another can be reduced to the flipping of a protection bit in the address space mappings of the sending and receiving modules (Figure 8.2). The method completely eliminates copying, while retaining a fairly standard operating system structure and protection boundaries.

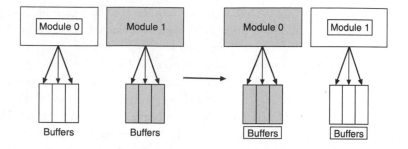

FIGURE 8.2. In Peterson's scheme, the buffers are in fact shared using virtual memory, exploiting protection features to avoid risk of corruption. To pass a buffer, access to it is enabled in the destination address space and disabled in the sender's address space. (In the figure, the white buffers represent real pointers and the gray ones represent invalid page-table entries pointing to the same memory regions but with access disabled.) When the buffer finally reaches the last module in the pipeline, it is freed and reallocated for a new message arriving from the left. Such an approach reduces the overhead of layering to the costs associated with manipulation of the page-table entries associated with the modules comprising the pipeline.

8.3 Active Messages

At the University of California, Berkeley, and Cornell University, researchers have explored techniques for fast message passing in parallel computing systems. Culler and von Eicken observed that operating system overheads are the dominant source of overhead in message-oriented parallel computing systems (see Thekkanth and Levy, von Eicken et al.) [1992]). Their work resulted in an extremely aggressive attack on communication costs, in which the application interacts directly with an I/O device and the overhead for sending or receiving a message can be reduced to as little as a few instructions. The CPU and latency overhead of an operating system is slashed in this manner, with important impact on the performance of parallel applications. Moreover, as we will see, similar ideas can be implemented in general-purpose operating systems.

An *active message* is a type of message generated within a parallel application that takes advantage of knowledge that the program running on the destination node of a parallel computer is precisely the same as the program on the source node to obtain substantial performance enhancements. In the approach, the sender is able to anticipate much of the work that the destination node would normally have to do if the source and destination were written to run on general-purpose operating systems. Moreover, because the source and destination are the same program, the compiler can effectively short circuit much of the work and overhead associated with mechanisms for general-purpose message generation and for dealing with heterogeneous architectures. Finally, because the communication hardware in parallel computers does not lose messages, active messages are designed for a world in which message loss and processor failure do not occur.

The basic approach is as follows. The sender of a message generates the message in a format that is preagreed between the sender and destination. Because the destination is running the same program as the sender and is running on the same hardware architecture, such a message will be directly interpretable by the destination without any of the overhead for describing data types and layout that one sees in normal RPC environments. Moreover, the sender places the address of a handler for this particular class of message into the header of the message—that is, a program running on machine A places an address of a handler that resides within machine B directly into the message. On the reception machine, as the message is copied out of the network interface, its first bytes are already sufficient to transfer control to a handler compiled specifically to receive messages of this type. This reduces the overhead of communication from the tens of thousands of instructions common on general-purpose machines to as few as five to ten instructions. In effect, the sender is able to issue a procedure call directly into the code of the destination process, with most of the overhead associated with triggering an interrupt on the destination machine and with copying data into the network on the sending side and out of the network on the receiving side. In some situations (e.g., when the destination node is idle and waiting for an incoming request) even the interrupt can be eliminated by having the destination wait in a tight polling loop.

Obviously, active messages make sense only if a single application is loaded onto multiple nodes of a parallel computer, such as the CM5 or SP2, and hence has complete trust in those programs and accurate knowledge of the memory layout of the nodes with which it communicates. In practice, the types of systems that use the approach normally have identical programs running on each node. One node is selected as the master and controls the computation, while the other nodes, its slaves, take actions on the orders of the master. The actual programming model visible to the user is one in which a sequential program initiates parallel actions by invoking parallel operations, or procedures, which have been programmed to distribute work among the slaves and then to wait for them to finish computing before taking the next step. This model is naturally matched to active messages, which can now be viewed as optimizing normal message passing to take advantage of the huge amount of detailed information available to the system regarding the way that messages will be handled. In these systems, there is no need for generality, and generality proves to be expensive. Active messages are a general way of optimizing to extract the maximum performance from the hardware by exploiting this prior knowledge. (See Figure 8.3.)

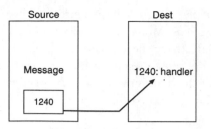

FIGURE 8.3. An active message includes the address of the handler to which it should be passed directly in the message header. In contrast with a traditional message-passing architecture, in which such a message would be copied repeatedly through successively lower layers of the operating system, an active message is copied directly into the network adapter by the procedure that generates it in the application program. It is effectively transferred directly to the application layer handler on the receiving side with no additional copying. Such a zero copy approach reduces communication latencies to a bare minimum and eliminates almost all overhead on the messages themselves. However, it also requires a high level of mutual trust and knowledge between source and destination, a condition that is more typical of parallel supercomputing applications than general distributed programs.

Active messages are useful in support of many programming constructs. The approach can be exploited to build extremely inexpensive RPC interactions, but it is also applicable to direct language support for data replication or parallel algorithms in which data or computation is distributed over the modes of a parallel processor. (See Figure 8.4.) Culler and von Eicken have explored a number of such options and reported particular success with language-based embedding of active messages within a parallel version of the C programming language, which they call "split C," and in a data-parallel language called ID-90.

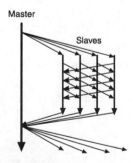

FIGURE 8.4. A typical parallel program employs a sequential master thread of control, which initiates parallel actions on slave processors and waits for them to complete before starting the next computational step. While computing, the slave nodes may exchange messages, but this too tends to be both regular and predictable. Such applications match closely with the approach to communication used in active messages, which trades generality for low overhead and simplicity.

8.4 Beyond Active Messages: U-Net

At Cornell University, von Eicken has continued the work begun in his study of active messages, looking for ways of applying the same optimizations in general-purpose operating systems connected to shared communication devices. U-Net is a communication architecture designed for use within a standard operating system such as UNIX or NT; it is intended to provide the standard protection guarantees taken for granted in these sorts of operating systems (see von Eicken et al. [1995]). These guarantees are provided, however, in a way that imposes extremely little overhead relative to the performance that can be attained in a dedicated application that has direct control over the communication device interface. U-Net gains this performance using an implementation that is split between traditional software functionality integrated into the device driver and nontraditional functionality implemented directly within the communication controller interfaced to the communication device. Most controllers are programmable—hence, the approach is more general than it may sound, although it should also be acknowledged that existing systems very rarely reprogram the firmware of device controllers to gain performance!

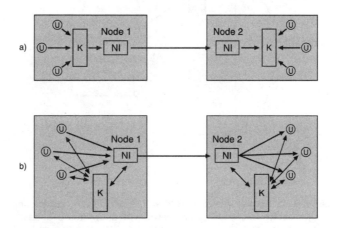

FIGURE 8.5. In a conventional communication architecture, all messages pass through the kernel before reaching the I/O device (a), resulting in high overheads. U-Net bypasses the kernel for I/O operations (b), while preserving a standard protection model.

The U-Net system (see Figure 8.5) starts with an observation we have made repeatedly in prior chapters, namely that the multiple layers of protocols and operating system software between the application and the communication wire represent a tremendous barrier to performance, impacting both latency and throughput. U-Net overcomes these costs by restructuring the core operating system layers that handle such communication so that channel setup and control functions can operate out of band, while the application interacts directly with the device itself. Such a direct path results in minimal latency for the transfer of data from source to destination, but it raises significant protection concerns: If an application can interact directly with the device, there is no obvious reason that it will not be able to subvert the interface to violate the protection on memory controlled by other applications or break into communication channels that share the device but were established for other purposes.

The U-Net architecture is based on a concept of a *communication segment*, which is a region of memory shared between the device controller and the application program. Each application is assigned a set of pages within the segment for use in sending and receiving messages and is prevented from accessing pages not belonging to it. Associated with the application are three queues: one pointing to received messages, one to outgoing messages, and one to free memory regions. Objects in the communication segment are of fixed size, simplifying the architecture at the cost of a small amount of overhead. (See Figure 8.6.)

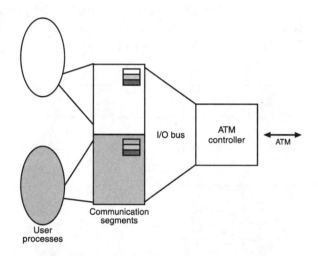

FIGURE 8.6. U-Net shared memory architecture permits the device controller to directly map a communication region shared with each user process. The send, receive, and free message queues are at known offsets within the region. The architecture provides strong protection guarantees and yet slashes the latency and CPU overheads associated with communication. In this approach, the kernel assists in setup of the segments but is not interposed on the actual I/O path used for communication once the segments are established.

Each of these communication structures is bound to a U-Net channel, which is a communication session for which permissions have been validated, linking a known source to a known destination over an established ATM communication channel. The application process plays no role in specifying the hardware communication channels to which its messages will be sent: It is restricted to writing in memory buffers that have been allocated for its use and to update the send, receive, and free queue appropriately. These restrictions are the basis of the U-Net protection guarantees cited earlier.

U-Net maps the communication segment of a process directly into its address space, pinning the pages into physical memory and disabling the hardware caching mechanisms so that updates to a segment will be applied directly to that segment. The set of communication segments for all the processes using U-Net is mapped to be visible to the device controller over the I/O bus of the processor used; the controller can thus initiate DMA or direct memory transfers in and out of the shared region as needed and without delaying any setup. A limitation of this approach is that the I/O bus is a scarce resource shared by all devices on a system, and the U-Net mapping excludes any other possible mapping for this region. However, on some machines (e.g., the cluster-style multiprocessors discussed in Chapter 24), there are no other devices contending for this mapping unit, and dedicating it to the use of the communication subsystem makes perfect sense.

The communication segment is directly monitored by the device controller. U-Net accomplishes this by reprogramming the device controller, although it is also possible to imagine an implementation in which a kernel driver would provide this functionality. The controller watches for outgoing messages on the send queue; if one is present, it immediately sends the message. The delay between when a message is placed on the send queue and when sending starts is never larger than a few microseconds. Incoming messages are automatically placed on the receive queue unless the pool of memory is exhausted; should that occur, any incoming messages are discarded silently. To accomplish this, U-Net needs only to look at the first bytes of the incoming message, which give the ATM channel number on which it was transmitted. These are used to index into a table maintained within the device controller that gives the range of addresses within which the communication segment can be found, and the head of the receive and free queues are then located at a fixed offset from the base of the segment. To minimize latency, the addresses of a few free memory regions are cached in the device controller's memory.

Such an approach may seem complex because of the need to reprogram the device controller. In fact, however, the concept of a programmable device controller is a very old one (IBM's channel architecture for the 370 series of computers already supported a similar programmable channel architecture nearly 20 years ago). Programmability such as this remains fairly common, and device drivers that download code into controllers are not unheard of today. Thus, although unconventional, the U-Net approach is not actually unreasonable. The style of programming required is similar to that used when implementing a device driver for use in a conventional operating system.

With this architecture, U-Net achieves impressive application-to-application performance. The technology easily saturates an ATM interface operating at the OC3 performance level of 155 MB/sec, and measured end-to-end latencies through a single ATM switch are as low as 2 6 µs for a small message. These performance levels are also reflected in higher-level protocols: Versions of UDP and TCP have been layered over U-Net and are capable of saturating the ATM for packet sizes as low as 1 KB; similar performance is achieved with a standard UDP or TCP technology only for very large packets of 8 KB or more. Overall, performance of the approach tends to be an order of magnitude or more better than with a conventional architecture for all metrics not limited by the raw bandwidth of the ATM: throughput for small packets, latency, and computational overhead of communication. Such results emphasize the importance of rethinking standard operating system structures in light of the extremely high performance that modern computing platforms can achieve.

Returning to the point made at the beginning of this chapter, a technology such as U-Net also improves the statistical properties of the communication channel. There are fewer places at which messages can be lost; hence reliability increases and, in well-designed applications, may approach perfect reliability. The complexity of the hands-off mechanisms employed as messages pass from application to controller to ATM and back to the receiver is greatly reduced—hence, the measured latencies are much tighter than in a conventional environment, where dozens of events could contribute toward variation in latency. Overall, then, U-Net is not just a higher-performance communication architecture; it is also one that is more conducive to the support of extremely reliable distributed software.

8.5 Protocol Compilation Techniques

U-Net seeks to provide very high performance by supporting a standard operating system structure in which a nonstandard I/O path is provided to the application program. A different direction of research, best known through the results of the SPIN project at the University of Washington (see Bershad et al.), is concerned with building operating systems that are dynamically extensible through application programs coded in a special type-safe language and linked directly into the operating system at run time. In effect, such a technology compiles the protocols used in the application into a form that can be executed close to the device driver. The approach results in speedups that are impressive by the standards of conventional operating systems, although less dramatic than those achieved by U-Net.

The key idea in SPIN is to exploit dynamically loadable code modules to place the communication protocol very close to the wire. The system is based on Modula-3, a powerful modern programming language similar to C++ or other modular languages, but type safe. Among other guarantees, type safety implies that a SPIN protocol module can be trusted not to corrupt memory or to leak dynamically allocated memory resources. This is in contrast with, for example, the situation for a stream module, which must be trusted to respect such restrictions.

SPIN creates a run-time context within which the programmer can establish communication connections, allocate and free messages, and schedule lightweight threads. These features are sufficient to support communication protocols such as the ones that implement typical RPC or stream modules, as well as for more specialized protocols such as those used to implement file systems or to maintain cache consistency. The approach yields latency and throughput improvements of as much as a factor of two when compared to a conventional user-space implementation of similar functionality. Most of the benefit is gained by avoiding the need to copy messages across address space boundaries and to cross protection boundaries when executing the short code segments typical of highly optimized protocols. Applications of SPIN include support for stream-style extensibility in protocols, but also less traditional operating system features, such as distributed shared memory and file system paging over an ATM between the file system buffer pools of different machines.

Perhaps more significant is the fact that a SPIN module has control over the conditions under which messages are dropped because of a lack of resources or time to process them. Such control, lacking in traditional operating systems, permits an intelligent and controlled degradation if necessary—a marked contrast with the more conventional situation in which, as load gradually increases, a point is reached where the operating system essentially collapses, losing a high percentage of incoming and outgoing messages, often without indicating that any error has occurred.

Like U-Net, SPIN illustrates that substantial performance gains in distributed protocol performance can be achieved by concentrating on the supporting infrastructure. Existing operating systems remain single-user centric in the sense of having been conceived and implemented with dedicated applications in mind. Although such systems have evolved successfully into platforms capable of supporting distributed applications, they are far from optimal in terms of overhead imposed on protocols, data loss characteristics, and length of the I/O path followed by a typical message on its way to the wire. As work such as this enters the mainstream, significant reliability benefits will spill over to end users, who often experience the side-effects of the high latencies and loss rates of current architectures as sources of unreliability and failure.

8.6 Related Reading

For work on kernel and microkernel architectures for high-speed communication: Amoeba (see Mullender et al., van Renesse [1988, 1989]), Chorus (see Armand et al., Rozier et al. [Fall 1988, December 1988]), Mach (see Rashid), QNX (see Hildebrand), Sprite (see Ousterhout et al. [1988]).

Issues associated with the performance of threads are treated in Anderson et al.

Packet filters are discussed in the context of Mach in Mogul et al.

The classic paper on RPC cost analysis is Shroeder and Burrows, but see also Clark and Tennenhouse.

TCP cost analysis and optimizations are presented in Clark et al., Jacobson (1988, 1990).

Lightweight RPC is treated in Bershad et al. (1989).

Fbufs and the *x*-Kernel are discussed in Abbott and Peterson, Drushel and Peterson, Peterson et al.

Active messages are covered in Thekkanth and Levy, von Eicken et al. (1992), and U-Net is discussed in von Eicken et al. (1995).

SPIN is treated in Bershad et al. (1995).

Part II

THE WORLD WIDE WEB

Part II focuses on the technologies that make up the World Wide Web, which, in a general sense, include Internet e-mail and news as well as the Mosaic-style of network document browser that has seized the public's attention. Our discussion is detailed enough to provide the reader with a good understanding of the key components of the technology base and the manner in which they are implemented—without going to such an extreme level of detail as to lose track of our broader agenda, which is to understand how reliable distributed computing services and tools can be introduced into the sorts of critical applications that may soon be placed on the Web.

CHAPTER 9 ✧ ✧ ✧ ✧

The World Wide Web

CONTENTS

9.1 The World Wide Web

As recently as 1993, it was common to read of a coming revolution in communication and computing technologies. Authors predicted a future information economy, the emergence of digital libraries and newspapers, the prospects of commerce over the network, and so forth. Yet the press was also filled with skeptical articles, suggesting that although there might well be a trend toward an information superhighway, it seemed to lack on-ramps accessible to normal computer users.

In an astonishingly short period of time, this situation has reversed itself. By assembling a relatively simple client/server application using mature, well-understood technologies, a group of researchers at CERN and at the National Center for Supercomputing Applications (NCSA) developed a system for downloading and displaying documents over a network. They employed an object-oriented approach in which their display system could be programmed to display various types of objects: audio, digitized images, text, hypertext documents represented using the HyperText Markup Language (HTML) (a standard for representing complex documents), and other data types. They agreed upon a simple resource location scheme, capable of encoding the information needed to locate an object on a server and the protocol with which it should be accessed. Their display interface integrated these concepts with easily used, powerful graphical user interface (GUI) tools. And suddenly, by pointing and clicking, a completely unsophisticated user could access a rich collection of data and documents over the Internet. Moreover, authoring tools for hypertext documents already existed, making it surprisingly easy to create elaborate graphics and sophisticated hypertext materials. By writing simple programs to track network servers, checking for changed content, and following hypertext links, substantial databases of Web documents were assembled, against which sophisticated information retrieval tools could be applied. Overnight, the long-predicted revolution in communications took place.

Two years later, there seems to be no end to the predictions for the potential scope and impact of the information revolution. One is reminded of the early days of the biotechnology revolution, during which dozens of companies were launched, fortunes were earned, and the world briefly overlooked the complexity of the biological world in its unbridled enthusiasm for a new technology. Of course, initial hopes can be unrealistic. A decade or so later, the biotechnology revolution is beginning to deliver on some of its initial promise, but the popular press and the individual in the street have long since become disillusioned.

The biotechnology experience highlights the gap that often forms between the expectations of the general populace and the deliverable reality of a technology area. We face a comparable problem in distributed computing today. On the one hand, the public seems increasingly convinced that the information society has arrived. Popular expectations for this technology are hugely inflated, and it is being deployed on a scale and rate that is surely unprecedented in the history of technology. Yet, the fundamental science underlying Web applications is in many ways very limited. The vivid graphics and ease with which hundreds of thousands of data sources can be accessed obscure more basic technical limitations, which may prevent the use of the Web for many of the uses that the popular press currently anticipates. Break-out 9.1 discusses several Web applications.

Web Applications

The network name service is structured like an inverted tree.

Web browser's system only needs to contact local name and Web services.

The Web operates like a postal service. Computers have names and addresses, and communication is by the exchange of electronic letters (messages) between programs. Individual systems don't need to know how to locate all the resources in the world. Instead, many services, such as the name service and Web document servers, are structured to pass requests via local representatives, which forward them to more remote ones, until the desired location or a document is reached.

To retrieve the Web document www.cs.cornell.edu/Info/Projects/HORUS a browser must first map the name of the Web server, www.cs.cornell.edu, to an address. If the address is unknown locally, the request will be forwarded to a central name server and then to one at Cornell (1–3 in figure above). The request to get the document will often pass through one or more Web proxies on its way to the Web server (4–9). These intermediaries save copies of frequently used information in short-term memory. Thus, if many documents are fetched from Cornell, the server address will be remembered by the local name service, and if the same document is fetched more than once, one of the Web proxies will respond rapidly using a saved copy. The term "caching" refers to the hoarding of reused information in this manner.

Our Web surfer looks irritated, perhaps because the requested server "is overloaded or not responding." This common error message is actually misleading because it can be provoked by many conditions, some of which don't involve the server at all—for example, the name service may have failed or become overloaded, or this may be true of a Web proxy,

as opposed to the Cornell Web server itself. The Internet addresses for any of these may be incorrect or stale (e.g., if a machine has been moved). The Internet connections themselves may have failed or become overloaded.

Although caching dramatically speeds response times in network applications, the Web does not track the locations of cached copies of documents, and it offers no guarantees that cached documents will be updated. Thus, a user may sometimes see a stale (outdated) copy of a document. If a document is complex, a user may even be presented with an inconsistent mixture of stale and up-to-date information.

With wider use of the Web and other distributed computing technologies, critical applications will require stronger guarantees. Such applications depend upon correct, consistent, secure, and rapid responses. If an application relies on rapidly changing information, stale responses may be misleading, incorrect, or even dangerous, as in the context of a medical display in a hospital or the screen image presented to an air traffic controller.

One way to address such concerns is to arrange for cached copies of vital information, such as resource addresses, Web documents, and other kinds of data, to be maintained consistently and updated promptly. By reliably replicating information, computers can guarantee rapid response to requests, avoid overloading the network, and avoid single points of failure. The same techniques also offer benefits from scalable parallelism, where incoming requests are handled cooperatively by multiple servers in a way that balances load to give better response times.

9.2 Web Security and Reliability

As we will see, the basic functionality of the Web can be understood in terms of a large collection of independently operated servers. A Web browser is little more than a graphical interface capable of issuing remote procedure calls to such a server, or using simple protocols to establish a connection to a server by which a file can be downloaded. The model is stateless: Each request is handled as a separate interaction, and if a request times out, a browser will simply display an error message. On the other hand, the simplicity of the underlying model is largely concealed from the user, who has the experience of a session and a strong sense of continuity and consistency when all goes well—for example, a user who fills in a graphical form seems to be in a dialog with the remote server, although the server, such as an NFS server, would not normally save any meaningful state for this dialog.

The reason that this should concern us becomes clear when we consider some of the uses to which Web servers are being put. Commerce over the Internet is being aggressively pursued by many companies. Such commerce will someday take many forms, including direct purchases and sales between companies, and direct sales of products and information to users. Today, the client of a Web server who purchases a product provides credit card billing information and trusts the security mechanisms of the browser and remote servers to protect these data from intruders. But, unlike a situation in which this information is provided by telephone, the Web is a shared packet-forwarding system in which a number of forms of intrusion are possible. For the user, interacting with a server over the Web may seem comparable to interacting to an agent over a telephone.

The introduction of encryption technologies will soon eliminate the most extreme deficiencies in this situation. Yet data security alone is just one element of a broader set of requirements. As the reader should recall from the first chapters of this book, RPC-based systems have the limitation that when a timeout occurs, it is often impossible for the user to determine if a request has been carried out, and if a server sends a critical reply just when the network malfunctions, the contents of that reply may be irretrievably lost. Moreover, there are no standard ways to guarantee that an RPC server will be available when it is needed, or even to be sure that an RPC server purporting to provide a desired service is in fact a valid representative of that service—for example, when working over the Web, how can a user be convinced that a remote server offering to sell jewelry at very competitive prices is not in fact fraudulent? Indeed, how can the user become convinced that the Web page for the bank down the street is in fact a legitimate Web page presented by a legitimate server, and not a fraudulent version that has been maliciously inserted onto the Web? At the time of this writing, the proposed Web security architectures embody at most partial responses to these concerns.

Full-service banking and investment support over the Web is likely to emerge in the near future. Moreover, many banks and brokerages are developing Web-based investment tools for internal use, in which remote servers price equities and bonds, provide access to financial strategy information, and maintain information about overall risk and capital exposure in various markets. Such tools also potentially expose these organizations to new forms of criminal activity: insider trading and fraud. Traditionally banks have kept their money in huge safes, buried deep underground. Here, one faces the prospect that billions of dollars will be protected primarily by the communication protocols and security architecture of the Web. We should ask ourselves if these are understood well enough to be trusted for such a purpose.

Web interfaces are extremely attractive for remote control of devices. How long will it be before such an interface is used to permit a plant supervisor to control a nuclear power plant from a remote location, or permit a physician to gain access to patient records or current monitoring status from home? Indeed, a hospital could potentially place all of its medical records onto Web servers, including everything from on-line telemetry and patient charts to x-rays, laboratory data, and even billing. But when this development occurs, how will we know whether or not hackers could also gain access to these databases, perhaps even manipulating the care plans for patients?

A trend toward critical dependence on information infrastructure and applications is already evident within many corporations. There is an increasing momentum behind the idea of developing corporate knowledge bases in which the documentation, strategic reasoning, and even records of key meetings would be archived for consultation and reuse. It is easy to imagine the use of a Web model for such purposes, and I am aware of several efforts directed to developing products based on this concept.

Taking the same idea one step further, the military sees the Web as a model for future information-based conflict management systems. Such systems would gather data from diverse sources, integrating these data and assisting all levels of the military command hierarchy in making coordinated, intelligent decisions that reflect the rapidly changing battlefield situation and that draw on continuously updated intelligence and analysis. The outcome of battles may someday depend on the reliability and integrity of information assets.

Libraries, as well as publishers of newspapers, journals, and books, are increasingly looking to the Web as a new paradigm for publishing the material they assemble. In this model, a subscriber to a journal or book would read it through some form of Web interface, either being charged on a per-access basis or provided with some form of subscription.

The list goes on. What is striking is the extent to which our society is rushing to make the transition, placing its most critical activities and valuable resources on the Web. A perception has been created that to be a viable company in the late 1990s, it will be necessary to make as much use of this new technology as possible. Obviously, such a trend presupposes that Web servers and interfaces are reliable enough to safely support the envisioned uses.

Many of the applications cited above have extremely demanding security and privacy requirements. Several involve situations in which human lives might be at risk if the envisioned Web application malfunctions by presenting the user with stale or incorrect data; in others, the risk is that great sums of money could be lost, a business might fail, or a battle lost. Fault tolerance and guaranteed availability are likely to matter as much as security: One wants these systems to protect data against unauthorized access, but also to guarantee rapid and correct access by authorized users.

Today, reliability of the Web is often taken as a synonym for *data security*. When this broader spectrum of potential uses is considered, however, it becomes clear that reliability, consistency, availability, and trustworthiness will be at least as important as data security if critical applications are to be safely entrusted to the Web or the Internet. Unfortunately, however, these considerations rarely receive attention when the decision to move an application to the Web is made. In effect, the enormous enthusiasm for the potential information revolution has triggered a great leap of faith that it has already arrived. And, unfortunately, it already seems to be too late to slow, much less reverse, this trend. Our only option is to understand how Web applications can be made sufficiently reliable to be used safely in the ways that society now seems certain to employ them.

Unfortunately, this situation seems very likely to deteriorate before any significant level of awareness that there is even an issue here will be achieved. As is traditionally the case in technology areas, reliability considerations are distinctly secondary to performance and user-oriented functionality in the development of Web services. If anything, the trend seems to be a form of latter-day gold rush, in which companies are stampeding to be first to introduce the critical servers and services on which Web commerce will depend. Digital cash servers, signature authorities, special-purpose Web search engines, and services that map from universal resource names to locations providing those services are a few examples of these new dependencies; they add to a list that already included such technologies as the routing and data transport layers of the Internet, the domain name service, and the Internet address resolution protocol. To a great degree, these new services are promoted to potential users on the basis of functionality, not robustness. Indeed, the trend at the time of this writing seems to be to stamp "highly available" or "fault tolerant" on more or less any system capable of rebooting itself after a crash. As we have already seen, recovering from a failure can involve much more than simply restarting the failed service.

The trends are being exacerbated by the need to provide availability for "hot Web sites," which can easily be swamped by huge volumes of requests from thousands or millions of potential users. To deal with such problems, Web servers are turning to a variety of ad hoc replication and caching schemes, in which the document corresponding to a particular Web request may be fetched from a location other than its ostensible home. The prospect thus created is of a world within which critical data are entrusted to Web servers, which replicate these data for improved availability and performance, but without necessarily providing strong guarantees that the information in question will actually be valid (or detectably stale) at the time it is accessed. Moreover, standards such as HTTP V1/0 remain extremely vague as to the conditions under which it is appropriate to cache documents and when they should be refreshed if they have become stale.

Broadly, the picture would seem to reflect two opposing trends. On the one hand, as critical applications are introduced into the Web, users may begin to depend on the correctness and accuracy of Web servers and resources, along with other elements of the Internet infrastructure, such as its routing layers, data transport performance, and so forth. To operate safely, these critical applications will often require a spectrum of behavioral *guarantees.* On the other hand, the modern Internet offers guarantees in none of these areas, and the introduction of new forms of Web services, many of which rapidly become indispensable components of the overall infrastructure, is only exacerbating the gap. Recalling our list of potential uses in commerce—banking, medicine, the military, and others—the potential for very serious failures becomes apparent. We are moving toward a world in which the electronic equivalents of the bridges that we traverse may collapse without warning, in which road signs may be out of date or intentionally wrong, and in which the agents with which we interact over the network may sometimes be clever frauds controlled by malicious intruders.

As a researcher, one can always adopt a positive attitude toward such a situation, identifying technical gaps as research opportunities or open questions for future study. Many of the techniques presented in this book could be applied to Web browsers and servers, and doing so would permit those servers to overcome some (not all!) of the limitations identified above. Yet it seems safe to assume that by the time this actually occurs, many critical applications will already be operational using technologies that are only superficially appropriate.

Short of some major societal pressure on the developers and customers for information technologies, it is very unlikely that the critical Web applications of the coming decade will achieve a level of reliability commensurate with the requirements of the applications. In particular, we seem to lack a level of societal consciousness of the need for a reliable technical base and a legal infrastructure that assigns responsibility for reliability to the developers and deployers of the technology. Lacking both the pressure to provide reliability and any meaningful concept of accountability, there is very little to motivate developers to focus seriously on reliability issues. Meanwhile, the prospect of earning huge fortunes overnight has created a near hysteria to introduce new Web-based solutions in every imaginable setting.

9.3 Related Reading

On the Web: (Berners-Lee et al. [1992, 1994, 1995], Gosling and McGilton [1995a, 1995b]).

There is also a large amount of on-line material concerning the Web—for example, in the archives maintained by Netscape Corporation (http://www.netscape.com).

CHAPTER 10 ✧ ✧ ✧ ✧

The Major Web Technologies

CONTENTS

10.1   Components of the Web

This chapter briefly reviews the component technologies of the World Wide Web (see Berners-Lee [1992, 1994]), but not associated technologies, such as e-mail and network bulletin boards, which are discussed in Chapter 11. The Web draws on the basic client/server and stream protocols discussed earlier—hence, the issue here is how those technologies can be applied to a distributed problem, not the development of a new or different technology base. In the case of the Web, there are three broad technology areas (see Figure 10.1). A *Web browser* is a program for interfacing to a *Web server*. There are various levels of browsers, but the most widely used are based on graphical windowing displays, which permit the display of textual material, including sophisticated formatting directives and graphical images, and implement access through hypertext links on behalf of the user. Web browsers also have a concept of an object type and will run the display program appropriate to a given type when asked to do so. This permits a user to download and replay a video image file, audio file, or other forms of sophisticated media. (Fancier display programs typically download access information only, then launch a viewer of their own, which pulls the necessary data and, for example, displays these data in real time.)

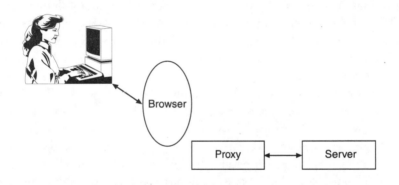

FIGURE 10.1. Components of a typical Web application. The user interacts with a graphical browser, which displays HTML documents and other graphical objects and issues HTTP commands to the servers on which objects needed by the user are stored. A proxy may be used to cache responses. Historically, HTTP applications have not fully specified what is or is not a cacheable response—hence, the use of this feature varies depending upon the origin of the proxy. Individual browsers may be capable of specialized display behaviors, such as rasterized display of graphical images or execution of pseudocode programs written in languages such as Java or Visual BASIC. Although not shown, there may be more than one level of proxy between the browser and server, and requests may "tunnel" through one or more firewalls before reaching the server. Moreover, messages passing over the Internet are relatively insecure and could be intercepted, read, and even modified on the path in either direction, if a Web security architecture is not employed.

Web servers and the associated concept of Web proxies (which are intermediaries that can act as servers by responding to queries using cached documents) represent the second major category of Web technologies. This is the level at which issues such as coherent replication and caching occur, and in which the Web authentication mechanisms are currently implemented.

The third major technology area underlying the Web consists of the *search engines,* which locate Web documents and index them in various ways, implementing query-style access on behalf of a Web user. These search engines have two sides to them: a user-interface side, in which they accept queries from a Web user and identify Web resources that match the specified request, and a document-finding side, which visits Web servers and follows hyperlinks to locate and index new documents. At present, few users think in terms of search engines as playing a critical role in the overall technology area. This could change, however, to the degree that individuals become dependent upon search engines to track down critical information and to report it promptly. One can easily imagine a future in which a financial analyst would become completely reliant upon such interfaces, as might a military mission planner, an air traffic controller, or a news analyst. If we believe that the Web will have the degree of impact that now seems plausible, such developments begin to seem very likely.

These technologies will soon be supplanted by others. Security and authentication services, provided by various vendors, are emerging to play a key role in establishing trustworthy links between Web users and companies from which they purchase services; these security features include data encryption, digital signatures with which the identity of a user can be validated, and tools for performing third-party validation of transactions whereby an intermediary trusted by two parties mediates a transaction between them. Digital cash and digital banks will surely emerge to play an important role in any future commercial digital market. Special-purpose telecommunication service providers will offer servers that can be used to purchase telecommunication connections with special properties for conferences; remote teleaccess to devices; and communication lines with guarantees of latency, throughput, error rate, and so forth. Web implementations of "auction" facilities will permit the emergence of commodities markets in which large purchases of commodities can be satisfied through a process of bidding and bid matching. Completely digital stock exchanges will follow soon after. Thus, while the early use of the Web is primarily focused on a paradigm of remote access and retrieval, the future of the Web will come closer and closer to creating a virtual environment, while also introducing new paradigms for social interaction and commerce. And these new electronic worlds will depend upon a wide variety of critical services to function correctly and reliably.

10.2 HyperText Markup Language

The HyperText Markup Language, or HTML, is a standard for representing textual documents and the associated formatting information needed to display them. HTML is quite sophisticated and includes such information as text formatting attributes (font, color, size, etc.), a means for creating lists and specifying indentation, and tools for implementing other standard formats. HTML also has conditional mechanisms, such as methods for displaying data in a concise form that can later be expanded upon request by the user. The HTML standard envisions various levels of compliance, and the most appropriate level for use in the Web has become a significant area of debate within the community.

HTML offers ways of *naming* locations in documents and specifying what are called *hypertext links* or *metalinks*. These links are textual representations of a document, a location in a document, or a service that the reader of a document can access. There are two forms of HTML links: those representing embedded documents, which are automatically retrieved and displayed when the parent document is displayed, and conditional links, which are typically shown in the form of a "button," which the user can select to retrieve the specified object. These buttons can be true buttons, regions within the document text (typically highlighted in color and underlined), or regions of a graphical image. This last approach is used to implement touch-sensitive maps and pictures.

10.3   Virtual Reality Markup Language

At the time of this writing, a number of proposals have emerged for Virtual Reality Markup Languages (VRML), which undertake to represent virtual reality worlds—three-dimensional or interactive data structures in which browsing has much of the look and feel of navigation in the real world. Although there will certainly be a vigorous debate in this area before standards emerge, it is easy to imagine an evolutionary path whereby interaction with the Web will become more and more like navigation in a library, a building, a city, or even the world.

It is entirely likely that by late in the 1990s, Web users who seek information about hotels in Paris, France, will simply fly there through a virtual reality interface, moving around animated scenes of Paris and even checking the rooms that they are reserving, all from a workstation or PC. An interactive agent, or "avatar," may welcome the visitor and provide information about the hotel, speaking much like the talking heads already seen on some futuristic television shows. Today, obtaining the same information involves dealing with flat two-dimensional Web servers presenting HTML documents to their users and with flat text-oriented retrieval systems; both are frequently cited as important impediments to wider use of the Web. Yet, a small number of innovative research groups and companies are already demonstrating VRML systems and language proposals.

Unfortunately, at the time of this writing, the degree of agreement on VRML languages and interfaces was still inadequate to justify any extended discussion in this book. Thus, although I am personally convinced that VRML systems may represent the next decisive event in the trend toward widespread adoption of the Web, there is little more that can be said about these systems except that they represent an extremely important development that merits close attention.

10.4   Uniform Resource Locators

When a document contains a hypertext link, that link takes the form of a *uniform resource locator,* or URL. A URL specifies the information needed by a Web server to track down a specified document. This typically consists of the protocol used to find that document (i.e., "FTP" or "HTTP," the hypertext transfer protocol), the name of the server on which the document resides (i.e., "www.cs.cornell.edu"), an optional Internet port number to use when contacting that server (otherwise the default port number is used), and a path name for the resource in question relative to the default for that server.

Cornell's Horus research project maintains a World Wide Web page with URL http://www.cs.cornell.edu/Info/Projects/Horus.html, meaning that the hypertext transfer protocol should be used over the Internet to locate the server www.cs.cornell.edu and to connect to it using the default port number. The document Info/Projects/Horus.html can be found there. The extension .html tells the Web browser that this document contains HTML information and should be displayed using the standard HTML display software. The "://" separator is a form of syntactic convention and has no special meaning. Variant forms of the URL are also supported— for example, if the protocol and machine name are omitted, the URL is taken to represent a path. Such a path can be a network path ("//" followed by a network location), an absolute path ("/" followed by a file name in the local file system), or a relative path (a file name that does not start with a "/" and that is interpreted relative to the directory from which the browser is running). In some cases a port number is specified after the host name; if it is omitted (as above), port number 80 is assumed.

Most Web users are familiar with the network path form of URL, because this is the form that is used to retrieve a document from a remote server. Within a document, however, the relative path notation tends to be used heavily, so that if a document and its subdocuments are all copied from one server to another, the subdocuments can still be found.

10.5   HyperText Transfer Protocol

The HyperText Transfer Protocol (HTTP) is one of the standard protocols used to retrieve documents from a Web server (see Berners-Lee [1995]). In current use, HTTP and FTP are by far the most commonly used file transfer protocols and are supported by all Web browsers. In the future, new transfer protocols implementing special features or exploiting special properties of the retrieved object may be introduced. HTTP was designed to provide lightness (in the sense of ease of implementation) and speed, which is clearly necessary in distributed, collaborative, hypermedia applications. However, as the scale of use of the Web has expanded, and the load upon it has grown, it has become clear that HTTP does not really provide either of these properties. This has resulted in a series of "hacks," which improve performance but also raise consistency issues, notably through the growing use of Web proxies that cache documents.

Web browsers typically provide extensible interfaces: New types of documents can be introduced, and new forms of display programs and transfer protocols are therefore needed to retrieve and display them. This requirement creates a need for flexibility at multiple levels: search, front-end update options, annotation, and selective retrieval. For this purpose, HTTP supports an extensible set of methods, which are typically accessed through different forms of URLs and different document types (extensions such as .txt, .html, etc.). The term, "URI" (Uniform Resource Indentifier) has become popular to express the idea that the URL may be a locator but may also be a type of name, indicating the form of abstract service that should be consulted to retrieve the desired document. As we will see shortly, this permits an HTTP server to construct documents upon demand, with content matched to the remote user's inquiry.

The hypertext transfer protocol itself is implemented using a very simple RPC-style interface, in which all messages are represented as user-readable ASCII strings, although often containing encoded or even encrypted information. Messages are represented in the same way that Internet mail passes data in messages. This includes text and also a form of encoded text called the Multipurpose Internet Mail Extensions, or MIME (the HTTP version is "MIME-like" in the sense that it extends a normal MIME scheme with additional forms of encoding). However, HTTP can also be used as a generic protocol for contacting other sorts of document repositories, including document caches (these are often called proxies), gateways that may impose some form of firewall between the user and the outside world, and other servers that handle such protocols as Gopher, FTP, NNTP, SMTP, and WAIS. When this feature is used, the HTTP client is expected to understand the form of data available from the protocol it employs and to implement the necessary mechanisms to convert the resulting data into a displayable form and to display it to the user.

In the normal case, when HTTP is used to communicate with a Web server, the protocol employs a client/server style of request-response, operating over a TCP connection that the client makes to the server and later breaks after its request has been satisfied. Each request takes the form of a request method or command, a URI, a protocol version identifier, and a MIME-like message containing special parameters to the request server. These may include information about the client, keys or other proofs of authorization, arguments that modify the way the request will be performed, and so forth. The server responds with a status line, which gives the message's protocol version and outcome code (success of one of a set of standard error codes), and then a MIME-like message containing the content associated with the reply. In normal use the client sends a single request over a single connection and receives a single response back from the server. More complicated situations can occur if a client interacts with an HTTP server over a connection that passes through proxies, which can cache replies, gateways, or other intermediaries; we return to these issues in Section 10.7.

HTTP messages can be compressed, typically using the UNIX compression tools "gzip" or "compress." Decompression is done in the browser upon receipt of a MIME-like message indicating that the body type has compressed content. The HTTP commands consist of:

▶ *Get:* The get command is used to retrieve a document from a Web server. Normally, the document URL is provided as an argument to the command, and the document itself is returned to the server in its response message. Thus, the command "GET // www.cs.cornell.edu/Info.html HTTP/1.0" could be used to request that the document "Info.html" be retrieved from "www.cs.cornell.edu," compressed and encoded into a MIME-like object, and returned to the requesting client. The origin of the resource is included but does not preclude caching: If a proxy sees this request it may be able to satisfy it out of a cache of documents that includes a copy of the Info.html previously retrieved from www.cs.cornell.edu. In such cases, the client will be completely unaware that the document came from the proxy and not from the server which keeps the original copy.

There are some special cases in which a get command behaves differently. First, there are cases in which a server should calculate a new HTML document for each request. These are handled by specifying a URL that identifies a program in a special area on the Web server called the cgi-bin area and encodes arguments to the program in the path-name suffix (the reader can easily observe this behavior by looking at the path name generated when a search request is issued to one of the major Web search engines, such as Lycos or Yahoo!). A Web server that is asked to retrieve one of these program objects will instead run the program, using the path-name suffix as an argument and creating a document as output in a temporary area that is then transmitted to the client. Many form-fill queries associated with Web pages use this approach, as opposed to the "post" command, which transmits arguments in a manner that requires slightly more sophisticated parsing and hence somewhat more effort on the part of the developer.

A second special case occurs if a document has moved; in this case, the get command can send back a redirection error code to the client that includes the URL of the new location. The browser can either reissue its request or display a short message indicating that this document has moved *here*. A conditional form of get, called *If-Modified-Since*, can be used to retrieve a resource only if it has changed some specified data. It is often used to refresh a cached object: If the object has not changed, minimal data are moved.

The get operation does not change the state of the server, and (in principle) the server will not need to retain any memory of the get operations that it has serviced. In practice, many servers cheat on the rules in order to prefetch documents likely to be needed in future get operations, and some servers keep detailed statistics about the access patterns of clients. We will return to this issue; it raises some serious concerns about both privacy and security of Web applications.

▶ *Head:* The head command is similar to get, but the server must not send any form of entity body in the response. The command is typically used to test a hypertext link for validity or to obtain accessibility and modification information about a document without actually retrieving the document. Thus, a browser that periodically polls a document for changes could use the head command to check the modification time of the document and only issue a get command if the document has changed.

▶ *Post:* The post command is used to request that the destination server accept the information included in the request as a new subordinate of the resource designated by the path.

This command is used for annotating existing resources (the client posts a note on the resource); posting of a conventional message to an e-mail destination, bulletin board, mailing list, or chat session; providing a block of data obtained through a form-fill; or extending a database or file through an append operation.

This set of commands can be extended by individual servers—for example, a growing number of servers support a subscription mechanism by which each update to a document will automatically be transmitted for as long as a connection to the server remains open. This feature is needed by services that dynamically send updates to displayed documents—for example, to provide stock market quotes to a display that shows the market feed in real time. However, unless such methods are standardized through the Internet Task Force they may only be supported by individual vendors. Moreover, special-purpose protocols may sometimes make more sense for such purposes: The display program that displays a medical record could receive updates to the EKG part of the displayed document, but it could also make a connection to a specified EKG data source and map the incoming data onto the part of the document that shows the EKG. The latter approach may make much more sense than one in which updates are received in HTTP format, particularly for data that are compressed in unusual ways or for which the desired quality of service of the communication channels involves unusual requirements or a special setup procedure.

Status codes play a potentially active role in HTTP. Thus, in addition to the standard codes ("created," "accepted," "document not found") there are codes that signify that a document has moved permanently or temporarily, providing the URL at which it can be found. Such a response is said to redirect the incoming request, but it can also be used in load-balancing schemes—for example, certain heavily used Web sites are implemented as clusters of computers. In these cases, an initial request will be directed to a load-balancing server, which redirects the request using a temporary URL to whichever of the servers in the cluster is presently least loaded. Because the redirection is temporary, a subsequent request will go back to the front-end server.

A curious feature of HTTP is that the client process is responsible both for opening and for closing a separate TCP connection for each command performed on the server. If retrieval of a document involves multiple get operations, multiple channels will be opened: one for each request. One might question this choice, since the TCP channel connection protocol represents a source of overhead that could be avoided if the browser were permitted to maintain connections for longer periods. Such an architecture is considered inappropriate, however, because of the potentially large number of clients that a server may be simultaneously handling. Thus, although it might seem that servers could maintain state associated with their attached channels, in practice this is not done. Even so, the solution can leave the server with a lot of resources tied up on behalf of channels. In particular, in settings where Internet latencies are high (or when clients fail), servers may be left with a large number of open TCP connections, waiting for the final close sequence to be executed by the corresponding clients. For a heavily loaded server, these open connections represent a significant form of overhead.

10.6 Representations of Image Data

Several standards are employed to compress image data for storage in Web servers. These include Graphics Interchange Format (GIF), an encoding for single images; Motion Picture Experts Group (MPEG) and Joint Photographic Experts Group (JPEG), which encode video data consisting of multiple frames; and a growing number of proprietary protocols. Text documents are normally represented using HTML, but PostScript is also supported by many browsers, as is the "rich text format" used by Microsoft's text processing products.

In the most common usage, GIF files are retrieved using a rasterized method in which a low-quality image can be rapidly displayed and then gradually improved as additional information is retrieved. The idea is to start by fetching just part of the data (perhaps every fourth raster of the image) and to interpolate between the rasters using a standard image interpolation scheme. Having finished this task, half of the remaining rasters will be fetched and the interpolation recomputed using these additional data; now, every other raster of the image will be based on valid data. Finally, the last rasters are fetched and the interpolation becomes unnecessary. The user is given the impression of a photographic image that gradually swims into focus. Depending on the browser used, this scheme may sweep from the top of the image to the bottom as a form of "wipe," or some sort of randomized scheme may be used. Most browsers permit the user to interrupt an image transfer before it finishes, so that a user who accidentally starts a very slow retrieval can work with the retrieved document even before it is fully available.

This type of retrieval is initiated using options to the get command and may require compatibility between the browser and the server. A less-sophisticated browser or server may not support rasterized retrieval, in which case the rasterization option to get will be ignored and the image displayed top to bottom in the standard manner. The most sophisticated browsers now on the market maintain a type of device driver, which is used to customize their style of retrieval to the type of Web server and code version number from which a document is retrieved.

In contrast to the approach used for GIF files, MPEG and JPEG files and documents represented in formats other than HTML are normally transferred to a temporary space on the user's file system for display by an appropriate viewer. In these cases, the file object will typically be entirely transferred before the viewer can be launched, potentially resulting in a long delay before the user is able to see the video data played back or the contents of the text document.

The Web is designed to be extensible. Each type of object is recognized by its file extension, and each Web server is configured with *viewer* programs for each of these types. It is expected that new file types will be introduced over time, and new types of viewers will be developed to display the corresponding data. However, although such viewers can often be downloaded over the network, users should be extremely cautious before doing so. A Web document viewer is simply a program that the user downloads and runs, and there is nothing to prevent that program from taking actions that have nothing at all to do with the ostensible display task. The program could be a form of virus or worm, or it could be designed to damage the user's computer system or to retrieve data from it and send it to third parties. For this reason, the major vendors of Web browsers are starting to offer libraries of certified viewers for the more important types of Web data. Their browsers will automatically download these types of viewers, which are in some ways similar to dynamically loaded executables in a standard operating system. When the user attempts to configure a new and nonstandard viewer, on the other hand, the browser may warn against this or even refuse to do so.

An important class of viewers is those that use their own data retrieval protocols to fetch complex image data. These viewers are typically launched using very small, compact image descriptions, which can be understood as domain-specific URLs. Once started, the viewer uses standard windowing primitives to discover the location of its display window on the screen, and it then begins to retrieve and display data into this location in real time. The advantage of such an approach is that it avoids the need to download the full image object before it can be displayed. Since an image object may be extremely large, there are enormous advantages to such an approach, and it is likely that this type of specialized image display will become more and more common in the future.

10.7   Authorization and Privacy Issues

Certain types of resources require that the Web browser authenticate its requests by including a special field, WWW authorization field, with the request. This field provides *credentials* containing the authentication information that will be used to decide if permission for the request should be granted. Credentials are said to be valid within a *realm*.

The basic HTTP authentication scheme is based on a model in which the user must present a user-ID and password to obtain credentials for access to a realm (see Berners-Lee [1995]). The user-ID and password are transmitted in a slightly obscured but insecure mode: They are translated to a representation called base64, encoded as an ASCII string of digits, and sent over the connection to the server. This approach is only secure to the degree that the communication channel to the server is secure; if an intruder were to capture such an authorization request in transit over the network (e.g., by installing a "packet sniffer" at a gateway), the same information could later be presented to the same realm and server to authenticate access by the intruder. Nonetheless, the basic authentication scheme is required from all servers, including those that can operate with stronger protection. Browsers that communicate with a server for which stronger security is available will often warn the user before sending a message that performs basic authentication.

When transferring genuinely sensitive information, Web applications typically make use of a trusted intermediary that provides session keys, using what is called public key encryption to authenticate channels and then a secret key encryption scheme to protect the data subsequently sent on that channel (the so-called *secure sockets layer* is described more fully in Internet Engineering Task Force and Denning and Branstad). At the core of this approach is a technology for publishing keys that can be used to encrypt data so that these data can be read only by a process that holds the corresponding private key. The basic idea is that the public keys for services to be used by a client can be distributed to that client in some way that is hard to disrupt or tamper with, and the client can then create messages that are illegible to any process other than the desired server. A client that has created a key pair for itself can similarly publish its public key, in which case it will be able to receive messages that only it can read. Because public key cryptography is costly, the recommended approach involves using a public key handshake to generate a secret key with which the data subsequently exchanged on the channel can be encrypted; in this manner, a faster protocol such as DES or RC4 can be employed for any large objects that need to be transferred securely. We discuss security architectures for distributed systems more fully in Chapter 19.

There are ways to attack this sort of security architecture, but they are potentially difficult to mount. If an intruder can break or steal the private keys used by the client or server, it may be possible for the intruder to misrepresent itself as one or the other and initiate secured transactions at leisure. Another option is to attack the stored public key information, in order to replace a public key with a falsified one that would permit a faked version of a server to mimic the real thing. Realistically, however, these would be very difficult types of attacks to engineer without some form of insider access to the systems on which the client and server execute or without have a fast way of breaking the cryptographic system used to implement the session keys. In practice, it is generally believed that although the basic authentication scheme is extremely fragile, the stronger Web security architecture should be adequate for most commercial transactions between individuals, provided, however, that the computer on which the client runs can be trusted. Whether the same schemes are adequate to secure transactions between banks, or military systems that transmit orders to the battlefield, remains an open question.

Web technologies raise a number of privacy issues that go beyond the concerns one may have about connection security. Many HTTP requests either include sensitive information, such as authentication credentials, or include fields that reveal the identity of the sender, URIs of documents being used by the sender, or software version numbers associated with the browser or server. These forms of information all can be misused. Moreover, many users employ the same password for all their authenticated actions; therefore, a single corrupt server that relies on the basic authentication scheme might reveal a password that can be used to attack secure servers that use the basic scheme.

Web servers are often considered to be digital analogs of libraries. In the United States, it is illegal for a library to maintain records of the documents that a client has examined in the past: Only current locations of documents may be maintained in the records. Web servers that keep logs of accesses may thus be doing something that would be illegal if the server were indeed the legal equivalent of a library. Nonetheless, it is widely reported that such logging of requests is commonly done, often to obtain information on typical request patterns. The concern, of course, is that information about the private reading habits of individuals is deemed to be personal and protected in the United States, and logs that are gathered for a purpose such as maintaining statistics on frequency of access to parts of a document base might be abused for some less-acceptable purpose.

Access patterns are not the only issue here. Knowledge of a URI for a document within which a pointer to some other document was stored may be used to gain access to the higher-level document by following the link backwards. This higher-level document may, however, be private and sensitive to the user who created it. With information about the version numbers of software on the browser or server, an intruder may be able to attack one or both using known security holes. A proxy could be subverted and modified to return incorrect information in response to get commands; to modify data sent in put commands; or to replay requests (even encrypted ones), which will then be performed more than once to the degree that the server was genuinely stateless. These are just a few of the most obvious concerns that one could raise about HTTP authentication and privacy.

These considerations point to the sense in which we tend to casually trust Web interfaces in ways that may be highly inappropriate. In a literal sense, use of the Web is a highly public activity today: Much of the information passed is basically insecure, and even the protection of passwords may be very limited. Although security is improving, the stronger security mechanisms are not yet standard. Even if one trusts the security protocol implemented by the Web, one must also trust many elements of the environment—for example, one may need to trust that the copy of a secure Web browser one has downloaded over the network wasn't modified in the network on the way to the user's machine or modified on the server itself from which it was retrieved. How can the user be sure that the browser he or she is using has not been changed in a way that will prevent it from following the normal security protocol?

One thinks of the network as anonymous, but user-ID information is present in nearly every message sent over it. Patterns of access can be tracked and intruders may be able to misrepresent a compromised server as one that is trusted by using techniques that are likely to be undetectable to the user. Yet the familiarity and comfort associated with the high quality of graphics and easily used interfaces to Web browsers and key services lull the user into a sense of trust. Because the system "feels" private, much like a telephone call to a mail-order sales department, one feels safe in revealing credit card information or other relatively private data. With the basic authentication scheme of the Web, doing so is little different from jotting it down on the wall of a telephone booth. The secure authentication scheme is considerably better, but it is not yet widely standard.

Within the Web community, the general view of these issues is that they represent fairly minor problems. The Web security architecture (the cryptographic one) is considered reasonably strong, and although the various dependencies cited above are widely recognized, it is also felt that they do not correspond to gaping exposures or "show stoppers" that could prevent digital commerce on the Web from taking off. The laws that protect private information are reasonably strong in the United States, and it is assumed that these offer recourse to users who discover that information about themselves is being gathered or used inappropriately. Fraud and theft by insiders is generally believed to be a more serious problem, and the legal system again offers the best recourse to such problems. For these reasons, most members of the Web community would probably feel more concerned about overload, denial of services due to failure, and consistency than about security.

I believe that the bottom line is not yet clear. It would be nice to believe that security is a problem of the past, but a bit more experience with the current Web security architecture will be needed before one can feel confident that it has no unexpected problems that clever intruders might be able to exploit. In particular, it is troubling to realize that the current security architecture of the Web depends upon the integrity of software that will increasingly be running on unprotected PC platforms and that may have been downloaded from unsecured sites on the Web. While Java and other interpreted languages could reduce this threat, it seems unlikely to go away soon. In the current environment, it would be surprising *not* to see the emergence of computer viruses that specialize in capturing private keys and revealing them to external intruders without otherwise damaging the host system. This sort of consideration (and we will see a related problem when we talk about non-PC systems that depend upon standard file systems such as NFS) can only engender some degree of skepticism about the near-term prospects for real security in the Web.

10.8   Web Proxy Servers

In Figure 10.1 a *server proxy* was shown between the browser and document server. Such proxies are a common feature of the World Wide Web, and they are widely perceived as critical to the eventual scalability of the technology. A proxy is any intermediary process through which HTTP operations pass on their way to the server specified in the document URL. Proxies are permitted to cache documents or responses to certain categories of requests, and in future systems they may even use cached information to dynamically construct responses on behalf of local users.

This leads to a conceptual structure in which each server can be viewed as surrounded by a ring of proxies that happen to be caching copies of documents associated with it (Figure 10.2). However, because the Web is designed as a stateless architecture, this structure is not typically represented: One could deduce a possible structure from the log of requests to the server, but information is not explicitly maintained in regard to the locations of copies of documents. Thus, a Web server would not typically have a means by which it could inform proxies that have cached documents when the primary copy changes. Instead, the proxies periodically refresh the documents they manage by using the head command to poll the server for changes or the conditional get command to simply pull an updated copy if one is available.

In Chapters 13 through 16 of this book, we will be looking at techniques for explicitly managing groups of processes that need to coherently replicate data, such as Web documents. These techniques could be used to implement coherent replication within a set of Web proxies, provided that one is prepared to relax the stateless system architecture normally used between the proxies and the primary server. It is likely that Web documents of the future will be more and more dynamic in many settings, making such coherency a problem of growing importance to the community selling Web-based information that must be accurate to have its maximum value.

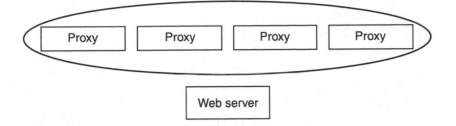

FIGURE 10.2. Conceptually, the proxies that cache a document form a distributed process group, although this group would not typically be explicitly represented a consequence of the stateless philosophy used in the overall Web architecture.

In the most common use of Web proxies today, however, their impact is to increase availability at the cost of visible inconsistency when documents are updated frequently. Such proxies reduce load on the Web server and are often able to respond to requests under conditions when a Web server might be inaccessible, crashed, or overloaded. However, unless a Web proxy validates every document before returning a cached copy of it, which is not a standard behavior, a proxy may provide stale data to its users for a potentially unbounded period of time, decreasing the perceived reliability of the architecture. Moreover, even if a proxy does refresh a cached record periodically, the Web potentially permits the use of multiple layers of proxy between the user and the server that maintains the original document. Thus, knowing that the local proxy has tried to refresh a document is not necessarily a strong guarantee of consistency. Head operations cannot be cached; if this command is used to test for freshness, there is a reasonable guarantee that staleness can be detected. But all types of get commands can be cached, so even if a document is known to be stale, there may be no practical way to force an uncooperative proxy to pass a request through to the primary server.

10.9 Java, HotJava, and Agent-Based Browsers

One way to think of an HTML document is as a form of program that the browser executes interpretively. Such a perspective makes it natural to take the next step and to consider sending a genuine program to the browser, which it could execute locally to the user. Doing so has significant performance and flexibility benefits and has emerged as a major area of research. One way to obtain this behavior is to introduce new application-specific document types. When a user accesses such a document, his or her browser will download the associated data file and then run a type-specific display program to display its contents. If the type of the file is a new one not previously known to the browser, it will also download the necessary display program, which is called an "agent." But this is clearly a risky proposition: The agent may well display the down-loaded data, but nothing prevents it from also infecting the host machine with viruses, scanning local files for sensitive data, or damaging information on the host.

Such considerations have resulted in research on new forms of *agent programming languages* (see Reiter [November 1994]) that are safe and yet offer the performance and flexibility benefits of downloaded display code. Best known among the programming languages available for use in programming such display agents is Sun Microsystem's HotJava browser, which downloads and runs programs written in an object-oriented language called Java (see Gosling and McGilton [1995a, 1995b]). (See Figure 10.3.). Other options also exist. The TCL/TK ("Tickle-Toolkit") language has become tremendously popular, and it can be used to rapidly prototype very sophisticated display applications (see Ousterhout [1994]). Many industry analysts predict that Visual BASIC, an extremely popular programming language for designing interactive PC applications, will rapidly emerge as a major alternative to Java. Interestingly, all of these are *interpreted* languages. The security problems associated with importing untrustworthy code are increasingly causing companies that see the Web as their future to turn to interpretation as a source of protection against hostile intrusion into a machine on which a browser is running.

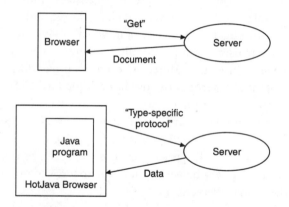

FIGURE 10.3. In a conventional Web interface, the user's requests result in retrieval of full documents (top). The browser understands HTML and can display it directly; for other types of documents it will copy the incoming object to a temporary location and then execute the appropriate display program. If an object type is unknown, the user may be forced to import potentially untrustworthy programs over the network. When using an agent language such as Java (bottom), the browser becomes an interpreter for programs that execute directly on the user's workstation and that can implement type-specific protocols for retrieving data over the network and displaying it. Not only does the security of the architecture improve (the HotJava browser is designed with many protection features), but the ability to execute a program on the user's workstation means that data transmission from server to client can be optimized in a type-specific way, interact with nonstandard servers (that may use protocols other than HTTP), and dynamically retrieve data over a period of time (unlike the approach used for HTML, which involves retrieving a document just once and then breaking the connection).

The Java language is designed to resemble C++, but it has built-in functions for interaction with a user through a graphical interface. These are called "applets" and consist of little graphical application objects that perform such operations as drawing a button or a box, providing a pull-down menu, and so forth. The language is touted as being robust and secure, although security is used here in the sense of protection against viruses and other forms of misbehavior by imported applications; the Java environment provides nothing new for securing the path from the browser to the server or authenticating a user to the server.

Interestingly, Java has no functions or procedures and no concept of data structures. The entire model is based on a very simple, pure object interface approach: Programmers work *only* with object classes and their methods. The argument advanced by the developers of Java is that this "functional" model is adequate and simple, and, by offering only one way to express a given task, the risk of programmer errors is reduced and the standardization of the resulting applications increased. Other missing features of Java include multiple inheritance (a problematic aspect of C++), operator overloading (in Java, an operator means just what it seems to mean—in C++, an operator can mean almost anything at all), automatic coercions (again, a costly C++ feature that is made explicit and hence controlled in Java), pointers, and goto statements. In summary, Java looks somewhat similar to C or C++, but it is in fact an extremely simplified subset, actually containing only the absolute minimum mechanisms needed to program sophisticated display applications without getting into trouble or somehow contaminating the client workstation on which the downloaded applet will execute.

Java is a multithreaded language, offering many of the same benefits as are seen in RPC servers that use threads. At the same time, however, Java is designed with features that protect against concurrency bugs. It supports dynamic memory allocation, but uses a memory management model that has no pointers or pointer arithmetic, eliminating one of the major sources of bugs for typical C and C++ programs. A background garbage collection facility quietly cleans up unreferenced memory, making memory leaks less likely than in C or C++ where memory can be lost while a program executes. The language provides extensive compile-time checking, and it uses a second round of run-time checking to prevent Java applications from attempting to introduce viruses onto a host platform or otherwise misbehaving in ways that could crash the machine. The latter can even protect against programs written to look like legitimate Java object codes but that were compiled using hostile compilers.

Although security of a Java application means "safe against misbehavior by imported agents," Java was designed with the secure sockets layer of the Web in mind. The language doesn't add anything new here, but it does include support for the available network security options such as firewalls and the security features of HTTP. Individual developers can extend these or use them to develop applications that are safe against intruders who might try to attack a server or steal data by snooping over a network. Java can thus claim to have closed the two major security holes in the Web: that of needing to import untrusted software onto a platform and that of the connection to the remote server.

The actual role of a Java program is to build a display for the user, perhaps using data solicited from a Java server, and to interact with the user through potentially sophisticated control logic.

Such an approach can drastically reduce the amount of data that a server must send to its clients—for example, a medical data server might need to send graphs, charts, images, and several other types of objects to the user interface. Using an HTML approach, the server would construct the necessary document upon request and send it to the display agent, which would then interactively solicit subdocuments needed to form the display. The graphical data would be shipped in the form of GIF, MPEG, or JPEG images, and the entire document might require the transmission of megabytes of information from server to display agent.

By writing a Java program for this purpose, the same interaction could be dramatically optimized. Unlike conventional browsers, the HotJava browser is designed without any built-in knowledge of the protocols used to retrieve data over the Internet. Thus, where a standard browser is essentially an expert in displaying HTML documents retrieved using HTTP, HotJava understands both HTML and HTTP through classes of display and retrieval objects, which implement code needed to deal with retrieving such documents from remote servers and displaying them to the user. Like any browser, HotJava includes built-in object classes for such standard Internet protocols and objects as HTTP, HTML, SMTP (the mail transfer protocol), URLs (making sense of Web addresses), GIF, NNTP (the news transfer protocol), FTP, and Gopher. However, the user can add new document types and new retrieval protocols to this list—in fact the user can even add new kinds of document addresses, if desired. At run time, the HotJava browser will ask the appropriate class of object to resolve the address, fetch the object, and display it.

Sometimes, the browser may encounter an object type that it doesn't know how to display—for example, a downloaded Java applet may contain references to other Java objects with which the browser is unfamiliar. In this case, the browser will automatically request that the server transfer the Java display code needed to display the unknown object class. The benefit of this approach is that the server can potentially maintain a tremendously large database of object types—within the limit, each object on the server can be a type of its own, or the server could actually construct a new object type for each request. Abstractly it would seem that the browser would need unlimited storage capacity to maintain the methods needed to display such a huge variety of objects, but, in practice, by downloading methods as they are needed, the actual use of memory in the browser is very limited. Moreover, this model potentially permits the server to revise or upgrade the display code for an object, perhaps to fix a bug or add a new feature. The next time that the browser downloads the method, the new functionality will immediately be available.

The developers of Java talk about the language as supporting dynamic content because new data types and the code needed to display them can be safely imported from a server, at run time, without concern about security violations. One could even imagine a type of server that would construct a new Java display program in response to each incoming request, compile it on the fly, and in this way provide new *computed* object classes dynamically. Such an approach offers intriguing new choices in the endless quest for generality without excess code or loss of performance.

Indeed, Java programs can potentially use nonstandard protocols to communicate with the server from which they retrieve data. Although this feature is somewhat limited by the need for the HotJava browser to maintain a trusted and secure environment, it still means that Java applications can break away from the very restricted HTTP protocol, implementing flexible protocols for talking to the server and providing functionality that would be hard to support directly over HTTP.

Returning to our medical example, these features make Java suitable for supporting specialized display programs that might be designed to compute medical graphics directly from the raw data. The display objects could also implement special-purpose data compression or decompression algorithms matched to the particular properties of medical image data. Moreover, the language can potentially support a much richer style of user interface than would otherwise be practical: If it makes sense to do so, a display object could permit its users to request that data be rescaled, that a graph be rotated, that certain features be highlighted, and so forth—all in an application-specific manner and without soliciting additional data from the server. Whatever makes sense to the application developer can be coded by picking an appropriate document representation and designing an appropriate interactive display program in the form of an interpreted Java object or objects. This is in contrast to a more standard approach, in which the browser has a very limited set of capabilities and any other behavior that one might require must be implemented on the server.

Although Java, when initiated, was lauded for its security features, it wasn't long before security concerns about Java surfaced. Many of these may have been corrected by the time this book goes to press, but it may be useful to briefly touch upon some examples of these concerns simply to make the points that reliability doesn't come easily in distributed systems and that complex technologies (such as HotJava), which are most promising in the long term, can often reduce reliability in the early period soon after they are introduced. Security threats associated with downloaded agents and other downloaded software may well become a huge problem in the late 1990s, because on the one hand we see enormous enthusiasm for rapid adoption of these technologies in critical settings, but, on the other hand, the associated security problems have yet to be fully qualified and are far from having been convincingly resolved.

I am aware of at least two issues that occurred early in the Java life cycle. The first of these was associated with a feature by which compiled Java code could be downloaded from Java servers to the HotJava browser. Although Java is intended primarily for interpretive use, this compilation technique is important for performance, and, in the long term, it is likely that Java will be an increasingly compiled language. Nonetheless, the very early versions of the object code downloading mechanism apparently had a bug that clever hackers could exploit to download malicious software that might cause damage to client file systems. This problem was apparently fixed soon after it appeared and before any major use had been made of it by the community that develops viruses. Yet one can only wonder how many early versions of the HotJava browser are still in use and hence still exposed to this bug.

A second problem was reported in spring 1996, and it was nicknamed the "Black Widow applet." Java Black Widow applets are hostile programs created to infect the systems of users who surf the Web, using Java as their technology (contact http://www.cs.princeton.edu/sip/pub/secure96.html for more information). These programs are designed to interfere with their host computers, primarily by consuming RAM and CPU cycles, so as to lower the performance available to the user. Some of these applets also make use of the compiled code problems of earlier Java servers, by making use of this ability available to a third party on the Internet, and, without the PC owner's knowledge, transfer information from the user's computer by subverting the HTTP protocol. Even sophisticated firewalls can be penetrated because the attack is launched from within the Java applet, which operates behind the firewall. Many users are surprised to realize that there may be untrustworthy Web sites that could launch an attack on a browser, and indeed many may not even be aware that they are using a browser that supports the Java technology and hence is at risk. Apparently, many of these problems will soon be fixed in new versions of the browsers offered by major vendors, but, again, one can only wonder how many older and hence flawed browsers will remain in the field, and for how long.

One can easily imagine a future in which such problems would lead vendors to create private networks, within which only trusted Web sites would be available, and to limit the ability of their browser technologies to download applications from untrusted sites. Without any doubt, such a world would appeal to the large network operators, since the user's PC would effectively be controlled by the vendor if this were to occur: In effect, the user's system would be able to download information only from the network service provider's servers and those of its affiliates. Yet one must also wonder if the promise of the Web could really be achieved if it is ultimately controlled by some small number of large companies. For the Web to emerge as a thriving economic force and a major contributor to the future information-based economy, it may be that only a free-enterprise model similar to the current Internet will work. If this is so, we can only hope that the security and reliability concerns that threaten the Internet today will be overcome to a sufficient degree to enable wider and wider use of the technology by users who have no particular restrictions imposed upon their actions by their network provider.

As an example, there is a great deal of research into what is called "sandbox" technologies, which consist of profiles that describe the expected behavior of an agent application and that can be enforced by the browser downloading them. To the degree that the profile itself is obtained from a trustworthy source and cannot be compromised or modified while being downloaded (perhaps a risky assumption!), one could imagine browsers that protect themselves against untrusted code by restricting the actions that the downloaded code can perform. The major vendors would then begin to play the role of certification authorities, providing (or signing) profile information, which is perhaps a more limited and hence less intrusive activity for them than to completely control some form of virtual private network and to restrict their browsers to operate only within its confines.

10.10 GUI Builders and Other Distributed CASE Tools

Java is currently the best known of the Web agent languages, but in time it may actually not be the most widely successful. As this book was being written, companies known for their graphical database access tools were hard at work on converting these into Web agent languages. Thus, languages such as Visual BASIC (the most widely used GUI language on PC systems) and PowerBuilder (a GUI building environment supporting a number of programming languages) are likely to become available in Java-like forms, supporting the development of graphical display agents that can be sent by a server to the user's Web browser, with the same sorts of advantages offered by Java. Database products such as Oracle's PowerObjects may similarly migrate into network-enabled versions over a short period of time. By offering tight integration with database systems, these developments are likely to make close coupling of database servers with the Web much more common than it was during the first few years after the Web phenomenon began.

Moreover, outright execution of downloaded programs may become more common and less risky over time. Recall that Java was introduced primarily as a response to the risks of downloading and executing special-purpose display programs for novel object types. If this sort of operation were less of a risk, there would be substantial advantages to a noninterpretive execution model. In particular, Java is unlikely to perform as well as compiled code, although it has a big advantage in being portable to a wide variety of architectures. Yet on relatively standard architectures, such as PCs, this advantage may not be all that important, and the performance issue could be critical to the success of the agent language.

Earlier in this book we discussed *object code editing* of the sort investigated by Lucco and Graham. These technologies offer a way to contain the potential actions of a piece of untrusted software, permitting a platform to import a function or program and yet to limit its actions to a set of operations that are considered safe. Object code editing systems represent a viable alternative to the Java model: One could easily imagine using these systems to download compiled code, encapsulating this code to eliminate any risk, and then directly executing it on the client's workstation or PC. Object code editors are potentially language independent: The program downloaded could be written in C, C++, assembly language, or BASIC. Thus, they have the benefit of not requiring the user to learn and work with a new programming language and model, as is the case for Java. It seems likely that object code editors will emerge to play an increasingly important role in the world of agents in the future, particularly if significant use of VRML applications begins to create a demanding performance problem on the client side.

10.11 TACOMA and the Agent Push Model

The agent languages described above, such as Java, are characterized by supporting a pull model of computation—that is, the client browser pulls the agent software from the server and executes it locally. However, there are applications in which one would prefer the converse model: One in which the browser builds an agent, which is then sent to the server to execute remotely. In particular, this would seem to be the case for applications in which the user needs to browse a very large database but only wishes to see a small number of selections that satisfy some property. In a conventional pull model such a user ultimately depends on the flexibility of the search interface and capabilities of the Web server; to the degree that the server is limited, the user will need to retrieve entire documents and study them.

Consider a situation in which the user is expected to pay for these documents—for example, the server might be an image archive, and the user may be purchasing specific images for use in a publication. The owners of the image archive won't want to release images for casual browsing, because the data are a valuable resource that they own. Each retrieved image may carry a hefty price and require some form of contractual agreement. In this case, short of somehow offering low-quality images to the browser and selling the high-quality ones (a viable model only for certain classes of application), there seems to be a serious obstacle to a Web-based solution.

The TACOMA language, developed by researchers at the University of Tromsö and at Cornell University, works to overcome these problems by offering an agent push model, in which the agent goes to the data, does some work, and may even migrate from server to server before ultimately returning results to the end user (see Asplin and Johansen, Johansen et al. [May 1995, June 1995, 1996]). TACOMA was originally designed for use in the StormCast system (see Johansen), a weather and environmental monitoring application.

The basic TACOMA problem area is easily described. StormCast collects and archives huge amounts of weather and environmental data in the far north, storing this information at a number of archive servers. The goal is to be able to use this sort of data to construct special-purpose weather forecasts, such as might be used by local airports or fishing vessels.

Not surprisingly, the extreme weather conditions of the Arctic make general weather prediction difficult. To predict the weather in a specific place, such as the fjords near Tromsö, one needs to combine local information about land topography and prevailing winds with remote information. If a storm is sweeping in from the north, data may be needed from a weather server offshore to the north; if the current prediction suggests that weather to the south is more important, the predictive software may need to extract data from an archive to the south. Moreover, predictions may need to draw on satellite data and periodically computed weather modeling information generated sporadically by supercomputing centers associated with the Norwegian Meteorology Organization. Thus, in the most general case, a local weather prediction could combine information extracted from dozens of archives containing gigabytes of potentially relevant information. It is immediately clear that the push model supported by Java-like languages cannot address this requirement. One simply doesn't want to move the data to the browser in this case.

Using TACOMA, the user writes programs in any of a number of supported languages, such as C, C++, TCL, or Perl. These programs are considered to travel from server to server carrying briefcases in which data are stored. Arriving at a server, a TACOMA agent will meet with an execution agent, which unpacks the program and the data from the briefcase, compiles the program, and executes it. A variety of security mechanisms are employed to limit any risk and to avoid the unintended proliferation of agents within the system. TACOMA itself implements the basic encapsulation mechanisms needed to implement briefcases (which are basically small movable file systems containing data files organized within folders) and provides support for the basic meet primitive by which an agent moves from place to place. Additional servers provide facilities for longer-term data storage, preprogrammed operations such as data retrieval from local databases, and navigation aids for moving within a collection of multiple servers.

Thus, to implement the weather prediction application described above, the application programmer would develop a set of agent programs. When a weather prediction is requested, these agents would be dispatched to seek out the relevant data, perhaps even doing computation directly at the remote data archive and sending back only the information actually needed by the user. The results of the search would then be combined and integrated into a single display object on the user's workstation.

Similarly, to overcome the image retrieval problems we discussed, a TACOMA agent might be developed that understands the user's search criteria. The agent could then be sent to the image archive to search for the desired images, sending back a list of images and their apparent quality relative to the search criteria.

It can be seen that the push model of agents raises a number of problems that don't arise in a pull setting: management of the team of agents that are executing on behalf of a given user, termination of a computation when the user has seen adequate results, garbage collection of intermediate results, limiting the resources consumed by agents, and even navigating the network. Moreover, it is unlikely that end users of a system will want to do any sort of programming, so TACOMA needs to be seen as a sort of agent middleware, used by an application programmer but hidden from the real user. Nonetheless, it is also clear that there are important classes of applications that the Java-style of pull agent will not be able to address and that a push style of agent could potentially solve. It seems very likely that as the Web matures, both forms of agent language will be of increasing importance.

10.12 Web Search Engines and Web Crawlers

An important class of Web servers is the *search engines,* which permit the user to retrieve URLs and short information summaries about documents of potential interest. Such engines typically have two components. A *Web crawler* is a program that hunts for new information on the Web and revisits sites for which information has been cached, revalidating the cache and refreshing information that has changed. Such programs normally maintain lists of Web servers and URLs. By retrieving the associated documents, the Web crawler extracts keywords and content information for use in resolving queries and also obtains new URIs that can be exploited to find more documents.

A Web search engine is a program that performs queries in a database of document descriptions maintained by a Web crawler. Such search engines accept queries in various forms (written language, often English, is the most popular query language) and then use document selection algorithms that attempt to match the words used in the query against the contents of the documents. Considerable sophistication within the search engine is required to ensure that the results returned to the user will be sensible ones and to guarantee rapid response. A consequence is that information retrieval, already an important research topic in computer science today, has become a key to the success of the Web.

Future Web search programs are likely to offer customizable search criteria by which documents can be located and presented to the user based on ongoing interests. An investor might have an ongoing interest in documents that predict future earnings for companies represented in a stock portfolio, a physician in documents relating to his or her specialization, and a lover of fine wines in documents that review especially fine wines or stores offering those wines at particularly good prices. Increasingly, Web users will be offered these services by the companies that today offer access to the Internet and its e-mail, chat, and bulletin board services.

10.13 Browser Extensibility Features: Plug-in Technologies

An important problem with existing Web browsers (including those that support Java or other agent languages) is that certain types of new features require that the browser itself be extended. Not all of these features can be expressed in the somewhat constrained style that Java permits. This has led the major Web browser providers to support browser extensibility. To cite two well-known examples, Netscape calls their extension feature "plug-ins," while Microsoft uses a catch-all term, "Active/X," to refer to these and other extensibility features in their browser, the Microsoft Internet Explorer.

The basic idea of a plug-in is similar to that of a dynamically linked and loaded module in a conventional operating system, except that the module is fetched over the network. Usually, the need for a plug-in is detected when a user downloads a Web page that makes use of an extensibility feature—for example, the Web page may contain a Java applet that imports the interface of a browser extension. Such an extension tells the browser to bind the applet to the interface if it is already loaded, but it also includes a URL, which can be used to fetch a copy if the extension is not available on the host machine. If the browser lacks the necessary extension, it asks the user to fetch and install it, after which the Web document can be displayed. Once downloaded, the extension becomes an executable component of the Web browser, providing new "native" interfaces that the Java applet can exploit. The same idea is also used to support new document types.

We will be discussing the Horus system in Chapter 18. Horus is a technology supporting group communication, which can be used (among other options) as a browser extension. By downloading Horus as a Netscape plug-on or Internet Explorer Active/X module, those browsers are enabled to directly support groupware applications. Other examples of plug-in technologies include support for the VRML language mentioned in Section 10.3, support for Internet audio and video display, and technologies that make use of special hardware features on the user's PC.

There are arguments for and against Web browser extensibility. Ideally, if all extensions could be expressed in Java or some other agent language, one would favor doing so because of the security benefits. However, these languages all have built-in limitations, and there are some types of extensions that are intrinsically insecure because they require direct access to the hardware of the computer or to the communication network. Technologies such as Horus cannot currently be coded in the Java language, for example, because Java does not support arbitrary use of the UDP communication protocol (for security reasons).

The ability to extend the Web browser itself overcomes this type of limitation. Indeed, the Web browser can be understood as a sort of remote operating system, which loads and executes arbitrary applications upon request. Moreover, plug-ins offer a way for a technology provider to sell technology: The sale is often integrated with the plug-in procedure, so that the user purchases and is billed for the technology in a single step, integrated with the act of downloading it. On the downside, plug-ins offer no security at all: A hostile plug-in has essentially unrestricted access both to the user's computer and to the network.

Microsoft's Active/X employs a user-mediated procedure to respond to this security issue. Under the argument that plug-ins will originate with mature, responsible companies, Active/X uses a signature scheme to authenticate the validity of any module that the user might download. Knowing that the technology provider and Microsoft have both signed off on the plug-in, the user can trust it to the same degree that he or she trusts other applications running on the same machine. Netscape is planning to support a similar capability.

In the short term, it seems likely that as long as the Web continues to explode with new applications and new functionality, heavy use of plug-in technology will be unavoidable. Indeed, over time, plug-ins could become the most common way to purchase new software for use on PCs and workstations! However, in the longer term, the security issues created by plug-ins may force many organizations to limit their use. It seems likely that the corporate network of the future will restrict the use of plug-ins to technologies that can be found in some sort of a corporate repository. Doing so would limit the types of applications that can be imported into the corporate network and allow the corporation to check imported technologies for potential threats. The legal protections that are now used to prosecute individuals who break into computing systems may eventually be found applicable to developers of malicious plug-ins and applications. This would also represent a step toward increased safety when downloading software over a network. However, as long as the overall technology area remains volatile, plug-ins are likely to be widely used, and serious security concerns are simply inevitable.

10.14 Important Web Servers

Although the Web can support a great variety of servers, business or enterprise use of the Web is likely to revolve around a small subset that supports a highly standardized commercial environment. Current thinking is that these would consist of the following:

▶ *Basic Web servers:* These would maintain the documents and services around which the enterprise builds its information model and applications.

- *Web commerce or merchant servers:* These would play the role of an on-line bank or checkbook. Financial transactions that occur over the network (buying and selling of products and services) would occur through the electronic analog of issuing a quote, responding with a purchase order, delivery of the product or service, billing, and payment by check or money transfer. Data encryption technologies and secure electronic transfer for credit card purchases will be crucial to ensuring that such commerce servers can be trusted and protected against third-party attack or other forms of intrusion.

- *Web exchange servers:* These are servers that integrate functionality associated with such subsystems as electronic mail, fax, chat groups and news groups, and multiplexing communication lines. Many such servers will also incorporate firewall technologies. Group scheduling, such as has been popularized by Lotus Notes, may also become a standard feature of such servers.

- *Web-oriented database servers:* As noted previously, early use of the Web has revolved around databases of HTML documents, but this is not likely to be the case in the long term. Over time, database servers will be increasingly integrated into the Web model. Indeed, it seems very likely that just as database and transactional systems are predominant in other client/server applications, they will ultimately dominate in Web environments.

10.15 Future Challenges

Although the explosive popularity of the Web makes it clear that the existing functionality of the system is more than adequate to support useful applications, evolution of the broader technology base will also require further research and development. Some of the major areas for future study include:

- *Improved GUI builders and browsers:* Although the first generation of browsers has already revolutionized the Web, the second wave of GUI builder technologies promises to open distributed computing to a vastly larger community. This development could revolutionize computing—increasing the use of networking by orders of magnitude and tremendously amplifying the existing trend toward critical dependency upon distributed computing systems.

  An interesting issue concerns the likely reliability impact of the widespread use of GUI builders, which might be characterized as the Computer-Aided Software Engineering (CASE) tools of distributed computing. On the positive side, GUI technologies encourage a tremendous degree of regularity in terms of application structure: Every Java application resembles every other Java application in terms of the basic computing model and the basic communication model, although this model allows considerable customization. But on the negative side, the weak intrinsic reliability of many GUI execution models and the consistency issues cited earlier are likely to become that much more visible as hundreds of thousands of application developers suddenly become Internet enabled. If these sorts of problems turn out to have common, visible consequences, the societal impact over time could be considerable.

Thus, we see a tradeoff here that may only become more clear with increased experience. My feeling is that wider use of the Web will create growing pressure to do something about network reliability and security and that the latter topic is receiving much more serious attention than the former. This could make reliability, in the sense of availability, consistency, manageability, and timely and guaranteed responsiveness, emerge as one of the major issues of the coming decade. But it is also possible that 90 percent of the market will turn out to be uninterested in such issues, with the exception of the need to gain sufficient security to support electronic commerce, and that vendors will simply focus on that 90 percent while largely overlooking the remaining 10 percent. This could lead us to a world of secure banking tools embedded in inconsistent and unreliable application software.

The danger, of course, is that if we treat reliability issues casually, we may begin to see major events in which distributed system unreliability has horrific consequences, in the sense of causing accidents, endangering health and privacy, bringing down banks, or other similarly frightening outcomes. Should terrorism ever become a major problem on the network, one could imagine scenarios in which that 10 percent exposure could suddenly loom as an immense problem. As discussed in the Introduction, we already face considerable risk through the increasing dependence of our telecommunication, power, banking, and air traffic control systems on the network, and this is undoubtedly just the beginning of a long-term trend.

If there is a light at the end of this particular tunnel, it is that a number of functionality benefits turn out to occur as side effects of the most effective reliability technologies, discussed in Part III. It is possible that the need to build better groupware and conferencing systems, better electronic stock markets and trading systems, and better tools for mundane applications such as document handling in large organizations will drive developers to look more seriously at the same technologies that also promote reliability. If this occurs, the knowledge base and tool base for integrating reliability solutions into GUI environments and elaborate agent programming languages could expand substantially, making reliability both easier and more transparent to achieve and more widely accessible.

▶ *Uniform Resource Locators (URLs):* Uniform resource locators suffer from excessive specificity: They tell the browser precisely where a document can be found. At the same time, they often lack information that may be needed to determine which version of a document is needed. Future developments will soon result in the introduction of uniform resource locators capable of uniquely identifying a document regardless of where it may be cached, including additional information to permit a user to validate its authenticity or to distinguish between versions. Such a uniform resource locator would facilitate increased use of caching within Web servers and proxies other than the originating server where the original copy of the document resides. Important issues raised by this direction of research include the management of consistency in a world of replicated documents that may be extensively cached.

- *Security and commerce issues:* The basic security architecture of the Web is very limited and rather trusting of the network. As noted earlier, a number of standards have been proposed in the areas of Web security and digital cash. These proposals remain difficult to evaluate and compare with one another, and considerable work will be needed before widely acceptable standards are available. These steps are needed, however, if the Web is to become a serious setting for commerce and banking.

- *Availability of critical Web data:* Security is only one of the reliability issues raised by the Web. Another important concern is availability of critical resources, such as medical documents that may be needed in order to treat patients in a hospital, banking records needed in secure financial applications, and decision-support documents needed for split-second planning in settings such as battlefields. Current Web architectures tend to include single points of failure, such as the Web server responsible for the original copy of a document or the authentication servers used to establish secure Web connections. When these resources are inaccessible or down, critical uses of the Web may be impossible. Thus, technologies permitting resources to be replicated for fault tolerance and higher availability will be of growing importance as critical applications are shifted to the Web.

- *Consistency and the Web:* Mechanisms for caching Web documents and replicating critical resources raise questions about the degree to which a user can trust a document to be a legitimate and current version of the requested document. With existing Web architectures, the only way to validate a document is to connect to its home server and use the head command to confirm that it has not changed since it was created. Moreover, there is no guarantee that a set of linked documents about to be retrieved will not be simultaneously updated on the originating server. Such a situation could result in a juxtaposition of stale and current documents, yielding a confusing or inconsistent result. More broadly, we need to understand what it means to say that a document or set of documents is seen in mutually consistent states, and how this property can be guaranteed by the Web. Where documents are replicated or cached, the same requirement extends to the replicas. In Chapter 17 we will consider solutions to these problems, but their use in the Web remains tentative and many issues will require further research, experimentation, and standardization.

10.16 Related Reading

On the Web: (see Berners-Lee et al. [1992, 1994, 1995]).

For Java: (see Gosling and McGilton [1995a, 1995b]).

For TACOMA: (see Asplin and Johansen, Johansen et al. [May 1995, June 1995, 1996]).

There is a large amount of on-line material concerning the Web—for example, in the archives maintained by Netscape Corporation, http://www.netscape.com.

CHAPTER 11 ✧ ✧ ✧ ✧

Related Internet Technologies

CONTENTS

The Web is just the latest of a series of Internet technologies that have gained extremely wide acceptance. In this chapter we briefly review some of the other important members of this technology family, including both old technologies such as mail and file transfer and new ones such as high-speed message bus architectures and security firewalls.

11.1 File Transfer Tools

The earliest networking technologies were those supporting file transfer in distributed settings. These typically consisted of programs for sending and receiving files, commands for initiating transfers and managing file transfer queues during periods when transfers backed up, utilities for administering the storage areas within which files were placed while a transfer was pending, and policies for assigning appropriate ownership and access rights to transferred files.

The most common file transfer mechanism in modern computer systems is that associated with the File Transfer Protocol (FTP), which defines a set of standard message formats and request types for navigating in a file system, searching directories, and moving files. FTP includes a security mechanism based on password authentication; however, these passwords are transmitted in an insecure way over the network, exposing them to potential attack by intruders. Many modern systems employ nonreusable passwords for this reason.

Other well-known file transfer protocols include the UNIX-to-UNIX copy program (UUCP) and the file transfer protocol standardized by the OSI protocol suite. Neither protocol is widely used, however, and FTP is a de facto standard within the Internet.

11.2 Electronic Mail

Electronic mail was the first of the Internet applications to gain wide popularity, and it remains a dominant technology at the time of this writing. Mail systems have become steadily easier to use and more sophisticated over time, and e-mail users are supported by increasingly sophisticated mail-reading and composition tools.

Underlying the e-mail system is a small collection of very simple protocols, of which the Simple Mail Transfer Protocol, or SMTP, is the most widely used and the most standard. The architecture of a typical mailing system is as follows. The user composes a mail message, which is encoded into ASCII (perhaps using a MIME representation) and then stored in a queue of outgoing e-mail messages. Periodically, this queue is scanned by a mail daemon program, which uses SMTP to actually transmit the message to its destinations. For each destination, the mail daemon establishes a TCP connection, delivers a series of e-mail messages, and receives acknowledgments of successful reception after the received mail is stored on the remote incoming mail queue. To determine the location of the daemon that will receive a particular piece of mail, the DNS for the destination host is queried. (See Figure 11.1.)

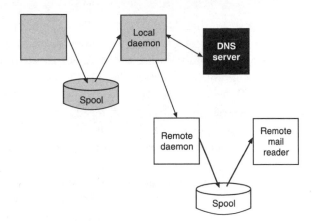

FIGURE 11.1. Steps in sending an e-mail. The user composes the e-mail and it is stored in a local spool for transmission by the local mail daemon. The daemon contacts the DNS to find the remote SMTP daemon's address for the destination machine, then transfers the mail file and, when an acknowledgment is received, deletes it locally. On the remote system, incoming mail is delivered to user mailboxes. The protocol is relatively reliable but can still lose mail if a machine crashes while transferring messages into or out of a spool or under other unusual conditions such as when there are problems forwarding an e-mail.

Users who depend upon network mail systems will be aware that the protocol is robust but not absolutely reliable. E-mail may be lost after it has been successfully acknowledged—for example, if the remote machine crashes just after receiving an e-mail. E-mail may be delivered incorrectly—for example, if the file system of a mail recipient is not accessible at the time the mail arrives and hence forwarding or routing instructions are not correctly applied. Further, there is the risk that e-mail will be inappropriately deleted if a crash occurs as the user starts to retrieve it from a mailbox. Thus, although the technology of e-mail is fairly robust, experienced users learn not to rely upon it in a critical way. To a limited degree, the return receipt mechanisms of modern mailers can overcome these difficulties, but heterogeneity prevents these from representing a completely satisfactory solution. Similarly, use of the more sophisticated e-mail mechanisms, such as e-mail with attached files, remains limited by incompatibilities between the mail reception processes that must interpret the incoming data.

11.3   Network Bulletin Boards (Newsgroups)

Network bulletin boards evolved in parallel with the e-mail system and hence share many of the same properties. Bulletin boards differ primarily in the way that messages are viewed and the way that they are distributed.

As most readers are aware, a bulletin board is typically presented to the user as a set of articles, which may be related to one another in what are called "conversations" or "threads." These are represented by special fields in the message headers that identify each message uniquely and permit one message to refer to another. Messages are typically stored in some form of directory structure, and the programs used to view them operate by displaying the contents of the directory and maintaining a simple database in which the messages each user has read are tracked.

The news distribution protocol, implemented by the NNTP daemon, is simple but highly effective. (See Figure 11.2.) It is based on a concept of flooding. Each news message is posted to a newsgroup. Associated with each newsgroup is a graph representing the connections between machines that wish to accept copies of postings to the group. To post a message, the user creates it and enqueues it in an outgoing news area, where the news daemon will eventually find it. The daemon for a given machine will periodically establish contact with a set of machines adjacent to it in the news distribution graph, exchanging messages that give the current set of available postings and their subjects. If a daemon connects to an adjacent daemon that has not yet received a copy of a particular posting, it forwards it over the link and vice versa.

This protocol is fairly reliable, but not absolutely so. Similar to the case of e-mail, an ill-timed crash can cause a machine to lose a copy of a recently received news posting after it has been confirmed, in which case there may be a gap in the news sequence for the corresponding group unless some other source happens to offer a copy of the same message on a different connection. Messages that are posted concurrently may be seen in different orders by different readers, and if a posting does not explicitly list *all* of the prior postings on which it is dependent, this can be visible to readers, because their display programs will not recognize that one message predates another. The display algorithms can also be fooled into displaying messages out of order by clock synchronization errors, which can erroneously indicate that one message is earlier than another.

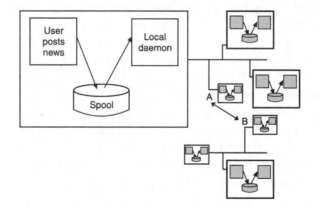

FIGURE 11.2. The network news protocol, NNTP, floods the network with gossip between machines that have news articles and machines that have yet to receive them. In this example, a message posted by a user reaches a forwarding node, *A*, which gossips with *B* by exchanging messages indicating the newsgroups for which each has recently received new postings. If *B* has not yet received the posting *A* just received, and is interested in the newsgroup, it will pull a copy from *A*. Failures of the network or of intermediate forwarding nodes can prevent articles from reaching their destinations quickly, in which case they will expire and may never reach some destinations. Thus, the protocol is quite reliable but not always reliable.

RELATED INTERNET TECHNOLOGIES   Chapter 11

The news protocol is known to suffer from a variety of security problems. It is trivial to forge a message by simply constructing what appears to be a legitimate news message and placing it in the reception area used by the news daemons. Such messages will be forwarded even if they misrepresent the name of the sender, the originating machine, or other information. Indeed, scripts for *spamming* newsgroups have become popular: These permit an individual to post a single message to a great number of newsgroups. To a very limited degree, the news distribution protocol has improved with time to resist such attacks, but for a user with even a small degree of sophistication, the technology is open to abuse.

Thus, while news systems are generally reliable, it would be inappropriate to use them in critical settings. In any use where the authenticity of messages is important, the context in which they were sent is significant, or the guarantee that postings will definitely reach their destinations is required, the technology is only able to provide partial solutions.

11.4 Message-Oriented Middleware Systems (MOMS)

Most major distributed system vendors offer products in what has become known as the message-oriented middleware, or MOMS, market. Typical of these products are Digital Equipment Corporation's MessageQ product line, IBM's MQSeries products, and the so-called "asynchronous message agent technology" available in some object-oriented computing systems—for example, CORBA Event Notification Services are likely to be positioned as MOMS products.

Broadly, these products fall into two categories. One very important area is concerned with providing network access to mainframe systems. IBM's MQSeries product is focused on this problem, as are perhaps a dozen comparable products from a variety of vendors, although MQSeries can also be useful in other settings. Technologies of this sort typically present the mainframe through a service interface abstraction, which permits the distributed system application developer to use a client/server architecture to develop applications. These architectures are frequently asynchronous in the sense that the sending of a request to the mainframe system is decoupled from the handling of its reply—much as if one were sending mail to the mainframe server, which will later send mail back containing the results of an inquiry. The message queuing system lives between the clients and the mainframe server, accepting the outgoing messages, transmitting them to the mainframe using the protocols appropriate for the mainframe operating system, arranging for the requests to be executed in a reliable manner (in some cases, even launching the associated application, if it is not already running), and then repeating the sequence in the opposite direction when the reply is sent back. Of course, these products are not confined to the mainframe connectivity problem, and many systems use them as front ends to conventional servers running on network nodes or workstations. However, the mainframe connectivity issue seems to be the driving force behind this market.

The second broad category of products uses a similar architecture but is intended more as a high-level message-passing abstraction for direct use in networked applications. In these products, of which DEC's MessageQ is perhaps typical, the abstraction presented to the user is of named mailboxes to which messages can be sent by applications on the network, much as user's send e-mail to one another. Unlike e-mail, the messages in question contain binary data, but the idea is very similar. Later, authorized applications dequeue the incoming messages for processing, sending back replies if desired or simply consuming them silently. As one might expect, these products contain extensive support for such options as priority levels (so that urgent messages can skip ahead of less-critical ones), flow control (so that message queues won't grow without limit if the consumer process or processes are slow), security, queue management, load-balancing (when several processes consume from the same queues), data persistence, and fault tolerance (for long-running applications).

If the model of this second category of message-queuing products is that of an e-mail system used at a program-to-program level, the performance is perhaps closer to that of a special-purpose file system. Indeed, many of these systems work very much as a file system would work: Adding a message to a queue is done by appending the message to a file representing the queue, and dequeueing a message is done by reading from the front of the file and freeing the corresponding disk space for reuse.

The growing popularity of message-oriented middleware products is typically due to their relative ease of use when compared to datagram-style message communication. Applications that communicate using RPC or datagrams need to have the producer and consumer processes running at the same time, and they must engage in a potentially complex binding protocol whereby the consumer or server process registers itself and the producer or client process locates the server and establishes a connection to it. Communication is, however, very rapid once this connection establishment phase has been completed. In contrast, a message-oriented middleware system does not require that the producer and consumer both be running at the same time or even that they be aware of one another: A producer may not be able to predict the process that will dequeue and execute its request, and a consumer process may be developed long before it is known what the various producers of messages it consumes will be. The downside of the model is that these products can be very slow in comparison to direct point-to-point communication over the network (perhaps by a factor of hundreds!) and that they can be hard to manage, because of the risk that a queue will leak messages and grow very large or that other subtle scheduling effects will cause the system to become overloaded and to thrash.

There is a good on-line source of additional information on middleware products, developed by the Message-Oriented Middleware Association, (MOMA): http://www.sbexpos.com/sbexpos/associations/moma/home.html. Information on specific products should be obtained from their vendors.

11.5 Message Bus Architectures

Starting with the V operating system in 1985 (see Cheriton and Zwaenepoel) and the Isis news application in 1987 (see Birman and Joseph [November 1987]), a number of distributed systems have offered a bulletin board style of communication directly to application programs; MIT's

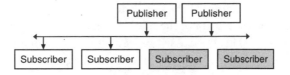

FIGURE 11.3. Message bus architectures (also known as publish/subscribe systems) originated as an application of process groups in the V system, and a fault-tolerant version was included in early versions of the Isis Toolkit. These technologies subsequently became commercially popular, particularly in financial and factory-floor settings. A benefit of the architecture is that the subscriber doesn't need to know who will publish on a given subject or vice versa. As shown, messages on the "white" subject will reach the "white" but not the "gray" subscribers—a set that can change dynamically and that may grow to include applications not planned at the time the publishers were developed. A single process can publish upon or subscribe to multiple subjects. This introduces desirable flexibility in a technology that can also achieve extremely high performance by using hardware multicast or broadcast, as well as fault tolerance or consistency guarantees if the technology base is implemented with reliability as a goal. An object-oriented encapsulation of message bus technology is provided as part of CORBA, through its Event Notification Service (ENS). In this example, the application is divided into publishers and subscribers, but in practice a single process can play both roles and can subscribe to many subjects. Wide area extensions normally mimic the gossip scheme used in support of network bulletin boards.

Zephyr system followed soon after (see DellaFera et al.). In this technology, which is variously called message bus communication, message queues, subject-based addressing, or process group addressing, processes register interest in message subjects by *subscribing* to them. The same or other processes can then send out messages by *publishing* them under one or more subjects. The message bus system is responsible for matching publisher to subscriber in a way that is efficient and transparent to both: The publisher is typically unaware of the current set of subscribers (unless it wishes to know) and the subscriber is typically not aware of the current set of publishers (again, unless it has some reason to ask for this information). (See Figure 11.3.)

Message bus architectures are in most respects very similar to network bulletin boards. An application can subscribe to many subjects, just as a user of a bulletin board system can monitor many bulletin board topics. Both systems typically support some form of hierarchical name space for subjects, and both typically allow one message to be sent to multiple subjects. The only significant difference is that message bus protocols tend to be optimized for high speed, using broadcast hardware if possible, and they typically deliver messages as soon as they reach their destination, through some form of *upcall* to the application process. In contrast, a bulletin board system usually requires a polling interface, in which the reader checks for new news and is not directly notified at the instant a message arrives.

The usual example of a setting in which a message bus system might be used is that of a financial trading application or stock exchange. In such systems, the subjects to which the messages are sent are typically the names of the financial instruments being traded: /equities/ibm or /bonds/at&t. Depending on the nature of the message bus, reliability guarantees may be nonexistent, weak, or very strong. The V system's process group technology illustrates the first approach: an application subscribed by joining a process group and published by sending to it, with the group name corresponding to the subject. V multicasts to process groups lacked any sort of strong reliability guarantees—hence, such an approach usually will deliver messages but not always. V transmitted these messages using hardware multicast features of the underlying transport technology, or point-to-point transport if hardware were not available.

The Teknekron Information Bus (TIB) (see Oki et al.) and Isis Message Distribution System (MDS) (see Glade) are good examples of modern technologies that support the V system with stronger reliability properties. TIB is extremely popular in support of trading floors and has had some success in factory automation environments. Isis is used in similar settings but where reliability is an especially important attribute: stock exchange systems, critical on-line control software in factories, and telecommunication applications.

Teknekron's TIB architecture is relatively simple, but it is still sufficient to provide high performance, scalability, and some degree of fault tolerance when publishers fail. In this system, messages are typically transmitted by the publisher using hardware broadcast with an overlaid retransmission mechanism, which ensures that messages will be delivered reliably and in the order they were published provided that the publisher doesn't fail. Point-to-point communication is used if a subject has only a small number of subscribers. Much as for a stream protocol, overload or transient communication problems can cause exceptional conditions in which messages could be lost. However, such conditions are uncommon in the settings where TIB is normally used.

The TIB system provides a fail-over capability if there are two or more equivalent publishers for a given class of subjects. In such a configuration, subscribers are initially connected to a primary publisher; the backup goes through the motions of publishing data, but TIB inhibits the transmission of any data. However, if the primary publisher fails, the TIB system will eventually detect this. Having done so, the system will automatically reconfigure so that the subscriber will start to receive messages from the other source. In this manner, TIB is able to guarantee that messages will normally be delivered in the order they are sent, and it will normally not have gaps or out-of-sequence delivery. These properties can, however, be violated if the network becomes severely overloaded, a failure occurs on the publisher site, or the subscriber becomes temporarily partitioned away from the network. In these cases a gap in the sequence of delivered messages can occur during the period of time required for the fail-over to the operational server.

The Isis Message Distribution System (MDS) is an example of a message bus that provides very strong reliability properties. This system is implemented using a technology based on reliable process groups (discussed in Chapters 13 through 18), in which agreement protocols are used to ensure that messages will be delivered to all processes that are subscribing to a subject, or to none. The approach also permits the active replication of publishers, so that a backup can provide precisely the same sequence of messages as the primary. By carefully coordinating the handling of failures, MDS is able to ensure that even if a failure does occur, the sequence of messages will not be disrupted: All subscribers that remain connected to the system will see the same messages, in the same order, even if a publisher fails, and this order will not omit any messages if the backup is publishing the same data as the primary.

MDS is implemented over hardware multicast but uses this feature only for groups with large fanout; point-to-point communication is employed for data transport when the number of subscribers to a subject is small or when the subscribers are not on the same branch of a local area network. The resulting architecture achieves performance comparable to that of TIB in normal cases, but under overload, when TIB can potentially lose messages, Isis MDS will typically choke back the publishers and, if the load becomes extreme, may actually drop slow receivers from the system as a way to catch up. These different design choices are both considered to represent reliable behavior by the vendors of the two products; clearly, the real issue for any given application will be the degree of match between the reliability model and the needs of the applications that consume the data. MDS would be preferable in a system where ordering and guaranteed delivery are very important to the end user; TIB might be favored in a setting where continued flow of information is ultimately valued more than the ordering and reliability properties of the system.

The reader may recall that the CORBA Event Notification Service (ENS) uses a message bus architecture. TIB is in fact notable for supporting an object-oriented interface similar to the one required for implementations of this service. The Isis MDS has been integrated into the Orbix + Isis product, and therefore can be used as an object-oriented CORBA ENS through Orbix (see Iona Ltd. and Isis Distributed Systems, Inc.).

Both systems also provide a form of message spooling and playback facility. In TIB, this takes the form of a subscriber that spools all messages on specified subjects to disk, replaying them later upon request. MDS also includes a spooling technology, which can store messages for future replay to a process. The MDS implementation of this playback technology preserves the ordering and reliability attributes of the basic publication mechanism and is carefully synchronized with the delivery of new messages so that a subscriber can obtain a seamless playback of spooled messages followed by the immediate delivery of new messages—in a correct order and without gaps or duplication.

Both the TIB and MDS architectures can be extended to large-scale environments using a protocol much like the one described in the section for network bulletin boards.

11.6 Internet Firewalls and Gateways

Internet firewalls have recently emerged to play a nearly ubiquitous role in Internet settings. We discuss firewalls in more detail in Chapter 19 and consequently limit ourselves to some very brief comments here.

A firewall is a message-filtering system. It resides at the perimeter of a distributed system, where messages enter and leave it from the broader Internet. The specific architecture used may be that of a true packet filter, or it may be one that permits application-level code to examine the incoming and outgoing packets (so-called application-level proxy technology). Although firewalls are not normally considered to be distributed programs, if a network has multiple access points, the firewall will be instantiated separately at each. Considered as a set, the collection of firewall programs will then be a distributed system, although the distributed aspect may be implicit.

Firewall programs typically operate by examining each message on the basis of source, destination, and authentication information (if enabled). Messages are permitted to pass through only if they satisfy some criteria controlled by the system administrator, or, in the case of an application-level proxy technology, if the application programs that compose the firewall consider the message to be acceptable. In this manner, a network can be made porous for network bulletin board and e-mail messages, but opaque to incoming FTP and remote log-in attempts. It can also be made to accept only packets digitally signed by an acceptable user or originating machine.

A *gateway* is a program placed outside of a protected domain that offers users a limited set of options for accessing information protected within the domain. A gateway may permit selected users to establish log-in sessions with the protected domain, for example, but only after challenging the user to provide a one-time password or some other form of authentication. Gateways often provide mechanisms for file transfer as well, again requiring authentication before transfers are permitted.

For typical Internet sites, the combination of firewalls and gateways provides the only real security. Although passwords are still employed within the firewall, such systems may be open to other forms of attack if an intruder manages to breach the firewall or to gain access permission from the gateway. Nonetheless, these represent relatively strong barriers to intrusion. Whereas systems that lack gateways and firewalls report frequent penetrations by hackers and other agents arriving over the network, firewalls and gateways considerably eliminate such attacks. They do not create a completely secure environment, but they do offer an inexpensive way to repel all but the most sophisticated attackers.

11.7 Related Reading

Most Internet technologies are documented through the so-called Request for Comments (RFC) reports, which are archived at various sites within the network, notably on a server maintained by SRI Corporation.

Part III

RELIABLE DISTRIBUTED COMPUTING

In this third and final part of the book, we ask how distributed computing systems can be made reliable—motivated by our review of servers used in Web settings, but seeking to generalize beyond these specific cases to include future servers that may be introduced by developers of new classes of critical distributed computing applications. Our focus is on communication technologies, but we do review persistent storage technologies based on the transactional computing model, particularly as it has been generalized to apply to objects in distributed environments.

CHAPTER 12 ✧ ✧ ✧ ✧
_____

How and Why Computer Systems Fail

CONTENTS

Throughout this part of the book, we will be concerned with technologies for making real distributed systems reliable (see Birman and van Renesse [1996]). Before undertaking this task, it will be useful to briefly understand the reasons that distributed systems fail. Although there are some dramatic studies of the consequences of failures (see, for example, Peterson [1995]), our discussion draws primarily on work by Jim Gray (see Gray [1990], Gray and Reuter, Gray et al.), who studied this issue while at Tandem Computers, and on presentations by Anita Borr (see Borr and Wilhelmy), who was a developer of Tandem's transactional system architecture (see Bartlett), and Ram Chilaragee, who has studied the same question at IBM (see Chilaragee). All three researchers focused on systems designed to be as robust as possible and might have drawn different conclusions had they looked at large distributed systems that incorporate technologies built with less-stringent reliability standards. Unfortunately, there seems to have been relatively little formal study of failure rates and causes in systems that were *not* engineered with reliability as a primary goal, despite the fact that a great number of systems used in critical settings include components with this property.

12.1 Hardware Reliability and Trends

Hardware failures were a dominant consideration in developing reliable systems until late in the 1980s. Hardware can fail in many ways, but as electronic packaging has improved and the density of integrated circuits increased, hardware reliability has grown enormously. This improved reliability reflects the decreased heat production and power consumption of smaller circuits, the reduction in the number of off-chip connections and wiring, and improved manufacturing techniques. A consequence is that hardware-related system downtime is fast becoming a minor component of the overall reliability concerns faced in a large, complex distributed system.

To the degree that hardware failures remain a significant reliability concern today, the observed problems are most often associated with the intrinsic limitations of connectors and mechanical devices. Thus, computer network problems (manifest through message loss or partitioning failures, where a component of the system becomes disconnected from some other component) are high on the list of hardware-related causes of failure for any modern system. Disk failures are also a leading cause of downtime in systems dependent upon large file or database servers, although RAID-style disk arrays can protect against such problems to a limited degree.

A common hardware-related source of downtime has very little to do with failures, although it can seriously impact system availability and perceived reliability. Any critical computing system will, over its life cycle, live through a series of hardware generations. These can force upgrades, because it may become costly and impractical to maintain old generations of hardware. Thus, routine maintenance and downtime for replacement of computing and storage components with more modern versions must be viewed as a planned activity, which can emerge as one of the more serious sources of system unavailability if not dealt with through a software architecture that can accommodate dynamic reconfiguration of critical parts of the system while the remainder of the system remains on-line. This issue of planning for future upgrading, expansion, and new versions of components extends throughout a complex system, encompassing all its hardware and software technologies.

12.2 Software Reliability and Trends

Earlier in this book, we observed that software reliability is best understood as a process, encompassing not just the freedom of a system from software bugs, but also such issues as the software design methodology, the testing and life-cycle quality-assurance process used, the quality of self-checking mechanisms and of user interfaces, the degree to which the system implements the intended application (i.e., the quality of match between system specification and problem specification), and the mechanisms provided for dealing with anticipated failures, maintenance, and upgrades. This represents a rich, multidimensional collection of issues, and few critical systems deal with them as effectively as one might wish. Software developers, in particular, often view software reliability in simplified terms, focusing exclusively on the software specification that their code must implement and on its correctness with regard to that specification.

Even this narrower issue of correctness remains an important challenge; indeed, many studies of system downtime in critical applications have demonstrated that even after rigorous testing, software bugs account for a substantial fraction of unplanned downtime (figures in the range of 25 to 35 percent are common), and this number is extremely hard to reduce (see, for example, Peterson [1995]). Jim Gray and Bruce Lindsey, who have studied reliability issues in transactional settings, once suggested that the residual software bugs in mature systems can be classified into two categories, which they called *Bohrbugs* and *Heisenbugs* (see Gray and Reuter, Gray et al.). (See Figure 12.1.)

FIGURE 12.1. Developers are likely to discover and fix Bohrbugs, which are easily localized and reproducible sources of errors. Heisenbugs are fuzzy and hard to pin down. Often, these bugs are actually symptoms of some other bug, which doesn't cause an immediate crash; the developer will tend to work around them but may find them extremely hard to fix in a convincing way. The frequency of such bugs diminishes very slowly over the life cycle of an application.

A Bohrbug is a solid, reproducible problem: If it occurs, and one takes note of the circumstances, the scenario can be reproduced and the bug will repeat itself. The name is intended to remind us of Bohr's model of the atomic nucleus: a small hard object, well localized in space. Gray and Lindsey found that as systems mature, the relative frequency of Bohrbugs drops steadily over time, although other studies (notably by Anita Borr) suggest that the population of Bohrbugs is periodically replenished when a system must be upgraded or maintained over its life cycle.

Heisenbugs are named for the Heisenberg model of the nucleus: a complex wave function that is influenced by the act of observation. These bugs are typically side-effects of problems that occurred much earlier in an execution, such as overrunning an array or accidentally dereferencing a pointer after the object to which it points has been freed. Such errors can corrupt the application in a way that will cause it to crash, but not until the corrupted data structure is finally referenced, which may not occur until long after the bug actually was exercised. Because such a bug is typically a symptom of the underlying problem, rather than an instance of the true problem itself, Heisenbugs are exquisitely sensitive to the order of execution. Even with identical inputs, a program that crashed once may run correctly back in the laboratory.

Not surprisingly, the major source of crashes in a mature software system turns out to be Heisenbugs. Anita Borr's work actually goes further, finding that most attempts to fix Heisenbugs actually make the situation worse than it was in the first place. This observation is not surprising to engineers of complex, large software systems: Heisenbugs correspond to problems that can be tremendously hard to track down, and they are often fixed by patching around them at run time. Nowhere is the gap between theory and practice in reliable computing more apparent than in the final testing and bug correction stages of a major software deployment that must occur under time pressure or a deadline.

12.3 Other Sources of Downtime

Jointly, hardware and software downtime, including downtime for upgrades, is typically said to account for some two-thirds of system downtime in critical applications. The remaining third of downtime is attributable to planned maintenance, such as making backups, and environmental factors, such as power outages, air conditioning or heating failures, leaking pipes, and other similar problems.

Although there may be little hope of controlling these forms of downtime, the trend is to try to treat them using software techniques that distribute critical functionality over sufficient numbers of computers, and separate them to a sufficient degree so that redundancy can overcome unplanned outages. Having developed software capable of solving such problems, downtime for hardware maintenance, backups, or other routine purposes can often be treated in the same way as other forms of outages. Such an approach tends to view system management, monitoring, and on-line control as a part of the system itself: A critical system should, in effect, be capable of modeling its own configuration and triggering appropriate actions if critical functionality is compromised for any reason. In the chapters that follow, this will motivate us to look at issues associated with having a system monitor its own membership (the set of processes that compose it) and, dynamically, adapt itself in a coordinated, consistent manner if changes are sensed. Although the need for brevity will prevent us from treating system management issues in the degree of detail that the problem deserves, we will develop the infrastructure on which reliable management technologies can be implemented, and we will briefly survey some recent work specifically on the management problem.

12.4 Complexity

Many developers would argue that the single most serious threat to distributed system reliability is the *complexity* of many large distributed systems. Indeed, distributed systems used in critical applications often interconnect huge numbers of components using subtle protocols, and the resulting architecture may be extremely complex. The good news, however, is that when such systems are designed for reliability, the techniques used to make them more reliable may also tend to counteract this complexity.

In the chapters that follow we will be looking at replication techniques that permit critical system data and services to be duplicated as a way to increase reliability. When this is done correctly, the replicas will be consistent with one another and the system as a whole can be thought of as containing just a single instance of the replicated object, but one that happens to be more reliable or more secure than any single object normally would be. If the object is active (a program), it can be *actively replicated* by duplicating the inputs to it and consolidating the outputs it produces. These techniques lead to a proliferation of components but also impose considerable regularity upon the set of components. They thus control the complexity associated with the robustness intervention.

We will also be looking at system management tools that monitor sets of related components, treating them as groups within which a common management, monitoring, or control policy can be applied. Again, by factoring out something that is true for all system components in a certain class or set of classes, these techniques reduce complexity. What were previously a set of apparently independent objects are now explicitly seen to be related objects that can be treated in similar ways, at least for purposes of management, monitoring, or control.

Broadly, then, we will see that although complexity is a serious threat to reliability, complexity can potentially be controlled by capturing and exploiting regularities in distributed system structure—regularities that are common when such systems are designed to be managed, fault tolerant, secure, or otherwise reliable. To the degree that this is done, the system structure becomes more explicit and hence complexity is reduced. In some ways, the effort of building the system will increase: This structure needs to be specified and needs to remain accurate as the system subsequently evolves. But in other ways, the effort is decreased: By managing a set of components in a uniform way, one avoids the need to do so on an ad hoc basis, which may be similar for the members of the set but not identical if the component management policies were developed independently.

These observations are a strong motivation for looking at technologies that can support grouping of components in various ways and for varied purposes. However, they also point to a secondary consideration: Unless such technologies are well integrated with system development software tools, they will prove to be irritating and hard to maintain as a system is extended over time. As we will see, researchers have been more involved with the former problem than the latter one, but this situation has now begun to change, particularly with the introduction of CORBA-based reliability solutions, which are well integrated with CORBA development tools.

12.5 Detecting Failures

Surprisingly little work has been done on the problem of building failure detection subsystems. A consequence is that many distributed systems detect failures using timeouts—an error-prone approach that forces the application to overcome inaccurate failure detections in software.

Recent work by Vogels (see Vogels) suggests that many distributed systems may be able to do quite a bit better. Vogels makes the analogy between detecting a failure and discovering that one's tenant has disappeared. If a landlord were trying to contact a tenant whose rent check is late, it would be a little extreme to contact the police after trying to telephone that tenant once, at an arbitrary time during the day, and not receiving any reply. More likely, the landlord would telephone several times, inquire of neighbors, check to see if the mail is still being collected and if electricity and water are being consumed, and otherwise check for indirect evidence of the presence or absence of the tenant.

Modern distributed systems offer a great number of facilities that are analogous to these physical options. The management information base of a typical computing node (its MIB) provides information on the active processes and their consumption of resources such as memory, computing time, and I/O operations. Often, the network itself is instrumented, and indeed it may sometimes be possible to detect a network partition in an accurate way by querying MIBs associated with network interface and routing nodes. If the operating system on which the application in question is running is accessible, one can sometimes ask it about the status of the processes it is supporting. And, in applications designed with fault tolerance in mind, there may be the option of integrating self-checking mechanisms directly into the code, so that the application will periodically verify that it is healthy and take some action, such as resetting a counter, each time the check succeeds. Through such a collection of tactics, one can potentially detect most failures rapidly and accurately and even distinguish partitioning failures from other failures such as crashes or application termination. Vogels has implemented a prototype of a failure investigator service that uses these techniques, yielding much faster and better failure detection than is traditionally assumed possible in distributed systems. Unfortunately, however, this approach is not at all standard. Many distributed systems rely entirely on timeouts for failures; as one might expect, this results in a high rate of erroneous detections and a great deal of complexity in order to overcome their consequences.

12.6 Hostile Environments

The discussion in this chapter has enumerated a great variety of reliability threats that a typical distributed system may need to anticipate and deal with. The problems considered, however, were all of a nature that might be considered "routine," in the sense that they all fall into the category of building software and hardware to be robust against anticipated classes of accidental failures and to be self-managed in ways that anticipate system upgrades and maintenance events.

Yet, it is sometimes surprising to realize that the Internet is a hostile environment, and growing more so. Modern computer networks are shared with a huge population of computer-literate users, whose goals and sense of personal ethics may differ tremendously from those of the system developer. Whether intentionally or otherwise, these network users represent a diffuse threat; they may unexpectedly probe a distributed system for weaknesses or even subject it to a well-planned and orchestrated assault without prior warning.

The intentional threat spectrum is as varied as the accidental threat spectrum reviewed earlier. The most widely known of the threats are computer viruses, which are software programs designed to copy themselves from machine to machine and to do damage to the machines on which they manage to establish themselves. (A benign type of virus that does no damage is called a *worm,* but because the mere presence of an unanticipated program can impact system reliability, it is perhaps best to take the view that all undesired intrusions into a system represent a threat to reliable behavior.) A virus may attack a system by violating assumptions it makes about the environment or the network, breaking through security codes and passwords, piggybacking a ride on legitimate messages, or any number of other routes. Attacks that exploit several routes at the same time are more and more common—for example, simultaneously compromising some aspect of the telecommunication infrastructure on which an application depends while also presenting the application with an exceptional condition that it can only handle correctly when the telecommunication subsystem is also functioning.

Other types of intentional threats include unauthorized users or authorized users who exceed their normal limitations. In a banking system, one worries about a rogue trader or an employee who seeks to divert funds without detection. A disgruntled employee may seek to damage the critical systems or data of an organization. In the most extreme case, one can imagine hostile actions directed at a nation's critical computing systems during a period of war or terrorism. Today, this sort of *information warfare* may seem like a suitable topic for science fiction writers, yet, as society shifts increasingly critical activities onto computing and communication technology, the potential targets for attack will eventually become rich enough to interest military adversaries.

Clearly, no computing system can be protected against every conceivable form of internal and external threat. Distributed computing can, however, offer considerable benefits against a well-known and fully characterized threat profile. By distributing critical functionality over sets of processes that must cooperate and coordinate their actions in order to perform sensitive functions, the barrier against external threats can be formidable. A terrorist who might easily overcome a system that effectively lacks any defenses at all would face a much harder problem overcoming firewalls, breaking through security boundaries, and interfering with critical subsystems designed to continue operating correctly even if some limited number of system components crash or are compromised. Later we will discuss virtual private network technologies, which take such approaches even further, preventing all communication within the network except that initiated by authenticated users. Clearly, if a system uses a technology such as this, it will be relatively hard to break into. However, the cost of such a solution may be higher than most installations can afford.

As the developer of a critical system, the challenge is to anticipate the threats that it must overcome and to do so in a manner that balances costs against benefits. Often, the threat profile that a component subsystem may face will be localized to that component—hence, the developer may need to go to great lengths in protecting some especially critical subsystems against reliability and security threats, while using much more limited and less-costly technologies elsewhere in the same system. The goal of this book involves a corresponding issue—that of understanding not just how a reliability problem can be solved, but also how the solution can be applied in a selective and localized manner, so that a developer who faces a specific problem in a specific context can draw on a solution tailored to that problem and context, without requiring that the entire system be reengineered to overcome a narrow threat.

Today, we lack a technology with these attributes. Most fault-tolerant and security technologies demand that the developer adopt a fault-tolerant or secure computing and communication architecture starting with the first lines of code entered into the system. With such an approach, fault tolerance and security become very hard to address late in the game, when substantial amounts of technology already exist. Unfortunately, however, most critical systems are built up out of preexisting technology, which will necessarily have been adapted to the new use and hence will necessarily be confronted with new types of reliability and security threats that were not anticipated in the original setting. What is needed, then, is a technology base that is flexible enough to teach us how to overcome a great variety of possible threats, but that is also flexible enough to be used in a narrow and selective manner (so that the costs of reliability are localized to the component being made reliable), efficient (so that these costs are as low as possible), and suitable for being introduced *late in the game,* when a system may already include substantial amounts of preexisting technology.

The good news, however, is that current research is making major strides in this direction. In the following chapters, we will be looking at many of the fundamental challenges that occur in overcoming various classes of threats. We will discuss computing models that are dynamic, self-managed, and fault tolerant, and we will see how a technology based on *wrapping* preexisting interfaces and components with look-alike technologies that introduce desired robustness features can be used to harden complex, preexisting systems, albeit with many limitations. Finally, we will consider some of the large-scale system issues raised when a complex system must be managed and controlled in a distributed setting. While it would be an overstatement to claim that all the issues have been solved, it is clear that considerable progress toward an integrated technology base for hardening critical systems is being made.

I have few illusions about reliability: Critical computing systems will continue to be less reliable than they should be until the customers and societal users of such systems demand reliability, and the developers begin to routinely concern themselves with understanding the threats to reliability in a given setting—planning a strategy for responding to those threats and for testing the response. However, there is reason to believe that in those cases where this process does occur, a technology base capable of rising to the occasion can be provided.

12.7 Related Reading

On dramatic system failures and their consequences: (see Gibbs, Peterson [1995]).

How and why systems fail and what can be done about it: (see Birman and van Renesse [1996], Borr and Wilhelmy, Chilaragee, Gray [1990], Gray and Reuter, Gray et al.).

On the failure investigator: (see Vogels).

On understanding failures: (see Cristian [1996]).

CHAPTER 13 ✧ ✧ ✧ ✧

Guaranteeing Behavior in Distributed Systems

CONTENTS

13.1 Consistent Distributed Behavior

It is intuitively natural to expect that a distributed system should be able to mimic the behavior of a nondistributed one. Such a system would benefit from its distributed architecture to gain better performance, fault tolerance, or other advantages, and yet it would implement a specification that may originally have been conceived with no particular attention to issues of distribution (see Birman and van Renesse [1996]). In effect, the specification of this type of distributed system describes the behavior of the system as if the system were a single entity.

Because system designers rarely think in terms of concurrent behaviors of a system, describing a system as if it were actually not distributed and only worrying about its distributed nature later is in fact a very natural way to approach distributed software development. A developer who adopts this approach would probably say that the intended interpretation of the design methodology is that the final system should behave *consistently with the specification*. Pressed to explain this more rigorously, such a developer would probably say that the behavior of a correct distributed implementation of the specification should be indistinguishable from a possible nondistributed behavior of a nondistributed program conforming with the specification. Earlier, we saw that transactional systems use such an approach when they require serializability, and our hypothetical developer might well cite that example as an illustration of how this concept might be put into practice. We will explore ramifications of such an approach—namely, the requirement that there be a way to explain distributed behaviors of the system in terms of simpler, nondistributed system specifications.

One can ask a number of questions about distributed consistency. The purpose of this chapter is to set down some of these questions and to start exploring at least some of the answers. Other answers, and in some cases refined questions, will appear in subsequent chapters. The broad theme of this chapter, however, starts with a recognition that the concept of reliability built into the standard technologies we have discussed until now is at best a very ad hoc one. At worst, it may be fundamentally meaningless. To do better, we need to start by associating meaning with at least some intuitively attractive distributed reliability goal.

Previously we asked if a reliable stream were ultimately more reliable than an RPC protocol. And, in some deep sense, the answer seems to be a negative one: A stream built using timeouts for failure detection is ultimately no more reliable than a protocol such as RPC, which retries each request several times before timing out. In both cases, data that have been acknowledged are known to have reached their destination: The RPC provides such acknowledgments in the form of a reply, while the stream does so implicitly by accepting data beyond the capacity of its window (if a stream has accepted more data than its window limitation, the sender can deduce that the initial portion of the data must have been acknowledged). But the status of an RPC in progress, or of data still in the window of a stream, is uncertain in very similar senses, and if an error is reported the sender will not have any way to know what happened to that data. We saw earlier that such an error may occur even if neither the sender nor the destination fails. Yet streams are often considered a reliable way of transferring data. Obviously, they are more reliable than datagrams most of the time, but it would be an overstatement to claim that a stream is reliable in some absolute sense.

To discuss a distributed system or protocol in depth, we need to understand the conditions under which consistent behavior is achievable and the conditions under which consistency is achievable. This will turn out to be a complex, but solvable, problem. The solution reveals a path to a surprisingly rich class of distributed computing systems in which all sorts of guarantees and properties can be achieved. Beyond merely offering rigorous ways of overcoming communication failures, these will include distributed security properties, self-management, load-balancing, and other forms of dynamic adaptation to the environment. All of these are properties that fall under the broad framework of consistent behaviors that are, to a degree, achievable in distributed settings.

We will also discover some hard limits on what can be done in a distributed system. Certain forms of consistency and fault tolerance are impossible, and although these results leave us a great deal of room in which to maneuver, they still circumscribe the achievable behaviors of distributed systems with strong properties.

13.2   Warning: Rough Road Ahead!

Before launching into this material, it may be useful to offer a few words of warning to the reader. The material in this chapter and the next two chapters is of a somewhat different nature than most of what we have covered up to this point. Previous chapters have surveyed the state of the art in distributed computing and to do this it has rarely been necessary to deal with abstractions. In these next few chapters, we'll also be surveying the state of the art, but in an area of distributed computing that is concerned primarily with abstractions and in which some of the protocols used are very subtle. This may make the treatment seem very abstract and difficult in comparison to what we have discussed up to now, and it also differs from the style of the remainder of the book, which returns largely to higher-level issues.

In the introduction to this book we commented that in a better world, distributed system engineers could reach over to a well-stocked shelf of tools and technologies for reliable distributed computing and find powerful, computer-aided design tools, which would make fault-tolerant computing or other reliability goals transparently achievable. But with the exception of a small number of products (the Electra system, which will be discussed in Chapter 18, and the Orbix+Isis product line), there are few technologies that offer a plug-and-play approach to reliability.

Despite the emergent character of reliability as a software product, we already know a great deal about building reliable systems, and there are some very powerful research systems that demonstrate how these concepts can be put to work. We'll review quite a few of them, and most of the systems we describe are available to potential users either for research efforts or as free, public-domain software. It seems very likely that the set of available products will expand in the future. Moreover, as we will see in Chapters 16 and 17, some of the existing groupware technologies are designed to be directly usable in their present form, even if better packaging would make them even easier to use.

Thus, there is really little alternative but for the potential technology consumer to approach this area by trying to acquire a somewhat deeper perspective on what can, and what cannot, be accomplished, viewing the research systems as prototypes of solutions for the area. To the degree that a developer has deep pockets, many of these technologies could be reimplemented in-house. Less well-heeled developers may be able to work with the existing products, and they will often be able to gain access to public-domain technologies or research prototypes. Finally, companies that offer products where reliability might represent a significant market advantage could size the development effort that would be required to develop and deploy new product lines, with reliability as a major feature, by looking at the protocols that are needed by such systems and viewing the research prototypes as proof of concepts that could be imitated at a known cost. In the absence of that shelf of reliability technologies in the local software store, this seems to be the most reasonable way to approach this situation.

Those readers whose interest in this book is primarily practical may find the remainder of this chapter and Chapters 14, 15, and 16 excessively abstract or theoretical. Every effort has been made to ensure that Chapters 17 through 25 can be read without detailed knowledge of the following material, and these readers should be able to skim any of these chapters without being lost in the remainder of the book.

13.3 Membership in a Distributed System

The first chapters of this book have glossed over what turns out to be one of the fundamental questions that we will ask about a distributed system—namely, that of determining what processes belong to the system. Before trying to solve this problem, it will be helpful to observe that there are two prevailing interpretations of what membership should mean in a distributed setting. The more widely used interpretation concerns itself with dynamically varying subsets of a static maximal set of processes.

This static view of membership corresponds to the behavior of transactional database systems and other systems in which a set of servers is associated with physical resources, such as the database files being managed, and collection of multimedia images or some sort of hardware. In such a system, the name of a process would not normally be its process identifier (pid), but rather would be derived from the name of the resource it manages. If there are three identical replicas of a database, we might say that the process managing the first replica is named a, the second b, and the third c, and we could reuse these same names even if replica a crashes and must be rebooted. One would model this type of system as having a fixed maximum membership (a,b,c), but operating with a dynamically varying subset of the members. We will call this a static membership model because the subsets are defined over a static population of processes.

Similarly, we can define a dynamic model of system membership. In this model, processes are created, execute for a period of time, and then terminate. Each process has a unique name, which will never be reused. In a dynamic membership model, the system is defined to be the set of processes that are operational at a given point in time. A dynamic system or set of processes would begin execution when some initial membership is booted and then would evolve as processes join the system (having been created) or leave the system (having terminated or crashed).

Abstractly, the space of possible process names is presumably finite, so one could argue that this dynamic approach to membership isn't really so different from the static one. In practice, however, a static system typically has such a small number of components, perhaps as few as three or four, that the problem is genuinely a very different one. After all, it would not be uncommon for a complex distributed system to spawn hundreds or thousands of processes per hour as it executes. Indeed, many systems of this sort are defined as if the space of processor identifiers were in fact infinite, under the assumption that the system is very unlikely to have any processor that remains operational long enough to actually start reusing identifiers. After all, even if a system spawns 100 processes per second, it would take one and one-half years to exhaust a 32-bit process identifier space.

One could imagine an argument in favor of a hybrid model of system membership—one in which a dynamically managed set of processes surrounds a more static core set of servers. Such a model would be more realistic than the static and dynamic models, because real computing networks do tend to have fixed sets of servers on which the client workstations and computers depend. The main disadvantage of using this mixed model as the overall system model is that it leads to very complex descriptions of system behavior. For this reason, although it is technically feasible to work with a mixed model, the results presented in this book are expressed in terms of the static and dynamic models of system membership. More specifically, we will work by generalizing results from a static model into a more dynamic one.

In designing systems, however, it is often helpful for the developer to keep in mind that even if a dynamic membership model were employed, information about static resources can still be useful. Indeed, by drawing on such information, it may be possible to solve problems that, in a purely dynamic model, could not be solved—for example, we will look at situations where groups of processes have membership that varies over time, increasing as processes join and decreasing as they depart or fail. Sometimes, if a failure has partitioned such a group, it is hard to know which component of the group owns some critical resource. However, suppose that we also notice that two out of three of the computers attached to this resource are accessible in one component of the partitioned group, and hence at most one of the three is accessible in the other component. Such information is sufficient to let the former component treat itself as the owner of the service associated with those computers: The other component will recognize that it must limit its actions to avoid conflict. Thus, even though we focus here on developing a dynamic membership model, by doing so we do not preclude the use of static information as part of the algorithms used by the resulting groups.

13.4   Time in Distributed Systems

In discussing the two views of system membership, we made casual reference to temporal properties of a system. Clearly, the concept of time represents a second fundamental component of any distributed computing model. In the simplest terms, a distributed system is any set of processes that communicates by message passing and carrying out desired actions over time.

Specifications of distributed behavior often include such terms as "when," "before," "after," and "simultaneously," and we will need to develop the tools to make this terminology rigorous.

In nondistributed settings, time has an obvious meaning—at least to nonphysicists. The world is full of clocks, which are accurate and synchronized to varying degrees. Something similar is true for distributed systems: All computers have some form of clock, and clock synchronization services are a standard part of any distributed computing environment. Moreover, just as in any other setting, these clocks have limited accuracy. Two different processes, reading their local clocks at the same instant in (real) time, might observe different values, depending on the quality of the clock synchronization algorithm. Clocks may also drift over long periods of time.

The use of time in a distributed system raises several problems. One obvious problem is to devise algorithms for synchronizing clocks accurately. In Chapter 20 we will look at several such algorithms. However, even given very accurate clocks, communication systems operate at such high speeds that the use of physical clocks for fine-grained temporal measurements can only make sense for processes sharing the same clock—for example, by operating on the same computer. This leads to something of a quandary: In what sense is it meaningful to say that one event occurs and then another does so, or that two events are concurrent, if no means are available by which a program could label events and compare their times of occurrence?

Looking at this question in 1978, Leslie Lamport proposed a model of logical time that answers this question (see Lamport [July 1978, 1984]). Lamport considered sets of processes (they could be static or dynamic) that interact by message passing. In his approach, the execution of a process is modeled as a series of atomic events, each of which requires a single unit of logical time to perform. More precisely, his model represents a process by a tuple $(E_p, <_p)$, where E_p is a set of events that occurred within process p, and $<_p$ is a partial order on those events. The advantage of this representation is that it captures any concurrency available within p. Thus, if a and b are events within p, $a <_p b$ means that a happens before b, in some sense meaningful to p—for example, b might be an operation that reads a value written by a, b could have acquired a lock that a released, or p might be executing sequential code in which operation b isn't initiated until after a has terminated.

Notice that there are many levels of granularity at which one might describe the events that occur as a process executes. At the level of the components from which the computer was fabricated, computation consists of concurrent events that implement the instructions or microinstructions executed by the user's program. At a higher level, a process might be viewed in terms of statements in a programming language, control-flow graphs, procedure calls, or units of work that make sense in some external frame of reference, such as operations on a database. Concurrency within a process may result from interrupt handlers, parallel programming constructs in the language or run-time system, or the use of lightweight threads. Thus, when we talk about the events that occur within a process, it is understood that the designer of a system will typically have a granularity of representation that seems natural for the distributed protocol or specification at hand and that events are encoded to this degree of precision. In this book, most examples will be at a very coarse level of precision, in which we treat all the local computation that occurs within a process, between when it sends or receives a first message, and when it sends or receives a second message, as a single event or as being associated with the send or receive event itself.

Lamport models the sending and receiving of messages as events. Thus, an event a could be the sending of a message m, denoted $snd(m)$; the reception of m, denoted $rcv(m)$; or the delivery of m to application code, denoted $deliv(m)$. When the process at which an event occurs is not clear from context, we will add the process identifier as a subscript: $snd_p(m)$, $rcv_p(m)$, and $deliv_p(m)$. The reasons for separating receiving events from delivery events are to enable us to talk about protocols that receive a message and do things to it, or delay it, before letting the application program see it. Not every message sent will necessarily be received, and not every message received will necessarily be delivered to the application; the former property depends on the reliability characteristics of the network, and the latter depends on the nature of the message.

Consider a process p with an event $snd(m)$ and a process q in which there is a corresponding event $rcv(m)$ for the same message m. Clearly, the sending of a message precedes its receipt. Thus, we can introduce an additional partial order that orders send and receive events for the same messages. Denote this communication ordering relation by $<_m$ so that we can write $snd_p(m) <_m rcv_q(m)$.

This leads to a definition of logical time in a distributed system as the transitive closure of the $<_p$ relations for the processes p that comprise the system and $<_m$. We will write $a \rightarrow b$ to denote the fact that a and b are ordered within this temporal relation, which is often called the potential causality relation for the system. In words, we will say that a happened before b. If neither $a \rightarrow b$ nor $b \rightarrow a$, we will say that a and b occur *concurrently*.

Potential causality is useful in many ways. First, it allows us to be precise when talking about the temporal properties of algorithms used in distributed systems—for example, when we have used phrasing such as "at a point in time" or "when" in relation to a distributed execution, it may not have been clear just what it means to talk about an instant in time that spans a set of processes composing the system. Certainly, the discussion at the start of this chapter, in which it was noted that clocks in a distributed system will often not be sufficiently synchronized to measure time, should have raised concerns about the concept of simultaneous events. An instant in time should correspond to a set of simultaneous events, one per process in the system, but the most obvious way of writing down such a set (namely, writing the state of each process as that process reaches some designated time) would not physically be realizable by any protocol we could implement as a part of such a system.

Consider, however, a set of concurrent events, one per process in a system. Such a set potentially represents an instantaneous snapshot of a distributed system, and even if the events did not occur at precisely the same instant in real time, there is no way to determine this from within the system. We will use the term *consistent cut* to refer to a set of events with this property (see Chandy and Lamport). A consistent snapshot is the full set of events that happen before or on a consistent cut. Note that a consistent snapshot will include the state of communication channels at the time of the consistent cut: The messages in the channels will be those for which the snapshot contains an *snd* event but lacks a corresponding *rcv* event.

Figure 13.1 illustrates this concept: The black cuts are inconsistent because they include message receive events but exclude the corresponding send events. The gray cuts satisfy the consistency property. If one thinks about process execution timelines as if they were made of rubber, the gray cuts correspond to possible distortions of the execution in which time never flows backward; the black cuts correspond to distortions that violate this property. (See also Figure 13.2.)

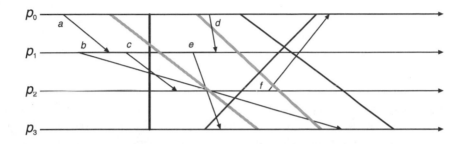

FIGURE 13.1. Examples of consistent (black) and inconsistent (gray) cuts. The gray cuts illustrate states in which a message receive event is included but the corresponding send event is omitted. Consistent cuts represent system states that could have occurred at a single instant in real time. Notice, however, that a consistent cut may not actually capture simultaneous states of the processes in question (i.e., a cut might be instantaneous in real time, but there are many consistent cuts that are not at all simultaneous) and that there may be many such cuts through a given point in the history of a process.

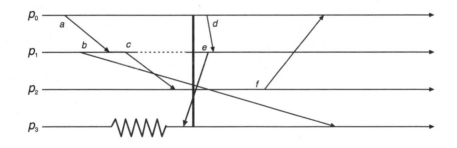

FIGURE 13.2. Distorted timelines that might correspond to faster or slower executions of the processes illustrated in Figure 13.1. Here we have redrawn the earlier execution to make an inconsistent cut appear to be physically instantaneous by slowing down process p_1 (dashed lines) and speeding up p_3 (jagged lines). But notice that to get the cut straight we now have message e traveling backwards in time—an impossibility! The gray cuts in Figure 13.1, in contrast, can all be straightened without such problems. This lends credibility to the idea that a consistent cut is a state that could have occurred at an instant in time, while an inconsistent cut is a state that could not have occurred in real time.

If a program or a person were to look at the state of a distributed system along an inconsistent cut (i.e., by contacting the processes one by one to check each individual state and then assembling a picture of the system as a whole from the data obtained), the results could be confusing and meaningless—for example, if a system manages some form of data using a lock, it could appear that multiple processes hold the lock simultaneously. To see this, imagine that process p holds the lock and then sends a message to process q in which it passes the lock to q. If our cut happened to show q after it received this message (and hence obtained the lock) but showed p before it sent it (and hence when it still held the lock), p and q would appear to both hold the lock. Yet in the real execution, this state never occurred. Were a developer trying to debug a distributed system, considerable time could be wasted in trying to sort out real bugs from these sorts of virtual bugs introduced as artifacts of the way the system state was collected!

The value of consistent cuts is that they represent states the distributed system might actually have been in at a single instant in real time. Of course, there is no way to know which of the feasible cuts for a given execution correspond to the actual real-time states through which the system passed, but Lamport's observation was that in a practical sense, to even ask this question reveals a basic misunderstanding of the nature of time in distributed systems. In his eyes, the consistent cuts for a distributed system are the *more meaningful* concept of simultaneous states for that system, while external time, being inaccessible within the system, is actually *less* meaningful. Lacking a practical way to make real-time clocks that are accurate to the resolution necessary to accurately timestamp events, he argued that real time is in fact not a very useful property for protocols that operate at this level. Of course, we can still use real time for other purposes that demand lesser degrees of accuracy, and we will reintroduce it later, but for the time being, we accept this perspective. For a discussion about some uses of consistent cuts, see Babaoglu and Marzullo.

Potential causality is a useful tool for reasoning about a distributed system, but it also has more practical significance. There are several ways to build logical clocks with which causal relationships between events can be detected to varying degrees of accuracy.

A very simple logical clock can be constructed by associating a counter with each process and message in the system. Let LT_p be the logical time for process p (the value of p's copy of this counter), and let LT_m be the logical time associated with message m (also called the logical timestamp of m). The following rules are used to maintain these counters.

1 If $LT_p < LT_m$, process p sets $LT_p = LT_m + 1$
2 If $LT_p \geq LT_m$, process p sets $LT_p = LT_p + 1$
3 For other events, process p sets $LT_p = LT_p + 1$

We will use the notation $LT(a)$ to denote the value of LT_p when event a occurred at process p. It can easily be shown that if $a \rightarrow b$, $LT(a) < LT(b)$: From the definition of the potential causality relation, we know that if $a \rightarrow b$, there must exist a chain of events $a \equiv e_0 \rightarrow e_1 \ldots \rightarrow e_k \equiv b$, where each pair is related either by the event ordering $<_p$ for some process p or by the event ordering $<_m$ on messages. By construction, the logical clock values associated with these events can only increase, establishing the desired result. On the other hand, $LT(a) < LT(b)$ does not imply that a $\rightarrow$ b, since concurrent events may have the same timestamps.

For systems in which the set of processes is static, logical clocks can be generalized in a way that permits a more accurate representation of causality. A vector clock is a vector of counters, one per process in the set (see Fidge, Mattern, Schiper et al.). Similar to the concept of logical clocks, we will say that VT_p and VT_m represent the vector times associated with process p and message m, respectively. Given a vector time VT, the notation $VT[p]$ denotes the entry in the vector corresponding to process p.

The rules for maintaining a vector clock are similar to the ones used for logical clocks, except that a process only increments its own counter. Specifically:

1 Prior to performing any event, process p sets $VT_p[p] = VT_p[p] + 1$

2 Upon delivering a message m, process p sets $VT_p = \max(VT_p, VT_m)$

In the second situation, the function max applied to two vectors is just the element-by-element maximum of the respective entries. We now define two comparison operations on vector times. If $VT(a)$ and $VT(b)$ are vector times, we will say that $VT(a) \leq VT(b)$ if $\forall I: VT(a)[i] \leq VT(b)[i]$. When $VT(a) \leq VT(b)$ and $\exists i: VT(a)[i] < VT(b)[i]$, we will write $VT(a) < VT(b)$.

In words, a vector time entry for a process p is just a count of the number of events that have occurred at p. If process p has a vector clock with $Vt_p[q]$ set to six, this means that some chain of events has caused p to hear (directly or indirectly) from process q subsequent to the sixth event that occurred at process q. Thus, the vector time for an event e tells us, for each process in the vector, how many events occurred at that process causally prior to when e occurred. If $VT(m) = [17,2,3]$, corresponding to processes $\{p, q, r\}$, we know that 17 events occurred at process p that causally precede the sending of m, two at process q, and three at process r.

It is easy to see that vector clocks accurately encode potential causality. If $a \rightarrow b$, then we again consider a chain of events related by the process or message ordering: $a \equiv e_0 \rightarrow e_1 \ldots \rightarrow e_k \equiv b$. By construction, at each event the vector time can only increase (i.e., $VT(e_i) < VT(e_{i+1})$), because each process increments its own vector time entry prior to each operation, and receive operations compute an element-by-element maximum. Thus, $VT(a) < VT(b)$. However, unlike a logical clock, the converse also holds: If $VT(a) < VT(b)$, then $a \rightarrow b$. To see this, let p be the process at which event a occurred, and consider $VT(a)[p]$. In the case where b also occurs at process p, we know that $\forall I: VT(a)[i] \leq VT(b)[i]$—hence, if a and b are not the same event, a must happen before b at p. Otherwise, suppose that b occurs at process q. According to the algorithm, process q only changes $VT_q[p]$ upon delivery of some message m for which $VT(m)[p] > VT_q[p]$ at the event of the delivery. If we denote b as e_k and $deliv(m)$ as e_{k-1}, the send event for m as e_{k-2}, and the sender of m by q', we can now trace a chain of events back to a process q'' from which q' received this vector timestamp entry. Continuing this procedure, we will eventually reach process p. We will now have constructed a chain of events $a \equiv e_0 \rightarrow e_1 \ldots \rightarrow e_k \equiv b$, establishing that $a \rightarrow b$, the desired result.

This tells us that if we have a fixed set of processes and use vector timestamps to record the passage of time, we can accurately represent the potential causality relationship for messages sent and received, and other events, within that set. Doing so will also allow us to determine when events are concurrent: This is the case if neither $a \rightarrow b$ nor $b \rightarrow a$.

There has been considerable research on optimizing the encoding of vector timestamps, and the representation presented above is far from the best possible in a large system (see Charron-Bost). For a very large system, it is considered preferable to represent causal time using a set of event identifiers, $\{e_0, e_1, \ldots e_k\}$ such that the events in the set are concurrent and causally precede the event being labeled (see Peterson [1987], Melliar-Smith and Moser [1993]). Thus, if $a \to b$, $b \to d$, and $c \to d$, one could say that event d took place at causal time $\{b, c\}$ (meaning "after events b and c"), event b at time $\{a\}$, and so forth. In practice the identifiers used in such a representation would be process identifiers and event counters maintained on a per-process basis—hence, this *precedence-order* representation is recognizable as a compression of the vector timestamp. The precedence-order representation is useful in settings where processes can potentially construct the full $\to$ relation and in which the level of true concurrency is fairly low. The vector timestamp representation is preferred in settings where the number of participating processes is fairly low and the level of concurrency may be high.

Logical and vector clocks will prove to be powerful tools in developing protocols for use in real distributed applications—for example, with either type of clock we can identify sets of events that are concurrent and that satisfy the properties required from a consistent cut. The method favored in a specific setting will typically depend upon the importance of precisely representing the potential causal order and on the overhead that can be tolerated. Notice, however, that while logical clocks can be used in systems with dynamic membership, this is not the case for a vector clock. All processes that use a vector clock must be in agreement upon the system membership used to index the vector. Thus, vector clocks, as formulated here, require a static concept of system membership. (Later we will see that they can be used in systems where membership changes dynamically as long as the places where the changes occur are well defined and no communication spans those membership change events.)

The remainder of this chapter focuses on problems for which logical time, represented through some form of logical timestamp, represents the most natural temporal model. In many distributed applications, however, some concept of real time is also required, and our emphasis on logical time in this section should not be taken as dismissing the importance of other temporal schemes. Methods for synchronizing clocks and for working within the intrinsic limitations of such clocks are the subject of Chapter 20.

13.5   Failure Models and Reliability Goals

Any discussion of reliability is necessarily phrased with respect to the reliability threats of concern in the setting under study—for example, we may wish to design a system so that its components will automatically restart after crash failures, which is called the *recoverability problem*. Recoverability does not imply continuous availability of the system during the periods before a faulty component has been repaired. Moreover, the specification of a recoverability problem would need to say something about how components fail: through clean crashes, which never damage persistent storage associated with them; in other limited ways; in arbitrary ways, which can cause unrestricted damage to the data directly managed by the faulty component; and so forth. These are the sorts of problems typically addressed using variations on the transactional computing technologies introduced in Section 7.5 and to which we will return in Chapter 21.

A higher level of reliability may entail *dynamic availability*, whereby the operational components of a system are guaranteed to continue providing correct, consistent behavior even in the presence of some limited number of component failures—for example, one might wish to design a system so that it will remain available provided that at most one failure occurs, under the assumption that failures are clean ones involving no incorrect actions by the failing component before its failure is detected and it shuts down. Similarly, one might want to guarantee reliability of a critical subsystem up to t failures involving arbitrary misbehavior by components of some type. The former problem would be much easier to solve, since the data available at operational components can be trusted; the latter would require a voting scheme in which data are trusted only when there is sufficient evidence as to their validity, so even if t arbitrary faults were to occur, the deduced value would still be correct.

Failure categories include the benign version, which would be an example of a *halting* failure, and the unrestricted version, which would fall into the *Byzantine* failure model. An extremely benign (and in some ways not very realistic) model is the *fail-stop* model, in which machines fail by halting and the failures are *reported* to all surviving members by a notification service (the challenge, needless to say, is implementing a means for accurately detecting failures and turning them into a reporting mechanism that can be trusted not to make mistakes).

In the following sections, we will provide precise definitions of a small subset of the problems that one might wish to solve in a static membership environment subject to failures. This represents a rich area of study and any attempt to exhaustively treat the subject could easily fill a book. However, as noted at the outset, our primary focus in this book is to understand the most appropriate reliability model for realistic distributed systems. For a number of reasons, a dynamic membership model is more closely matched to the properties of typical distributed systems than the static one; even when a system uses a small hardware base that is relatively static, we will see that availability goals frequently make a dynamic membership model more appropriate for the application itself. Accordingly, we will confine ourselves here to a small number of particularly important problems and to a very restricted class of failure models.

13.6 Reliable Computing in a Static Membership Model

The problems we now focus on are concerned with replicating information in a static environment subject to fail-stop failures and with solving the same problem in a Byzantine failure model. By replication, we mean supporting a variable that can be updated or read and that behaves like a single, nonfaulty variable even when failures occur at some subset of the replicas. Replication may also involve supporting a locking protocol, so that a process needing to perform a series of reads and updates can prevent other processes from interfering with its computation;

in the most general case this problem becomes the transactional one discussed in Chapter 7. We'll use replication as a sort of gold standard against which various approaches can be compared in terms of cost, complexity, and properties.

Replication turns out to be a fundamental problem for other reasons as well. As we begin to look at tools for distributed computing in the coming chapters, we will see that even when these tools do something that can seem very far from replication per se, they often do so by replicating other forms of state that permit the members of a set of processes to cooperate implicitly by looking at their local copies of this replicated information.

Some examples of replicated information will help make this point clear. The most explicit form of replicated datum is simply a replicated variable of some sort. In a bank, one might want to replicate the current holdings of yen as part of a distributed risk-management strategy seeking to avoid overexposure to yen fluctuations. Replication of this information means that it is made locally accessible to the traders (perhaps worldwide): Their computers don't need to fetch these data from a central database in New York; the traders can have these data directly accessible at all times. Obviously, such a model entails supporting updates from many sources, but it should also be clear why one might want to replicate information this way. Notice also that by replicating these data, the risk that they will be inaccessible when needed (because lines to the server are overloaded or the server itself is down) is greatly reduced.

Similarly, a hospital might want to view a patient's medication record as a replicated data item, with copies posted on the workstation of the patient's physician, displayed on a virtual chart at the nursing station, visible next to the bed on a status display, and available on the pharmacy computer. One could, of course, build such a system to use a central server and design all of these other applications as clients of the server that poll it periodically for updates, similar to the way that a Web proxy refreshes cached documents by polling their home server. But it may be preferable to view these data as replicated if, for example, each of the applications needs to represent these data in a different way and needs to guarantee that its version is up to date. In such a setting, data are replicated in the conceptual sense, and, although one might choose to implement the replication policy using a client/server architecture, doing so is basically an implementation decision. Moreover, such a client/server architecture would create a single point of failure for the hospital that can be highly undesirable.

An air traffic control system needs to replicate information about flight plans and current trajectories and speeds. This information resides in the database of each air traffic control center that tracks a given plane, and it may also be visible on the workstation of the controller. If plans to develop "free flight" systems advance, such information will also need to be replicated within the cockpits of planes that are close to one another. Again, one could implement such a system with a central server, but doing so in a setting as critical as air traffic control makes little sense: The load on a central server would be huge, and the single-point failure concerns would be impossible to overcome. The alternative is to view the system as one in which these sorts of data are replicated.

We previously saw that Web proxies can maintain copies of Web documents, caching them to satisfy get requests without contacting the document's home server. Such proxies form a group that replicates the document—although in this case, the Web proxies typically would not know anything about each other, and the replication algorithm depends upon the proxies polling the main server and noticing changes. Thus, document replication in the Web is not able to guarantee that data will be consistent. However, one could imagine modifying a Web server so that when contacted by caching proxy servers of the same "make," it would track the copies of its documents and explicitly refresh them if they change. Such a step would introduce consistent replication into the Web, an issue we will discuss further in Sections 17.3 and 17.4.

Distributed systems also replicate more subtle forms of information. Consider, for example, a set of database servers on a parallel database platform. Each is responsible for some part of the load and backs up some other server, taking over for it in the event that it should fail (we'll see how to implement such a structure shortly). These servers replicate information concerning which servers are included in the system, which server is handling a given part of the database, and what the status of the servers (operational or failed) is at a given point in time. Abstractly, this is replicated data used by the servers to drive their individual actions. One could imagine designating one special server as the master, which would distribute the rules based on what the other servers are doing. This would be one way of implementing the replication scheme.

Finally, if a server is extremely critical, one can actively replicate it by providing the same inputs to two or more replicas (see Birman [1993], Birman and Joseph, Birman and van Renesse [1994, 1996], Cooper and Birman, Schneider [1990], van Renesse et al.). If the servers are deterministic, they will now execute in lockstep, taking the same actions at the same time and providing tolerance of limited numbers of failures. A checkpoint/restart scheme can then be introduced to permit additional servers to be launched as necessary.

Thus, while replication is an important problem in itself, it also underlies a great many other distributed behaviors. One could, in fact, argue that replication is the most fundamental of the distributed computing paradigms. By understanding how to solve replication as an abstract problem, we will also gain insight into how these other problems can be solved.

13.6.1 The Distributed Commit Problem

We begin by discussing a classical problem that occurs as a subproblem in several of the replication methods that follow. This is the *distributed commit problem* and involves performing an operation in an all-or-nothing manner (see Gray [1979], Gray and Reuter).

The commit problem occurs when we wish to have a set of processes that all agree on whether or not to perform some action that may not be possible for some of the participants. To overcome this initial uncertainty, it is necessary to first determine whether or not all the participants will be able to perform the operation and then communicate the outcome of the decision to the participants in a reliable way (the assumption is that once a participant has confirmed that it can perform the operation, this remains true even if it subsequently crashes and must be restarted). We say that the operation can be *committed* if the participants can all perform it. Once a commit decision is reached, this requirement will hold even if some participants fail and later recover. On the other hand, if one or more participants are unable to perform the operation when initially queried, or some can't be contacted, the operation as a whole *aborts*, meaning that no participant should perform it.

Consider a system composed of a static set S containing processes $\{p_0,\ p_1,\ \dots p_n\}$ that fail by crashing and that maintain both *volatile* data, which are lost if a crash occurs, and *persistent* data, which can be recovered after a crash in the same state they had at the time of the crash. An example of persistent data would be information in a disk file; volatile data are any information in a processor's memory, on some sort of scratch area, that will not be preserved if the system crashes and must be rebooted. It is frequently much cheaper to store information in volatile data—hence, it would be common for a program to write intermediate results of a computation to volatile storage. The commit problem will now occur if we wish to arrange for all the volatile information to be saved persistently. The all-or-nothing aspects of the problem reflect the possibility that a computer might fail and lose the volatile data it held; in this case the desired outcome would be that no changes to any of the persistent storage areas occur.

As an example, we might want all of the processes in S to write some message into their persistent data storage. During the initial stages of the protocol, the message would be sent to the processes, which would each store it in their volatile memory. When the decision is made to try to commit these data, the processes clearly cannot just modify the persistent area, because some process might fail before doing so. Consequently, the commit protocol involves first storing the volatile information into a persistent but temporary region of storage. Having done so, the participants would signal their ability to commit.

If all the participants are successful, it is safe to begin transfers from the temporary area to the real data storage region. Consequently, when these processes are later told that the operation as a whole should commit, they would copy their temporary copies of the message into a permanent part of the persistent storage area. If the operation aborts, they would not perform this copy operation. As should be evident, the challenge of the protocol will be to handle the recovery of a participant from a failed state; in this situation, the protocol must determine whether any commit protocols were pending at the time of failure and, if so, whether they terminated in a commit or an abort state.

A distributed commit protocol is normally initiated by a process that we will call the *coordinator;* assume that this is process p_0. In a formal sense, the objective of the protocol is for p_0 to solicit votes for or against a commit from the processes in S and then to send a *commit* message to those processes only if all the votes are in favor of commit; otherwise an *abort* message is sent. To avoid a trivial solution in which p_0 always sends an abort, we would ideally like to require that if all processes vote for commit and no communication failures occur, the outcome should be commit. Unfortunately, however, it is easy to see that such a requirement is not really meaningful because communication failures can prevent messages from reaching the coordinator. Thus, we are forced to adopt a weaker, nontriviality requirement, which states that if all processes vote for commit and all the votes reach the coordinator, the protocol should commit.

A commit protocol can be implemented in many ways—for example, RPC could be used to query the participants and later to inform them of the outcome, or a token could be circulated among the participants that they would each modify before forwarding, indicating their vote. The most standard implementations, however, are called two- and three-phase commit protocols, often abbreviated as 2PC and 3PC.

Two-Phase Commit

A 2PC protocol operates in rounds of multicast communication. Each phase is composed of one round of messages to the participants and one round of replies from the recipients to the sender. The coordinator initially selects a unique identifier for this run of the protocol—for example, by concatenating its own process ID to the value of a logical clock. The protocol identifier will be used to distinguish the messages associated with different runs of the protocol that happen to execute concurrently, and in the remainder of this section we will assume that all the messages under discussion are labeled by this initial identifier.

The coordinator starts by sending out a first round of messages to the participants. These messages normally contain the protocol identifier, the list of participants (so that all the participants will know who the other participants are), and a message "type" indicating that this is the first round of a 2PC protocol. In a static system, where all the processes in the system participate in the 2PC protocol, the list of participants can be omitted because it has a well-known value. Additional fields can be added to this message depending on the situation in which the 2PC was needed—for example, it could contain a description of the action the coordinator wishes to take (if this is not obvious to the participants), a reference to some volatile information the coordinator wishes to have copied to a persistent data area, and so forth. 2PC is thus a very general tool, which can solve any number of specific problems sharing the attribute of needing an all-or-nothing outcome as well as the requirement that participants must be queried as to whether or not they will be able to perform the operation before it is safe to assume that they can do so.

Each participant, upon receiving the first message, takes such local actions as are needed to decide if it can vote in favor of commit—for example, a participant may need to set up some sort of persistent data structure, recording that the 2PC protocol is underway and saving the information that will be needed to perform the desired action if a commit occurs. In the previous example, the participant would copy its volatile data to the temporary persistent region of the disk and then force the records to the disk. Having done this (which may take some time), the participant sends back its vote. The coordinator collects votes, but also uses a timer to limit the duration of the first phase (the initial round of outgoing messages and the collection of replies). If a timeout occurs before the first-phase replies have all been collected, the coordinator aborts the protocol. Otherwise, it makes a commit or abort decision according to the votes it collects.[1]

Now we enter the second phase of the protocol, in which the coordinator sends out commit or abort messages in a new round of communication. Upon receipt of these messages, the participants take the desired action or, if the protocol is aborted, they delete the associated information from their persistent data stores. Figure 13.3 illustrates this basic skeleton of the 2PC protocol.

---

[1] As described, this protocol already violates the nontriviality goal we discussed earlier. No timer is really safe in an asynchronous distributed system, because an adversary could set the minimum message latency to the timer value plus one second, and, in this way, cause the protocol to abort despite the fact that all processes voted to commit and all messages will reach the coordinator. Concerns such as these can seem unreasonably narrow-minded, but they are actually important in trying to pin down the precise conditions under which commit is possible. The practical community tends to be fairly relaxed about such issues, while the theory community tends to take problems of this sort very seriously. It is regrettable but perhaps inevitable that some degree of misunderstanding results from these different points of view.

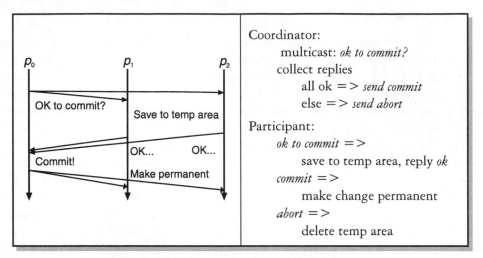

Coordinator:
 multicast: *ok to commit?*
 collect replies
  all ok => *send commit*
  else => *send abort*

Participant:
 ok to commit =>
  save to temp area, reply *ok*
 commit =>
  make change permanent
 abort =>
  delete temp area

In the figure: p₀, p₁, p₂ — OK to commit? — Save to temp area — OK... OK... — Commit! — Make permanent

FIGURE 13.3. Skeleton of two-phase commit protocol

Several failure cases need to be addressed. The coordinator could fail before starting the protocol, during the first phase, while collecting replies, after collecting replies but before sending the second-phase messages, or during the transmission of the second-phase messages. The same is true for a participant. For each case we need to specify a recovery action, which will lead to successful termination of the protocol with the desired all-or-nothing semantics.

In addition to this, the protocol described above omits consideration of the storage of information associated with the run. In particular, it seems clear that the coordinator and participants should not need to keep any form of information indefinitely in a correctly specified protocol. Our protocol makes use of a protocol identifier, and we will see that the recovery mechanisms require that some information be saved for a period of time, indexed by protocol identifier. Thus, rules will be needed for garbage collection of information associated with terminated 2PC protocols. Otherwise, the information base in which these data are stored might grow without limit, ultimately posing serious storage and management problems.

We start by focusing on participant failures, then turn to the issue of coordinator failure, and finally discuss the question of garbage collection.

Suppose that a process p_i fails during the execution of a 2PC protocol. With regard to the protocol, p_i may be any of several states. In its initial state, p_i will be unaware of the protocol. In this case, p_i will not receive the initial vote message; therefore, the coordinator aborts the protocol. The initial state ends when p_i has received the initial vote request and is prepared to send back a vote in favor of commit (if p_i doesn't vote for commit, or isn't yet prepared, the protocol will abort in any case). We will now say that p_i is *prepared to commit*. In the prepared to commit state, p_i is compelled to learn the outcome of the protocol even if it fails and later recovers. This is an important observation because the applications that use 2PC often must lock critical resources or limit processing of new requests by p_i while they are prepared to commit. This means that until p_i learns the outcome of the request, it may be unavailable for other types of processing.

Such a state can result in denial of services. The next state entered by p_i is called the *commit* or *abort* state, in which it knows the outcome of the protocol. Failures occurring at this stage must not be allowed to disrupt the termination actions of p_i, such as the release of any resources that were tied up during the prepared state. Finally, p_i returns to its initial state, garbage collecting all information associated with the execution of the protocol and retaining only the effects of any committed actions.

From this discussion, we see that a process recovering from a failure will need to determine whether or not it was in a prepared to commit, commit, or abort state at the moment of the failure. In a prepared to commit state, the process will need to find out whether the 2PC protocol terminated in a commit or abort, so there must be some form of system service or protocol outcome file in which this information is logged. Having entered a commit or abort state, the process needs a way to complete the commit or abort action even if it is repeatedly disrupted by failures in the act of doing so. We say that the action must be *idempotent*, meaning that it can be performed repeatedly without ill effects. An example of an idempotent action would be copying a file from one location to another: Provided that access to the target file is disallowed until the copying action completes, the process can copy the file once or many times with the same outcome. In particular, if a failure disrupts the copying action, it can be restarted after the process recovers.

Not surprisingly, many systems that use 2PC are structured to take advantage of this type of file copying. In the most common approach, information needed to perform the commit or abort action is saved in a *log* on the persistent storage area. The commit or abort state is represented by a bit in a table, also stored in the persistent area, describing pending 2PC protocols and indexed by protocol identifier. Upon recovery, a process first consults this table to determine the actions it should take, and it then uses the log to carry out the action. Only after successfully completing the action does a process delete its knowledge of the protocol and garbage collect the log records that were needed to carry it out. (See Figure 13.4.)

Coordinator:	Participant:
multicast: *ok to commit?*	*ok to commit* $=>$
collect replies	save to temp area, reply *ok*
all ok $=>$ *log commit to outcomes table*	*commit* $=>$
send commit	make change permanent
else $=>$ *send abort*	*abort* $=>$
collect acknowledgments	delete temp area
garbage collect protocol outcome information	
	After failure:
	for each pending protocol
	contact coordinator to learn
	outcome

FIGURE 13.4. 2PC extended to handle participant failures

Up to now, we have not considered coordinator failure—hence, it would be reasonable to assume that the coordinator itself plays the role of tracking the protocol outcome and saving this information until all participants are known to have completed their commit or abort actions. The 2PC protocol thus needs a final phase in which messages flow back from participants to the coordinator, which must retain information about the protocol until all such messages have been received.

Consider next the case where the coordinator fails during a 2PC protocol. If we are willing to wait for the coordinator to recover, the protocol requires a few changes to deal with this situation. The first change is to modify the coordinator to save its commit decision to persistent storage *before* sending commit or abort messages to the participants.[2] Upon recovery, the coordinator is now guaranteed to have available the information needed to terminate the protocol, which it can do by simply retransmitting the final commit or abort message. A participant not in the precommit state would acknowledge such a message but take no action; a participant waiting in the precommit state would terminate the protocol upon receipt of it.

One major problem with this solution to 2PC is that if a coordinator failure occurs, the participants are blocked, waiting for the coordinator to recover. As noted earlier, precommit often ties down resources or involves holding locks—hence, blocking in this manner can have serious implications for system availability. (See Figure 13.3.) Suppose that we permit the participants to communicate among themselves. Could we increase the availability of the system so as to guarantee progress even if the coordinator crashes?

Again, there are three stages of the protocol to consider. If the coordinator crashes during its first phase of message transmission, a state may result in which some participants are prepared to commit, others may be unable to commit (they have voted to abort and know that the protocol will eventually do so), and still other processes may not know anything at all about the state of the protocol. If it crashes during its decision, or before sending out all the second-phase messages, there may be a mixture of processes left in the prepared state and processes that know the final outcome.

Suppose that we add a timeout mechanism to the participants: In the prepared state, a participant that does not learn the outcome of the protocol within some specified period of time will timeout and seek to complete the protocol on its own. Clearly, there will be some unavoidable risk of a timeout occurring because of a transient network failure, much as in the case of RPC failure-detection mechanisms discussed earlier in the book. Thus, a participant that takes over in this case cannot safely conclude that the coordinator has actually failed. Indeed, any mechanism

---

[2]It is actually sufficient for the coordinator to save only commit decisions in persistent storage. After failure, a recovering coordinator can safely presume the protocol to have aborted if it finds no commit record; the advantage of such a change is to make the abort case less costly, by removing a disk I/O operation from the critical path before the abort can be acted upon. The elimination of a single disk I/O operation may seem like a minor optimization, but in fact it can be quite significant in light of the tenfold latency difference between a typical disk I/O operation (10–25 ms) and a typical network communication operation (perhaps 1–4 ms latency). One doesn't often have an opportunity to obtain an order of magnitude performance improvement in a critical path—hence, these are the sorts of engineering decisions that can have very important implications for overall system performance!

for takeover will need to work even if the timeout is set to 0 and even if the participants try to run the protocol to completion starting from the instant they receive the phase-one message and enter a prepared to commit state!

Accordingly, let p_i be some process that has experienced a protocol timeout in the prepared to commit state. What are p_i's options? The most obvious would be for it to send out a first-phase message of its own, querying the state of the other p_j. From the information gathered in this phase, p_i may be able to deduce that the protocol either committed or aborted. This would be the case if, for example, some process p_j had received a second-phase outcome message from the coordinator before it crashed. Having determined the outcome, p_i can simply repeat the second phase of the original protocol. Although participants may receive as many as n copies of the outcome message (if all the participants timeout simultaneously), this is clearly a safe way to terminate the protocol.

On the other hand, it is also possible that p_i would be unable to determine the outcome of the protocol. This would occur, for example, if all processes contacted by p_i, as well as p_i itself, were in the prepared state, with a single exception: process p_j, which does not respond to the inquiry message. Perhaps p_j has failed, or perhaps the network is temporarily partitioned. The problem now is that only the coordinator and p_j can determine the outcome, which depends entirely on p_j's vote. If the coordinator is itself a participant, as is often the case, a single failure can thus leave the 2PC participants blocked until the failure is repaired! This risk is unavoidable in a 2PC solution to the commit problem. (See Figure 13.5.)

Coordinator:
multicast: *ok to commit?*
collect replies
  all *ok* => *log commit to outcomes table*
    wait until safe on persistent store
    send commit
  else => *send abort*
collect acknowledgments
garbage collect protocol outcome
  information

After failure:
for each pending protocol in outcomes table
  send outcome (commit or abort)
  wait for acknowledgments
  garbage collect outcome information

Participant: first time message received
ok commit =>
  save to temp area, reply *ok*
commit =>
  make change permanent
abort =>
  delete temp area
Message is a duplicate (recovering
  coordinator)
send acknowledgment

After failure:
for each pending protocol
  contact coordinator to learn
  outcome

FIGURE 13.5. 2PC protocol extended to overcome coordinator failures

Earlier, we discussed the garbage collection issue. Notice that in this extension to 2PC, participants must retain information about the outcome of the protocol until they are certain that all participants know the outcome. Otherwise, if a participant p_j were to commit but forgot that it had done so, it would be unable to assist some other participant p_i in terminating the protocol after a coordinator failure.

Garbage collection can be done by adding a third phase of messages from the coordinator (or a participant who takes over from the coordinator) to the participants. This phase would start after all participants have acknowledged receipt of the second-phase commit or abort message, and it would simply tell participants that it is safe to garbage collect the protocol information. The handling of coordinator failure can be similar to that during the pending state. A timer is set in each participant that has entered the final state but not yet seen the garbage collection message. Should the timer expire, such a participant can simply echo the commit or abort message, which all other participants acknowledge. Once all participants have acknowledged the message, a garbage collection message can be sent out and the protocol state safely garbage collected.

Notice that the final round of communication, for purposes of garbage collection, can often be delayed for a period of time and then run once in a while, on behalf of many 2PC protocols at the same time. When this is done, the garbage collection protocol is itself best viewed as a 2PC protocol executing perhaps once per hour. During its first round, a garbage collection protocol would solicit from each process in the system the set of protocols for which they have reached the final state. It is not difficult to see that if communication is first in, first out (FIFO) in the system, then 2PC protocols—even if failures occur—will complete in FIFO order. This being the case, each process needs only to provide a single protocol identifier, per protocol coordinator, in response to such an inquiry: the identifier of the last 2PC initiated by the coordinator to have reached its final state. The process running the garbage collection protocol can then compute the minimum over these values. For each coordinator, the minimum will be a 2PC protocol identifier, which has fully terminated at all the participant processes and can be garbage collected throughout the system.

We now arrive at the final version of the 2PC protocol, shown in Figure 13.6. Notice that this protocol has a potential message complexity, which increases as $O(n^2)$ with the worst case occurring if a network communication problem disrupts communication during the three basic stages of communication. Further, notice that although the protocol is commonly called a two-phase commit, a true two-phase version will always block if the coordinator fails. The version shown in Figure 13.6 gains a higher degree of availability at the cost of additional communication for purposes of garbage collection. However, although this protocol may be more available than our initial attempt, it can still block if a failure occurs at a critical stage. In particular, participants will be unable to terminate the protocol if a failure of both the coordinator and a participant occurs during the decision stage of the protocol.

Coordinator:	Participant: first time message received
multicast: *ok to commit?*	*ok to commit* =>
collect replies	save to temp area, reply *ok*
all *ok* => *log commit to outcomes table*	*commit* =>
wait until safe on persistent store	log outcome, make change
send commit	permanent
else => *send abort*	*abort* =>
collect acknowledgments	log outcome, delete temp area
After failure:	Message is a duplicate (recovering
for each pending protocol in outcomes table	coordinator)
send outcome (commit or abort)	*send acknowledgment*
wait for acknowledgments	After failure:
Periodically:	*for each pending protocol*
query each process: *terminated protocols?*	contact coordinator to learn
for each coordinator: determine *fully*	outcome
terminated protocols	After timeout in *prepare to commit* state:
2PC to garbage collect outcome	query other participants about state
information	outcome can be deduced =>
	run coordinator/recovery protocol
	outcome uncertain =>
	must wait

FIGURE 13.6. Final version of 2PC commit: Participants attempt to terminate protocol without blocking periodic 2PC protocol used to garbage collect outcome information saved by participants and coordinators for recovery.

Three-Phase Commit

In 1981, Skeen and Stonebraker studied the cases in which 2PC can block (see Skeen [June 1982]). Their work resulted in a protocol called *three-phase commit* (3PC), which is guaranteed to be nonblocking provided that only fail-stop failures occur. Before we present this protocol, it is important to stress that the fail-stop model is not a very realistic one: This model requires that processes fail only by crashing and that such failures *be accurately detectable* by other processes that remain operational. As we will see, inaccurate failure detections and network partition failures continue to pose the threat of blocking in this protocol. In practice, these considerations limit the utility of the protocol (because we lack a way to accurately sense failures in most systems, and network partitions are a real threat in most distributed environments). Nonetheless, the protocol sheds light both on the issue of blocking and on the broader concept of consistency in distributed systems; therefore, it is presented here.

As in the case of the 2PC protocol, 3PC really requires a fourth phase of messages for purposes of garbage collection. However, this problem is easily solved using the same method presented in

Figure 13.6 for the case of 2PC. For brevity, we focus on the basic 3PC protocol and overlook the garbage collection issue.

Recall that 2PC blocks under conditions in which the coordinator crashes and one or more participants crash, such that the operational participants are unable to deduce the protocol outcome without information that is only available to the coordinator and/or these participants. The fundamental problem is that in a 2PC protocol, the coordinator can make a commit or abort decision, which would be known to some participant p_j and even acted upon by p_j, but would be totally unknown to other processes in the system. The 3PC protocol prevents this from occurring by introducing an additional round of communication and delaying the prepared state until processes receive this phase of messages. By doing so, the protocol ensures that the state of the system can always be deduced by a subset of the operational processes, provided that the operational processes can still communicate reliably among themselves.

Coordinator:	Participant: logs state on each message
multicast: *ok to commit?*	*ok to commit* =>
collect replies	save to temp area, reply *ok*
all *ok* => log *precommit*	*precommit* =>
send *precommit*	enter precommit state,
else => send *abort*	*acknowledge*
collect acks from nonfailed participants	*commit* =>
all *ack* => log *commit*	make change permanent
send *commit*	*abort* =>
collect acknowledgments	delete temp area
garbage collect protocol outcome information	After failure:
	collect participant state information
	all *precommit, or any committed* =>
	push forward to commit
	else =>
	push back to abort

FIGURE 13.7. Outline of a three-phase commit protocol

A typical 3PC protocol operates as shown in Figure 13.7. As in the case of 2PC, the first-round message solicits votes from the participants. However, instead of entering a prepared state, a participant that has voted for commit enters an *ok to commit* state. The coordinator collects votes and can immediately abort the protocol if some votes are negative or if some votes are missing. Unlike for a 2PC, it does not immediately commit if the outcome is unanimously positive. Instead, the coordinator sends out a round of *prepare to commit* messages, receipt of which causes all participants to enter the prepare to commit state and to send an acknowledgment. After receiving acknowledgments from all participants, the coordinator sends *commit* messages and the participants commit. Notice that the *ok to commit* state is similar to the *prepared* state in the 2PC protocol, in that a participant is expected to remain capable of committing even if failures and recoveries occur after it has entered this state.

If the coordinator of a 3PC protocol detects failures of some participants (recall that in this model, failures are accurately detectable) and has not yet received their acknowledgments to its *prepare to commit* messages, the 3PC can still be committed. In this case, the unresponsive participants can be counted upon to run a recovery protocol when the cause of their failure is repaired, and that protocol will lead them to eventually commit. The protocol thus has the property of only committing if all operational participants are in the *prepared to commit* state. This observation permits any subset of operational participants to terminate the protocol safely after a crash of the coordinator and/or other participants.

The 3PC termination protocol is similar to the 2PC protocol, and it starts by querying the state of the participants. If any participant knows the outcome of the protocol (commit or abort), the protocol can be terminated by disseminating that outcome. If the participants are all in a prepared to commit state, the protocol can safely be committed.

Suppose, however, that some mixture of states is found in the state vector. In this situation, the participating processes have the choice of driving the protocol forward to a commit or back to an abort. This is done by rounds of message exchange that either move the full set of participants to *prepared to commit* and then to *commit* or that back them to *ok to commit* and then abort. Again, because of the fail-stop assumption, this algorithm runs no risk of errors. Indeed, the processes have a simple and natural way to select a new coordinator at their disposal: Since the system membership is assumed to be static, and since failures are detectable crashes (the fail-stop assumption), the operational process with the lowest process identifier can be assigned this responsibility. It will eventually recognize the situation and will then take over, running the protocol to completion.

Notice also that even if additional failures occur, the requirement that the protocol only commit once and that all operational processes are in a *prepared to commit* state and only abort when all operational processes have reached an *ok to commit state* (also called *prepared to abort*) eliminates many possible concerns. However, this is true only because failures are accurately detectable and because processes that fail will always run a recovery protocol upon restarting. (See Figure 13.8.)

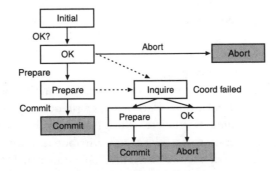

FIGURE 13.8. States for a nonfaulty participant in 3PC protocol

It is not hard to see how this recovery protocol should work. A recovering process is compelled to track down an operational process, which knows the outcome of the protocol, and to learn the outcome from that process. If all processes fail, the recovering process must identify the subset of processes that were the last to fail (see Skeen [1985]), learning the protocol outcome from them. In the case where the protocol had not reached a commit or abort decision when all processes failed, it can be resumed using the states of the participants that were the last to fail, together with any other participants that have recovered in the interim.

Unfortunately, however, the news for 3PC is actually not quite as good as this protocol may make it seem, because real systems do not satisfy the fail-stop failure assumption. Although there may be some specific conditions under which failures are detectable by crashes, these most often depend upon special hardware. In a typical network, failures are only detectable using timeouts, and the same imprecision that makes reliable computing difficult over RPC and streams also limits the failure-handling ability of the 3PC.

The problem that occurs is most easily understood by considering a network partitioning scenario, in which two groups of participating processes are independently operational and trying to terminate the protocol. One group may see a state that is entirely *prepared to commit* and would want to terminate the protocol by commit. The other, however, could see a state that is entirely *ok to commit* and would consider abort to be the only safe outcome: After all, perhaps some unreachable process voted against commit! Clearly, 3PC will be unable to make progress in settings where partition failures can occur. We will return to this issue in Section 13.8, when we discuss a basic result by Fisher, Lynch, and Paterson; the inability to terminate a 3PC protocol in settings that don't satisfy fail-stop failure assumptions is one of many manifestations of the so-called "FLP impossibility" result (see Fisher et al., Ricciardi [1996]). For the moment, though, we find ourselves in the uncomfortable position of having a solution to a problem that is similar to, but not quite identical to, the one that occurs in real systems. One consequence of this is that few systems make use of 3PC commit protocols today: Given a situation in which 3PC is less likely to block than 2PC, but may nonetheless block when certain classes of failures occur, the extra cost of the 3PC is not generally seen as bringing a return commensurate with its cost.

13.6.2 Reading and Updating Replicated Data with Crash Failures

The 2PC protocol represents a powerful tool for solving end-user applications. In this section, we focus on the use of 2PC to implement a data replication algorithm in an environment where processes fail by crashing. Notice that we have returned to a realistic failure model here—hence, the 3PC protocol would offer few advantages.

Accordingly, consider a system composed of a static set S containing processes $\{p_0, p_1, \dots p_n\}$, which fail by crashing and which maintain volatile and persistent data. Assume that each process p_i maintains a local replica of some data object, which is updated by operation $update_i$ and read using operation $read_i$. Each operation, both local and distributed, returns a value for the replicated data object. Our goal is to define the distributed operations $UPDATE$ and $READ$, which remain available even when $t < n$ processes have failed and return results indistinguishable from those that might be returned by a single, nonfaulty process. Secondary goals are to understand the relationship between t and n and to determine the maximum level of availability that can be achieved without violating the one-copy behavior of the distributed operations.

The best-known solutions to the static replication problem are based on *quorum* methods (see Gifford, Skeen [February 1982], Thomas). In these methods, both *UPDATE* and *READ* operations can be performed on less than the full number of replicas, provided however that there is a guarantee of overlap between the replicas at which any successful *UPDATE* is performed and those at which any other *UPDATE* or any successful *READ* is performed. Let us denote the number of replicas that must be read to perform a *READ* operation by q_r and the number to perform an *UPDATE* by q_u. Our quorum overlap rule states that we need $q_r + q_u > n$ and that $q_u + q_u > n$.

An implementation of a quorum replication method associates a *version number* with each data item. The version number is just a counter, which will be incremented by each attempted update. Each replica will include a copy of the data object, together with the version number corresponding to the update that wrote that value into the object.

To perform a *READ* operation, a process reads q_r replicas and discards any replicas with version numbers smaller than those of the others. The remaining values should all be identical, and the process treats any of these as the outcome of its *READ* operation.

To perform an *UPDATE* operation, the 2PC protocol must be used. The updating process first performs a *READ* operation to determine the current version number and, if desired, the value of the data item. It calculates the new value of the data object, increments the version number, and then initiates a 2PC protocol to write the value and version number to q_u or more replicas. In the first stage of this protocol, a replica votes to abort if the version number it already has stored is larger than the version number proposed in the update. Otherwise, it locks out read requests to the same item and waits in an *ok to commit* state. The coordinator will commit the protocol if it receives only *commit* votes and if it is successful in contacting at least q_u or more replicas; otherwise, it aborts the protocol. If new read operations occur during the *ok to commit state*, they are delayed until the commit or abort decision is reached. On the other hand, if new updates arrive during the *ok to commit state*, the participant votes to abort them.

Our solution raises several issues. First, we need to be convinced that it is correct and to understand how it would be used to build a replicated object tolerant of t failures. A second issue is to understand the behavior of the replicated object if recoveries occur. The last issue to be addressed concerns concurrent systems: As stated, the protocol may be prone to livelock (cycles in which one or more updates are repeatedly aborted).

With regard to correctness, notice that the use of 2PC ensures that an *UPDATE* operation either occurs at q_u replicas or at none. Moreover, *READ* operations are delayed while an *UPDATE* is in progress (see Figure 13.9). Making use of the quorum overlap property, it is easy to see that if an *UPDATE* is successful, any subsequent *READ* operation must overlap with at least one replica, and the *READ* will therefore reflect the value of that *UPDATE* or of a subsequent one. If two *UPDATE* operations occur concurrently, one or both will abort. Finally, if two *UPDATE* operations occur in some order, then, since the *UPDATE* starts with a *READ* operation,

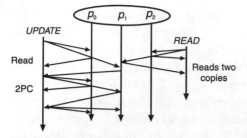

FIGURE 13.9. Quorum update algorithm uses a quorum read followed by a 2PC protocol for updates.

the later *UPDATE* will use a larger version number than the earlier one, and its value will be the one that persists.

To tolerate *t* failures, it will be necessary that the *UPDATE* quorum, q_u, be no larger than $n - t$. It follows that the *READ* quorum, q_r, must have a value larger than *t*—for example, in the common case where we wish to guarantee availability despite a single failure, *t* will equal 1. The *READ* quorum will therefore need to be at least 2, implying that a minimum of three copies are needed to implement the replicated data object. If three copies are in fact used, the *UPDATE* quorum would also be set to 2. We could also use extra copies: With four copies, for example, the *READ* quorum could be left at 2 (one typically wants reads to be as fast as possible and hence would want to read as few copies as possible), and the *UPDATE* quorum increased to 3, guaranteeing that any *READ* will overlap with any prior *UPDATE* and that any pair of *UPDATE* operations will overlap with one another. Notice, however, that with four copies, 3 is the smallest possible *UPDATE* quorum.

Our replication algorithm places no special constraints on the recovery protocol beyond those associated with the 2PC protocol itself. Thus, a recovering process simply terminates any pending 2PC protocols and can then resume participation in new *READ* and *UPDATE* algorithms.

Turning finally to the issue of concurrent *UPDATE* operations, it is evident that there may be a real problem here. If concurrent operations of this sort are required, they can easily force one another to abort. Presumably, an aborted *UPDATE* would simply be reissued, creating a livelock. One solution to this problem is to protect the *UPDATE* operation using a locking mechanism, permitting concurrent *UPDATE* requests only if they access independent data items. Another possibility is to employ some form of back-off mechanism, similar to the one used by an Ethernet controller. Later, when we consider dynamic process groups and atomic multicast, we will see additional solutions to this problem.

What should the reader conclude about this replication protocol? One important conclusion is that the protocol does not represent a very good solution to the problem and will perform very poorly in comparison with some of the dynamic methods introduced in Section 13.9. Limitations include the need to read multiple copies of data objects in order to ensure that the quorum overlap rule is satisfied despite failures, which makes *READ* operations costly. A second limitation is the extensive use of 2PC, itself a costly protocol, when doing *UPDATE* operations. Even a modest application may issue large numbers of *READ* and *UPDATE* requests, leading to a tremendous volume of I/O. This is in contrast with dynamic membership solutions, which turn out to be extremely sparing in I/O, permitting completely local *READ* operations and *UP-DATE* operations costing as little as one message per replica, and yet are able to guarantee very strong consistency properties. Perhaps for these reasons, quorum data management has seen relatively little use in commercial products and systems.

There is one setting in which quorum data management is found to be less costly: transactional replication schemes, typically as part of a replicated database. In these settings, database concurrency control eliminates the concerns raised earlier in regard to livelock or thrashing, and the overhead of the 2PC protocol can be amortized into a single 2PC protocol executing at the end of the transaction. Moreover, *READ* operations can sometimes cheat in transactional settings, accessing a local copy and later confirming that the local copy was a valid one as part of the first phase of the 2PC protocol terminating the transaction. Such a read can be understood as using a form of optimism, similar to that of an optimistic concurrency control scheme. The ability to abort thus makes possible significant optimizations in the solution.

However, few transactional systems have incorporated quorum replication. If one discusses the option with database companies, the message that emerges is clear: Transactional replication is perceived as being extremely costly, and 2PC represents a huge burden when compared with transactions that run entirely locally on a single, nonreplicated database. Transaction rates are approaching 10,000 per second for top-of-the-line commercial database products on nonreplicated, high-performance computers; rates of 100 transactions per second would be impressive for a replicated transactional product. The two orders of magnitude performance loss are more than the commercial community can readily accept, even if they confer increased product availability. We will return to this point in Chapter 21.

13.7 Replicated Data with Nonbenign Failure Modes

The discussion in the previous sections assumed a crash-failure model that is approximated in most distributed systems, but may sometimes represent a risky simplification. Consider a situation in which the actions of a computing system have critical implications, such as the software responsible for adjusting the position of an aircraft wing in flight or for opening the cargo door of the Space Shuttle. In settings such as these, the designer may hesitate to simply assume that the only failures that will occur will be benign ones.

There has been considerable work on protocols for coordinating actions under extremely pessimistic failure models, centering on what is called the Byzantine generals' problem, which explores a type of agreement protocol under the assumption that failures can produce arbitrarily incorrect behavior, but that the *number* of failures is known to be bounded. Although this assumption may seem more realistic than the assumption that processes fail by clean crashes, the model also includes a second type of assumption, which some might view as unrealistically benign: It assumes that the processors participating in a system share perfectly synchronized clocks, permitting them to exchange messages in rounds, which are triggered by the clocks (e.g., once every second). Moreover, the model assumes that the latencies associated with message exchange between correct processors is accurately known.

Thus, the model permits failures of unlimited severity, but at the same time assumes that the *number* of failures is limited and that operational processes share a very simple computing environment. Notice in particular that the round model would only be realistic for a very small class of modern parallel computers and is remote from the situation on distributed computing networks. The usual reasoning is that by endowing the operational computers with extra power (in the form of synchronized rounds), we can only make their task easier. Thus, understanding the minimum cost for solving a problem in this model will certainly teach us something about the minimum cost of overcoming failures in real-world settings.

The Byzantine generals' problem is as follows (see Lynch). Suppose that an army has laid siege to a city and has the force to prevail in an overwhelming attack. However, if divided, the army might lose the battle. Moreover, the commanding generals suspect that there are traitors in their midst. Under what conditions can the loyal generals coordinate their action so as to either attack in unison or not attack at all? The assumption is that the generals start the protocol with individual opinions on the best strategy: to attack or to continue the siege. They exchange messages to execute the protocol, and, if they decide to attack during the ith round of communication, they will all attack at the start of round $i + 1$. A traitorous general can send out any messages he or she likes and can lie about his or her own state, but can never forge the message of a loyal general. Finally, to avoid trivial solutions, it is required that if all the loyal generals favor attacking, an attack will result, and if all favor maintaining the siege, no attack will occur.

To see why this is difficult, consider a simple case in which three generals surround the city. Assume two are loyal, but that one favors attack and the other prefers to hold back. The third general is a traitor. Moreover, assume it is known that there is at most one traitor. If the loyal generals exchange their votes, they will both see a tie: one vote for attack, one opposed. Now suppose that the traitor sends an attack message to one general and tells the other to hold back. The loyal generals now see inconsistent states: One is likely to attack while the other holds back. With the forces divided, they would be defeated in battle. The Byzantine generals' problem is thus seen to be impossible for $t = 1$ and $n = 3$.

With four generals and at most one failure, the problem is solvable, but not trivially so. Assume that two loyal generals favor attack, the third favors retreat, and the fourth is a traitor. Again, it is known that there is at most one traitor. The generals exchange messages, and the traitor sends a retreat message to one and an attack message to two others. One loyal general will now have a tied vote: two votes to attack, two to retreat. The other two generals will see three votes for attack and one for retreat. A second round of communication will clearly be needed before this protocol can terminate! Accordingly, we now imagine a second round in which the generals circulate messages concerning their state in the first round. Two loyal generals will start this round knowing that it is safe to attack: On the basis of the messages received in the first round, they can deduce that even with the traitor's vote, the majority of loyal generals favored an attack. The remaining loyal general simply sends out a message that he or she is still undecided. At the end of this round, all the loyal generals will have one undecided vote, two votes that it is safe to attack, and one message from the traitor. Clearly, no matter what the traitor votes during the second round, all three loyal generals can deduce that it is safe to attack. Thus, with four generals and at most one traitor, the protocol terminates after two rounds.

By using this model, one can prove what are called lower bounds and upper bounds on the Byzantine generals' problem. A lower bound would be a limit to the quality of a possible solution to the problem—for example, one can prove that any solution to the problem capable of overcoming t traitors requires a minimum of $3t + 1$ participants (hence: $2t + 1$ or more loyal generals). The intuition into such a bound is fairly clear: The loyal generals must somehow be able to deduce a common strategy even with t participants whose votes cannot be trusted. For the others there must be a way to identify a majority decision. However, it is surprisingly difficult to prove that this must be the case. For our purposes, such a proof would represent a digression and is omitted, but interested readers are referred to the excellent treatment in Fisher et al. (August 1985). Another example of a lower bound concerns the minimum number of messages required to solve the problem: No protocol can overcome t faults with fewer than $t + 1$ rounds of message exchange, and hence $O(t * n^2)$ messages, where n is the number of participating processes.

In practical terms, these represent costly findings: Recall that our 2PC protocol is capable of solving a problem much like the Byzantine generals' problem in two rounds of message exchange requiring only $3n$ messages, albeit for a simpler failure model. Moreover, the quorum methods permit data to be replicated using as few as $t + 1$ copies to overcome t failures. Later in the book we will discuss even cheaper replication schemes with slightly weaker guarantees. Thus, a Byzantine protocol is very costly, and the best solutions are also fairly complex.

An upper bound on the problem would be a demonstration of a protocol that actually solves the Byzantine generals' problem and an analysis of its complexity (number of rounds of communication required or messages required). Such a demonstration is an upper bound because it rules out the need for a more costly protocol to achieve the same objectives. Clearly, one hopes for upper bounds that are as close as possible to the lower bounds, but unfortunately no such protocols have been found for the Byzantine generals' problem. The simple protocol illustrated here can easily be generalized into a solution for t failures that achieves the lower bound for rounds of message exchange, although not for numbers of messages required.

Suppose that we wanted to use the Byzantine generals' problem to solve a static data replication problem in a very critical or hostile setting. To do so, it would be necessary that the setting somehow correspond to the setup of the Byzantine generals' problem itself—for example, one could imagine using this problem to control an aircraft wing or the Space Shuttle cargo door by designing hardware that carries out voting through some form of physical process. The hardware would be required to implement the mechanisms needed to write software that executes in rounds, and the programs would need to be carefully analyzed to be sure that when operational, all the computing they do in each round can be completed before that round terminates.

On the other hand, one would not want to use a Byzantine protocol in a system where, at the end of the protocol, some single program will take the output of the protocol and perform a critical action. In that sort of a setting (unfortunately, far more typical of real computer systems), all we will have done is transfer complete trust in the set of servers within which the agreement protocol runs into complete trust in the single program that carries out their decision.

The practical use of the Byzantine protocol raises another concern: The timing assumptions built into the model are not realizable in most computing environments. While it is certainly possible to build a system with closely synchronized clocks and to approximate the synchronous rounds used in the model, the pragmatic reality is that few existing computer systems offer such a feature. Software clock synchronization, on the other hand, is subject to intrinsic limitations of its own, and for this reason is a poor alternative to the real thing. Moreover, the assumption that message exchanges can be completed within known, bounded latency is very hard to satisfy in general-purpose computing environments.

Continuing in this vein, one could also question the extreme pessimism of the failure model. In a Byzantine setting the traitor can act as an adversary, seeking to force the correct processes to malfunction. For a worst-case analysis this makes a good deal of sense. But having understood the worst case, one can also ask whether real-world systems should be designed to routinely assume such a pessimistic view of the behavior of system components. After all, if one is this negative, shouldn't the hardware itself also be suspected of potential misbehavior, as well as the compiler and the various prebuilt system components that implement message passing? In designing a security subsystem or implementing a firewall, such an analysis makes a lot of sense. But when designing a system that merely seeks to maintain availability despite failures, and is not expected to come under active and coordinated attack, an extremely pessimistic model would be both unwieldy and costly.

From these considerations, one sees that a Byzantine computing model may be applicable to certain types of special-purpose hardware, but it will rarely be directly applicable to more general distributed computing environments where we might raise a reliability goal. As an aside, it should be noted that Rabin has introduced a set of probabilistic Byzantine protocols that are extremely efficient, but that accept a small risk of error (the risk diminishes exponentially with the number of rounds of agreement executed) (see Rabin). Developers who seek to implement Byzantine-based solutions to critical problems would be wise to consider using these elegant and efficient protocols.

13.8 Reliability in Asynchronous Environments

At the other side of the spectrum is what we call the *asynchronous* computing model (see Breakout 13.1), in which a set of processes cooperates by exchanging messages over communication links that are arbitrarily slow and balky. The assumption here is that the messages sent on the links eventually get through, but that there is no meaningful way to measure progress except by the reception of messages. Clearly such a model is overly pessimistic, but in a way that is different from the pessimism of the Byzantine model, which extended primarily to failures—here we are pessimistic about our ability to measure time or to predict the amount of time actions will take. A message that arrives after a century of delay would be processed no differently than a message received within milliseconds of being transmitted. At the same time, this model assumes that processes fail by crashing, taking no incorrect actions and simply halting silently.

One might wonder why the asynchronous system completely eliminates any physical concept of time. We have seen that real distributed computing systems lack ways to closely synchronize clocks and are unable to distinguish network partitioning failures from processor failures, so there is a sense in which the asynchronous model isn't as unrealistic as it may initially appear. Real systems do have clocks and use these to establish timeouts, but generally they lack a way to ensure that these timeouts will be accurate, as we saw when we discussed RPC protocols and the associated reliability issues in Chapter 4. Indeed, if an asynchronous model can be criticized as specifically unrealistic, this is primarily in its assumption of reliable communication links: Real systems tend to have limited memory resources, and a reliable communication link for a network subject to extended partitioning failures will require unlimited spooling of the messages sent. This represents an impractical design point: A better model would state that when a process is *reachable*, messages will be exchanged reliably with it, but if it becomes *inaccessible*, messages to it will be lost and its state, faulty or operational, cannot be accurately determined. In Italy, Babaoglu and his colleagues are studying such a model, but this is recent work and the full implications of this design point are not yet fully understood (see Babaoglu et al. [1994]). Other researchers, such as Cristian, are looking at models that are partially asynchronous: They have time bounds, but the bounds are large compared to typical message-passing latencies (see Cristian [1996]). Again, it is too early to say whether or not this model represents a good choice for research on realistic distributed systems.

Within the purely asynchronous model, a classical result limits what we can hope to accomplish. In 1985, Fisher, Lynch, and Paterson proved that the asynchronous consensus problem (similar to the Byzantine generals' problem, but posed in an asynchronous setting) is impossible if even a single process can fail. Their proof revolves around the use of a type of message scheduler that delays the progress of a consensus protocol and holds regardless of the way that the protocol itself works. Basically, they demonstrate that any protocol guaranteed to produce only correct outcomes in an asynchronous system can be indefinitely delayed by a complex pattern of network partitioning failures. More recent work has extended this result to some of the communication protocols we will discuss in the rest of this chapter (see Chandra et al. [1996], Ricciardi [1996]).

The Fisher, Lynch, and Paterson (FLP) proof is short but quite sophisticated, and it is common for practitioners to conclude that it does not correspond to any scenario that would be expected to occur in a real distributed system—for example, recall that 3PC is unable to make progress when failure detection is unreliable because of message loss or delays in the network. The FLP result predicts that if a protocol such as 3PC is capable of solving the consensus problem, it can be prevented from terminating. However, if one studies the FLP proof, it turns out that the type of partitioning failure exploited by the proof is at least superficially very remote from the pattern of crashes and network partitioning that forces the 3PC to block.

Thus, it is a bit facile to say that FLP predicts that 3PC will block in this specific way, because the proof constructs a scenario that seems to have relatively little to do with the one that causes problems in a protocol such as 3PC. At the very least, one would be expected to relate the FLP scheduling pattern to the situation when 3PC blocks, and I am not aware of any research that has made this connection concrete. Indeed, it is not entirely clear that 3PC *could* be used to solve the consensus problem: Perhaps the latter is actually a more difficult problem, in which case the inability to solve consensus might not imply that 3PC cannot be solved in asynchronous systems.

As a matter of fact, although it is obvious that 3PC cannot be solved when the network is partitioned, if we carefully study the model used in FLP we realize that network partitioning is

The Asynchronous Computing Model

Although we refer to our model as asynchronous, it is in fact more constrained. In the asynchronous model, as used by distributed system theoreticians, processes communicate entirely by message passing and there is no concept of time. Message passing is reliable but individual messages can be delayed indefinitely, and there is no meaningful concept of failure except for a process that crashes (taking no further actions) or that violates its protocol by failing to send a message or discarding a received message. Even these two forms of communication failure are frequently ruled out.

The form of asynchronous computing environment used in this chapter, in contrast, is intended to be "realistic." This implies that there are clocks on the processors and expectations regarding typical round-trip latencies for messages. Such temporal data can be used to define a concept of reachability or to trigger a failure-detection mechanism. The detected failure may not be attributable to a specific component (in particular, it will be impossible to know if a *process* failed or just the *link* to it), but the fact that some sort of problem has occurred will be detected, perhaps very rapidly. Moreover, in practice, the frequency with which failures are erroneously suspected can be kept low.

Jointly, these properties make the asynchronous model used in this book different from the one used in most theoretical work. And this is a good thing, too: In the fully asynchronous model, it is known that the group membership problem cannot be solved, in the sense that any protocol capable of solving the problem may encounter situations in which it cannot make progress. In contrast, these problems are always solvable in asynchronous environments, which satisfy sufficient constraints on the frequency of correctly or incorrectly detected failures and on the quality of communication.

not actually considered in the model: The FLP result assumes that every message sent will eventually be received, in FIFO order. Thus, FLP essentially requires that every partition eventually be fixed and that every message eventually get through. The tendency of 3PC to block during partitions, which concerned us above, is not captured by FLP because FLP is willing to wait until such a partition is repaired (and implicitly assumes that it will be), while we wanted 3PC to make progress even while the partition was present (whether or not it will eventually be repaired).

To be more precise, FLP tells us that any asynchronous consensus decision can be *indefinitely delayed*, not merely delayed, until a problematic communication link is fixed. Moreover, it says that this is true even if every message sent in the system eventually reaches its destination. During this period of delay the processes may thus be quite active. Finally, and in some sense most surprising of all, the proof doesn't require that any process fail at all: It is entirely based on a pattern of message delays. Thus, FLP not only predicts that we would be unable to develop a 3PC protocol guaranteeing progress despite failures, but that, in actuality, there is no 3PC protocol that can terminate at all, even if no failures actually occur and the network is merely subject to unlimited numbers of network partitioning events. We convinced ourselves that 3PC would need to block (wait) in a single situation; FLP tells us that if a protocol such as 3PC can be used to solve the consensus, then there is a sequence of communication failures that would prevent it from reaching a commit or abort point regardless of how long it executes!

13.8.1 Three-Phase Commit and Consensus

To see that 3PC solves consensus, we should be able to show how to map one problem to the other and back—for example, suppose that the inputs to the participants in a 3PC protocol are used to determine their vote, for or against commit, and that we pick one of the processes to run the protocol. Superficially, it may seem that this is a mapping from 3PC to consensus. But recall that consensus of the type considered by FLP is concerned with protocols that tolerate a single failure, which would presumably include the process that starts the protocol. Moreover, although we didn't get into this issue, consensus has a nontriviality requirement, which is that if all the inputs are 1 the decision will be 1, and if all the inputs are 0 the decision will be 0. As stated, our mapping of 3PC to consensus might not satisfy nontriviality while also overcoming a single failure. Thus, while it would not be surprising to find that 3PC is equivalent to consensus, neither is it obvious that the correspondence is an exact one.

But assume that 3PC is in fact equivalent to consensus. In a *theoretical* sense, FLP would represent a very strong limitation on 3PC. In a *practical* sense, however, it is unclear whether it has direct relevance to developers of reliable distributed software. Previously, we commented that even the scenario that causes 2PC to block is extremely unlikely unless the coordinator is also a participant; thus, 2PC (or 3PC when the coordinator actually is a participant) would seem to be an adequate protocol for most real systems. Perhaps we are saved from trying to develop some other protocol to evade this limitation: FLP tells us that any such protocol will sometimes block. But once 2PC or 3PC has blocked, one could argue that it is of little practical consequence whether this was provoked by a complex sequence of network partitioning failures or by something simple and blunt, such as the simultaneous crash of a majority of the computers in the network. Indeed, we would consider that 3PC has failed to achieve its objectives as soon as the first partitioning failure occurs and it ceases to make *continuous* progress. Yet the FLP result, in some sense, hasn't even kicked in at this point: It relates to *ultimate* progress. In the FLP work, the issue of a protocol being blocked is not really modeled in the formalism at all, except in the sense that such a protocol has not yet reached a decision state.

We thus see that although FLP tells us that the asynchronous consensus problem cannot *always* be solved, it says nothing at all about when problems such as this actually *can* be solved. As we will see, more recent work answers this question for asynchronous consensus. However, unlike an impossibility result, to apply this new result one would need to be able to relate a given execution model to the asynchronous one and a given problem to consensus.

FLP is frequently misunderstood as having proven the impossibility of building fault-tolerant distributed software for realistic environments. This is not the case at all! FLP doesn't say that one cannot build a consensus protocol tolerant of one failure or of many failures. It simply says that if one does build such a protocol, and then runs it in a system with no concept of global time whatsoever and no timeouts, there will be a pattern of message delays that prevents it from terminating. The pattern in question may be extremely improbable, meaning that one might still be able to build an asynchronous protocol that would terminate with overwhelming probability. Moreover, realistic systems have many forms of time: timeouts, loosely synchronized global clocks, and (often) a good idea of how long messages should take to reach their destinations and to be acknowledged. This sort of information allows real systems to evade the limitations imposed by FLP or at least to create a run-time environment that differs in fundamental ways from the FLP-style of asynchronous environment.

This brings us to the more recent work in this area, which presents a precise characterization of the conditions under which a consensus protocol can terminate in an asynchronous environment. Chandra and Toueg have shown how the consensus problem can be expressed using what they call "weak failure detectors," which are a mechanism for detecting that a process has failed without necessarily doing so accurately (see Chandra and Toueg [1991], Chandra et al. [1992]). A weak failure detector can make mistakes and change its mind; its behavior is similar to what might result by setting some arbitrary timeout—declaring a process faulty if no communication is received from it during the timeout period, and then declaring that it is actually operational after all if a message subsequently turns up (the communication channels are still assumed to be reliable and FIFO). Using this model, Chandra and Toueg prove that consensus can be solved provided that a period of execution occurs during which all genuinely faulty processes are suspected as faulty, and during which at least one operational process is never suspected as faulty by any other operational process. One can think of this as a constraint on the quality of the communication channels and the timeout period: If communication works well enough, and timeouts are accurate enough, for a long enough period of time, a consensus decision can be reached. Interested readers should also refer to Babaoglu et al. (1995), Friedman et al., Guerraoui and Schiper, Ricciardi (1996). Two very recent papers in this area are by Babaoglu et al. (1996) and Neiger.

What Chandra and Toueg have done has general implications for the developers of other forms of distributed systems that seek to guarantee reliability. We learn from this result that to guarantee progress, the developer may need to guarantee a higher quality of communication than in the classical asynchronous model, a degree of clock synchronization (lacking in the model), or some form of accurate failure detection. With any of these, the FLP limitations can be evaded (they no longer hold). In general, it will not be possible to say "my protocol always terminates" without also saying "when such and such a condition holds" on the communication channels, the timeouts used, or other properties of the environment.

This said, the FLP result does create a quandary for practitioners who hope to be rigorous about the reliability properties of their algorithms by making it difficult to talk in rigorous terms about what protocols for asynchronous distributed systems actually guarantee. We would like to be able to talk about one protocol being more tolerant of failures than another, but now we see that such statements will apparently need to be made about protocols in which one can only guarantee fault tolerance in a conditional way and where the conditions may not be simple to express or to validate.

What seems to have happened here is that we lack an appropriate concept of what it means for a protocol to be live in an asynchronous setting. The FLP concept of liveness is rigorously defined and not achievable, but in any case this concept does not address the more relative concept of liveness that we seek when developing a nonblocking commit protocol. As it happens, even this more relative form of liveness is not always achievable, and this coincidence has sometimes led practitioners and even theoreticians to conclude that the forms of liveness are the same, since neither is always possible. This subtle but very important point has yet to be treated adequately by the theoretical community. We need a model in which we can talk about 3PC making progress under conditions when 2PC would not do so without getting snarled in the impossibility of guaranteeing progress for all possible runs in the asynchronous model.

Returning to our data replication problem, these theoretical results do have some practical implications. In particular, they suggest that there may not be much more that can be accomplished in a static computing model. The quorum methods give us a way to overcome failures or damage to limited numbers of data objects within a set of replicas; although expensive, such methods clearly work. While they would not work with a very serious type of failure in which processes behave maliciously, the Byzantine protocol and consensus literature suggest that one cannot always solve this problem in an asynchronous model, and the synchronous model is sufficiently specialized as to be largely inapplicable to standard distributed computing systems.

Our best hope, in light of these limitations, will be to focus on the poor performance of the style of replication algorithm arrived at above. Perhaps a less-costly algorithm would represent a viable option for introducing tolerance to at least a useful class of failures in realistic distributed environments. Moreover, although the FLP result tells us that for certain categories of objectives availability must always be limited, the result does not speak directly to the sorts of tradeoffs between availability and cost seen in 2PC and 3PC. Perhaps we should talk about optimal progress and identify the protocol structures that result in the best possible availability without sacrificing consistency, even if we must accept that our protocols will (at least theoretically) remain exposed to scenarios in which they are unable to make progress.

13.9 The Dynamic Group Membership Problem

If we want to move beyond the limitations of the protocols presented previously, it will be necessary to identify some point of leverage offering a practical form of freedom that can be exploited to reduce the costs of reliability, measured in terms of messages exchanged to accomplish a goal, such as updating replicated data, or rounds of messages exchanged for such a purpose. What options are available to us?

Focusing on the model itself, it makes sense to examine the static nature of the membership model. In light of the fact that real programs are started, run for a while, and then terminated, one might ask if a static membership model really makes sense in a distributed setting. The most apparent response is that the hardware resources on which the programs execute will normally be static and that such a model actually may not be *required* for many applications. Moreover, the hardware base used in many systems does change over time, albeit slowly and fairly infrequently. If one assumes that certain platforms are always started with the same set of programs running on them, a static membership model is closely matched to the environment. However, if one looks at the same system over a time scale long enough to encompass periods of hardware reconfiguration and upgrades, or looks at the applications themselves, the static model seems less natural.

Suppose that one concedes that a given application is closely matched to the static model. Even in this case, what really makes that system static is the memory its component programs retain of their past actions; otherwise, even though the applications tend to run on the same platforms, this is of limited importance outside of the implications for the maximum number of applications that might be operational at any time. Suppose that a computer has an attached disk on it, and some program is the manager for a database object stored on that disk. Each time this program interacts with other programs in the system, its entire past history of interactions is evidenced by the state of the database at the time of the interaction. It makes sense to model such a system as having a set of processes that may be inaccessible for periods of time but that are basically static and predefined, because the data saved on disk lets them prove that this was the case. Persistence, then, is at the core of the static model.

Another example of a setting in which the resources are fundamentally static would be a system in which a set of computers controls a collection of external devices. The devices will presumably not represent a rapidly changing resource—hence, it makes a lot of sense to model the programs that manage them as a static group of processes. Here, the external world is persistent to the degree that it retains memory of the actions taken in the past. Without such memory, however, one might just as well call this a dynamic system model.

In fact, there are a great many types of programs that are launched and execute with no particular memory of the past activities on the computer used to run them. These programs may tend to run repeatedly on the same set of platforms, and one certainly could apply a static system membership model to them. Yet in a deeper sense they are oblivious to the place at which they run. Such a perspective gives rise to what we call the *dynamic membership model*, in which the set of processes comprising the system at a given time varies. New processes are started and *join* the system, while active processes *leave* the system when they terminate, fail, or simply choose to disconnect themselves. While such a system may also have a static set of processes associated with it, such as a set of servers shared by the dynamically changing population of transient processes, it makes sense to adopt a dynamic membership model to deal with this collection of transient participants.

Even the process associated with a static resource such as a database can potentially be modeled as a dynamic component of the system to which it belongs. In this perspective there is a need for the system to behave in a manner consistent with the externally persistent actions taken by previous members. These actions may be recorded in databases or have had external effects on the environment. The process that manages such a resource is treated as a dynamic component of the system. It must be launched when the database server is booted and terminate when the server crashes. Its state is volatile: The database persists, but not the memory contents of the server at the time it shuts down. Thus, even if one believes that a static system model is physically reasonable, it may still be acceptable to model this with a dynamic model—provided that attention is paid to externally persistent actions initiated by system members. The advantage of doing this is that the model will not be tied to the system configuration, so hardware upgrades and other configuration changes do not necessarily have to be treated outside of the model in which reliability was presented. To the degree that systems need to be self-managing, such freedom can be extremely useful.

The dynamic model poses new challenges—not the least of which is to develop mechanisms for tracking the current membership of the system, since this will no longer be a well-known quantity. It also offers significant potential for improvements in our protocols, particularly if processes that leave the system are considered to have terminated. The advantage, as we will see shortly, is that agreement within the operational group of system members is an easier problem to solve than the sorts of consensus and replication problems examined above for a static set of possibly inaccessible members.

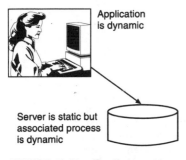

Application is dynamic

Server is static but associated process is dynamic

FIGURE 13.10. Trading system

In the rest of this section, we explore some of the fundamental problems raised by dynamic membership. The treatment distinguishes two cases: Those in which actions must be *dynamically uniform*, meaning that any action taken by a process must be consistent with subsequent actions by the operational part of the system (see Amir et al. [July 1992], Malkhi et al., Schiper and Sandoz), and those in which dynamic uniformity is not required, meaning that the operational part of the system is taken to define the system, and the states and actions of processes that subsequently fail can be discarded. Dynamic uniformity captures the case of a process that, although executed within a dynamic system model, may perform externally visible actions. In our database server example, the server process can perform such actions, by modifying the contents of the database it manages. (See Figure 13.10.)

Dynamic uniformity may seem similar to the property achieved by a commit protocol, but there are important differences. In a commit protocol, we require that if any process commits some action, all processes will commit it. This obligation holds within a statically defined set of processes: A process that fails may later recover, so the commit problem involves an indefinite obligation, with regard to a set of participants, that is specified at the outset. In fact, the obligation holds even if a process reaches a decision and then crashes without telling any other process what that decision was.

Dynamic uniformity says that if any process performs some action, all processes that remain operational will also perform it (Figure 13.11)—that is, the obligation to perform an action begins as soon as any process in the system performs that action, and it then extends to processes that remain operational, but not to processes that fail. This is because, in a dynamic membership model, we adopt the view that if a process is excluded from the membership it has failed, and a process that fails never rejoins the system. In practice, a process that fails may reconnect as a new process, but when we consider join events, we normally say that a process that joins the system is only required to perform those actions that are initiated after its join event (in a causal sense). The idea is that a process joining the system, whether it is new or an old process reconnecting after a long delay, learns the state of the system as it first connects and is then guaranteed to perform all the dynamically uniform actions that any process performs, until such time as it fails or (equivalently) is excluded from the system membership set.

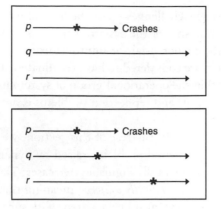

FIGURE 13.11. Nonuniform (top) and dynamically uniform (bottom) executions. In the nonuniform case, process _p_ takes an action (denoted "*") that no other process takes. This is acceptable because _p_ crashes. In the dynamically uniform case, all processes must take the action if any one process does. This implies that before _p_ took the action, its intention to do so must have been communicated to the remainder of the system.

Real-world applications often have a need for dynamic uniformity, but this is not invariably the case. Consider a desktop computing system used in a financial application. The role of the application program might be to display financial information of importance to the trader in a timely manner, to compute financial metrics such as theoretical prices for specified trading instruments, and to communicate new trades to the exchanges where they will be carried out. Such systems are common and represent a leading-edge example of distributed computing used in support of day-to-day commerce. These systems are also quite critical to their users, so they are a good example of an application in which reliability is important.

Is such an application best viewed as static or dynamic, and must the actions taken be dynamically uniform ones? If we assume that the servers providing trading information and updates are continuously operational, there is little need for the application on the trader's desktop to retain significant state. It may make sense to cache frequently used financial parameters or to keep a local log of submitted trading requests as part of the application, but such a system would be unlikely to provide services when disconnected from the network and is unlikely to have any significant form of local state information that is not also stored on the databases and other servers that maintain the firm's accounts.

Thus, while there are some forms of information that may be stored on disk, it is reasonable to view this as an example of a dynamic computing application. The trader sits down at his or her desk and begins to work, launching one or more programs that connect to the back-end computing system and that start providing data and support services. At the end of the day, or when no longer needed, the programs shut down. In the event of a network communication problem that disconnects the desktop system from the system servers, the trader is more likely to move to a different station than to try to continue using it while disconnected. Any state that was associated with the disconnected station will rapidly become stale, and when the station is reconnected, it may make more sense to restart the application programs as if from a failure rather than try to reintegrate them into the system in order to exploit local information not available on the servers.

This type of dynamicism is common in client/server architectures. Recall the discussion of caching when we considered Web proxies in Part II. The cached documents reside within a dynamically changing set of processes surrounding a static set of servers on which the originals reside. The same could be said for the file systems we examined in Part I, which cache buffers or whole files.

It is not often required that actions taken by a user-interface program satisfy a dynamic uniformity property. Dynamic uniformity occurs when a single action is logically taken at multiple places in a network—for example, when a replicated database is updated or when traders are asked to take some action and it is important that *all* of them do so. In our trading system, one could imagine the server sending information to multiple client systems, or there could be a client entering a trading strategy that impacts multiple other clients and servers. More likely, however, the information sent by servers to the client systems will be pricing updates and similar data, and if the client system and server system fail just as the information is sent, there may not be any great concern that the same message must be delivered to every other client in the network. Similarly, if the client sends a request for information to the server, which fails at the instant the request is sent, it may not be terribly important if the request is seen by the server or not. These are examples of nondynamic uniform actions.

In contrast, although the process that handles the database can be viewed as a dynamic member of the system, its actions do need to satisfy the dynamically uniform property. If a set of such servers maintains replicated data or has any sort of consistency property that spans them as a set, it will be important that an action taken by one process also be taken by the others.

Thus, our example can be modeled as a dynamic system. Some parts of the system may require dynamic uniformity, but others can manage without this property. As we will see shortly, this choice has significant performance implications—hence, we will work toward a family of protocols in which dynamic uniformity is available on an operation-by-operation basis, depending on the needs of the application and the price of such a guarantee.

13.10 The Group Membership Problem

The role of a group membership service (GMS) is to maintain the membership of a distributed system on behalf of the processes that compose it. As previously described, processes join and leave the system dynamically over its lifetime. We will adopt a model in which processes wishing to join do so by first contacting the GMS, which updates the list of system members and then grants the request. Once admitted to the system, a process may interact with other system members. Finally, if the process terminates, the GMS service will again update the list of system members.

13.10.1 GMS Controversy

My work concerning the GMS problem included an effort to break out the problem from the context in which we use it, specifying it as a formal abstract question using temporal logic, and then using the temporal logic properties to reason about the correctness and optimality of the protocols used to implement GMS. Unfortunately, some of this work was later found to have flaws, specifically in the temporal logic formulas that were proposed as the rigorous specification of the GMS. Readers who become interested in this area should be aware that the GMS specification used in papers by Ricciardi and Birman does have some fairly serious problems. Better specifications have subsequently been proposed by Malkhi (unpublished 1995), Babaoglu et al. (1996), and Neiger.

The essential difficulty is that the behavior of a GMS implementation depends upon future events. Suppose that a process p suspects that process q is faulty. If p itself remains in the system, q will eventually be excluded from it. But there are cases in which p itself might be excluded from the system, in which both p and q might be excluded, and in which the system as a whole is prevented from making progress because less than a majority of the processes that participated in the previous system view have remained operational. Unfortunately, it is not clear which of these cases will apply until *later in the execution* when the system's future has become definite. In the specification of the GMS, one wants to say that if p suspects the failure of q then q will eventually be excluded from the system, unless p is excluded from the system. However, formalizing this type of uncertain action is extremely difficult in the most popular forms of temporal logics.

A further problem was caused by what is ultimately a misunderstanding of terminology. In the theory community, one says that a problem is "impossible" if the problem is "not always solvable" in the target environment. Thus, because there can be sequences of failures for which GMS is unable to make progress, the GMS problem is an "impossible one." The theory community deals with this by defining special types of failure detectors and proving that a problem such as GMS is (always) solvable given such a failure detector. They do not necessarily tackle the issues associated with implementing such a failure detector.

In our specification of GMS, Ricciardi and I used a similar approach, characterizing GMS as "live provided that the frequency of failures was sufficiently low," but using English rather than temporal logic formulas to express these conditions. There were some who were critical of this approach, arguing that from the formulas alone it was impossible to know precisely what the conditions were for which GMS is guaranteed to be live. Moreover, they proved that in the absence of some form of restrictive condition (some form of failure detector), the GMS problem was "impossible" in the sense that it cannot always be solved in the general asynchronous model. This is not surprising and reflects the same limitations we encountered when looking at the multiphase commit protocol.

In summary, then, there exists an attempt to formalize the GMS problem, and there are some known and serious problems with the proposed formalization. However, the proposed protocols are generally recognized to work correctly, and the concerns that have been raised revolve around relatively subtle theoretical issues having to do with the way the formal specification of the problem was expressed in temporal logic. At the time of this writing, there has been encouraging progress on these issues, and we are closer than ever before to knowing precisely when GMS can be solved and precisely how this problem relates to the asynchronous consensus problem. These research topics, however, seem to have no practical implications at all for the design of distributed software systems such as the ones we present here: They warn us that we will not always be able to guarantee progress, consistency and high availability, but they do not imply that we will not "usually" be successful in all of these goals.

13.10.2 GMS and Other System Processes

The interface between the GMS and other system processes provides three operations, shown in Table 13.1. The *join* operation is invoked by a process that wishes to become a system member. The *monitor* operation is used by a process to register its interest in the status of some other process; should that process be dropped from the membership, a callback will occur to notify it of the event. Such a callback is treated as the equivalent of a failure notification in the fail-stop computing model: The process is considered to have crashed, all communication links with it are severed, and messages subsequently received from it are rejected. Finally, the *leave* operation is used by a process that wishes to disconnect itself from the system, or by some other system component that has detected a fault and wishes to signal that a particular process has failed. We assume throughout this section that failure detections are inaccurate in the sense that they may result from partitioning of the network, but they are otherwise of sufficiently good quality as to rarely exclude an operational process as faulty.

The GMS itself will need to be highly available—hence, it will typically be implemented by a set of processes that cooperate to implement the GMS abstraction. Although these processes would normally reside on a statically defined set of server computers, so that they can readily be located by a process wishing to join the system, the actual composition of the group may vary over time due to failures within the GMS service itself, and one can imagine other ways of tracking down representatives (files, name services, the use of hardware broadcast to poll for a member, etc.). Notice that in order to implement the GMS abstraction on behalf of the rest of the system, a GMS server needs to solve the GMS problem on its own behalf. We will say that it uses a *group membership protocol*, or GMP, for this purpose. Thus, the GMP deals with the membership of a small service, the GMS, which the rest of the system (a potentially large set of processes) employs to track the composition of the system as a whole.

Similar to the situation for other system processes that don't comprise the GMS, the GMP problem is defined in terms of *join* and *leave* events; the latter being triggered by the inability of the GMS processes to communicate with one another. Clearly, such an environment creates the threat of a partitioning failure, in which a single GMS might split into multiple GMS subinstances, each of which considers the other to be faulty. What should our goals be when such a partitioned scenario occurs?

Suppose that our distributed system is being used in a setting such as air traffic control. If the output of the GMS is treated as being the logical equivalent of a failure notification, one would

TABLE 13.1. GMS Operations

Operation	Function	Failure handling
join (process-ID, callback) returns (time, GMS list)	Calling process is added to membership list of system, returns logical time of the join event and a list giving the membership of the GMS service. The callback function is invoked whenever the core membership of the GMS changes.	Idempotent: can be reissued to any GMS process with same outcome
leave (process-ID) returns void	Can be issued by any member of the system. GMS drops the specified process from the membership list and issues notification to all members of the system. If the process in question is really operational, it must rejoin under a new process-ID.	Idempotent: fails only if the GMS process that was the target is dropped from the GMS membership list
monitor (process-ID, callback) returns callback-ID	Can be issued by any member of the system. GMS registers a callback and will invoke callback (process-ID) later if the designated process fails.	Idempotent: as for *leave*

expect the system to reconfigure itself after such notification to restore full air traffic control support within the remaining set of processes. If some component of the air traffic control system is responsible for advising controllers about the status of sectors of the airspace (free or occupied), and the associated process fails, the air traffic system would probably restart it by launching a new status manager process.

Now, a GMS partition would be the likely consequence of a network partition, raising the prospect that two air traffic sector services could find themselves simultaneously active, both trying to control access to the same portions of the airspace, and neither aware of the other! Such an inconsistency would have disastrous consequences. While the partitioning of the GMS might be permissible, it is clear that at most one of the resulting GMS components should be permitted to initiate new actions.

From this example we see that although we might want to allow a system to remain operational during partitioning of the GMS, we also need a way to pick one component of the overall system as the primary one, within which authoritative decisions can be taken on behalf of the system as a whole (see Malkhi, Ricciardi [1993]). Nonprimary components might report information on the basis of their state as of the time a partitioning occurred, but would not permit potentially conflicting actions (such as routing a plane into an apparently free sector of airspace) to be initiated. Such an approach clearly generalizes: One can imagine a system in which some applications would be considered primary within a component and considered nonprimary for other purposes. Moreover, there may be classes of actions that are safe even within a nonprimary component; an example would be the reallocation of air traffic within sectors of the air traffic service already owned by the partition at the time the network failed. But it is clear that any GMP solution should at least track a primary component so that actions can be appropriately limited.[3]

The key properties of the primary component of the GMS are that its membership should overlap with the membership of a previous primary component of the GMS and that there should only be one primary component of the GMS within any partitioning of the GMS as a whole. We will then say that the primary component of a partition of the distributed system as a whole is determined by the primary component of the GMP to which its processes are connected.

In the beginning of this chapter, we discussed concepts of time in distributed settings. In defining the primary component of a partitioned GMS we used temporal terms without making it clear exactly what form of time was intended. In the following text, we have *logical* time in mind. In particular, suppose that process p is a member of the primary component of the GMS, but then suddenly becomes partitioned away from the remainder of the GMS, executing for an arbitrarily long period of time without sending or receiving any additional messages and finally shutting down. From the discussion up to now, it is clear that we would want the GMS to reconfigure itself to exclude p, if possible, forming a new primary GMS component, which can permit further progress in the system as a whole. But now the question occurs as to whether or not p would be aware that this has happened. If not, p might consider itself a member of the previous primary component of the GMS, and we would now have two primary components of the GMS active simultaneously.

There are two ways in which we could respond to this issue. The first involves a limited introduction of time into the model. Where clocks are available, it would be useful to have a mechanism whereby any process that ceases to be a member of the primary component of a partitioned GMS can detect this situation within a bounded period of time—for example, it would be helpful to know that within two seconds of being excluded from the GMS, p would know that it is no longer a member of the primary component. If we assume that clocks are synchronized to a specified degree, we would ideally like to be able to compute the smallest time constant δ, such that it is meaningful to say that p will detect this condition within time $t + \delta$ of when it occurs.

In addition to this, we will need a way to capture the sense in which it is legal for p to lag behind the GMS in this manner, albeit for a limited period of time. Notice that because we wish

---

[3]Later, we will present some work by Idit Keidar and Danny Dolev, in which a partitioned distributed system is able to make progress despite the fact that no primary component is ever present. However, in this work a static system membership model is used, and no action can be taken until a majority of the processes in the system as a whole are known to be aware of the action. This is a costly constraint and seems likely to limit the applicability of the approach.

to require that primary components of the GMS have overlapping membership, if we are given two different membership lists for the GMS, a and b, then either $a \rightarrow b$, or $b \rightarrow a$. Thus, rather than say that there should be at most one primary component of the GMS active simultaneously, we will say that any two concurrently active membership lists for the GMS (in the sense that each is considered current by some process) should be ordered by causality. Equivalently, we could now say that there is at most a single sequence of GMS membership lists that is considered to represent the primary component of the GMS. We will use the term "view" of the GMS membership to denote the value of the membership list that holds for a given process within the GMS at a specified point in its execution.

If the GMS can experience a partitioning failure, it can also experience the *merging* of partitions (see Amir et al. [1992], Malkhi, Moser et al. [1994]). The GMP should therefore include a merge protocol. Finally, if all the members of the GMS fail, or if the primary partition is somehow lost, the GMP should provide for a restart from complete failure or for identification of the primary partition when the merge of two nonprimary partitions makes it possible to determine that there is no active primary partition within the system. We'll discuss this issue at some length in Chapter 15.

The protocol that we now present is based on one that was developed as part of the Isis system in 1987 (see Birman and Joseph [February 1987]), but was substantially extended by Ricciardi in 1991 as part of her Ph.D. dissertation (see Ricciardi [1992, 1993], Ricciardi and Birman). A slightly more elaborate version of this protocol has been proved optimal, but we present a simpler version for clarity. The protocol has an interesting property: All GMS members see exactly the same sequence of join and leave events. The members use this property to obtain an unusually efficient protocol execution.

To avoid placing excessive trust in the correctness or fault tolerance of the clients, our goal will be to implement a GMS for which all operations are invoked using a modified RPC protocol. Our solution should allow a process to issue requests to any member of the GMS server group with which it is able to establish contact. The protocol implemented by the group should stipulate that *join* operations are idempotent: If a joining process times out or otherwise fails to receive a reply, it can reissue its request, perhaps to a different server. Having joined the system, clients that detect apparent failures merely report them to the GMS. The GMS itself will be responsible for all forms of failure notification, both for GMS members and other clients. Thus, actions that would normally be triggered by timeouts (such as reissuing an RPC or breaking a stream connection) will be triggered in our system by a GMS callback notifying the process doing the RPC or maintaining the stream that the party it is contacting has failed. Table 13.1 summarizes this interface.

13.10.3 Protocol Used to Track GMS Membership

We start by presenting the protocol used to track the core membership of the GMS service itself. These are the processes responsible for implementing the GMS abstraction, but not their clients. We assume that the processes all watch one another using some form of network-level ping operation, detecting failures by timeout.

Both the addition of new GMS members and the deletion of apparently failed members are handled by the GMS coordinator, which is the GMS member that has been operational for the longest period of time. As we will see, although the GMS protocol permits more than one process to be added or deleted at a time, it orders all add and delete events so that this concept of oldest process is well defined and consistent throughout the GMS. If a process believes the GMS coordinator has failed, it treats the next highest ranked process (perhaps itself) as the new coordinator.

Our initial protocol will be such that any process suspected of having failed is subsequently *shunned* by the system members that learn of the suspected failure. Upon detection of an apparent failure, a GMS process immediately ceases to accept communication from the failed process. It also immediately sends a message to every other GMS process with which it is communicating, informing them of the apparent failure; they then shun the faulty process as well. If a shunned process is actually operational, it will learn that it is being shunned when it next attempts to communicate with a GMS process that has heard of the fault; at this point it is expected that the shunned process will rejoin the GMS under a new process identifier. In this manner, a suspected failure can be treated as if it were a real one.

Having developed this initial protocol, we will discuss extensions that allow partitions to form and later merge in Section 13.10.6, and again in Chapter 15, where we present an execution model that makes use of this functionality.

Upon learning of a failure or an addition request, the GMS coordinator starts a protocol that will lead to the updating of the membership list, which is replicated among all GMS processes. The protocol requires two phases when the processes being added or deleted do not include the old GMS coordinator; a third phase is used if the coordinator has failed and a new coordinator is taking over. Any number of add operations can be combined into a single round of the protocol. A single round can also perform multiple delete operations, but here there is a limit: At most a minority of the processes present in a given view can be dropped from the subsequent view (more precisely, a majority of the processes in a given view must acknowledge the next view; obviously, this implies that the processes in question must be alive).

In the two-phase case, the first round of the protocol sends the list of add and delete events to the participants, including the coordinator itself. All acknowledge receipt. The coordinator waits for as many replies as possible, but also requires a majority response from the current membership. If less than a majority of processes are reachable it waits until communication is restored before continuing. If processes have failed and only a minority are available, a special protocol is executed.

Unless additional failures occur at this point in the protocol, which would be very unlikely, a majority of processes acknowledge the first-round protocol. The GMS coordinator now commits the update in a second round, which also carries with it notifications of any failures that were detected during the first round. Indeed, the second-round protocol can be compacted with the first round of a new instance of the deletion protocol, if desired. The GMS members update their membership view upon reception of the second-round protocol messages.

In what one hopes will be an unusual condition, it may be that a majority of the previous membership cannot be contacted because too many GMS processes have crashed. In this case, a

GMS coordinator still must ensure that the failed processes did not acquiesce in a reconfiguration protocol of which it was not a part. In general, this problem may not be solvable—for example, it may be that a majority of GMS processes have crashed, and prior to crashing they could have admitted any number of new processes and deleted the ones now trying to run the protocol. Those new processes could now be anywhere in the system. In practice, however, this problem is often easy to solve: The GMS will most often execute within a static set of possible server hosts, and even if this set has some small degree of dynamicism, it is normally possible to track down any GMS server by checking a moderate number of nodes for a representative.

Both of these cases pertained to the case where the coordinator did not fail. A three-phase protocol is required when the current coordinator is suspected of having failed and some other coordinator must take over. The new coordinator starts by informing at least a majority of the GMS processes listed in the current membership that the coordinator has failed and then collects their acknowledgments and current membership information. At the end of this first phase, the new coordinator may have learned of pending add or delete events that were initiated by the prior coordinator before it was suspected of having failed. The first-round protocol also has the effect of ensuring that a majority of GMS processes will start to shun the old coordinator. The second and third rounds of the protocol are exactly as for the normal case: The new coordinator proposes a new membership list, incorporating any add events it has learned about, as well as all the delete events, including those it learned about during the initial round of communication and those from the prior coordinator. It waits for a majority to acknowledge this message and then commits it, piggybacking suspected failure information for any unresponsive processes.

Ricciardi has provided detailed proof that the above protocol results in a single, ordered sequence of process add and leave events for the GMS and that it is immune to partitioning (see Ricciardi [1992]). The key to her proof is the observation that any new membership list installed successfully necessarily must be acknowledged by a majority of the previous list, and therefore any two concurrent protocols will be related by a causal path. One protocol will learn of the other, or both will learn of one another, and this is sufficient to prevent the GMS from partitioning. Ricciardi shows that if the ith round of the protocol starts with n processes in the GMS membership, an arbitrary number of processes can be added to the GMS and at most $[n/2] - 1$ processes can be excluded (this is because of the requirement that a majority of processes agree with each proposed new view). In addition, she shows that even if a steady stream of join and leave or failure events occurs, the GMS should be able to continuously output new GMS views provided that the number of failures never rises high enough to prevent majority agreement on the next view. In effect, although the protocol may be discussing the proposed $i + 2$nd view, it is still able to commit the $i + 1$st view.

13.10.4 GMS Protocol to Handle Client Add and Join Events

We now turn to the issues that occur if a GMS server is used to manage the membership of a larger number of client processes, which interact with it through the interface given earlier.

In this approach, a process wishing to join the system will locate an operational GMS member. It then issues a *join* RPC to that process. If the RPC times out, the request can simply be reissued to some other member. When the join succeeds, it learns its ranking (the time at which the join took place) and the current membership of the GMS service, which is useful in setting up subsequent monitoring operations. Similarly, a process wishing to report a failure can invoke the *leave* operation in any operational GMS member. If that member fails before confirming that the operation has been successful, the caller can detect this by receiving a callback reporting the failure of the GMS member itself and then can reissue the request.

To solve these problems, we could now develop a specialized protocol. Before doing so, however, it makes sense to ask if the GMS is not simply an instance of a service that manages replicated data on behalf of a set of clients; if so, we should instead develop the most general and efficient solutions possible for the replicated data problem, and then use them within the GMS to maintain this specific form of information. And, indeed, it is very natural to adopt this point of view.

To transform the one problem into the other, we need to understand how an RPC interface to the GMS can be implemented such that the GMS would reliably offer the desired functionality to its clients, using data replication primitives internally for this purpose. Then we can focus on the data replication problem separately and convince ourselves that the necessary primitives can be developed and can offer efficient performance.

The first problem that needs to be addressed concerns the case where a client issues a request to a representative of the GMS that fails before responding. This can be solved by ensuring that such requests are *idempotent*, meaning that the same operation can be issued repeatedly and will repeatedly return the identical result—for example, an operation that assigns the value 3 to a variable x is idempotent, whereas an operation that increments x by adding 1 to it would not be. We can make the client join operation idempotent by having the client uniquely identify itself and repeat the identifier each time the request must be reissued. Recall that the GMS returns the time of the join operation; this can be made idempotent by arranging it so that if a client join request is received from a client already listed as a system member, the time currently listed is returned and no other action is taken.

The remaining operations are all initiated by processes that belong to the system. These, too, might need to be reissued if the GMS process contacted to perform the operation fails before responding (the failure would be detected when a new GMS membership list is delivered to a process waiting for a response, and the GMS member it is waiting for is found to have been dropped from the list). It is clear that exactly the same approach can be used to solve this problem. Each request need only be uniquely identifiable—for example, using the process identifier of the invoking process and some form of counter (request 17 from process p on host h).

The central issue is thus reduced to replication of data within the GMS or within similar groups of processes. We will postpone this problem momentarily, returning later when we give a protocol for implementing replicated data within dynamically defined groups of processes.

13.10.5 GMS Notifications with Bounded Delay

If the processes within a system possess synchronized clocks, it is possible to bound the delay before a process becomes aware that it has been partitioned from the system. Consider a system in which the health of a process is monitored by the continued reception of some form of "still alive" messages received from it; if no such message is received after delay σ, any of the processes monitoring that process can report it as faulty to the GMS. (Normally, such a process would also cease to accept incoming messages from the faulty process and would also gossip with other processes to ensure that if p considers q to have failed, then any process that receives a message from p will also begin to shun messages from q.) Now, assume further that all processes receiving a "still alive" message acknowledge it.

In this setting, p will become aware that it may have been partitioned from the system within a maximum delay of $2 * \varepsilon + \sigma$, where ε represents the maximum latency of the communication channels. More precisely, p will discover that it has been partitioned from the system $2 * \varepsilon + \sigma$ time units after it last had contact with a majority of the previous primary component of the GMS. In such situations, it would be appropriate for p to break any channels it has to system members and to cease taking actions on behalf of the system as a whole.

Thus, although the GMS may run its protocol to exclude p as early as $2 * \varepsilon$ time units before p discovers that it has been partitioned from the main system, there is a bound on this delay. The implication is that the new primary system component can safely break locks held by p or otherwise take over actions for which p was responsible after $2 * \varepsilon$ time units have elapsed. (See Figure 13.12.)

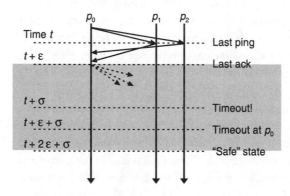

FIGURE 13.12. If channel delays are bounded, a process can detect that it has been partitioned from the primary component within a bounded time interval, making it safe for the primary component to take over actions from it even if externally visible effects may be involved. The gray region denotes a period during which the new primary process will be unable to take over because there is some possibility that the old primary process is still operational in a nonprimary component and may still be initiating authoritative actions. At the end of the gray period a new primary process can be appointed within the primary component. There may be a period of real time during which no primary process is active, but there is no risk that two processes could be simultaneously active. One can also bias a system in the other direction, so that there will always be at least one primary process active provided that the rate of failures is limited.

Reasoning such as this is only possible in systems where clocks are synchronized to a known precision and in which the delays associated with communication channels are also known. In practice, such values are rarely known with any accuracy, but coarse approximations may exist. Thus, in a system where message-passing primitives provide expected latencies of a few milliseconds, one might take ε to be a much larger number: one second or ten seconds. Although extremely conservative, such an approach would in practice be quite safe. Later, we will examine real-time issues more closely, but it is useful to keep in mind that very coarse-grained real-time problems are often simple in distributed systems where the equivalent fine-grained real-time problems would be very difficult or impossible. At the same time, even a coarse-grained rule such as this one would only be safe if there were good reason to believe that the value of ε was a safe approximation. Some systems provide no guarantees of this sort at all, in which case incorrect behavior could result if a period of extreme overload or some other unusual condition caused the ε limit to be exceeded.

To summarize, the core primary partition GMS protocol must satisfy the following properties:

- *C-GMS-1:* The system membership takes the form of system views. There is an initial system view, which is predetermined at the time the system starts. Subsequent views differ by the addition or deletion of processes.

- *C-GMS-2:* Only processes that request to be added to the system are added. Only processes that are suspected of failure or that request to leave the system are deleted.

- *C-GMS-3:* A majority of the processes in view i of the system must acquiesce in the composition of view $i + 1$ of the system.

- *C-GMS-4:* Starting from an initial system view, subsequences of a single sequence of system views are reported to system members. Each system member observes such a subsequence starting with the view in which it was first added to the system and continuing until it fails, when it either leaves the system or is excluded from the system.

- *C-GMS-5:* If process p suspects process q of being faulty, and if the core GMS service is able to report new views, either q will be dropped from the system, or p will be dropped, or both.

- *C-GMS-6:* In a system with synchronized clocks and bounded message latencies, any process dropped from the system view will know that this has occurred within bounded time.

As noted previously, the core GMS protocol will not always be able to make progress: There are patterns of failures and communication problems that can prevent it from reporting new system views. For this reason, *C-GMS-5* is a conditional liveness property: If the core GMS is able to report new views, then it eventually acts upon process add or delete requests. It is not yet clear what conditions represent the weakest environment within which liveness of the GMS can always be guaranteed. For the protocol given above, the core GMS will make progress provided that at most a minority of processes from view i fail or are suspected of having failed during the period needed to execute the two- or three-phase commit protocol used to install new views. Such a characterization may seem evasive, since such a protocol may execute extremely rapidly in some settings and extremely slowly in others. However, unless the timing properties of the system are sufficiently strong to support estimation of the time needed to run the protocol, this seems to be as strong a statement as can be made.

We note that the failure detector called ◊W in Chandra and Toueg's work is characterized in terms somewhat similar to this (see Chandra and Toueg [1991], Chandra et al. [1992]). Very recent work by several researchers (see Babaoglu et al. [1995], Friedman et al., Guerraoui [1995]) has shown that the ◊W failure detector can be adapted to asynchronous systems in which messages can be lost during failures or processes can be killed because the majority of processes in the system consider them to be malfunctioning. Although fairly theoretical in nature, these studies are shedding light on the conditions under which problems such as membership agreement can always be solved and those under which agreement may not always be possible (the theoreticians are fond of calling the latter problems "impossible"). To present this work here, however, is beyond the scope of this book.

13.10.6   Extending the GMS to Allow Partition and Merge Events

Research on the Transis system, at Hebrew University in Jerusalem, has yielded insights into the extension of protocols, such as the one used to implement our primary component GMS, so that they can continue operation during partitionings that leave no primary component, or allow activity in a nonprimary component, reconciling the resulting system state when partitions later remerge (see Amir et al. [1992], Malkhi). Some of this work was done jointly with the Totem project at the University of California, Santa Barbara (see Moser et al.).

Briefly, the approach is as follows. In Ricciardi's protocols, when the GMS is unable to obtain a majority vote in favor of a proposed new view, the protocol ceases to make progress. In the extended protocol, such a GMS can continue to produce new views, but no longer considers itself to be the primary partition of the system. Of course, there is also a complementary case in which the GMS encounters some other GMS and the two merge their membership views. It may now be the case that one GMS or the other was the primary component of the system, in which case the new merged GMS will also be primary for the system. On the other hand, perhaps a primary component fragmented in such a way that none of the surviving components considers itself to be the primary one. When this occurs, it may be that later, such components will remerge and the primary component can then be deduced by study of the joint histories of the two components. Thus, one can extend the GMS to make progress even when partitioning occurs.

Some recent work at the University of Bologna, on a system named Relacs, has refined this approach into one that is notable for its simplicity and clarity. Ozalp Babaoglu, working with Alberto Bartoli and Gianluca Dini, has demonstrated that a very small set of extensions to a view-synchronous environment suffice to support EVS-like functionality. They call their model Enriched View Synchrony and describe it in a technical report that appeared shortly before this book went to press (see Babaoglu et al. [1996]). Very briefly, Enriched View Synchrony arranges to deliver only *nonoverlapping* group views within different components of a partitioned system. The reasoning behind this is that overlapping views can cause applications to briefly believe that the same process or site resides on both sides of a partition, leading to inconsistent behavior. Then, they provide a set of predicates by which a component can determine whether or not it has a quorum that would permit direct update of the global system state, as well as algorithmic tools for assisting in the state merge problem that occurs when communication is reestablished. I am not aware of any implementation of this model yet, but the primitives are simple and an implementation in a system such as Horus (Chapter 18) would not be difficult.

Having described these approaches, an important question remains: whether or not it is *desirable* to allow a GMS to make progress in this manner. We defer this point until Chapter 16. In addition, as was noted in footnote 3, Keidar and Dolev have shown that there are cases in which no component is ever the primary one for the system, and yet dynamically uniform actions can still be performed through a type of gossip that occurs whenever the network becomes reconnected and two nonminority components succeed in communicating. Although interesting, this protocol is costly: Prior to taking any action, a majority of all the processes in the system must be known to have seen the action. Indeed, Keidar and Dolev developed their solution for a static membership model, in which the GMS tracks subsets of a known maximum system membership. The majority requirement makes this protocol costly—hence, although it is potentially useful in the context of wide area systems that experience frequent partition failures, it is not likely that one would use it directly in the local area communication layers of a system. We will return to this issue in Chapter 15 in conjunction with the model called *Extended Virtual Synchrony*.

13.11 Dynamic Process Groups and Group Communication

When the GMS is used to ensure systemwide agreement on failure and join events, the illusion of a fail-stop computing environment is created (see Sabel and Marzullo). If one were to implement the 3PC protocol using the notifications of the GMS service as a failure-detection mechanism, the 3PC protocol would be nonblocking—provided, of course, that the GMS service itself were able to remain active and continue to output failure detections. The same power that the GMS brings to the 3PC problem can also be exploited to solve other problems, such as data replication, and offers us ways to do so that can be remarkably inexpensive relative to the quorum update solutions presented previously. Yet, we can say that these systems are reliable in a strong sense: Under conditions when the GMS can make progress, such protocols will also make progress and will maintain their consistency properties continuously, at least when permission to initiate new actions is limited to the primary component in the event that the system experiences a partitioning failure.

In the following sections, we develop this idea into an environment for computing with what are called *virtually synchronous process groups*. We begin by focusing on a simpler problem, closely related to 2PC—namely the reliable delivery of a message to a statically defined group of processes. Not surprisingly, our solution will be easily understood in terms of the 2PC protocol: delivering messages in the first phase if internal consistency is all that we require and doing so in the second phase if dynamic uniformity (external consistency) is needed. We will then show how this solution can be extended to provide ordering for the delivery of messages; later, such ordered and reliable communication protocols will be used to implement replicated data and locking. Next, we show how the same protocols can also be used to implement dynamic groups of processes. In contrast to the dynamic membership protocols used in the GMS, however, these protocols will be quite a bit simpler and less costly. We then introduce a synchronization mechanism, which allows us to characterize these protocols as failure-atomic with respect to group membership changes; this implements a model called the *view synchrony* model. Finally, we show how view synchrony can support a more extensive execution model called *virtual synchrony*, which supports a particularly simple and efficient style of fault-tolerant computing. Thus, step by step, we will show how to built up a reliable and consistent computing environment starting with the protocols embodied in the group membership service.

Up to the present we have focused on protocols in terms of a single group of processes at a time, but the introduction of sophisticated protocols and tools in support of process group computing also creates the likelihood that a system will need to support a great many process groups simultaneously and that a single distributed application may embody considerable numbers of groups—perhaps many groups for each process. Such developments have important performance implications and will motivate us to reexamine our protocols.

Finally, we will turn to the software engineering issues associated with support for process group computing. This topic, which is addressed in the next chapter, will center on a specific software system developed by me (and my colleagues), called Horus. The chapter also reviews a number of other systems, and, in fact, one of the key goals of Horus is to be able to support the features of some of these other systems within a common framework.

13.11.1   Group Communication Primitives

A *group communication primitive* is a procedure for sending a message to a set of processes that can be addressed without knowledge of the current membership of the set. Recall that we discussed the concept of a hardware *broadcast* capable of delivering a single message to every computer connected to some sort of communication device. Group communication primitives would normally transmit to subsets of the full membership of a computing system, so we use the term "multicast" to describe their behavior. A multicast is a protocol that sends a message from one *sender* process to multiple *destination* processes, which *deliver* it.

Suppose we know that the current composition of some group G is $\{p_0 \dots p_k\}$. What properties should a multicast to G satisfy?

The answer to this question will depend upon the application. As will become clear in Chapters 16 and 17, there are a great number of reliable applications for which a multicast with relatively weak properties would suffice—for example, an application simply seeking information that any of the members of G can provide might multicast an inquiry to G as part of an RPC-style protocol requiring a single reply, taking the first one that is received. Such a multicast would ideally avoid sending the message to the full membership of G, resorting instead to heuristics for selecting a member that is likely to respond quickly (such as one on the same machine as the sender) and implementing the multicast as an RPC to this member, falling back to some other strategy if no local process is found or if it fails to respond before a timeout elapses. One might argue that this is hardly a legitimate implementation of a multicast, since it often behaves like an RPC protocol, but there are systems that implement precisely this functionality and find it extremely useful.

A multimedia system might use a similar multicast, but with real-time rate control or latency properties corresponding to the requirements of the display software (see Rowe and Smith). As groupware uses of distributed systems become increasingly important, one can predict that vendors of Web browsers will focus on offering such functions and that telecommunication service providers will provide the corresponding communication support. Such support would probably need to be aware of the video encoding that is used (e.g., MPEG), so that it can recognize and drop data selectively if a line becomes congested or a date frame is contaminated or will arrive too late to be displayed.

Distributed systems that use groups as a structuring construct may require guarantees of a different nature, and we now focus on those. A slightly more ambitious multicast primitive that could be useful in such a group-oriented application might work by sending the message to the full membership of the destination set, but without providing reliability, ordering, flow control, or any form of feedback in regard to the outcome of the operation. Such a multicast could be implemented by invoking the IP multicast (or UDP multicast) transport primitives discussed in Chapter 3. The user whose application requires any of these properties would implement some form of end-to-end protocol to achieve them.

A more elaborate form of multicast would be a *failure-atomic* multicast, which guarantees that for a specified class of failures, the multicast will either reach all of its destinations or none of them. As we just observed, there are really two forms of failure atomicity that might be interesting, depending on the circumstance. A failure-atomic multicast is *dynamically uniform* if it guarantees that if one process delivers the multicast, then all processes that remain operational will do so, regardless of whether or not the initial recipient remains operational subsequent to delivering the message. A failure-atomic multicast that is not dynamically uniform would only guarantee that if one waits long enough, one will find either that all the destinations that remained operational delivered the message or that none did so. To avoid trivial outcomes, both primitives require that the message will be delivered eventually if the sender doesn't fail.[4]

To reiterate a point made earlier, the key difference between a dynamically uniform protocol and one that is merely failure-atomic but nonuniform has to do with the obligation incurred by the first delivery event. From the perspective of a recipient process p, if m is sent using a protocol that provides dynamic uniformity, then when p delivers m it also knows that any future execution

---

[4]Such a definition leaves open the potential for another sort of trivial solution: One in which the act of invoking the multicast primitive causes the sender to be excluded from the system as faulty. A rigorous nontriviality requirement would also exclude this sort of behavior, and there may be other trivial cases that I am not aware of. However, as was noted earlier in the book, our focus here is on reliability as a practical engineering discipline and not on the development of a mathematics of reliability.

of the system in which a set of processes remains operational will also guarantee the delivery of m within its remaining destinations among that set of processes, as illustrated in Figure 13.13. (We state it this way because processes that join after m was sent are not required to deliver m.) On the other hand, if process p receives a nonuniform multicast m, p knows that if both the sender of m and p crash or are excluded from the system membership, m may not reach its other destinations, as seen in Figure 13.14.

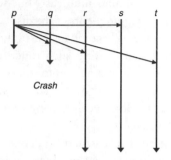

FIGURE 13.13. A dynamically uniform multicast involves a more costly protocol, but if the message is delivered to one destination, the system guarantees that the remaining destinations will also receive it. This is sometimes called a safe delivery, in the sense that it is safe to take actions that leave externally visible effects with respect to which the remainder of the system must be consistent. However, a nonuniform multicast is often safe for applications in which the action taken upon receipt of the message has only internal effects on the system state or when consistency with respect to external actions can be established in other ways—for example, from the semantics of the application.

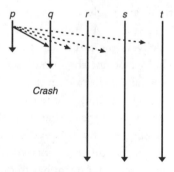

FIGURE 13.14. Nondynamic uniform message delivery: The message is delivered to one destination, q, but then the sender and q crash and the message is delivered to none of the remaining destinations.

Dynamic uniformity is a costly property to provide, but p would want this guarantee if its actions upon receiving m will leave some externally visible trace that the system must know about, such as redirecting an airplane or issuing money from an automatic teller. A nondynamic failure-atomicity rule would be adequate for most internal actions, such as updating a data structure maintained by p, and even for some external ones, such as displaying a quote on a financial analyst's workstation or updating an image in a collaborative work session. In these cases, one may want the highest possible performance and not be willing to pay a steep cost for the dynamic uniformity property because the guarantee it provides is not actually a necessary one. Notice that neither property ensures that a message will reach *all* of its destinations, because no protocol can be sure that a destination will not crash before it has an opportunity to deliver the message, as seen in Figure 13.15.

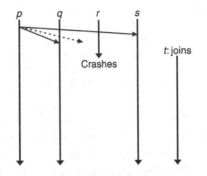

FIGURE 13.15. Neither form of atomicity guarantees that a message will actually be delivered to a destination that fails before the message is delivered or that joins the system after the message is sent.

In this section of the book, the word "failure" is understood to refer to the reporting of failure events by the GMS (see Sabel and Marzullo). Problems that result in the detection of an apparent failure by a system member are reported to the GMS but do not directly trigger any actions (except that messages from apparently faulty processes are ignored) until the GMS officially notifies the full system that the failure has occurred.

13.12 Delivery Ordering Options

Turning now to multicast delivery ordering, let us start by considering a multicast that offers no guarantees whatsoever. Using such a multicast, a process that sends two messages, m_0 and m_1, concurrently would have no assurances at all about their relative order of delivery or relative atomicity—that is, suppose that m_0 was the message sent first. Not only might m_1 reach any destinations that it shares with m_0 first, but a failure of the sender might result in a scenario where m_1 was delivered atomically to all its destinations, but m_0 was not delivered to any process that remains operational (Figure 13.20). Such an outcome would be atomic on a per-multicast basis, but it might not be a very useful primitive from the perspective of the application developer! Thus, while we should ask what forms of order a multicast primitive can guarantee, we should also ask how order is connected to atomicity in our failure-atomicity model.

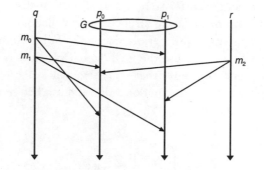

FIGURE 13.16. An unordered multicast provides no guarantees. Here, m_0 was sent before m_1, but is received after m_1 at destination p_0. The reception order for m_2, sent concurrently by process r, is different at each of its destinations.

We will be studying a hierarchy of increasingly ordered delivery properties. The weakest of these is usually called "sender order" or "FIFO order" and requires that if the same process sends m_0 and m_1, then m_0 will be delivered before m_1 at any destinations they have in common (see Figure 13.17). A slightly stronger ordering property is called "causal delivery order," and says that if $send(m_0) \rightarrow send(m_1)$, then m_0 will be delivered before m_1 at any destinations they have in common (Figure 13.18). Still stronger is an order whereby any processes that receive the same two messages receive them in the same order: If, at process p, $deliv(m_0) \rightarrow deliv(m_1)$, then m_0 will be delivered before m_1 at all destinations they have in common (see Figure 13.18). This is sometimes called a totally ordered delivery protocol, but this is something of a misnomer, since one can imagine a number of ordering properties that would be total in this respect without necessarily implying the existence of a single systemwide total ordering on all the messages sent in the system. The reason for this is that our definition focuses on delivery orders where messages overlap, but it doesn't actually relate these orders to an acyclic systemwide ordering. The Transis project calls this type of locally ordered multicast an "agreed" order, and we like this term too: The destinations agree on the order, even for multicasts that may have been initiated concurrently and that may be unordered by their senders (Figure 13.19). However, the agreed order is more commonly called a "total" order or an "atomic" delivery order in the systems that support multicast communication and in the literature.

One can extend the agreed order into a causal agreed order (now one requires that if the sending events were ordered by causality, the delivery order will respect the causal send order) or into a systemwide agreed order (one requires that there exists a single systemwide total order on messages, such that the delivery ordering used at any individual process is consistent with the message ordering in this system's total order). Later we will see why these are not identical orderings. Moreover, in systems that have multiple process groups, the issue of how to extend ordering properties to span multiple process groups will occur.

It has been proposed that total ordering be further classified as *weak* or *strong* in terms analogous to the dynamically uniform and nonuniform delivery properties. A weak total ordering

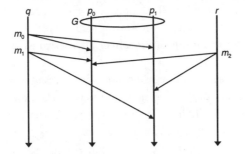

FIGURE 13.17. Sender order or FIFO multicast. Notice that m_2, which is sent concurrently, is unordered with respect to m_0 and m_1.

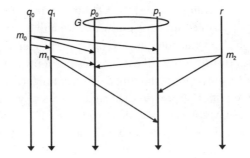

FIGURE 13.18. Causally ordered multicast delivery. Here m_0 is sent before m_1 in a causal sense, because a message is sent from q_0 to q_1 after m_0 was sent and before q_1 sends m_1. Perhaps q_0 has requested that q_1 send m_1. m_0 is consequently delivered before m_1 at destinations that receive both messages. Multicast m_2 is sent concurrently and no ordering guarantees are provided. In this example, m_2 is delivered after m_1 by p_0 and before m_1 by p_1.

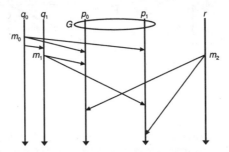

FIGURE 13.19. When using a totally ordered multicast primitive, p_0 and p_1 receive exactly the same multicasts, and the messages are delivered in identical order. Here, the order happens to also be causal, but this is not a specific guarantee of the primitive.

13.12 DELIVERY ORDERING OPTIONS

property would be one guaranteed to hold only at *correct* processes, namely those remaining operational until the protocol terminates. A strong total ordering property would hold even at faulty processes, namely those failing after delivering messages but before the protocol as a whole has terminated.

Suppose that a protocol fixes the delivery ordering for messages m_1 and m_2 at process p, delivering m_1 first. If p fails, a weak total ordering would permit the delivery of m_2 before m_1 at some other process q that survives the failure, even though this order is not the one seen by p. Like dynamic uniformity, the argument for strong total ordering is that this may be required if the ordering of messages may have externally visible consequences, which could be noticed by an external observer interacting with a process that later fails, and then interacting with some other process that remained operational. Naturally, this guarantee has a price, and one would prefer to use a less costly weak protocol in settings where such a guarantee is not required.

Let us now return to the issue raised briefly above, concerning the connection between the ordering properties for a set of multicasts and their failure-atomicity properties. To avoid creating an excessive number of possible multicast protocols, we will assume here that the developer of a reliable application will also want the specified ordering property to extend into the failure-atomicity properties of the primitives used. That is, in a situation where the ordering property of a multicast would imply that message m_0 should be delivered before m_1 if they have any destinations in common, we will require that if m_1 is delivered successfully, then m_0 must also be delivered successfully, whether or not they actually do have common destinations. This is sometimes called a *gap-freedom* guarantee: It is the constraint that failures cannot leave holes or gaps in the ordered past of the system. Such a gap is seen in Figure 13.20.

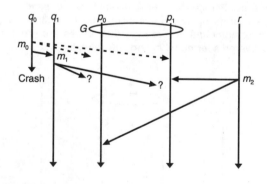

FIGURE 13.20. In this undesirable scenario, the failure of q_0 leaves a causal gap in the message delivery order, preventing q_1 from communicating with members of G. If m_1 is delivered, the causal ordering property would be violated, because $send(m_0) \rightarrow send(m_1)$. But m_0 will never be delivered. Thus, q_1 is logically partitioned from G.

Notice that this rule is stated so that it would apply even if m_0 and m_1 have no destinations in common. The reason for this is that ordering requirements are normally transitive: If m_0 is before m_1 and m_1 is before m_2, then m_0 is also before m_2, and we would like both delivery ordering obligations and failure-atomicity obligations to be guaranteed between m_0 and m_2. Had we instead required that "in a situation where the ordering property of a multicast implies that message m_0 should be delivered before m_1, then if they have any destinations in common, we will also require that if m_1 is delivered successfully, then m_0 must be too," the delivery atomicity requirement might not apply between m_0 and m_2.

Lacking a gap-freedom guarantee, one can imagine runs of a system that would leave orphaned processes, which are technically prohibited from communicating with one another—for example, in Figure 13.20, q_1 sends message m_1 to the members of group G causally after m_0 was sent by q_0 to G. The members of G are now required to deliver m_0 before delivering m_1. However, if the failure-atomicity rule is such that the failure of q_0 could prevent m_0 from ever being delivered, this ordering obligation can only be satisfied by *never* delivering m_1. One could say that q_1 has been partitioned from G by the ordering obligations of the system! Thus, if a system provides ordering guarantees and failure-atomicity guarantees, it should normally extend the latter to encompass the former.

Yet an additional question arises if a process sends multicasts to a group while processes are joining or leaving it. In these cases the membership of the group will be in flux at the time that the message is sent, and one can imagine several ways of interpreting how a system could implement group atomicity.

13.12.1 Nonuniform Failure-Atomic Group Multicast

Consider the following simple, but inefficient group multicast protocol. The sender adds a header to its message listing the membership of the destination group at the time that it sends the message. It now transmits the message to the members of the group, perhaps taking advantage of a hardware multicast feature if one is available, and otherwise transmitting the message over stream-style reliable connections to the destinations. (However, unlike a conventional stream protocol, here we will assume that the connection is only broken if the GMS reports that one of the end points has left the system.)

Upon receipt of a message, the destination processes deliver it immediately, but then resend it to the remaining destinations. Again, each process uses reliable stream-style channels for this retransmission stage, breaking the channel only if the GMS reports the departure of an end point. A participant will now receive one copy of the message from the sender and one from each nonfailed participant other than itself. After delivery of the initial copy, it therefore discards any duplicates. We will now argue that this protocol is failure-atomic, although not dynamically uniform.

To see that it is failure-atomic, assume that some process p_i receives and delivers a copy of the message and remains operational. Failure atomicity tells us that all other destinations that remain operational must also receive and deliver the message. It is clear that this will occur, since the only condition under which p_i would fail to forward a message to p_j would be if the GMS reports that p_i has failed, or if it reports that p_j has failed. But we assume that p_i does not fail, and the output of the GMS can be trusted in this environment. Thus, the protocol achieves failure atomicity. To see that the protocol is not dynamically uniform, consider the situation if the sender sends a copy of the message only to process p_i and then both processes fail. In this case, p_i may have delivered the message and then executed for some extended period of time before crashing or detecting that it has been partitioned from the system. The message has thus been delivered to one of the destinations and that destination may well have acted on it in a visible way; however, none of the processes that remain operational will ever receive it. As we noted earlier, this often will not pose a problem for the application, but it is a behavior that the developer must anticipate and treat appropriately.

As can be seen in Figure 13.21, this simple protocol is a costly one: To send a message to n destinations requires $O(n^2)$ messages. Of course, with hardware broadcast functions, or if the network is not a bottleneck, the cost will be lower, but the protocol still requires each process to send and receive each message approximately n times.

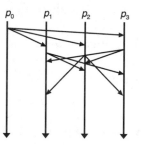

FIGURE 13.21. A very simple reliable multicast protocol. The initial round of messages triggers a second round of messages as each recipient echoes the incoming message to the other destinations.

But now, suppose that we delay the retransmission stage of the protocol, doing this only if the GMS informs the participants that the sender has failed. This change yields a less-costly protocol, which requires n messages (or just one, if hardware broadcast is an option), but in which the participants may need to save a copy of each message indefinitely. They would do this just in case the sender fails.

Recall that we are transmitting messages over a reliable stream. It follows that within the lower levels of the communication system, there is an occasional acknowledgment flowing from each participant back to the sender. If we tap into this information, the sender will know when the participants have all received copies of its message. It can now send a second-phase message, informing the participants that it is safe to delete the saved copy of each message, although they must still save the message identification information to reject duplicates if the sender happens to crash midway through this stage. At this stage the participants can disable their retransmission logic and discard the saved copy of the message (although not its identification information), since any retransmitted message would be a duplicate. Later, the sender could run still a third phase, telling the participants that they can safely delete even the message identification information, because after the second phase there will be no risk of a failure that would cause the message to be retransmitted by the participants.

But now a further optimization is possible. There is no real hurry to run the third phase of this protocol, and even the second phase can be delayed to some degree. Moreover, most processes that send a multicast will tend to send a subsequent one soon afterwards: This principle is well known from all forms of operating systems and database software. It can be summarized by this maxim: The most likely action by any process is to repeat the same action it took most recently. Accordingly, it makes sense to delay sending out messages for the second and third phase of the protocol in the hope that a new multicast will be initiated; this information can be piggybacked onto the first stage of an outgoing message associated with that subsequent protocol!

In this manner, we arrive at a solution, illustrated in Figure 13.22, that has an average cost of n messages per multicast, or just one if hardware broadcast can be exploited, plus some sort of background cost associated with the overhead to implement a reliable stream channel. When a failure does occur, any pending multicast will suddenly generate as many as n^2 additional messages, but even this effect can potentially be mitigated. Since the GMS provides the same membership list to all processes and the message itself carries the list of its destinations, the participants can delay briefly in the hope that some jointly identifiable lowest-ranked participant will turn out to have received the message and will terminate the protocol on behalf of all. We omit the details of such a solution, but any serious system for reliable distributed computing would implement a variety of such mechanisms to keep costs down to an absolute minimum and to maximize the value of each message actually transmitted using piggybacking, delaying tactics, and hardware broadcast.

13.12.2 Dynamically Uniform Failure-Atomic Group Multicast

We can extend the above protocol to one that is dynamically uniform, but doing so requires that no process deliver the message until it is known the processes in the destination group all have a copy. (In some cases it may be sufficient to know that a majority have a copy, but we will not concern ourselves with these sorts of special cases now, because they are typically limited to the processes that actually run the GMS protocol.)

We could accomplish this with the original inefficient protocol of Figure 13.21, by modifying the original nonuniform protocol to delay the delivery of messages until a copy has been received from every destination that is still present in the membership list provided by the GMS. However, such a protocol would suffer from the inefficiencies that led us to optimize the original protocol into the one shown in Figure 13.22. Accordingly, it makes more sense to focus on that improved protocol.

Here, it can be seen that an additional round of messages will be needed before the multicast can be delivered initially; the rest of the protocol can then be used without change (Figure 13.23). Unfortunately, though, this initial round also delays the delivery of the messages to their destinations. In the original protocol, a message could be delivered as soon as it reached a

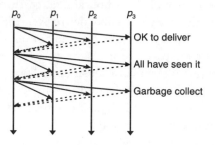

FIGURE 13.22. An improved three-phase protocol. Ideally, the second and third phases would be piggybacked onto other multicasts from the same sender to the same set of destinations and would not require extra messages.

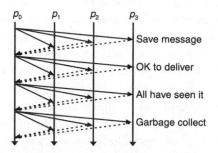

FIGURE 13.23. A dynamically uniform version of the optimized, reliable multicast protocol. Latency to delivery may be much higher, because no process can deliver the message until all processes have received and saved a copy. Here, the third and fourth phases can piggyback on other multicasts, but the first two stages may need to be executed as promptly as possible to avoid increasing the latency still further. Latency is often a key performance factor.

destination for the first time—thus, the latency to delivery is precisely the latency from the sender to a given destination for a single hop. Now the latency might be substantially increased: For a dynamically uniform delivery, we will need to wait for a round trip to the slowest process in the set of destinations, and then one more hop until the sender has time to inform the destinations that it is safe to deliver the messages. In practice, this may represent an increase in latency of a factor of ten or more. Thus, while dynamically uniform guarantees are sometimes needed, the developer of a distributed application should request this property only when it is genuinely necessary, or performance (to the degree that latency is a factor in performance) will suffer badly.

13.12.3 Dynamic Process Groups

When we introduced the GMS, our system became very dynamic, allowing processes to join and leave at will. But not all processes in the system will be part of the same application, and the protocols presented in the previous section are therefore assumed to be sent to groups of processes that represent subsets of the full system membership. This is seen in Figure 13.24, which illustrates the structure of a hypothetical trading system in which services (replicated for improved performance or availability) implement theoretical pricing calculations. Here we have one big system, with many small groups in it. How should the membership of such a subgroup be managed?

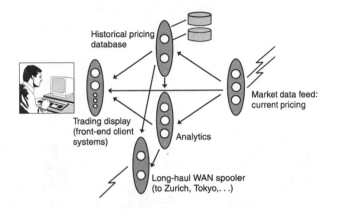

FIGURE 13.24. Distributed trading systems may have both static and dynamic uses for process groups. The historical database, replicated for load-balancing and availability, is tied to the databases themselves and can be viewed as static. This is also true of the market data feeds, which are often redundant for fault tolerance. Other parts of the system, however, such as the analytics (replicated for parallelism) and the client interface processes (one or more per trader), are highly dynamic groups. For uniformity of the model, it makes sense to adopt a dynamic group model, but to keep in mind that some of these groups manage physical resources.

In this section, we introduce a membership management protocol based on the idea that a single process within each group will serve as the coordinator for membership changes. If a process wishes to join the group, or voluntarily leaves the group, this coordinator will update the group membership accordingly. (The role of coordinator will really be handled by the layer of software that implements groups, so this won't be visible to the application process itself.) Additionally, the coordinator will monitor the members (through the GMS and by periodically pinging them to verify that they are still healthy), excluding any failed processes from the membership (much as in the case of a process that leaves voluntarily).

In the approach we present here, all processes that belong to a group maintain a local copy of the current membership list. We call this the "view" of the group and say that each time the membership of the group changes, a new view of the group is reported to the members. Our protocol will have the property that all group members see the identical sequence of group views within any given component of a partitioned system. In practice, we will mostly be interested in primary component partitions, and, in these cases, we will simply say that all processes either see identical views for a group or, if excluded from the primary component, cease to see new views and eventually detect that they are partitioned, at which point a process may terminate or attempt to rejoin the system much as a new process would.

The members of a group depend upon their coordinator for the reporting of new views and consequently monitor the liveness of the coordinator by periodically pinging it. If the coordinator appears to be faulty, the member or members that detect this report the situation to the GMS in the usual manner, simultaneously cutting off communication to the coordinator and starting to piggyback or gossip this information on messages to other members, which similarly cut their channels to the coordinator and, if necessary, relay this information to the GMS. The GMS will eventually report that the coordinator has failed, at which point the lowest ranked of the remaining members takes over as the new coordinator.

Interestingly, we have now solved our problem, because we can use the nondynamic uniform multicast protocol to distribute new views within the group. In fact, this hides a subtle point, to which we will return momentarily—namely, the way to deal with ordering properties of a reliable multicast, particularly in the case where the sender fails and the protocol must be terminated by other processes in the system. However, we will see that the protocol has the necessary ordering properties when it operates over stream connections that guarantee FIFO delivery of messages and when the failure-handling mechanisms introduced earlier are executed in the same order that the messages themselves were initially seen (i.e., if process p_i first received multicast m_0 before multicast m_1, then p_i retransmits m_0 before m_1).

13.12.4 View-Synchronous Failure Atomicity

We have now created an environment within which a process that joins a process group will receive the membership view for that group as of the time it was added to the group. It will subsequently observe any changes that occur until it crashes or leaves the group, provided only that the GMS continues to report failure information. Such a process may now wish to initiate multicasts to the group using the reliable protocols presented earlier. But suppose that a process belonging to a group fails while some multicasts from it are pending? When can the other members be certain that they have seen all of its messages, so that they can take over from it if the application requires that they do so?

Up to now, our protocol structure would not provide this information to a group member—for example, it may be that process p_0 fails after sending a message to p_1 but to no other member. It is entirely possible that the failure of p_0 will be reported through a new process group view before this message is finally delivered to the remaining members. Such a situation would create difficult problems for the application developer, and we need a mechanism to avoid it. This is illustrated in Figure 13.25.

It makes sense to assume that the application developer will want failure notification to represent a final state with regard to the failed process. Thus, it would be preferable for all messages initiated by process p_0 to have been delivered to their destinations before the failure of p_0 is reported through the delivery of a new view. We will call the necessary protocol a *flush* protocol, meaning that it flushes partially completed multicasts out of the system, reporting the new view only after this has been done.

In the example shown in Figure 13.25, we did not include the exchange of messages required to multicast the new view of group G. Notice, however, that the figure is probably incorrect if

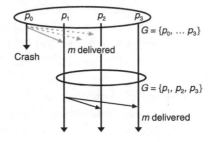

FIGURE 13.25. Although *m* was sent when p_0 belonged to *G*, it reaches p_2 and p_3 after a view change reporting that p_0 has failed. The earlier (gray) and later (black) delivery events thus differ in that the recipients will observe a different view of the process group at the time the message arrives. This can result in inconsistency if, for example, the membership of the group is used to subdivide the incoming tasks among the group members.

the new-view coordinator for group *G* is actually process p_1. To see this, recall that the communication channels are FIFO and that the termination of an interrupted multicast protocol requires only a single round of communication. Thus, if process p_1 simply runs the completion protocol for multicasts initiated by p_0 before it starts the new-view multicast protocol that will announce that p_0 has been dropped by the group, the pending multicast will be completed first. This is shown in Figure 13.26.

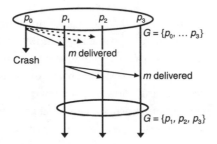

FIGURE 13.26. Process p_1 flushes pending multicasts before initiating the new-view protocol.

We can guarantee this behavior even if multicast *m* is dynamically uniform, simply by delaying the new-view multicast until the outcome of the dynamically uniform protocol has been determined.

On the other hand, the problem becomes harder if p_1 (which is the only process to have received the multicast from p_0) is not the coordinator for the new-view protocol. In this case, it will be necessary for the new-view protocol to operate with an additional round, in which the members of *G* are asked to flush any multicasts that are as yet unterminated, and the new-view protocol runs only when this flush phase has finished. Moreover, even if the new-view protocol is being executed to drop p_0 from the group, it is possible that the system will soon discover that some other process, perhaps p_2, is also faulty and must also be dropped. Thus, a flush protocol should flush messages *regardless of their originating process*, with the result that all multicasts will have been flushed out of the system before the new view is installed.

These observations lead to a communication property that Babaoglu and his colleagues have called *view synchronous communication*, which is one of several properties associated with the *virtual synchrony model* introduced by Thomas Joseph and me in 1985. A view-synchronous communication system ensures that any multicast initiated in a given view of some process group will be failure-atomic with respect to that view and will be terminated before a new view of the process group is installed.

One might wonder how a view-synchronous communication system can prevent a process from initiating new multicasts while the view installation protocol is running. If such multicasts are locked out, there may be an extended delay during which no multicasts can be transmitted, causing performance problems for the application programs layered over the system. But if such multicasts are permitted, the first phase of the flush protocol will not have flushed *all* the necessary multicasts!

A solution for this problem was suggested independently by Ladin and Malkhi, working on systems called Harp and Transis, respectively. In these systems, if a multicast is initiated while a protocol to install view i of group G is running, the multicast destinations are taken to be the future membership of G when that new view has been installed—for example, in Figure 13.26, a new multicast might be initiated by process p_2 while the protocol to exclude p_0 from G is still running. Such a new multicast would be addressed to $\{p_1, p_2, p_3\}$ (not to p_0), and it would be delivered only after the new view is delivered to the remaining group members. The multicast can thus be initiated while the view change protocol is running and would only be delayed if, when the system is ready to deliver a copy of the message to some group member, the corresponding view has not yet been reported. This approach will often avoid delays completely, since the new-view protocol was already running and will often terminate in roughly the same amount of time as will be needed for the new multicast protocol to start delivering messages to destinations. Thus, at least in the most common case, the view change can be accomplished even as communication to the group continues unabated. Of course, if multiple failures occur, messages will still queue up on receipt and will need to be delayed until the view flush protocol terminates, so this desirable behavior cannot always be guaranteed.

13.12.5 Summary of GMS Properties

The following is an informal (English-language) summary of the properties that a group membership service guarantees to members of subgroups of the full system membership. We use the term "process group" for such a subgroup. When we say "guarantees" the reader should keep in mind that a GMS service does not, and in fact cannot, guarantee that it will remain operational despite all possible patterns of failures and communication outages. Some patterns of failure or of network outages will prevent such a service from reporting new system views and will consequently prevent the reporting of new process group views. Thus, the guarantees of a GMS are relative to a constraint—namely, that the system provides a sufficiently reliable transport of messages and that the rate of failures is sufficiently low.

▶ *GMS-1:* Starting from an initial group view, the GMS reports new views that differ by addition and deletion of group members. The reporting of changes is by the two-stage interface described previously, which gives protocols an opportunity to flush pending communication from a failed process before its failure is reported to application processes.

▶ *GMS-2:* The group view is not changed capriciously. A process is added only if it has started and is trying to join the system, and it is deleted only if it has failed or is suspected of having failed by some other member of the system.

▶ *GMS-3:* All group members observe continuous subsequences of the same sequence of group views, starting with the view during which the member was first added to the group and ending either with a view that registers the voluntary departure of the member from the group or with the failure of the member.

▶ *GMS-4:* The GMS is fair in the sense that it will not indefinitely delay a view change associated with one event while performing other view changes. That is, if the GMS service itself is live, join requests will eventually cause the requesting process to be added to the group, and leave or failure events will eventually cause a new group view to be formed that excludes the departing process.

▶ *GMS-5:* Either the GMS permits progress only in a primary component of a partitioned network, or, if it permits progress in nonprimary components, all group views are delivered with an additional Boolean flag indicating whether or not the group view resides in the primary component of the network. This single Boolean flag is shared by all the groups in a given component: The flag doesn't indicate whether a given view of a group is primary for that group, but rather it indicates whether a given view of the group resides in the primary component of the encompassing network.

Although we will not pursue these points here, it should be noted that many networks have some form of critical resources on which the processes reside. Although the protocols given above are designed to make progress when a majority of the processes in the system remain alive after a partitioning failure, a more reasonable approach would also take into account the resulting resource pattern. In many settings, for example, one would want to define the primary partition of a network to be the one that retains the majority of the servers after a partitioning event. One can also imagine settings in which the primary should be the component within which access to some special piece of hardware remains possible, such as radar in an air traffic control application. These sorts of problems can generally be solved by associating weights with the processes in the system and redefining the majority rule as a weighted majority rule. Such an approach recalls work in the 1970s and early 1980s by Bob Thomas of BBN on weighted majority voting schemes and weighted quorum replication algorithms (see Gifford, Thomas).

13.12.6 Ordered Multicast

Earlier, we observed that our multicast protocol would preserve the sender's order if executed over FIFO channels and if the algorithm used to terminate an active multicast was also FIFO. Of course, some systems may seek higher levels of concurrency by using non-FIFO-reliable channels or by concurrently executing the termination protocol for more than one multicast, but, even so, such systems could potentially number multicasts to track the order in which they should be delivered. Freedom from gaps in the sender order is similarly straightforward to ensure.

This leads to a broader issue of what forms of multicast ordering are useful in distributed systems and how such orderings can be guaranteed. In developing application programs that make use of process groups, it is common to employ what Leslie Lamport and Fred Schneider call a *state machine* style of distributed algorithm (see Schneider [1990]). Later, we will see reasons why one might want to relax this model, but the original idea is to run identical software at each member of a group of processes and to use a failure-atomic multicast to deliver messages to the members in identical order. Lamport's proposal stated that Byzantine protocols should be used for this multicast, and, in fact, he also uses Byzantine protocols on messages output by the group members. The result of this is that the group as a whole gives the behavior of a single ultrareliable process, in which the operational members behave identically and the faulty behaviors of faulty members can be tolerated up to the limits of the Byzantine protocols. Clearly, the method requires deterministic programs and thus could not be used in applications that are multithreaded or that accept input through an interrupt-style of event notification. Both of these are common in modern software, so this restriction may be a serious one.

As we will use the concept, however, there is really only one aspect of the approach that is exploited—namely that of building applications that will remain in identical states if presented with identical inputs in identical orders. Here we may not require that the applications actually be deterministic, but merely that they be designed to maintain identically replicated states. This problem, as we will see, is solvable even for programs that may be very nondeterministic in other ways and very concurrent. Moreover, we will not be using Byzantine protocols, but will substitute various weaker forms of multicast protocols. Nonetheless, it has become usual to refer to this as a variation on Lamport's state machine approach, and it is certainly the case that his work was the first to exploit process groups in this manner.

FIFO Order

The FIFO multicast protocol is sometimes called *fbcast* (the "b" comes from the early literature, which tended to focus on static system membership and hence on "broadcasts" to the full membership; "fmcast" might make more sense here, but would be nonstandard). Such a protocol can be developed using the methods previously discussed, provided that the software used to implement the failure-recovery algorithm is carefully designed to ensure that the sender's order will be preserved, or at least tracked to the point of message delivery.

There are two variants on the basic *fbcast*: a normal *fbcast*, which is nonuniform, and a safe *fbcast*, which guarantees the dynamic uniformity property at the cost of an extra round of communication.

The costs of a protocol are normally measured in terms of the latency before delivery can occur, the message load imposed on each individual participant (which corresponds to the CPU usage in most settings), the number of messages placed on the network as a function of group size (this may or may not be a limiting factor, depending on the properties of the network), and the overhead required to represent protocol-specific headers. When the sender of a multicast is also a group member, there are really two latency metrics that may be important: latency from when a message is sent to when it is delivered, which is usually expressed as a multiple of the communication latency of the network and transport software, and the latency from when the sender initiates the multicast to when it learns the delivery ordering for that multicast. During this period, some algorithms will be waiting—in the sender case, the sender may be unable to proceed until it knows when its own message will be delivered (in the sense of ordering with respect to other concurrent multicasts from other senders). And in the case of a destination process, it is clear that until the message is delivered, no actions can be taken.

In all of these cases, *fbcast* and *safe fbcast* are inexpensive protocols. The latency seen by the sender is minimal: In the case of *fbcast*, as soon as the multicast has been transmitted, the sender knows that the message will be delivered in an order consistent with its order of sending. Still focusing on *fbcast*, the latency between when the message is sent and when it is delivered to a destination is exactly that of the network itself: Upon receipt, a message is immediately deliverable. (This cost is much higher, of course, if the sender fails while sending.) The protocol requires only a single round of communication, and other costs are hidden in the background and often can be piggybacked on other traffic. The header used for *fbcast* needs only to identify the message uniquely and capture the sender's order—information that may be expressed in a few bytes of storage.

For the safe version of *fbcast*, of course, these costs would be quite a bit higher, because an extra round of communication is needed to find out if all the intended recipients have a copy of the message. Thus, *safe fbcast* has a latency at the sender of roughly twice the maximum network latency experienced in sending the message (to the slowest destination and back) and a latency at the destinations of roughly three times this figure. Notice that even the fastest destinations are limited by the response times of the slowest destinations, although one can imagine partially safe implementations of the protocol in which a majority of replies would be adequate to permit progress, and the view change protocol would be changed correspondingly.

The *fbcast* and *safe fbcast* protocols can be used in a state machine style of computing under conditions where the messages transmitted by different senders are independent of one another, and hence the actions taken by recipients will commute—for example, suppose that sender p is reporting trades on a stock exchange and sender q is reporting bond pricing information. Although this information may be sent to the same destinations, it may or may not be combined in a way that is order sensitive. When the recipients are insensitive to the order of messages that originate in different senders, *fbcast* is a strong enough ordering to ensure that a state machine style of computing can safely be used. However, many applications are more sensitive to ordering than this, and the ordering properties of *fbcast* would not be sufficient to ensure that group members remain consistent with one another in such cases.

Causal Order

An obvious question to ask concerns the maximum amount of order that can be provided in a protocol that has the same cost as *fbcast*. At the beginning of this chapter, we discussed the causal ordering relation, which is the transitive closure of the message send/receive relation and the internal ordering associated with processes. In 1985, Thomas Joseph and I developed a causally ordered protocol with costs similar to that of *fbcast* and showed how it could be used to implement replicated data. We named the protocol *cbcast*. Shortly thereafter, Schmuck was able to show that causal order is a form of maximal ordering relation among *fbcast*-like protocols (see Schmuck). More precisely, he showed that any ordering property that can be implemented using an asynchronous protocol can be represented as a subset of the causal ordering relationship. This proves that causally ordered communication is the most powerful protocol possible with costs similar to that of *fbcast*.

The basic idea of a causally ordered multicast is easy to express. Recall that a FIFO multicast is required to respect the order in which any single sender sends a sequence of multicasts. If process p sends m_0 and then later sends m_1, a FIFO multicast must deliver m_0 before m_1 at any overlapping destinations. The ordering rule for a causally ordered multicast is almost identical: if $send(m_0) \rightarrow send(m_1)$, then a causally ordered delivery will ensure that m_0 is delivered before m_1 at any overlapping destinations. In some sense, causal order is just a generalization of the FIFO send order. For a FIFO order, we focus on events that happen in some order at a single place in the system. For the causal order, we relax this to events that are ordered under the "happens before" relationship, which can span multiple processes but is otherwise essentially the same as the send order for a single process. A causally ordered multicast simply guarantees that if m_0 is sent before m_1, then m_0 will be delivered before m_1 at destinations they have in common.

The first time one encounters the concept of causally ordered delivery, it can be confusing because the definition doesn't look at all like a definition of FIFO ordered delivery. In fact, however, the concept is extremely similar. Most readers will be comfortable with the idea of a thread of control that moves from process to process when RPC is used by a client process to ask a server to take some action on its behalf. We can think of the thread of computation in the server as being part of the thread of the client. In some sense, a single computation spans two address spaces. Causally ordered multicasts are simply multicasts ordered along such a thread of computation. When this perspective is adopted, one sees that FIFO ordering is in some ways the less natural concept: It artificially tracks ordering of events only when they occur in the same address space. If process p sends message m_0 and then asks process q to send message m_1, it seems natural to say that m_1 was sent after m_0. Causal ordering expresses this relation, but FIFO ordering only does so if p and q are in the same address space.

There are several ways to implement multicast delivery orderings, which are consistent with the causal order. We will now present two such schemes, both based on adding a timestamp to the message header before it is initially transmitted. The first scheme uses a logical clock; the resulting change in header size is very small but the protocol itself has high latency. The second scheme uses a vector timestamp and achieves much better performance. Finally, we discuss several ways of compressing these timestamps to minimize the overhead associated with the ordering property.

Causal ordering with logical timestamps. Suppose that we are interested in preserving causal order within process groups and in doing so only during periods when the membership of the group is fixed (the flush protocol that implements view synchrony makes this a reasonable goal). Finally, assume that all multicasts are sent to the full membership of the group. By attaching a logical timestamp to each message, maintained using Lamport's logical clock algorithm, we can ensure that if $send(m_1) \rightarrow send(m_2)$, then m_1 will be delivered before m_2 at overlapping destinations. The approach is extremely simple: Upon receipt of a message m_i a process p_i waits until it knows that there are no messages still in the channels to it from other group members, p_j, that could have a timestamp smaller than $LT(m_i)$.

How can p_i be sure of this? In a setting where process group members continuously emit multicasts, it suffices to wait long enough. Knowing that m_i will eventually reach every other group member, p_i can reason that eventually every group member will increase its logical clock to a value at least as large as $LT(m_i)$ and will subsequently send out a message with that larger timestamp value. Since we are assuming that the communication channels in our system preserve FIFO ordering, as soon as any message has been received with a timestamp greater than or equal to that of m_i from a process p_j, all future messages from p_j will have a timestamp strictly greater than that of m_i. Thus, p_i can wait long enough to have the full set of messages that have timestamps less than or equal to $LT(m_i)$, and then deliver the delayed messages in timestamp order. If two messages have the same timestamp, they must have been sent concurrently, and p_i can either deliver them in an arbitrary order or can use some agreed-upon rule (e.g., by breaking ties using the process-ID of the sender or its ranking in the group view) to obtain a total order. With this approach, it is no harder to deliver messages in an order that is causal and total than to do so in an order that is only causal.

Of course, in many (if not most) settings, some group members will send to the group frequently while others send rarely or participate only as message recipients. In such environments, p_i might wait in vain for a message from p_j, preventing the delivery of m_i. There are two obvious solutions to this problem: Group members can be modified to send a periodic multicast simply to keep the channels active, or p_i can ping p_j when necessary—in this manner flushing the communication channel between them.

Although simple, this causal ordering protocol is too costly for most settings. A single multicast will trigger a wave of n^2 messages within the group, and a long delay may elapse before it is safe to deliver a multicast. For many applications, latency is the key factor that limits performance, and this protocol is a potentially slow one because incoming messages must be delayed until a suitable message is received on every other incoming channel. Moreover, the number of messages that must be delayed can be very large in a big group, creating potential buffering problems.

Causal ordering with vector timestamps. If we are willing to accept a higher overhead, the inclusion of a vector timestamp in each message permits the implementation of a much more accurate message-delaying policy. Using the vector timestamp, we can delay an incoming message m_i precisely until any missing causally prior messages have been received. This algorithm, like the previous one, assumes that all messages are multicast to the full set of group members.

Again, the idea is simple. Each message is labeled with the vector timestamp of the sender as of the time when the message was sent. This timestamp is essentially a count of the number of causally prior messages that have been delivered to the application at the sender process, broken down by source. Thus, the vector timestamp for process p_1 might contain the sequence $[13,0,7,6]$ for a group G with membership $\{p_0, p_1, p_2, p_3\}$ at the time it creates and multicasts m_i. Process p_1 will increment the counter for its own vector entry (here we assume that the vector entries are ordered in the same way as the processes in the group view), labeling the message with timestamp $[13,1,7,6]$. The meaning of such a timestamp is that this is the first message sent by p_1, but that it has received and delivered 13 messages from p_0: seven from p_2 and six from p_3. Presumably, these received messages created a context within which m_i makes sense, and if some process delivers m_i without having seen one or more of them, it may run the risk of misinterpreting m_i. A causal ordering avoids such problems.

Now, suppose that process p_3 receives m_i. It is possible that m_i would be the very first message that p_3 has received up to this point in its execution. In this case, p_3 might have a vector timestamp as small as $[0,0,0,6]$, reflecting only the six messages it sent before m_i was transmitted. Of course, the vector timestamp at p_3 could also be much larger: The only real upper limit is that the entry for p_1 is necessarily 0, since m_i is the first message sent by p_1. The delivery rule for a recipient such as p_3 is now clear: It should delay message m_i until both of the following conditions are satisfied:

1 Message m_i is the *next* message, in sequence, from its sender

2 Every causally prior message has been received and delivered to the application

We can translate rule 2 into the following formula: If message m_i sent by process p_i is received by process p_j, then we delay m_i until, for each value of k different from i and j, $VT(p_j)[k] \geq VT(m_i)[k]$. Thus, if p_3 has not yet received any messages from p_0, it will not deliver m_i until it has received at least 13 messages from p_0. Figure 13.27 illustrates this rule in a simpler case, involving only two messages.

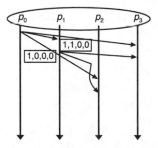

FIGURE 13.27. Upon receipt of a message with vector timestamp [1,1,0,0] from p_1, process p_2 detects that it is too early to deliver this message and delays it until a message from p_0 has been received and delivered.

We need to convince ourselves that this rule really ensures that messages will be delivered in a causal order. To see this, it suffices to observe that when m_i was sent, the sender had already received and delivered the messages identified by $VT(m_i)$. Since these are precisely the messages causally ordered before m_i, the protocol only delivers messages in an order consistent with causality.

The causal ordering relationship is acyclic—hence, one would be tempted to conclude that this protocol can never delay a message indefinitely. But, in fact, it can do so if failures occur. Suppose that process p_0 crashes. Our flush protocol will now run, and the 13 messages that p_0 sent to p_1 will be retransmitted by p_1 on its behalf. But if p_1 also fails, we could have a situation in which m_i, sent by p_1 causally after having received 13 messages from p_0, will never be safely deliverable, because no record exists of one or more of these prior messages! The point here is that although the communication channels in the system are FIFO, p_1 is not expected to forward messages on behalf of other processes until a flush protocol starts when one or more processes have left or joined the system. Thus, a dual failure can leave a gap such that m_i is causally orphaned.

The good news, however, is that this can only happen if the *sender of* m_i fails, as illustrated in Figure 13.28. Otherwise, the sender will have a buffered copy of any messages that it received and that are still unstable, and this information will be sufficient to fill in any causal gaps in the message history prior to when m_i was sent. Thus, our protocol can leave individual messages that are orphaned, but it cannot partition group members away from one another in the sense that concerned us earlier.

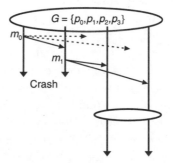

FIGURE 13.28. When processes p_0 and p_1 crash, message m_1 is causally orphaned. This would be detected during the flush protocol that installs the new group view. Although m_1 has been received by the surviving processes, it is not possible to deliver it while still satisfying the causal ordering constraint. However, this situation can only occur if the sender of the message is one of the failed processes. By discarding m_1 the system can avoid causal gaps. Surviving group members will never be logically partitioned (prevented from communicating with each other).

Our system will eventually discover any such causal orphan when flushing the group prior to installing a new view that drops the sender of m_i. At this point, there are two options: m_i can be delivered to the application with some form of warning that it is an orphaned message preceded by missing causally prior messages, or m_i can simply be discarded. Either approach leaves the system in a self-consistent state, and surviving processes are never prevented from communicating with one another.

Causal ordering with vector timestamps is a very efficient way to obtain this delivery ordering property. The overhead is limited to the vector timestamp itself and to the increased latency associated with executing the timestamp ordering algorithm and with delaying messages that genuinely arrive too early. Such situations are common if the machines involved are overloaded, channels are backlogged, or the network is congested and lossy, but otherwise they would rarely be observed. In the best case, when none of these conditions is present, the causal ordering property can be assured with essentially no additional cost in latency or messages passed within the system! On the other hand, notice that the causal ordering obtained is definitely not a total ordering, as was the case in the algorithm based on logical timestamps. Here, we have a genuinely cheaper ordering property, but it is also less ordered.

Timestamp compression. The major form of overhead associated with a vector-timestamp causality is that of the vectors themselves. This has stimulated interest in schemes for compressing the vector-timestamp information transmitted in messages. Although an exhaustive treatment of this topic is well beyond the scope of this book, there are some specific optimizations that are worth mentioning.

Suppose that a process sends a burst of multicasts—a common pattern in many applications. After the first vector timestamp, each subsequent message will contain a nearly identical timestamp, differing only in the timestamp associated with the sender itself, which will increment for each new multicast. In such a case, the algorithm could be modified to omit the timestamp: A missing timestamp would be interpreted as being the previous timestamp, incremented in the sender's field only. This single optimization can eliminate most of the vector-timestamp overhead seen in a system characterized by bursty communication. More accurately, what has happened here is that the sequence number used to implement the FIFO channel from source to destination makes the sender's own vector-timestamp entry redundant. We can omit the vector timestamp because none of the other entries were changing and the sender's sequence number is represented elsewhere in the packets being transmitted.

An important case of this optimization occurs if all the multicasts to some group are sent along a single causal path—for example, suppose that a group has some form of token, which circulates within it, and only the token holder can initiate multicasts to the group. In this case, we can implement *cbcast* using a single sequence number: the first *cbcast*, the second *cbcast*, and so forth. Later, this form of *cbcast* will turn out to be important. Notice, however, that if there are concurrent multicasts from different senders (i.e., if senders can transmit multicasts without waiting for the token), the optimization is no longer able to express the causal ordering relationships on messages sent within the group.

A second optimization is to reset the vector-timestamp fields to 0 each time the group changes its membership and to sort the group members so that any passive receivers are listed last in the group view. With these steps, the vector timestamp for a message will tend to end in a series of zeros, corresponding to those processes that have not sent a message since the previous view change event. The vector timestamp can then be truncated: The reception of a short vector would imply that the missing fields are all zeros. Moreover, the numbers themselves will tend to stay smaller and hence can be represented using shorter fields (if they threaten to overflow, a flush protocol can be run to reset them). Again, a single very simple optimization would be expected to greatly reduce overhead in typical systems that use this causal ordering scheme.

A third optimization involves sending only the difference vector, representing those fields that have changed since the previous message multicast by this sender. Such a vector would be more complex to represent (since we need to know which fields have changed and by how much) but much shorter (since, in a large system, one would expect few fields to change in any short period of time). This generalizes into a run-length encoding.

This third optimization can also be understood as an instance of an ordering scheme introduced originally in the Psync, Totem, and Transis systems. Rather than representing messages by counters, a precedence relation is maintained for messages: a tree of the messages received and the causal relationships between them. When a message is sent, the leaves of the causal tree are transmitted. These leaves are a set of concurrent messages, all of which are causally prior to the message now being transmitted. Often, there will be very few such messages, because many groups would be expected to exhibit low levels of concurrency.

The receiver of a message will now delay it until those messages it lists as causally prior have been delivered. By transitivity, no message will be delivered until all the causally prior messages have been delivered. Moreover, the same scheme can be combined with one similar to the logical time-stamp ordering scheme of the first causal multicast algorithm to obtain a primitive that is both causally and totally ordered. However, doing so necessarily increases the latency of the protocol.

Causal multicast and consistent cuts. At the outset of this chapter we discussed concepts of logical time, defining the causal relation and introducing the definition of a consistent cut. Notice that the delivery events of a multicast protocol such as *cbcast* are concurrent and can be thought of as occurring at the same time in all the members of a process group. In a logical sense, *cbcast* delivers messages at what may look to the recipients like a single instant in time. Unfortunately, however, the delivery events for a single *cbcast* do not represent a consistent cut across the system, because communication that was concurrent with the *cbcast* could cross it. Thus, one could easily encounter a system in which a *cbcast* is delivered at process p, which has received message m, but where the same *cbcast* was delivered at process q (the eventual sender of m) before m had been transmitted.

With a second *cbcast* message, it is actually possible to identify a true consistent cut, but to do so we need to either introduce a concept of an epoch number or inhibit communication briefly. The inhibition algorithm is easier to understand. It starts with a first *cbcast* message, which tells the recipients to inhibit the sending of new messages. The process group members receiving this message send back an acknowledgment to the process that initiated the *cbcast*. The initiator, having collected replies from all group members, now sends a second *cbcast* telling the group members that they can stop recording incoming messages and resume normal communication. It is easy to see that all messages in the communication channels when the first *cbcast* was received will now have been delivered and that the communication channels will be empty. The recipients now resume normal communication. (They should also monitor the state of the initiator, in case it fails!) The algorithm is very similar to the one for changing the membership of a process group, discussed previously.

Noninhibitory algorithms for forming consistent cuts are also known. One way to solve this problem is to add *epoch numbers* to the multicasts in the system. Each process keeps an *epoch counter* and tags every message with the counter value. In the consistent cut protocol, the first-phase message now tells processes to increment the epoch counters (and not to inhibit new messages). Thus, instead of delaying new messages, they are sent promptly but with epoch number $k + 1$ instead of epoch number k. The same algorithm now works to allow the system to reason about the consistent cut associated with its kth epoch even as it exchanges new messages during epoch $k + 1$. Another well-known solution takes the form of what is called an *echo protocol,* in which two messages traverse every communication link in the system (see Chandy and Lamport). For a system with all-to-all communication connectivity, such protocols will transmit $O(n^2)$ messages, in contrast with the $O(n)$ required for the inhibitory solution.

This *cbcast* provides a relatively inexpensive way of testing the distributed state of the system to detect a desired property. In particular, if the processes that receive a *cbcast* compute a predicate or write down some element of their states at the moment the message is received, these states will fit together cleanly and can be treated as a glimpse of the system as a whole at a single instant in time. To count the number of processes for which some condition holds, it is sufficient to send a *cbcast* asking processes if the condition holds and to count the number that return *true.* The result is a value that could in fact have been valid for the group at a single instant in real time. On the negative side, this guarantee only holds with respect to communication that uses causally ordered primitives. If processes communicate with other primitives, the delivery events of the *cbcast* will not necessarily be prefix closed when the send and receive events for these messages are taken into account. Marzullo and Sabel have developed optimized versions of this algorithm.

Some examples of properties that could be checked using our consistent cut algorithm include the current holder of a token in a distributed locking algorithm (the token will never appear to be lost or duplicated), the current load on the processes in a group (the states of members will never be accidentally sampled at different times yielding an illusory load that is unrealistically high or low), the wait-for graph of a system subject to infrequent deadlocks (deadlock will never be detected when the system is in fact not deadlocked), and the contents of a database (the database will never be checked at a time when it has been updated at some locations but not others). On the other hand, because the basic algorithm inhibits the sending of new messages in the group, albeit briefly, there will be many systems for which the performance impact is too high and a solution that sends more messages but avoids inhibition states would be preferable. The epoch-based scheme represents a reasonable alternative, but we have not treated fault-tolerance issues; in practice, such a scheme works best if all cuts are initiated by some single member of a group, such as the oldest process in it, and a group flush is known to occur if that process fails and some other takes over from it. We leave the details of this algorithm as a small problem for the reader!

Exploiting topological knowledge. Many networks have topological properties, which can be exploited to optimize the representation of causal information within a process group that implements a protocol such as cbcast. Within the NavTech system, developed at INESC in Portugal, wide area applications operate over a communication transport layer implemented as part of NavTech. This structure is programmed to know the location of wide area network links and to make use of hardware multicast where possible (see Rodrigues and Verissimo, Rodrigues et al). A consequence is that if a group is physically laid out with multiple subgroups interconnected over a wide area link, as seen in Figure 13.29, the message need only be sent once over each link.

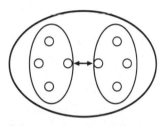

FIGURE 13.29. In a complex network, a single process group may be physically broken into multiple subgroups. With knowledge of the network topology, the NavTech system is able to reduce the information needed to implement causal ordering.

In a geographically distributed system, it is frequently the case that all messages from some subset of the process group members will be relayed to the remaining members through a small number of relay points. Rodrigues exploits this observation to reduce the amount of information needed to represent causal ordering relationships within the process group. Suppose that message m_1 is causally dependent upon message m_0 and that both were sent over the same communication link. When these messages are relayed to processes on the other side of the link, they will appear to have been sent by a single sender and the ordering relationship between them can be compressed into the form of a single vector-timestamp entry. In general, this observation permits any set of processes that route through a single point to be represented using a single sequence number on the other side of that point.

Stephenson explored the same question in a more general setting involving complex relationships between overlapping process groups (the multigroup causality problem) (see Stephenson). His work identifies an optimization similar to this one, as well as others that take advantage of other regular layouts of overlapping groups, such as a series of groups organized into a tree or some other graph-like structure.

The reader may wonder about causal cycles, in which message m_2, sent on the right of a linkage point, becomes causally dependent on m_1, sent on the left, which was in turn dependent upon m_0, also sent on the left. Both Rodrigues and Stephenson made the observation that as m_2 is forwarded back through the link, it emerges with the old causal dependency upon m_1 reestablished. This method can be generalized to deal with cases where there are multiple links (overlap points) between the subgroups that implement a single process group in a complex environment.

Total Order

In developing our causally ordered communication primitive, we really ended up with a family of such primitives. Cheapest of these are purely causal in the sense that concurrently transmitted multicasts might be delivered in different orders to different members. The more costly ones combined causal order with mechanisms that resulted in a causal, total order. We saw two such primitives: One was the causal ordering algorithm based on logical timestamps, and the second (introduced very briefly) was the algorithm used for total order in the Totem and Transis systems, which extend the causal order into a total one using a canonical sorting procedure, but in which latency is increased by the need to wait until multicasts have been received from all potential sources of concurrent multicasts.[5] In this section we discuss totally ordered multicasts, known by the name *abcast*, in more detail.

When causal ordering is not a specific requirement, there are some very simple ways to obtain total order. The most common of these is to use a sequencer process or token (see Chang and Maxemchuk, Kaashoek). A sequencer process is a distinguished process, which publishes an ordering on the messages of which it is aware; all other group members buffer multicasts until the ordering is known and then deliver them in the appropriate order. A token is a way to move the sequencer around within a group: While holding the token, a process may put a sequence number on outgoing multicasts. Provided that the group only has a single token, the token ordering results in a total ordering for multicasts within the group. This approach was introduced in a very early protocol by Chang and Maxemchuk and remains popular because of its simplicity and low overhead. Care must be taken, of course, to ensure that failures cannot cause the token to be lost, briefly duplicated, or result in gaps in the total ordering that orphan subsequent messages. We saw this solution as an optimization to *cbcast* in the case where all the communication to a group originates along a single causal path within the group. From the perspective of the application, *cbcast* and *abcast* are indistinguishable in this case, which turns out to be a common and important one.

It is also possible to use the causally ordered multicast primitive to implement a causal and totally ordered token-based ordering scheme. Such a primitive would respect the delivery ordering property of *cbcast* when causally prior multicasts are pending in a group, similar to *abcast* when two processes concurrently try to send a multicast. Rather than present this algorithm here, however, we defer it until Chapter 16, when we present it in the context of a method for

---

[5]Most ordered of all is the flush protocol used to install new views: This delivers a type of message (the new view) in a way that is ordered with respect to all other types of messages. In the Isis Toolkit, there was actually a *gbcast* primitive, which could be used to obtain this behavior at the request of the user, but it was rarely used and more recent systems tend to use this protocol only to install new process group views.

implementing replicated data with locks on the data items. We do this because, in practice, token-based total ordering algorithms are more common than the other methods. The most common use of causal ordering is in conjunction with the specific replication scheme presented in Chapter 16; therefore, it is more natural to treat the topic in that setting.

Yet an additional total ordering algorithm was introduced by Leslie Lamport in his very early work on logical time in distributed systems (see Lamport [July 1978]) and later adapted to group communication settings by Skeen during a period when he and I collaborated on an early version of the Isis totally ordered communication primitive. The algorithm uses a two-phase protocol in which processes vote on the message ordering to use, expressing this vote as a logical timestamp.

The algorithm operates as follows. In the first phase of communication, the originator of the multicast (we'll call it the coordinator) sends the message to the members of the destination group. These processes save the message but do not yet deliver it to the application. Instead, each proposes a delivery time for the message using a logical clock, which is made unique by appending the process-ID. The coordinator collects these proposed delivery times, sorts the vector, and designates the maximum time as the *committed* delivery time. It sends this time back to the participants. They update their logical clocks (and hence will never propose a smaller time) and reorder the messages in their pending queue. If a pending message has a committed delivery time, and this time is smallest among the proposed and committed times for other messages, it can be delivered to the application layer.

This solution can be seen to deliver messages in a total order, since all the processes base the delivery action on the same committed timestamp. It can be made fault tolerant by electing a new coordinator if the original sender fails. One curious property of the algorithm, however, is that it has a nonuniform ordering guarantee. To see this, consider the case where a coordinator and a participant fail and that participant also proposed the maximum timestamp value. The old coordinator may have committed a timestamp that could be used for delivery to the participant, but that will not be reused by the remaining processes, which may therefore pick a different delivery order. Thus, just as dynamic uniformity is costly to achieve as an atomicity property, one sees that a dynamically uniform ordering property may be quite costly. It should be noted that dynamic uniformity and dynamically uniform ordering tend to go together: If delivery is delayed until it is known that all operational processes have a copy of a message, it is normally possibly to ensure that all processes will use identical delivery orderings.

This two-phase ordering algorithm and a protocol called the "born-order" protocol, which was introduced by the Transis and Totem systems (messages are ordered using unique message identification numbers that are assigned when the messages are first created, or "born"), have advantages in settings with multiple overlapping process groups, a topic to which we will return in Chapter 14. Both provide what is called "globally total order," which means that even *abcast* messages sent in different groups will be delivered in the same order at any overlapping destinations they may have.

The token-based ordering algorithms provide locally total order, which means that *abcast* messages sent in different groups may be received in different orders even at destinations that they share. This infers that one should use the globally total algorithms; such reasoning could be carried further to justify a decision to only consider globally total ordering schemes that also guarantee dynamic uniformity. However, this line of reasoning leads to more and more costly solutions. For most of my work, the token-based algorithms have been adequate, and I have never seen an application for which globally total dynamically uniform ordering was a requirement.

Unfortunately, the general rule seems to be that stronger ordering is more costly. On the basis of the known protocols, the stronger ordering properties tend to require that more messages be exchanged within a group, and they are subject to longer latencies before message delivery can be performed. We characterize this as unfortunate, because it suggests that in the effort to achieve greater efficiency, the designer of a reliable distributed system may be faced with a tradeoff between complexity and performance. Even more unfortunate is the discovery that the differences are extreme. When we look at Horus, we will find that its highest performance protocols (which include a locally total multicast that is nonuniform) are nearly three orders of magnitude faster than the best-known dynamically uniform and globally total ordered protocols (measured in terms of latency between when a message is sent and when it is delivered).

By tailoring the choice of protocol to the specific needs of an application, far higher performance can be obtained. On the other hand, it is very appealing to use a single, very strong primitive systemwide in order to reduce the degree of domain-specific knowledge needed to arrive at a safe and correct implementation. The designer of a system in which multicasts are infrequent and far from the critical performance path is fortunate indeed: Such systems can be built on a strong, totally ordered, and dynamically uniform communication primitive—the high cost will probably not be noticeable. The rest of us are faced with a more challenging design problem.

13.13 Communication from Nonmembers to a Group

Up to now, all of our protocols have focused on the case of group members communicating with one another. However, in many systems there is an equally important need to provide reliable and ordered communication from nonmembers into a group. This section presents two solutions to the problem—one for a situation in which the nonmember process has located a single member of the group but lacks detailed membership information about the remainder of the group, and one for the case of a nonmember that nonetheless has cached group membership information.

In the first case, our algorithm will have the nonmember process ask some group member to issue the multicast on its behalf, using an RPC for this purpose. In this approach, each such multicast is given a unique identifier by its originator, so that if the forwarding process fails before reporting on the outcome of the multicast, the same request can be reissued. The new forwarding process would check to see if the multicast was previously completed, issue it if not, and then return the outcome in either case. Various optimizations can then be introduced, so that a separate RPC will not be required for each multicast. The protocol is illustrated in Figure 13.30 for the normal case, when the contact process does not fail. Not shown is the eventual garbage collection phase needed to delete status information accumulated during the protocol and saved for use in the case where the contact eventually fails.

Our second solution uses what is called an *iterated* approach, in which the nonmember processes cache possibly inaccurate process group views. Specifically, each group view is given a unique identifier, and client processes use an RPC or some other mechanism to obtain a copy of the group view (e.g., they may join a larger group within which the group reports changes in its core membership to interested nonmembers). The client then includes the view identifier in its message and multicasts it directly to the group members. Again, the members will retain some limited history of prior interactions using a mechanism such as the one for the multiphase commit protocols. (See Figure 13.31.)

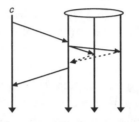

FIGURE 13.30. Nonmember of a group uses a simple RPC-based protocol to request that a multicast be done on its behalf. Such a protocol becomes complex when ordering considerations are added, particularly because the forwarding process may fail during the protocol run.

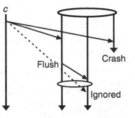

FIGURE 13.31. An iterated protocol. The client sends to the group as its membership is changing (to drop one member). Its multicast is terminated by the flush associated with the new-view installation (message just prior to the new view), and when one of its messages arrives late (dashed line), the recipient detects it as a duplicate and ignores it. Had the multicast been so late that all the copies were rejected, the sender would have refreshed its estimate of group membership and retried the multicast. Doing this while also respecting ordering obligations can make the protocol complex, although the basic idea is quite simple. Notice that the protocol is cheaper than the RPC solution: The client sends directly to the actual group members, rather than indirectly sending through a proxy. However, while the figure may seem to suggest that there is no acknowledgment from the group to the client, this is not the case: The client communicates over a reliable FIFO channel to each member—hence, acknowledgments are implicitly present. Indeed, some effort may be needed to avoid an implosion effect, which would overwhelm the client of a large group with a huge number of acknowledgments.

There are now three cases that may occur. Such a multicast can arrive in the correct view, it can arrive partially in the correct view and partially late (after some members have installed a new group view), or it can arrive entirely late. In the first case, the protocol is considered successful. In the second case, the group flush algorithm will push the partially delivered multicast to a view-synchronous termination; when the late messages finally arrive, they will be ignored as duplicates by the group members that receive them, since these processes will have already delivered the message during the flush protocol. In the third case, all the group members will recognize the message as a late one that was not flushed by the system and all will reject it. Some or all should also send a message back to the nonmember warning it that its message was not successfully delivered; the client can then retry its multicast with refreshed membership information. This last case is said to iterate the multicast. If it is practical to modify the underlying reliable transport protocol, a convenient way to return status information to the sender is by attaching it to the acknowledgment messages such protocols transmit.

This protocol is clearly quite simple, although its complexity grows when one considers the issues associated with preserving sender order or causality information in the case where iteration is required. To solve such a problem, a nonmember that discovers itself to be using stale group view information should inhibit the transmission of new multicasts while refreshing the group view data. It should then retransmit, in the correct order, all multicasts that are not known to have been successfully delivered while it was sending using the previous group view. Some care is required in this last step, however, because new members of the group may not have sufficient information to recognize and discard duplicate messages.

To overcome this problem, there are basically two options. The simplest case occurs when the group members transfer information to joining processes, including the record of multicasts successfully received from nonmembers prior to when the new member joined. Such a state transfer can be accomplished using a mechanism discussed in the next chapter. Knowing that the members will detect and discard duplicates, the nonmember can safely retransmit any multicasts that are still pending, in the correct order, followed by any that may have been delayed while waiting to refresh the group membership. Such an approach minimizes the delay before normal communication is restored.

The second option is applicable when it is impractical to transfer state information to the joining member. In this case, the nonmember will need to query the group, determining the status of pending multicasts by consulting with surviving members from the previous view. Having determined the precise set of multicasts that was dropped upon reception, the nonmember can retransmit these messages and any buffered messages and then resume normal communication. Such an approach is likely to have higher overhead than the first one, since the nonmember (and there may be many of them) must query the group after each membership change. It would not be surprising if significant delays were introduced by such an algorithm.

13.13.1 Scalability

The preceding discussion of techniques and costs did not address questions of scalability and limits. Yet clearly, the decision to make a communication protocol reliable will have an associated cost, which might be significant. We treat this topic in more detail in Section 18.8. The reader is cautioned to keep in mind that reliability does have a price, and many of the most demanding distributed applications, which generate extremely high message-passing loads, must be split into a reliable subsystem, which experiences lower loads and provides stronger guarantees for the subset of messages that pass through it, and a concurrently executed unreliable subsystem, which handles the bulk of communication but offers much weaker guarantees to its users. Reliability and properties can be extremely valuable, but one shouldn't make the mistake of assuming that reliability properties are *always* desirable or that such properties should be provided everywhere in a distributed system. Used selectively, these technologies are very powerful; used blindly, they may actually compromise reliability of the application by introducing undesired overheads and instability in those parts of the system that have strong performance requirements and weaker reliability requirements.

13.14 Communication from a Group to a Nonmember

The discussion in the preceding section did not consider the issues raised by transmission of replies from a group to a nonmember. These replies, however, and other forms of communication outside of a group, raise many of the same reliability issues that motivated the ordering and gap-freedom protocols presented previously—for example, suppose that a group is using a causally ordered multicast internally, and one of its members sends a point-to-point message to some process outside the group. In a logical sense, that message may now be dependent upon the prior causal history of the group, and if that process now communicates with other members of the group, issues of causal ordering and freedom from causal gaps will occur.

This specific scenario was studied by Ladin and Liskov, who developed a system in which vector timestamps could be exported by a group to its clients; the client later presented the timestamp back to the group when issuing requests to other members, and in this way the client was protected against causal ordering violations. The protocol proposed in that work used stable storage to ensure that even if a failure occurred, no causal gaps could occur.

Other researchers have considered the same issues using different methods. Work by Schiper, for example, explored the use of an $n \times n$ matrix to encode point-to-point causality information (see Schiper et al.), and the Isis Toolkit introduced mechanisms to preserve causal order when point-to-point communication was done in a system. We will present some of these methods in Chapter 16.

13.15 Summary

When we introduced the sender-ordered multicast primitive, we noted that it is often called *fbcast* in systems that explicitly support it; the causally ordered multicast primitive is known as *cbcast;* and the totally ordered one is called *abcast*. These names are traditional ones, and are obviously somewhat at odds with terminology in this book. More natural names might be *fmcast*, *cmcast*, and *tmcast*. However, since a sufficiently large number of papers and systems use the terminology of broadcasts and call the totally ordered primitive *atomic*, it would confuse many readers if we did not at least adopt the standard acronyms for these primitives.

Table 13.2 summarizes the most important terminology and primitives defined in this chapter.

TABLE 13.2. Terminology

Concept	Brief description
abcast	View-synchronous totally ordered group communication. If processes p and q both receive m_1 and m_2, then either both deliver m_1 prior to m_2, or both deliver m_2 prior to m_1. As noted earlier, *abcast* comes in several versions. Throughout the remainder of this book, we will assume that *abcast* is a locally total and nondynamic uniform protocol—that is, we focus on the least costly of the possible *abcast* primitives, unless we specifically indicate otherwise.
cabcast	Causally and totally ordered group communication. The delivery order is as for *abcast*, but it is also consistent with the causal sending order.
cbcast	View-synchronous causally ordered group communication. If $send(m_1) \rightarrow send(m_2)$, then processes that receive both messages deliver m_1 prior to m_2.
fbcast	View-synchronous FIFO group communication. If the same process p sends m_1 prior to sending m_2, then processes that receive both messages deliver m_1 prior to m_2.
gap freedom	The guarantee that if message m_i should be delivered before m_j and some process receives m_j and remains operational, m_i will also be delivered to its remaining destinations. A system that lacks this property can be exposed to a form of logical partitioning, where a process that has received m_j is prevented from (ever) communicating to some process that was supposed to receive m_i but will not because of a failure.
gbcast	A group communication primitive based upon the view-synchronous flush protocol. Supported as a user-callable API in the Isis Toolkit, but very costly and not widely used. *gbcast* delivers a message in a way that is totally ordered relative to all other communication in the same group.
group client	A nonmember of a process group that communicates with it and that may need to monitor the membership of that group as it changes dynamically over time.
member (of a group)	A process belonging to a process group.
process group	A set of processes that have joined the same group. The group has a *membership list,* which is presented to group members in a data structure called the *process group view*. This lists the members of the group and other information, such as their ranking.

TABLE 13.2. Terminology *(continued)*

Concept	Brief description
safe multicast	A multicast having the property that if any group member delivers it, then all operational group members will also deliver it. This property is costly to guarantee and corresponds to a *dynamic uniformity* constraint. Most multicast primitives can be implemented in a safe or an unsafe version; the less costly one is preferable. In this book, we are somewhat hesitant to use the term "safe," because a protocol lacking this property is not necessarily "unsafe." Consequently, we will normally describe a protocol as being dynamically uniform (safe) or nonuniform (unsafe). If we do not specifically say that a protocol needs to be dynamically uniform, the reader should assume that we intend the nonuniform case.
view-synchronous multicast	A way of sending a message to a process group so all the group members that don't crash will receive the message between the same pair of group views—that is, a message is delivered entirely before or entirely after a given view of the group is delivered to the members. If a process sends a multicast when the group membership consists of $\{p_0, \dots p_k\}$ and doesn't crash, the message will be delivered while the group view is still $\{p_0, \dots p_k\}$.
virtual synchrony	A distributed communication system in which process groups are provided, supporting view-synchronous communication and gap freedom, and in which algorithms are developed using a style of closely synchronous computing in which all group members see the same events in the same order and consequently can closely coordinate their actions. Such synchronization becomes virtual when the ordering properties of the communication primitive are weakened in ways that do not change the correctness of the algorithm. By introducing such weaker orderings, a group can be made more likely to tolerate failure and can gain a significant performance improvement.

13.16 Related Reading

On logical concepts of time: (see Lamport [July 1978, 1984]).

Causal ordering in message delivery: (see Birman and Joseph [February 1987, November 1987]).

Consistent cuts: (see Babaoglu and Marzullo, Chandy and Lamport).

Vector clocks: (see Fidge, Mattern).

Vector clocks used in message delivery: (see Birman et al., Ladin et al. [1992], Schiper et al.).

Optimizing vector clock representations: (see Charron-Bost, Melliar-Smith and Moser [1993]).

Compression using topological information about groups of processes: (see Birman et al. [1991], Rodrigues and Verissimo, Rodrigues et al.).

Static groups and quorum replication: (see Bernstein et al., Birman and Joseph [November 1987], Cooper [1985]).

Two-phase commit: (see Bernstein et al., Gray [1979], Gray and Reuter).

Three-phase commit: (see Skeen [June 1982, 1985]).

Byzantine protocol: (see Ben-Or, Coan and Thomas, Coan et al., Cristian et al., Rabin, Schneider [1984]).

Asynchronous consensus: (see Chandra and Toueg [1991], Fisher et al.); but see also Babaoglu et al. (1995), Friedman et al., Guerraoui and Schiper, Ricciardi (1996).

The method of Chandra and Toueg: (see Babaoglu et al. [1995], Chandra and Toueg [1991], Chandra et al. [1992, 1996], Friedman et al.)

Group membership: (see Birman and Joseph [February 1987, November 1987], Chandra et al. [1996], Cristian [April 1991], Melliar-Smith et al. [1991], Mishra [1991], Ricciardi and Birman); but see also Agarwal, Anceaume et al., Babaoglu et al. (1994, 1995), Birman and Glade, Chandra et al. (1996), Cristian and Schmuck, Friedman et al., Golding (1992), Guerraoui and Schiper, Reiter (1994), Ricciardi (1992, 1993, 1996), Rodrigues et al.

Partitionable membership: (see Amir et al. [1992], Moser et al. [1994]).

Fail-stop illusion: (see Sabel and Marzullo).

Token-based total order: (see Chang and Maxemchuk, Kaashoek).

Lamport's method: (see Birman and Joseph [February 1987], Lamport [July 1978]).

Communication from nonmembers of a group: (see Birman and Joseph [February 1987], Wood [1991]).

Point-to-point causality: (see Schiper et al.).

CHAPTER 14

Point-to-Point and Multigroup Considerations

CONTENTS

Until now, we have considered settings in which all communication occurs within a process group, and although we did discuss protocols by which a client can multicast into a group, we did not consider issues raised by replies from the group to the client. Primary among these is the question of preserving the causal order if a group member replies to a client, which was discussed in Section 13.14. We now turn to issues involving multiple groups, including causal order, total order, causal and total ordering domains, and coordination of the view-flush algorithms where more than one group is involved.

Even before starting to look at these topics, however, a broader philosophical issue must be considered. When one develops an idea, such as the combination of properties with group communication, there is always a question concerning just how far one wants to take the resulting technology. Process groups, as treated in the previous chapter, are localized and self-contained entities. In this chapter, we are concerned with extending this local model into an encompassing systemwide model. One can easily imagine a distributed system in which the fundamental communication abstraction is in fact the process group, with communication to a single process being viewed as a special case of the general process. In such a setting, one might well try to extend ordering properties so that they would apply systemwide and achieve an elegant and highly uniform programming abstraction.

There is a serious risk associated with this whole line of thinking—namely, that it will result in systemwide costs and systemwide overhead of a potentially unpredictable nature. Recall the end-to-end argument of Saltzer et al.: In most systems, given a choice between paying a cost where and when it is needed or paying that cost systemwide, one should favor the end-to-end solution, where the cost is incurred only when the associated property is desired. By and large, the techniques we present in this chapter should only be considered when there is a very clear and specific justification for using them. Any system that uses these methods casually is likely to perform poorly and exhibit unpredictable behavior.

14.1 Causal Communication outside of a Process Group

Although there are sophisticated protocols guaranteeing that causality will be respected for arbitrary communication patterns, the most practical solutions generally confine concurrency and associated causality issues to the interior of a process group—for example, at the end of Section 13.14, we briefly cited the replication protocol of Ladin and Liskov (see Ladin et al. [1992], Liskov et al.). This protocol transmits a timestamp to the client, and the client later includes the most recent of the timestamps it has received in any requests it issues to the group. The group members can detect causal ordering violations and delay such a request until causally prior multicasts have reached their destinations, as seen in Figure 14.1.

An alternative is to simply delay messages sent out of a group until any causally prior multicasts sent within the group have become stable—in other words, have reached their destinations.

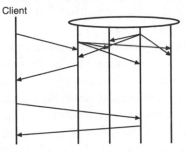

Client

FIGURE 14.1. In the replication protocol used by Ladin and Liskov in the Harp system, vector timestamps are used to track causal multicasts within a server group. If a client interacts with a server in that group, it does so using a standard RPC protocol. However, the group timestamp is included with the reply and can be presented with a subsequent request to the group. This permits the group members to detect missing prior multicasts and to appropriately delay a request, but omits the client's point-to-point messages from the causal state of the system. Such tradeoffs between properties and cost seem entirely appropriate, because an attempt to track causal order systemwide can result in significant overheads. A system such as the Isis Toolkit, which enforces causal order even for point-to-point message passing, generally does so by delaying after sending point-to-point messages until they are known to be stable—a simple and conservative solution that avoids the need to represent ordering information for such messages.

Since there is no remaining causal ordering obligation in this case, the message does not need to carry causality information. Moreover, such an approach may not be as costly as it sounds, for the same reason that the *flush* protocol introduced earlier turns out not to be terribly costly in practice: Most asynchronous *cbcast* or *fbcast* messages become stable shortly after they are issued—long before any reply is sent to the client. Thus, any latency is associated with the very last multicasts that were initiated within the group, and will normally be small. We will see a similar phenomenon (in more detail) in Section 17.5, which discusses a replication protocol for stream protocols.

There has been some work on the use of causal order as a systemwide guarantee, applying to point-to-point communication as well as multicasts. Unfortunately, representing such ordering information requires a matrix of size $O(n^2)$ in the size of the system. Moreover, this type of ordering information is only useful if messages are sent asynchronously (without waiting for replies). But, if this is done in systems that use point-to-point communication, there is no obvious way to recover if a message is lost (when its sender fails) after subsequent messages (to other destinations) have been delivered. Cheriton and Skeen discuss this form of all-out causal order in a well-known paper and conclude that it is probably not desirable (see Birman [1994], Cheriton and Skeen, Cooper [1994], Schiper et al., van Renesse [1993]). If point-to-point messages are treated as being causally prior to other messages, it is best to wait until they have been received before sending causally dependent messages to other destinations.[1] (We'll discuss Cheriton and Skeen's paper in more detail in Chapter 16.)

Early versions of the Isis Toolkit solved this problem without actually representing causal information at all, although later work replaced this scheme with one that waits for point-to-point messages to become stable (see Birman and Joseph [February 1987], Birman et al. [1993]). The approach was to piggyback pending messages (those that are not known to have reached all their destinations) on *all* subsequent messages, regardless of their destination (Figure 14.2)—that is, if process p has sent multicast m_1 to process group G and now wishes to send a message m_2 to any destination other than group G, a copy of m_1 is included with m_2. By applying this rule systemwide, p can be certain that if any route causes a message m_3, causally dependent upon m_1, to reach a destination of m_1, a copy of m_1 will be delivered too. A background garbage collection algorithm is used to delete these spare copies of messages when they do reach their destinations, and a simple duplicate suppression scheme is employed to avoid delivering the same message more than once if it reaches a destination several times.

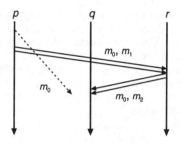

FIGURE 14.2. After sending m_0 asynchronously to q, p sends m_1 to r. To preserve causality, a copy of m_0 is piggybacked on this message, and on message r when it sends m_3 to q. This ensures that q will receive m_0 by the first causal path to reach it. A background garbage collection algorithm cleans up copies of messages that have become stable by reaching all of their destinations. To avoid excessive propagation of messages, the system always has the alternative of sending a message directly to its true destination and waiting for it to become stable, or it can simply wait until the message reaches its destination and becomes stable.

This scheme may seem wildly expensive, but it rarely sends a message more than once in applications that operate over Isis. One important reason for this is that Isis has other options available for use when the cost of piggybacking becomes too high—for example, instead of sending m_0 piggybacked to some destination far from its true destination, q, any process can

---

[1]Notice that this issue doesn't occur for communication to the same destination as the point-to-point message: One can send any number of point-to-point messages or individual copies of multicasts to a single process within a group without delaying. The requirement is that messages to *other* destinations must be delayed until these point-to-point messages are stable.

simply send m_0 to q, in this way making it stable. The system can also wait for stability to be detected by the original sender, at which point garbage collection will remove the obligation. Additionally, notice that m_0 only needs to be piggybacked once to any given destination. In Isis, which typically runs on a small set of servers, this means that the worst case was just to piggyback the message once to each server. For all of these reasons, the cost of piggybacking is never excessive in Isis. The Isis algorithm also has the benefit of avoiding any potential gaps in the causal communication order: If q has received a message that was causally ordered after m_1, then q will retain a copy of m_1 until m_1 is safe at its destination.

I am not aware of any system other than Isis that has used this approach. Perhaps the strongest argument against the approach is that it has an *unpredictable* overhead: One can imagine patterns of communication for which its costs would be high, such as a client/server architecture in which the server replies to a high rate of incoming RPCs. In principle, each reply will carry copies of some large number of prior but unstable replies, and the garbage collection algorithm will have a great deal of work to do. Moreover, the actual overhead imposed on a given message is likely to vary depending on the amount of time since the garbage collection mechanism was last executed. Recent group communication systems, such as Horus, seek to provide extremely predictable communication latency and bandwidth and steer away from mechanisms that are difficult to analyze in a straightforward manner.

14.2 Extending Causal Order to Multigroup Settings

Additional issues occur when groups can overlap. Suppose that a process sends or receives multicasts in more than one group—a pattern that is commonly observed in complex systems that make heavy use of group computing. Just as we asked how causal order can be guaranteed when a causal path includes point-to-point messages, we can also ask how causal and total order can be extended to apply to multicasts sent in a series of groups.

Consider first the issue of causal ordering. If process p belongs to groups g_1 and g_2, one can imagine a chain of multicasts that includes messages sent asynchronously in both groups—for example, perhaps we will have $m_1 \rightarrow m_2 \rightarrow m_3$, where m_1 and m_3 are sent asynchronously in g_1 and m_2 is sent asynchronously in g_2. Upon receipt of a copy of m_3, a process may need to check for and detect causal ordering violations, delaying m_3 if necessary until m_1 has been received. Actually, this example illustrates two problems, since we also need to be sure that the delivery atomicity properties of the system extend to sequences of multicasts sent in a different group. Otherwise, scenarios can occur whereby m_3 becomes causally orphaned and can never be delivered.

In Figure 14.3, for example, if a failure causes m_1 to be lost, m_3 can never be delivered. There are several possibilities for solving the atomicity problem, which lead to different possibilities for dealing with causal order. A simple option is to delay a multicast to group g_2 while there are causally prior multicasts pending in group g_1. In the example, m_2 would be delayed until m_1 becomes stable. Most existing process group systems use this solution, which is called the *conser-*

vative scheme. It is simple to implement and offers acceptable performance for most applications. To the degree that overhead is introduced, it occurs within the process group itself and hence is both localized and readily measured.

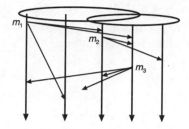

FIGURE 14.3. Message m_3 is causally ordered after m_1, and therefore may need to be delayed upon reception.

Less-conservative schemes are riskier in the sense that safety can be compromised when certain types of failures occur and they require more overhead; this overhead is less localized and consequently harder to quantify—for example, a *k-stability* solution might wait until m_1 is known to have been received at $k + 1$ destinations. The multicast will now be atomic provided that no more than k simultaneous failures occur in the group. However, we now need a way to detect causal ordering violations and to delay a message that arrives prematurely to overcome them.

One option is to annotate each multicast with multiple vector timestamps. This approach requires a form of piggybacking: Each multicast carries with it only timestamps that have changed or (if timestamp compression is used) only those with fields that have changed. Stephenson has explored this scheme and related ones and has shown that they offer general enforcement of causality at low average overhead. In practice, however, I am not aware of any systems that implement this method, apparently because the conservative scheme is so simple and because of the risk of a safety violation if a failure causes k processes to fail simultaneously.

Another option is to use the Isis style of piggybacking *cbcast* implementation. Early versions of the Isis Toolkit employed this approach, and, as noted earlier, the associated overhead turns out to be fairly low. The details are essentially identical to the method presented in Section 14.1. This approach has the advantage of also providing atomicity, but it has the disadvantage of having unpredictable costs.

In summary, there are several possibilities for enforcing causal ordering in multigroup settings. One should ask whether the costs associated with doing so are reasonable. The consensus of the community has tended to accept costs that are limited to within a single group (i.e., the conservative mode delays) but not costs that are paid systemwide (such as those associated with piggybacking vector timestamps or copies of messages). Even the conservative scheme, however, can be avoided if the application doesn't actually *need* the guarantee that this provides. Thus, the application designer should start with an analysis of the use and importance of multigroup causality before deciding to assume this property in a given setting.

14.3 Extending Total Order to Multigroup Settings

The total ordering protocols presented in Chapter 13 guarantee that messages sent in any one group will be totally ordered with respect to one another. However, even if the conservative stability rule is used, this guarantee does not extend to messages sent in different groups but received at processes that belong to both. Moreover, the local versions of total ordering permit some surprising global ordering problems. Consider, for example, multicasts sent to a set of processes that form overlapping groups, as shown in Figure 14.4. If one multicast is sent to each group, we could easily have process p receive m_1 followed by m_2, process q receive m_2 followed by m_3, process r receive m_3 followed by m_4, and process s receive m_1 followed by m_4. Since only a single multicast was sent in each group, such an order is total if only the perspective of the individual group is considered. Yet this ordering is clearly a cyclic one in a global sense.

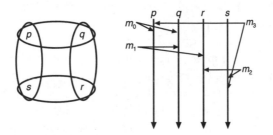

FIGURE 14.4. Overlapping process groups, seen from above and in a time-space diagram. Here, m_0 was sent to $\{p,q\}$, m_1 to $\{q,r\}$ and so forth, and since each group received only one message, there is no ordering requirement within the individual groups. Thus, an *abcast* protocol would never delay any of these messages. But one can deduce a global ordering for the multicasts. Process p sees m_0 after m_3, q sees m_0 before m_1, r sees m_1 before m_2, and s sees m_2 before m_3. This global ordering is thus cyclic, illustrating that many of our *abcast* ordering algorithms provide locally total ordering but not globally total ordering.

A number of schemes for generating a globally acyclic total ordering are known, and indeed one could have qualms about the use of the term "total" for an ordering that now turns out to sometimes admit cycles. Perhaps it would be best to say that previously we identified a number of methods for obtaining *locally total* multicast ordering whereas now we consider the issue of *globally total* multicast ordering.

The essential feature of the globally total schemes is that the groups within which ordering is desired must share some resource that is used to obtain the ordering property—for example, if a set of groups shares the same ordering token, the ordering of messages assigned using the token can be made globally total as well as locally total. Clearly, however, such a protocol could be costly, since the token will now be a single bottleneck for ordered multicast delivery.

In the Psync system an ordering scheme that uses multicast labels was introduced (see Peterson [1987], Peterson et al.); soon after, variations of this were proposed by the Transis and Totem systems (see Amir et al. [July 1992], Melliar-Smith and Moser [1989]). All of these methods work by using some form of unique label to place the multicasts in a total order determined by their labels. Before delivering such a multicast, a process must be sure it has received all other multicasts that could have smaller labels. The latency of this protocol is thus prone to rise with the number of processes in the aggregated membership of groups to which the receiving process belongs.

Each of these methods, and in fact all methods with which I am familiar, has performance that degrades as a function of scale. The larger the set of processes over which a globally total ordering property will apply, the more costly the ordering protocol. When deciding if globally total ordering is warranted, it is therefore useful to ask what sort of applications might be expected to notice the cycles that a local ordering protocol would allow. The reasoning is that if a cheaper protocol is still adequate for the purposes of the application, most developers would favor the cheaper protocol. In the case of globally total ordering, there are very few applications that really need this property.

Indeed, the following explanation may be the only widely cited example of a problem for which locally total order is inadequate and globally total order is consequently needed. Suppose that we wish to solve the dining philosophers' problem. In this problem, which is a classical synchronization problem well known to the distributed system community, a group of philosophers gather around a table. Between each pair of philosophers is a single shared fork, and at the center of the table is a plate of pasta. In order to eat, a philosopher must have one fork in each hand. The life of a philosopher is an infinite repetition of the sequence *think, pick up forks, eat, put down forks*. Our challenge is to implement a protocol that solves this problem and avoids deadlock.

Suppose that the processes in our example are the forks and that the multicasts originate in philosopher processes arrayed around the table. The philosophers can now request their forks by sending totally ordered multicasts to the process group of forks to their left and right. It is easy to see that if forks are granted in the order that the requests arrive, a globally total order avoids deadlock, but a locally total order is deadlock prone. Presumably, there is a family of multigroup locking and synchronization protocols for which similar results would hold. However, to repeat the point made above, I have never encountered a real-world application in which globally total order is needed. This being the case, such strong ordering should perhaps be held in reserve as an option for applications that specifically request it, but not as a default. If globally total order were as cheap as locally total order, the conclusion would be reversed.

14.4 Causal and Total Ordering Domains

We have seen that when ordering properties are extended to apply to multiple heavyweight groups, the costs of achieving ordering can rise substantially. Sometimes, however, such properties really are needed, at least in subsets of an application. If this occurs, one option may be to provide the application with control over these costs by introducing what are called *causal and total ordering domains.* Such a domain would be an attribute of a process group: At the time a group is created, it would be bound to an ordering domain identifier, which remains constant thereafter. We can then implement the rule that when two groups are in different domains, multicast ordering properties do not need to hold across them—for example, if group g_1 and group g_2 are members of different ordering domains, the system could ignore causal ordering between multicasts sent in g_1 and multicasts sent in g_2. More general still would be a scheme in which a domain is provided for each type of ordering: Two groups could then be in the same causal ordering domain but be in different total ordering domains. Implementation of ordering domains is simple if the corresponding multigroup ordering property is available within a system—for example, if group g_1 and group g_2 are members of different causal ordering domains, the conservative rule would be overlooked when a process switched from sending or receiving in one group to sending in the other. Delays would only occur when two groups are explicitly placed in the same ordering domain, presumably because the application actually requires multigroup ordering in this case.

It can be argued that the benefits associated with preserving causally total order systemwide are significantly greater than those for supporting globally total order. The reasoning is that causal order is needed to implement asynchronous data replication algorithms, and, since these have such a large performance advantage over other schemes, the benefits outweigh the costs of needing to enforce causal order across group boundaries. However, the conservative causality scheme is an adequate solution to this particular problem, and has the benefit of providing a systemwide guarantee with a local method. When combined with causal domains, such a mechanism has a highly selective cost. This said, however, it should also be noted that the *flush* primitive proposed earlier offers the same benefits and is quite easy to use. Thus, many real systems opt for causal ordering, do not delay when sending messages outside of a group, and provide a *flush* primitive for use by the application itself when causal ordering is needed over group boundaries. Such a compromise is visible to the user and is easily understood.

Similar reasoning seems to argue against globally total order: The primitive has a significant cost (mostly in terms of latency) and limited benefit. Thus, my work has stopped providing this property, after initially doing so in the early versions of the Isis Toolkit. The costs were simply too high to make globally total ordering the default, and the complexity of supporting a very rarely used mechanism argued against having the property at all.

14.5 Multicasts to Multiple Groups

An additional multigroup issue concerns the sending of a single multicast to a set of process groups in a single atomic operation. Until now, such an action would require that the multicast be sent to one group at a time, raising issues of nonatomic delivery if the sender fails midway. One can imagine solving this problem by implementing a multigroup multicast as a form of nonblocking commit protocol; Schiper and Raynal have proposed such a protocol in conjunction with their work on the Phoenix system (see Schiper and Raynal). However, there is another option, which is to create a new process group superimposed on the multiple destination groups and to send the multicast in that group. Interestingly, the best implementations of a group-created protocol require a single *fbcast*—hence, if one creates a group, issues a single multicast in it, and then deletes the group, this will incur a cost comparable to doing a multiphase commit over the same set of processes and then garbage collecting after the protocol has terminated!

This last observation argues against explicit support for sending a multicast to several groups at the same time, except in settings where the set of groups to be used cannot be predicted in advance and is very unlikely to be reused for subsequent communication—that is, although the application process can be presented with an interface that allows multicasts to be sent to sets of groups, it may be best to implement such a mechanism by creating a group in the manner described previously. In the belief that most group communication patterns will be reused shortly after they are first employed, such a group could then be retained for a period of time in the hope that a subsequent multicast to the same destinations will reuse its membership information. The group can then be torn down after a period during which no new multicasts are transmitted. Only if such a scheme is impractical would one need a multicast primitive capable of sending to many groups at the same time, and I am not familiar with any setting in which such a scheme is clearly not viable.

14.6 Multigroup View Management Protocols

A final issue that occurs in systems where groups overlap heavily is that our view management and flush protocol will run once for each group when a failure or join occurs, and our state transfer protocol will only handle the case of a process that joins a single group at a time. Clearly, these will be sources of inefficiency (in the first case) and inconvenience (in the second case) if group overlap is common. This observation, combined with the delays associated with conservative ordering algorithms and the concerns noted above in regard to globally total order, has motivated research on methods of collapsing heavily overlapped groups into smaller numbers of larger groups. Such approaches are often described as resulting in *lightweight* groups, because the groups seen by the application typically map onto some enclosing set of *heavyweight* groups.

Glade has explored this approach in Isis and Horus (see Glade et al.). His work supports the same process group interfaces as for a normal process group, but maps multicasts to lightweight groups into multicasts to the enclosing heavyweight groups. Such multicasts are filtered on arrival, so that an individual process will only be given copies of messages actually destined for it. The approach essentially maps the fine-grained membership of the lightweight groups to a coarser-grained membership in a much smaller number of heavyweight groups.

The benefit of Glade's approach is that it avoids the costs of maintaining large numbers of groups (the membership protocols run just once if a process joins or leaves the system, updating multiple lightweight groups in one operation). Moreover, the causal and total ordering guarantees of our single-group solutions will now give the illusion of multigroup causal and total ordering, with no changes to the protocols themselves. Glade argues that when a system produces very large numbers of overlapping process groups, there are likely to be underlying patterns that can be exploited to efficiently map the groups to a small number of heavyweight ones. In applications I am familiar with, Glade's point seems to hold. Object-oriented uses of Isis (e.g., Orbix+Isis) can generate substantial overlap when a single application uses multiple object groups. But this is also precisely the case where a heavyweight group will turn out to have the least overhead, because it precisely matches the membership of the lightweight groups.

Glade's algorithms in support of lightweight process groups are relatively simple. A multicast to such a group is mapped to a multicast to the corresponding heavyweight group and filtered on arrival. Membership changes can be implemented either by using an *abcast* to the lightweight group or, in more extreme cases, by using a flushed multicast, similar to the one used to install view changes in the heavyweight group. For most purposes, the *abcast* solution is sufficient.

14.7 Related Reading

On the timestamp technique used in Harp: (see Ladin et al. [1992], Liskov et al.).

On preserving causality in point-to-point message-passing systems: (see Schiper et al.).

On the associated controversy: (see Cheriton and Skeen), and on the responses (see Birman [1994], Cooper [1994], van Renesse [1993]).

On multiple groups in Isis: (see Birman and Joseph [February 1987], Birman et al.).

On communication from a nonmember of a group to a group: (see Birman and Joseph [February 1987], Wood [1993]).

On graph representations of message dependencies: (see Amir et al. [July 1992], Melliar-Smith and Moser [1989], Peterson [1987], Peterson et al.).

On lightweight process groups: (see Glade et al.).

CHAPTER 15 ✧ ✧ ✧ ✧

The Virtually Synchronous Execution Model

CONTENTS

The process group communication primitives introduced in the previous chapters create a powerful framework for algorithmic development. When the properties and primitives are combined for this purpose, we will say that a *virtually synchronous* execution environment results (see Birman and Joseph [February 1987, November 1987], Birman and van Renesse [1994]). However, although we built up our primitives from basic message passing, it is probably easier to understand the idea behind virtual synchrony in a top-down treatment. We'll then use the approach to develop an extremely high performance replicated data algorithm, as well as several other tools for consistent distributed computing.

15.1 Virtual Synchrony

Suppose that we want to use a process group (or a set of process groups) as a building block in a distributed application. The group members will join that group for the purpose of cooperation, perhaps to replicate data or to perform some operation fault tolerantly. The issue now occurs of how to design such algorithms with a high degree of confidence that they will operate correctly.

Recall the discussion of transactional serializability from Section 7.5. In that context, we encountered a similar problem: a set of concurrently executed programs that share files or a database and want to avoid interference with one another. The basic idea was to allow the developer to code these applications as if they would run in isolation from one another, one by one. The database itself is permitted to interleave operations for greater efficiency, but only in ways that preserve the illusion that each transaction executes without interruption. The results of a transaction are visible only after it commits; a transaction that aborts is automatically and completely erased from the memory of the system. As we noted at the time, transactional serializability allows the developer to use a simple programming model, while offering the system an opportunity to benefit from high levels of concurrency and asynchronous communication.

Virtual synchrony is not based on transactions, but introduces an approach similar to programming with process groups. In the virtual synchrony model, the simplifying abstraction seen by the developer is that of a set of processes (the group members) which all see the same events in the same order. These events are incoming messages to the group and group membership changes. The key insight, which is not a very deep one, is that since all the processes see the same inputs, they can execute the same algorithm and in this manner stay in consistent states. This is illustrated in Figure 15.1, which shows a process group receiving messages from several nonmembers. It then has a new member join and transfers the state of the group to this new member, which results in a failure of one of the old members. Notice that the group members see identical sequences of events while they belong to the group. The members differ, however, in their relative ranking within the group membership. There are many possible ways to rank the members of a group, but the most common one, which is used in this chapter, assumes that the rank is based on when members joined—the oldest member having the lowest ranking and so forth.

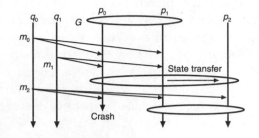

FIGURE 15.1. Closely synchronous execution: All group members see the same events (incoming messages and new group views) in the same order. In this example only the nonmembers of the group send multicasts to it, but the group members can also multicast to one another and can send point-to-point messages to the nonmembers—for example, a group RPC could be performed by sending a multicast to the group, to which one or more members could reply.

The State Machine approach of Lamport and Schneider first introduced this approach as part of a proposal for replicating objects in settings subject to Byzantine failures (see Schneider [1988, 1990]). Their work made a group of identical replicas of the object in question, and used the Byzantine protocol for all interactions with the group and to implement its interactions with the external world. However, the State Machine approach saw little use when it was first proposed for the same reason that the Byzantine protocol sees little practical use: Few computing environments satisfy the necessary synchronous computing and communication requirements, and it is difficult to utilize a service that employs a Byzantine fault model without extending the same approach to other aspects of the environment, such as any external objects with which the service interacts and the operating system software used to implement the communication layer.

A further concern about the State Machine approach is that all copies of the program see identical inputs in the identical order. If one program crashes because of a software bug, so will all the replicas of that program. Unfortunately, as we saw earlier, studies of real-world computing installations reveal that even in mature software systems, bugs that crash the application remain a proportionately important cause of failures. Thus, by requiring correct processes that operate deterministically and in lockstep, the State Machine approach is unable to offer protection against software faults.

Virtual synchrony is similar to the State Machine abstraction, but it moves outside the original Byzantine setting, while also introducing optimizations that overcome the concerns mentioned previously. The idea is to view the State Machine as a sort of reference model but to allow a variety of optimizations so that the true execution of a process group may be very far from lockstep synchrony. In effect, just as transactional executions allow operations to be interleaved, provided that the behavior is indistinguishable from some serial execution, a virtually synchronous execution also allows operations to be interleaved, provided that the result is indistinguishable from some closely synchronous (State Machine) execution. Nonetheless, the executions of the different replicas will be different enough to offer some hope that software bugs will not be propagated across the entire group.

To take a very simple example, suppose that we wish to support a process group whose members replicate some form of database and perform load-balanced queries upon it. The operations on the service will be queries and updates, and we will overlook failures (for the time being) to keep the problem as simple as possible.

Next, suppose that we implement both queries and database updates as totally ordered multicasts to the group membership. Every member will have the same view of the membership of the group, and each will see the same updates and queries in the same order. By simply applying the updates in the order they were received, the members can maintain identically replicated copies of the database. As for the queries, an approximation of load-balancing can be had using the ranking of processes in the group view. Suppose that the process group view ranks the members in $[0 \ldots n-1]$. Then the ith incoming query can be assigned to the process whose rank is $(i \bmod n)$. Each query will be handled by exactly one process.

We'll call this a *closely synchronous* execution. Frank Schmuck was the first to propose this term, observing that the actions of the receiving processes were closely synchronized but might be spread over a significant period of real time. The synchronous model, as discussed previously, would normally require real-time bounds on the time period over which an action is performed by the different processes. Notice that a closely synchronous execution does not require identical actions by identical processes: If we use the load-balancing idea outlined above, actions will be quite different at the different copies. Thus, a closely synchronous group is similar to a group that uses State Machine replication, but it is not identical.

Having developed this solution, however, there will often be ways to weaken the ordering properties of the protocols it uses—for example, it may be the case that updates are only initiated by a single source, in which case an *fbcast* protocol would be sufficient to provide the desired ordering. Updates will no longer be ordered with respect to queries if such a change is made, but in an application where a single process issues an update and follows it with a query, the update would always be received first and hence the query will reflect the result of doing the update. In a slightly fancier setting, *cbcast* might be needed to ensure that the algorithm will operate correctly—for example, with *cbcast* one would know that if an application issues an update and then tells some other process to do a query, that second process will see the effects of the causally prior updates. Often, an analysis such as this one can be carried very far.

Having substituted *fbcast* or *cbcast* for the original *abcast*, however, the execution will no longer be closely synchronous, since different processes may see different sequences of updates and queries and hence perform the same actions but in different orders. The significant point is that if the original analysis was performed correctly, the actions will produce an effect indistinguishable from what might have resulted from a closely synchronous execution. Thus, the execution appears closely synchronous, even though it is not. It is *virtually synchronous* in much the same sense that a transactional system creates the illusion of a serial execution even though the database server is interleaving operations from different transactions to increase concurrency.

Our transformation has the advantage of delivering inputs to the process group members in different orders, at least some of the time. Moreover, as we saw earlier, the process groups themselves are dynamically constructed, with processes joining them at different times. Also, the ranking of the processes within the group differs. Thus, there is substantial room for processes to execute in slightly different ways, affording a degree of protection against software bugs that could crash some of the members.

Recall the Gray/Lindsey characterization of Bohrbugs and Heisenbugs from Chapter 12. It is interesting to note that virtually synchronous replication can protect against Heisenbugs (see Birman and van Renesse [1994, 1996]). If a replica crashes because such a bug has been exercised, the probability that other group members will crash simultaneously is reduced by the many aspects of the execution that differ from replica to replica. Our transformation from a closely synchronous system to a virtually synchronous system increases the natural resiliency of the group, assuming that its constituent members are mature, well-debugged code. Nonetheless, some exposure to correlated failures is unavoidable, and the designer of a critical system should keep this in mind.

Additionally, notice that the *cbcast* primitive can be used asynchronously: There is no good reason for a process that issues a *cbcast* to perform an update to wait until the update has been completed by the full membership of the group. The properties of the *cbcast* protocol ensure that these asynchronously transmitted messages will reliably reach their destinations and that any causally subsequent actions by the same or different processes will see the effects of the prior *cbcasts*. In an intuitive sense, one could say that these *cbcast* protocols look as if they were performed instantly, even when they actually execute over an extended period of time.

In practice, the most common transformation that we will make is precisely this one: the replacement of a totally ordered *abcast* primitive with an asynchronous, causally ordered *cbcast* primitive. In the following sections, this pattern will occur repeatedly. Such transformations have been studied by Schmuck in his Ph.D. dissertation and are extremely general: What we do in the following sections can be done in many settings.

Thus, we have transformed a closely synchronous group application, in which the members operate largely in lockstep, into a very asynchronous implementation in which some members can pull ahead and others can lag behind, communication can occur concurrently with other execution, and the group may be able to tolerate software bugs that crash some of its members. These are important advantages and account for the appeal of this approach.

When an application has multiple process groups in it, an additional level of analysis is often required. As we saw in the previous chapter, multigroup causal (and total) ordering is expensive. When one considers real systems, it also turns out that multigroup ordering is often unnecessary: Many applications that need multiple groups use them for purposes that are fairly independent of one another. Operations on such independent groups can be thought of as commutative, and it may be possible to use *cbcast* to optimize such groups independently without taking the next step of enforcing causal orderings across groups. Where multigroup ordering is needed, it will often be confined to small sets of groups, which can be treated as an ordering domain. In this manner, a general solution results, which can scale to large numbers of groups while still preserving the benefits of the single asynchronous communication pattern seen in the *cbcast* protocol but not in the *abcast* protocols.

Our overall approach, it should be noted, is considerably less effective when the dynamic uniformity guarantees of the safe multicast protocols are required. The problem is that whereas asynchronous *cbcast* is a very fast protocol, which delivers messages during its first phase of communication, any dynamically uniform protocol will delay delivery until a second phase. The benefit of replacing *abcast* with *cbcast* in such a case is lost. Thus, one begins to see a major split between the algorithms that operate fairly synchronously, requiring more than a single phase of message passing before delivery can occur, and those that operate asynchronously, allowing the sender of a multicast to continue computing while multicasts that update the remainder of a group or that inform the remainder of the system of some event propagate concurrently to their destinations.

The following is a summary of the key elements of the virtual synchrony model:

▶ *Support for process groups:* Processes can join groups dynamically and are automatically excluded from a group if they crash.

▶ *Identical process group views and mutually consistent rankings:* Members of a process group are presented with identical sequences of group membership, which we call *views* of that process group. If a nonprimary component of the system forms after a failure, any process group views reported to processes in that component are identified as nonprimary, and the view sequence properties will otherwise hold for all the processes in a given component. The view ranks the components, and all group members see identical rankings for identical group views.

▶ *State transfer to the joining process:* A process that joins a group can obtain the group's current state from some prior member or from a set of members.

▶ *A family of reliable, ordered multicast protocols:* We have seen a number of these, including *fbcast*, *cbcast*, *abcast*, *cabcast*, the *safe* (dynamically uniform) versions of these, and the group flush protocol, which is sometimes given the name *gbcast*.

▶ *Gap-freedom guarantees:* After a failure, if some message, m_j, is delivered to its destinations, then any message, m_i, that the system is obliged to deliver prior to m_j will also have been delivered to its destinations.

▶ *View-synchronous multicast delivery:* Any pair of processes that are both members of two consecutive group views receive the same set of multicasts during the period between those views.[1]

▶ *Use of asynchronous, causal, or FIFO multicast:* Although algorithms will often be developed using a closely synchronous computing model, a systematic effort is made to replace synchronous, totally ordered, and dynamically uniform (safe) multicasts with less-costly alternatives—notably the asynchronous *cbcast* primitive in its nondynamic uniform (unsafe) mode.

---

[1]In some systems this is interpreted so that if a process fails, but its failure is not reported promptly, it is considered to have received multicasts that would have been delivered to it had it still been operational.

15.2 Extended Virtual Synchrony

As discussed in the previous section, the virtual synchrony model is inherently intolerant of partitioning failures: The model is defined in terms of a single system component within which process groups reside. In this primary component approach, if a network partitioning failure occurs and splits a group into fragments, only the fragment that resides in the primary component of the system is able to continue operation. Fragments that find themselves in the nonprimary component(s) of the system are typically forced to shut down, and the processes within them must reconnect to the primary component when communication is restored.

The basis of the primary component approach lies in a subtle issue, which we first saw when discussing commit protocols. In a dynamic distributed environment there can be symmetric failure modes resulting from communication problems that mimic process failures. In such a situation perhaps process p will consider that process q has failed while process q believes the opposite to be true. To make progress, one or the other (or perhaps both) of these events must become official. In a partitioned run of the system, only one of these conflicting states can become official.

At the core of the problem is the observation that if a system experiences a partitioning failure, it is impossible to guarantee that multiple components can remain operational (in the sense of initiating new actions, delivering messages, and new group views) with guarantees that also span both sides of the partition. To obtain strong systemwide guarantees a protocol must always wait for communication to be reestablished under at least some executions in at least one side of the partition. When we resolve this problem using the protocols discussed in the previous chapters, the primary component is permitted to make progress at the expense of inconsistency relative to other components: Within these other components, the set of messages delivered may be different from the set in the primary component, and the order may also be different. In the case of the dynamically uniform protocols the guarantees are stronger, but nonprimary components may be left in a state where some dynamically uniform multicasts are still undelivered and where new dynamically uniform ones are completely blocked. The primary component, in contrast, can make progress so long as its GMS protocol is able to make progress.

Some researchers, notably those involved with the Transis and Totem projects, have pointed out that there are applications that can tolerate inconsistency of the sort that could occur if progress were permitted in a nonprimary component of a partitioned system (see Agarwal, Dolev et al., Malkhi). In these systems, any component that can reach internal agreement on its membership is permitted to continue operation. However, only a single component of the system is designated as the primary one. An application that is safe only in the primary component would simply shut down in nonprimary components. Other applications, however, might continue to be available in nonprimary components, merging their states back into the primary component when the partitioning failure ends.

Carrying this observation even further, the Transis group has shown that there are distributed systems in which no component ever can be identified as the primary one, and yet every action initiated within the system can eventually be performed in a globally consistent manner (see Dolev et al., Keidar and Dolev). However, this work involves both a static system model and a very costly protocol, which delays performing an action until a majority of the processes in the system as a whole have acknowledged receipt of it. The idea is that actions can be initiated within dynamically defined components, which represent subsets of the true maximal system configuration, but these actions remain in a pending state until a sufficient number of processes are known to have seen them, which occurs when communication is restored between components. Eventually, knowledge of the actions reaches enough processes so that it becomes safe to perform them. But the protocol is clearly intended for systems that operate in a partitioned mode over very long periods of time and where there is no special hurry to perform actions. Yair Amir has extended this approach to deal with more urgent actions, but his approach involves weakening the global consistency properties (see Amir). Thus, one is faced with a basic tradeoff between ensuring that actions will occur quickly and providing consistency between the primary component of the system and other components. We can have one or the other, but not both at once.

Cornell's most recent system, Horus, supports an extended model of the former sort (see Malkhi). (In fact, this part of Horus was actually implemented by Malkhi, who ported the associated code from Transis into Horus.) However, it is quite a bit harder to work with than the primary partition model. The merge of states when an arbitrary application resumes contact between a nonprimary and a primary component can be very difficult and cannot, in general, be solved automatically. In practice, such an application would normally remember any update actions it has taken and save these on a queue. When a merge becomes possible, it would replace its state with that of the primary component and then reapply these updates, if any. But it is not clear how large a class of applications can operate this way. Moreover, unless dynamically uniform protocols are employed for updates, the nonprimary component's state may be inconsistent with the primary one in significant ways.

On the other hand, the primary component model is awkward in wide area networks where partitioning events occur frequently (see Figure 15.2). Here, the model will in effect shut down parts of the overall system that are physically remote from the main part of the system. Each time these parts manage to restart after a communication failure, a new communication problem will soon cut them off again. (See Figure 15.3.)

Recent work, which we will not discuss in detail, points to yet a third possible mode of operation. In this mode, a computing system would be viewed as a wide area network composed of interconnected local area networks, as was first proposed in the Transis project. Within each of the LAN systems one would run a local subsystem: a complete primary-component system with its own sets of process groups and a self-sufficient collection of services and applications. The WAN layer of the system would be built up by superimposing a second communication structure on the aggregate of LANs and would support its own set of WAN services. At this higher level of the system, one would use a true asynchronous communication model: If a partitioning event does occur, such a WAN system would wait until the problem is resolved. The WAN system would then be in a position to make use of protocols that don't attempt to make progress while communication is disrupted, but rather wait as long as necessary until the exchange of

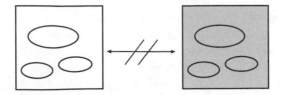

FIGURE 15.2. When a partitioning failure occurs, an application may be split into two or more fragments, each complete in the sense that it may potentially have a full set of processes and groups. In the primary component model, however, only one set is permitted to remain operational—hopefully one that has a full complement of processes and groups. In this figure, the white component might thus be alive after the link breaks, while the members of the gray component are prevented from continuing execution. The rationale underlying this model is that it is impossible to guarantee consistency if both sides of a partitioning failure are permitted to remain available while a communication failure is pending. Thus, we could allow both to run if we sacrifice consistency, but then we face a number of problems: Which side owns critical resources? How can the two sides overcome potential inconsistencies in their states as of the time of the partition failure event? There are no good general answers to these questions.

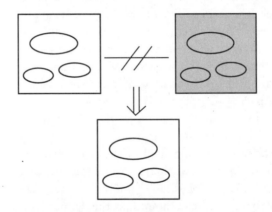

FIGURE 15.3. The extended virtual synchrony model allows both white and gray partitions to continue progress despite the inconsistencies that may occur between their states. However, only one of the components is considered to be the primary one. Thus, the white partition might be considered to be authoritative for the system, while the gray partition is permitted to remain alive but is known to be potentially stale. Later, when the communication between the components is restored, the various process group components merge, resulting in a single, larger system component with a single instance of each process group (shown at bottom of figure). The problem, however, is that merging must somehow overcome the inconsistencies that may have occurred during the original partitioning failure, and this may not always be possible. Working with such a model is potentially challenging for the developer. Moreover, one must ask what sorts of applications would be able to continue operating in the gray partition knowing that the state of the system at that point may be inconsistent—for example, it may reflect the delivery of messages in an order that differs from the order in the main partition, having atomicity errors or gaps in the message-delivery ordering.

messages resumes and the protocol can be pushed forward. The consensus protocol of Chandra and Toueg is a good example of a protocol one could use at the WAN level of a system structured in this manner, while the virtual synchrony model would be instantiated multiple times separately: once for each LAN subsystem.

In this two-tiered model (see Figure 15.4), an application would typically be implemented as a local part designed to remain available in the local component and to reconfigure itself to continue progress despite local failures. The primary component virtual synchrony model is ideal for this purpose. When an action is taken that has global implications, the local part would initiate a global action by asking the WAN architecture to communicate this message through the WAN system. The WAN system would use potentially slow protocols, which offer strong global ordering and atomicity properties at the expense of reduced progress when partitioning failures occur, delivering the resulting messages back into the various local subsystems. The local subsystems would then apply these updates to their global states.

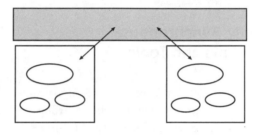

FIGURE 15.4. In a two-tiered model, each LAN has its own complete subsystem and runs using its own copy of the primary-component virtual synchrony model. A WAN system (gray) spans the LANs and is responsible for distributing global updates. The WAN layer may block while a partitioning failure prevents it from establishing the degree of consensus needed to safely deliver updates, but the local systems continue running even if global updates are delayed. Such a mixed approach splits the concerns of the application into local ones, where the focus is on local consistency and high availability, and global ones, where the focus is on global consistency even if partitioning failures can introduce long delays. I favor this approach, which is used in the Isis Toolkit's long-haul subsystem and has been applied successfully in such Isis applications as its wide area news facility.

Danny Dolev has suggested the following simple way to understand such a two-tiered system. In his view, the LAN subsystems run applications that are either entirely confined to the LAN (and have no interest in global state) or that operate by reading the *global* state but updating the *local* state. These applications do not directly update the global system state. Rather, if an action requires that the global state must be updated, the LAN subsystem places the associated information on a WAN action queue, perhaps replicating this for higher availability. From that queue, the WAN protocols will eventually propagate the action into the WAN level of the system, from where it will filter back down into the LAN level in the form of an update to the global system state. The LAN layer will then update its local state to reflect the fact that the requested action has finally been completed. The LAN layer of such a system would use the primary-component virtual synchrony model, while the WAN layer employs protocols based on the method discussed by Keidar and Dolev.

First introduced in the Isis system's long-haul service by Makpangou, and then extended through Dolev and Malkhi's work on the Transis architecture (which has a "LANsys" and a "WANsys" subsystem), two-tiered architectures such as this have received attention in many projects and systems. They are now used in Transis, Horus, NavTech, Phoenix, and Relacs. By splitting the application into the part that can be done locally with higher availability and the part that must be performed globally even if availability is limited, two-tiered architectures do not force a black or white decision on the developer. Moreover, a great many applications seem to fit well with this model. It seems likely that we will soon begin to see programming tools that encapsulate this architecture into a simple-to-use, object-oriented framework, making it readily accessible to a wide community of potential developers.

15.3 Virtually Synchronous Algorithms and Tools

In the following sections, we will develop a set of simple algorithms to illustrate the power and limitations of reliable multicast within dynamic process groups. These algorithms are just a small subset of the ones that can be developed using the primitives, and the sophisticated system designer may sometimes find a need for a causal and total multicast primitive (*cabcast*) or one with some other slight variation on the properties we have focused on here. Happily, the protocols we have presented are easily modified for special needs, and modern group communication systems, such as the Horus system, are designed precisely to accommodate such flexibility and fine-tuning. The following algorithms, then, should be viewed as a form of template upon which other solutions might be developed through a process of gradual refinement and adaptation.

15.3.1 Replicated Data and Synchronization

When discussing the static process group model, we put it to the test by using it to implement replicated data. The reader will recall from Section 13.7 that this approach was found to have a number of performance problems. The algorithm that resulted would have forced group members to execute nearly in lockstep, and the protocols themselves were costly in terms of both latency and messages required. Virtual synchrony, on the other hand, offers a solution to this problem that is inexpensive in all of these aspects, provided that dynamic uniformity is not required. When dynamic uniformity is required, the cost is still lower than for the static, quorum-replication methods, although the advantage is less pronounced.

We start by describing our replication and synchronization algorithm in terms of a closely synchronous execution model. We will initially focus on the nondynamic uniform case. Suppose that we wish to support *READ, UPDATE,* and *LOCK* operations on data replicated within a process group. As a first approximation to a solution, we would use *abcast* to implement the *UPDATE* and *LOCK* operations, while allowing any group member to perform *READ* operations using its local replica of the data maintained by the group.

Specifically, we will require each group member to maintain a private replica of the group data. When joining a group, the state transfer algorithm (developed below) must be used to initialize the replica associated with the joining process. Subsequently, all members will apply the same sequence of updates by tracking the order in which *UPDATE* messages are delivered and respecting this order when actually performing the updates. *READ* operations, as suggested above, are performed using the local replica (this is in contrast to the quorum methods, where a read must access multiple copies).

An *UPDATE* operation can be performed without waiting for the group members to actually complete the individual update actions. Instead, an *abcast* is issued asynchronously (without waiting for the message to be delivered), and the individual replicas perform the update when the message arrives.

Many systems make use of nonexclusive read locks. If necessary, these can also be implemented locally. The requesting process will be granted the lock immediately unless an exclusive (write) lock is registered at this copy. Finally, exclusive (write) *LOCK* operations are performed by issuing an *abcast* to request the lock and then waiting for each group member to grant it. A recipient of such a request waits until there are no pending read locks and then grants the request in the order it was received. The lock will later be released either with another *abcast* message or upon receipt of a new view of the process group reporting the failure of the process that holds the lock.

One would hope it is obvious that this implementation of replicated data will be tolerant of failures and guarantee the consistency of the copies. The individual replicas start in identical states because the state transfer to a joining member copies the state of the replicated data object from some existing member. Subsequent updates and lock operations behave identically at all copies. Thus, all see the same events in the same order, and all remain in identical states.

Now, let us ask how many of these *abcast* operations can be replaced with asynchronous *cbcast* operations. In particular, suppose that we replace *all* of the *abcast* operations with asynchronous *cbcast* operations. Remarkably, with just two small changes, the modified algorithm will be correct. The first change is that all updates must be guarded by a lock with appropriate granularity—that is, if any update might be in conflict with a concurrent update, we will require that the application must ensure that the update provides for some form of mutual exclusion. On the other hand, updates that are known to be independent commute and hence can be issued concurrently—for example, updates to two different variables maintained by the same process group can be issued concurrently, and, in groups where all the updates for a specific type of data originate with a single process, no locks may be required at all.

The second change is a bit more subtle: It has to do with the way that ordering is established when a series of write locks are requested within the group. The change is as follows. We will say that the first process to join the group, when the group is first created, is its *initial writer*. This process is considered to control write access to all the data items managed by the group.

Now, before doing an update, a process will typically request a lock, sending a *cbcast* to inform the group members of its *LOCK* request and waiting for the members to grant the request. In our original closely synchronous algorithm, a recipient of such a request granted it in first-come, first-served order when no local read-lock was pending. Our modified algorithm, however, will wait before granting lock requests. They simply pile up in a queue, ordered in whatever order the *cbcast* messages were delivered.

When the writer for a given lock no longer needs that lock, we will say that it becomes *prepared to pass the lock.* This process will react to incoming lock requests by sending out a *cbcast* that *grants* the first lock request on its copy of the queue. The grant message will be delivered to all group members. Once the grant message is received, a member dequeues the corresponding lock request (the causal ordering properties ensure that the request will indeed be found on the pending lock-request queue) and then grants it when any read locks present for the same item have been released. A writer thus grants the lock to the process that issued the oldest of the pending lock requests on its version of the lock queue.

Having obtained a grant message for its lock request, as well as individual confirmation messages from each group member that the lock has been acquired locally, the writer may begin issuing updates. In many systems the local read-lock mechanism will not be required, in which case the members need not confirm write-lock acquisition, and the writer need not wait for these messages. The members simply dequeue the pending lock request when the grant message arrives, and the writer proceeds to issue updates as soon as it receives the grant message.

It may at first seem surprising that this algorithm can work: Why should it ensure that the group members will perform the same sequence of updates on their replicas? To understand this, start by noticing that the members actually might not perform identical sequences of updates (Figure 15.5). However, any sequence of *conflicting updates* will be identical at all replicas, for the following reason: Within the group, there can be only one writer that holds a given lock. That writer uses *cbcast* (asynchronously) to issue updates and uses *cbcast* (again asynchronously) to grant the write lock to the subsequent writer. This establishes a total order on the updates: One can imagine a causal path traced through the group, from writer to writer, with the updates neatly numbered along it—the first update, the second, the granting of the lock to a new writer, the third update, the granting of the lock to a new writer, the fourth update, and so forth. Thus, when *cbcast* enforces the delivery order, any set of updates covered by the same lock will be delivered in the same order to all group members.

As for the nonconflicting updates: These commute with the others and hence would have had the same effect regardless of how they were ordered. The ordering of updates is thus significant only with respect to other conflicting updates.

Finally, the reader may have noticed that lock requests are not actually seen in the same order by each participant. This is not a problem, however, because the lock-request order isn't actually used in the algorithm. As long as the grant operation is reasonably fair and grants a request that is genuinely pending, the algorithm will work.

The remarkable thing about this new algorithm is that it is almost entirely asynchronous. Recall that our *cbcast* protocol delivers messages in the same view of the process group as the one

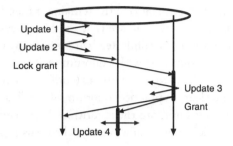

FIGURE 15.5. A set of conflicting updates is ordered because only one process can write at a time. Each update, and the lock-granting message, is an asynchronous *cbcast.* **Because the causal order is in fact a total one along this causal path (shown in bold), all group members see the same updates in the same order. Lock requests are not shown, but they too would be issued using an asynchronous** *cbcast.* **Notice that lock requests will not be seen in the same order by all processes, but this is not required for the algorithm to behave correctly. All that matters is that the grant operation grant a currently pending lock request, and, in this algorithm, all processes do have a way to track the pending requests, even though they may learn about those requests in different orders.**

that was in place when the *cbcast* was initiated. This implies that the sender of a *cbcast* can always deliver its own copy of the multicast as soon as it initiates the message. After all, by definition, a causally prior *cbcast* will already have been delivered to the sender, and the flush protocol enforces the view synchrony and causal gap-freedom guarantees. This means that a process wanting to issue an update can perform the update locally, sending off a *cbcast* that will update the other replicas without delaying local computation. Clearly, a lock request will block the process that issues it—unless that process happens to hold the write lock already, as is often the case in systems with bursty communication patterns. But it is clear that this minimal delay—the time needed to request permission to write and for the grant message to travel back to the requesting process—is necessary in any system.

Moreover, the algorithm can be simplified further. Although we used *cbcast* here, one could potentially replace this with *fbcast* by employing a sequence number: The *i*th update would be so labeled, and all group members would simply apply updates in sequence order. The token would now represent permission to initiate new updates (and the guarantee that the values a process reads are the most current ones). Such a change eliminates the vector-timestamp overhead associated with *cbcast*, and it is also recognizable as an implementation of one of the *abcast* protocols we developed earlier!

From the perspective of an application, this asynchronous replication and locking scheme may seem astonishingly fast. The only delays imposed upon the application are when it requests a new write lock. During periods when it holds the lock, or if it is lucky enough to find the lock already available, the application is never delayed at all. Read operations can be performed locally, and write operations respond as soon as the local update has been completed and the *cbcast* or *fbcast* (we'll just call it a *cbcast* for simplicity) to the remaining members has been handed off to the communication subsystem. Later, we will see that the Horus system achieves performance that can reach 85,000 such updates per second. Reads are essentially free—hence, millions could be done per second. When this is compared with a quorum read and update technology, in which it would be surprising to exceed 100 reads and updates (combined) in one second, the benefits of an asynchronous *cbcast* become clear. In practice, quorum schemes are often considerably slower than this because of the overheads built into the algorithm. Moreover, a quorum read or update forces the group members into lockstep, while our asynchronous replication algorithm encourages them to leap ahead of one another, buffering messages to be transmitted in the background.

However, we must not delude ourselves into viewing this algorithm as identical to the quorum replication scheme, because that scheme provides the equivalent of a dynamic uniformity guarantee and a strong total ordering. The algorithm described above could be modified to provide such a guarantee by using a safe *cbcast* in place of the standard *cbcast*. But such a change will make the protocol dramatically slower, because each *UPDATE* will now be delayed until at least a majority of the group members acknowledge receipt of the update message. Thus, although the algorithm would continue to perform *READ* operations from local replicas, *UPDATE* operations will now be subject to the same performance limits as for a quorum update. The benefit may still be considerable, but the advantage of this scheme over a quorum scheme would be much reduced.

In my experience, dynamic uniformity is rarely needed. If an application is about to take an externally visible action and it is important that in the event of a failure the other replicas of the application are in a consistent state with that of the application taking the action, this guarantee becomes important. In such cases, it can be useful to have a way to *flush* communication within the group, so that any prior asynchronous multicasts are forced out of the communication channels and delivered to their destinations. A *cbcast* followed by a *flush* is thus the equivalent of a safe *cbcast* (stronger, really, since the *flush* will flush all prior *cbcast*s, while a safe *cbcast* might not provide this guarantee). Many process group systems, including Horus, adopt this approach rather than one based on a safe *cbcast*. The application developer is unlikely to use *flush* very frequently— hence, the average performance may approximate that of our fully asynchronous algorithm, with occasional short delays when a *flush* pushes a few messages through the channels to their destinations. Unless large backlogs develop within the system, long delays are unlikely to occur. Thus, such a compromise can be very reasonable from the perspective of the application designer.

By way of analogy, many system developers are familiar with the behavior of operating systems that buffer disk I/O. In such settings, to increase performance, it is common to permit the application to continue operation as soon as a disk write is reflected in the cache contents— without waiting for the data to be flushed to the disk itself. When a stronger guarantee is required, the application explicitly requests that the disk buffer be flushed by invoking an appropriate primitive, such as the UNIX *fsync* system call. The situation created by the asynchronous *cbcast* is entirely analogous, and the role of the *flush* primitive is precisely the same as that of *fsync*.

What about a comparison with the closely synchronous algorithm from which ours was derived? Interestingly, the story here is not so clear. Suppose that we adopt the same approach to the dynamic uniformity issue, by using a *flush* primitive with this property if required. Now, the performance of the closely synchronous *abcast* algorithm will depend entirely on the way *abcast* is implemented. In particular, one could implement *abcast* using the *cbcast*-based lock and update scheme described in this section or using a rotating token (with very similar results). Such an *abcast* solution would push the logic of our algorithm into the communication primitive itself. In principle, performance could be the same as for an algorithm using *cbcast* explicitly—this level of performance has been achieved in experiments with the Horus system. The major issue is that to do this, one needs to use an *abcast* algorithm well matched to the communication pattern of the user, and this is not always possible: Many systems lack the knowledge to predict such patterns accurately.

15.3.2 State Transfer to a Joining Process

As we have noted several times, there is often a need to transfer information about the current state of a process group to a joining member at the instant it joins. In the iterated protocol by which a nonmember of a group communicates with a group, for example, it was necessary to ensure that the group members could learn how many messages had been successfully delivered from each client prior to the join event. In a replicated data algorithm, there is clearly a need to transfer a current copy of the data in question to the joining process.

The most appropriate representation of the state of the group, however, will be highly dependent on the application. Some forms of state may be amenable to extreme compression or may be reconstructable from information stored on files or logs using relatively small amounts of information at the time the process joins. Accordingly, we adopt the view that a state transfer should be done by the application itself. Such a transfer is requested at the time of the join and operates much like a remote procedure call.

Given this point of view, it is easy to introduce state transfer into the protocol for group flush. At the time a process first requests that it be added to the group, it should signal its intention to solicit state from the members. The associated information is passed in the form of a message to the group members and is carried along with the join protocol to be reported with the new group view after the members perform the flush operation.

Each member now faces a choice: It can stop processing new requests at the instant of the flush, or it can make a copy of its state as of the time of the flush for possible future use, in which case it can resume processing. The joining process will solicit state information RPC-style, pulling it from one or more of the prior members. If state information is needed from all members, they can send it without waiting for it to be solicited (Figure 15.6), although this can create a burst of communication load just at the moment when the flush protocol is still running, with the risk of momentarily overloading some processes or the network. At the other extreme, if a transfer is needed from just a single member, the joining process should transmit an asynchronous multicast, *terminating* the transfer after it has successfully pulled the state from some member. The remaining members can now resume processing requests or discard any information saved for use during the state transfer protocol.

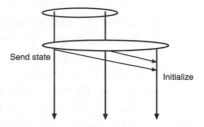

Send state

Initialize

FIGURE 15.6. One of several state transfer mechanisms. In this very simple scheme, the group members all send their copies of the state to the joining member and then resume computing. The method may be a good choice if the state is known to be small, since it minimizes delay, is fault tolerant (albeit sending redundant information), and is very easy to implement. If the state is large, however, the overhead could be substantial.

Perhaps the best among these options, if one single approach is desired as a default, is for the joining process to pull state from a single existing member, switching to a second member if a failure disrupts the transfer. The members should save the state in a buffer for later transfer, and they should use some form of out-of-band transfer (e.g., over a specially created TCP channel) to avoid sending large state objects over the same channels used for other forms of group communication and request processing. When the transfer is completed, the joining process should send a multicast telling the other members it is safe to delete their saved state copies. This is illustrated in Figure 15.7.

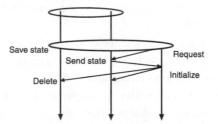

Save state

Send state

Request

Delete

Initialize

FIGURE 15.7. A good state transfer mechanism for cases where the state is of unknown size—the joining member solicits state from some existing member and then tells the group as a whole when it is safe to delete the saved data.

Having had considerable experience with state transfer mechanisms, I should warn developers of a frequently encountered problem. In many systems, the group state can be so large that transferring it represents a potentially slow operation—for example, in a file system application, the state transferred to a joining process might need to contain the full contents of every file modified since that process was last operational. Clearly, it would be a terrible idea to shut down the request processing by existing group members during this extended period of time!

Such considerations lead to three broad recommendations. First, if the state is very large, it is advisable to transfer as much of it as possible before initiating the join request. A mechanism can then be implemented by which any last-minute changes are transferred to the joining process—without extended delays. Second, the state transfer should be done asynchronously—in a manner that will not lead to congestion or flow-control problems impeding the normal processing of requests by the service. A service that technically remains available, but actually is inaccessible because its communication channels are crammed with data, to a joining process may seem very unavailable to other users. Finally, where possible, the approach of jotting down the state is preferable to one that shuts down a server even briefly during the transfer. Again, this reflects a philosophy whereby every effort is made to avoid delaying the response by the server to ongoing requests during the period while the join is still in progress.

15.3.3 Load-Balancing

One of the more common uses of process groups is to implement some form of load-balancing algorithm, whereby the members of a group share the workload presented to them in order to obtain a speedup from parallelism. It is no exaggeration to say that parallelism of this sort may represent the most important single property of process group computing systems: The opportunity to gain performance while also obtaining fault-tolerant benefits on relatively inexpensive cluster-style computing platforms is obviously of tremendous potential importance to the size of the market for process group server architectures.

There are two broad styles of load-balancing algorithms. The first style involves multicasting the client's request to the full membership of the group; the decision as to how the request should be processed is left for the group members to resolve. This approach has the advantage of requiring little trust in the client, but it has the disadvantage of communicating the full request (which may involve a large amount of data) to more processes than really need to see this information. In the second style, the client either makes a choice among the group members or is assigned a preferred group member to which its requests are issued. Here, some degree of trust in the behavior of the clients is accepted in order to reduce the communication load on the system. In this second style, the client may also need to implement a *fail-over policy* by which it reissues a request if the server to which it was originally issued turns out to be faulty or fails while processing it.

Load-balancing algorithms of the first sort require some form of deterministic rule by which incoming requests can be assigned within the server group. As an example, if incoming requests are issued using an *abcast* protocol, the group can take advantage of the fact that all members see the requests in the same order. The ith request can now be assigned to the server whose rank within the group is $i \bmod n$, or the servers can use some other deterministic algorithm for assigning the incoming work. (See Figure 15.8.)

If group members periodically send out load reports to one another, also using *abcast*, these load measures can be used to balance work in the following manner. Suppose that the servers in

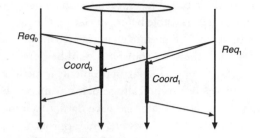

FIGURE 15.8. Load-balancing based on a coordinator scheme using ranking. Ideally, the load on the members will be fairly uniform.

the group measure their load on a simple numerical scale, with 0 representing an unloaded server, 1 representing a server currently handling a single request, and so forth. The load on a group of n servers now can be represented as a vector $[l_0, \ldots l_n]$. Think of these load values as intervals within a line segment of total length $L = (l_0 + \ldots l_n)$ and assume that the group members employ an algorithm for independently but identically generating pseudorandom numbers—the seed of which is transferred as part of the state passed to a joining process when it joins the group. Then, as each new request is received, the group members can independently pick the same random number on the segment $[0,L]$, assigning the request to the process corresponding to the interval within which that number falls. Such an approach will tend to equalize load by randomly spreading it within the group, and it has the benefit of working well even if the load values are approximate and may be somewhat inaccurate.

The same methods can be used as the basis for client affinity load-balancing schemes. In these, the group members provide the client with information that it uses to select the server to which requests will be sent—for example, the group can statically assign clients to particular members at the time the client first interacts with the group. Such an approach risks overloading a server whose clients happen to be unusually active, but it can also be advantageous if caching is a key determinate of request processing performance, since this server is more likely to benefit from the use of a caching algorithm. Alternatively, the client can randomly select a server for each new request within the group membership, or it can use the same load-balancing scheme outlined above to spread requests over the group membership using approximate load information, which the members would periodically broadcast to the clients. Any of these methods represents a viable option for distributing work, and the best choice for a given setting will depend on other information available only to the application designer, such as the likely size of the data associated with each request, fault-tolerant considerations (discussed in the next section), or issues such as the balance between queries (which can often be load-balanced) and update requests (which generally cannot).

In the Isis system, some use was made of load-balancing methods within which the members of a group subdivide the processing of individual requests among themselves (e.g., one process does half the work of processing a request and a second does the other half, as might be done to speed the search of a large database). In practice, I am not aware that much use was ever made of these methods and consequently they are not included in this book.

15.3.4 Primary-Backup Fault Tolerance

Earlier, we illustrated the concept of primary-backup fault tolerance, in which a pair of servers are used to implement a critical service. Virtually synchronous process groups offer a good setting within which such an approach can be used (see Budhiraja et al.).

Primary-backup fault tolerance is most easily understood if one assumes that the application is completely deterministic—that is, the behavior of the server program will be completely determined by the order of inputs to it and is therefore reproducible by simply replaying the same inputs in the same order to a second copy. Under this (admittedly unrealistic!) assumption, a backup server can track the actions of a primary server by simply arranging that a totally ordered broadcast be used to transmit incoming requests to the primary-backup group. The client processes should be designed to detect and ignore duplicate replies to requests (by numbering requests and including the number in the reply). The primary server can simply compute results for incoming requests and reply normally, periodically informing the backup of the most recent replies known to have been received safely. The backup mimics the primary, buffering replies and garbage collecting them when such a status message is received. If the primary fails, the backup resends any replies in its buffer.

Most primary-backup schemes employ some form of checkpoint method to launch a new replica if the primary process actually does fail. At some convenient point soon after the failure, the backup turned primary makes a checkpoint of its state, and simultaneously launches a new backup process.[2] The new process loads its initial state from the checkpoint and joins a process group with the primary. State transfer can also be used to initialize the backup, but this is often harder to implement because many primary-backup schemes must operate with old code, which is not amenable to change and in which the most appropriate form of state is hard to identify. Fortunately, it is just this class of server that is most likely to support a checkpoint mechanism.

The same approach can be extended to work with nondeterministic primary servers, but doing so is potentially much harder. The basic idea is to find a way to *trace* (keep a record of) the nondeterministic actions of the primary, so that the backup can be forced to repeat those actions in a trace-driven mode—for example, suppose that the only nondeterministic action taken by the primary is to request the time of day from the operating system. This system call can be modified to record the value so obtained, sending it in a message to the backup. If the backup pauses each time it encounters a time-of-day system call, it will either see a copy of the value used by the primary (in which case it should use that value and ignore the value of its local clock), or

---

[2]Interested readers may also want to read about log-based recovery techniques, which we do not cover in this book because these techniques have not been applied in many real systems. Alvisi presents a very general log-based recovery algorithm and reviews other work in this area both in his Ph.D. dissertation and in a paper with Marzullo.

it will see the primary fail (in which case it takes over as primary and begins to run off its local clock). Unfortunately, there can be a great many sources of nondeterminism in a typical program, and some will be very hard to deal with: lightweight thread scheduling, delivery of interrupts, shared memory algorithms, I/O ready notifications through system calls such as "select," and so forth. Moreover, it is easily seen that to operate a primary-backup scheme efficiently, the incoming requests, the corresponding replies, and these internal trace messages will need to be transmitted as asynchronously as possible, while respecting causality (see Figure 15.9). Our causal ordering algorithms were oriented toward group multicast, and this particular case would demand nontrivial analysis and optimization. Thus, in practice, primary-backup replication can be very hard to implement when using arbitrary servers.

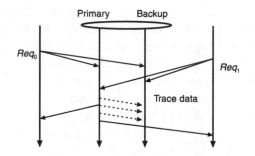

FIGURE 15.9. Primary-backup scheme for nondeterministic servers requires that trace information reach the backup. The fundamental requirement is a causal gap-freedom property: If a reply or some other visible consequence of a primary's actions is visible to a client or the outside world, all causally prior inputs to the primary, as well as trace information, must also be delivered to the backup. The trace data contain information about how nondeterministic actions were performed in the primary. The ordering obligation is ultimately a fairly weak one, and the primary could run far ahead of the backup, giving good performance and masking the costs of replication for fault tolerance. The complexity of the scheme is fairly high, because it can be hard to generate and use trace information—hence, it is rare to see primary-backup fault tolerance in nondeterministic applications.

Another drawback to the approach is that it may fail to overcome software bugs. As we can see, primary-backup replication is primarily appealing for deterministic applications. But these are just the ones in which Heisenbugs would be carefully repeated by a primary-backup solution, unless the fact of starting the backup from a state checkpoint introduces some degree of tolerance to this class of failures. Thus, the approach is likely to be exposed to correlated failures of the primary and backup in the case where it can be most readily applied.

15.3.5 Coordinator-Cohort Fault Tolerance

The coordinator-cohort approach to fault tolerance generalizes the primary-backup approach in ways that can help overcome the limitations previously mentioned. In this fault-tolerant method, the work of handling requests is shared by the group members. (The same load-sharing mechanisms discussed previously are used to balance the load.) The handler for a given request is said to be the *coordinator* for processing that request and is responsible for sending any updates or necessary trace information to the other members, which are termed the *cohorts* for that request. As in the primary-backup scheme, if the coordinator fails, one of the cohorts takes over.

Unlike the primary-backup method, there may be many coordinators active in the same group for many different requests. Moreover, the trace information in a primary backup scheme normally contains the information needed for the backup to duplicate the actions of the primary, whereas the trace data of a coordinator-cohort scheme will often consist of a log of updates the coordinator applied to the group state. In this approach, the cohorts do not actively replicate the actions of the coordinator, but merely update their states to reflect its updates. Locking must be used for concurrency control. In addition, the coordinator will normally send some form of a copy of the reply to its cohorts, so that they can garbage collect information associated with the pending requests for which they are backups. The approach is illustrated in Figure 15.10.

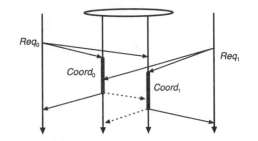

FIGURE 15.10. Coordinator-cohort scheme. The work of handling requests is divided among the processes in the group. Notice that as each coordinator replies, it also (atomically) informs the other group members that is has terminated. This permits them to garbage collect information about pending requests that other group members are handling. In the scheme, each process group member is actively handling some requests while passively acting as backup for other members on other requests. The approach is best suited for deterministic applications, but it can also be adapted to nondeterministic ones.

Some practical cautions limit the flexibility of this style of load-balanced and fault-tolerant computing (which is quite popular among users of systems such as the Isis Toolkit and Horus, we should add!). First, it is important that the coordinator selection algorithm do a good job of load-balancing, or some single group member may become overloaded with the lion's share of the requests. In addition to this, the method can be very complex for requests that involve nontrivial updates to the group state or that involve nondeterministic processing that the cohort may be expected to reproduce. In such cases, it can be necessary to use an atomic protocol for sending the reply to the requesting client and the trace information or termination information to the cohorts. Isis implements a protocol for this purpose: It is atomic and can send to the members of a group plus one additional member. However, such protocols are not common in most systems for reliable distributed computing. Given appropriate protocol support, however, and a reasonably simple server (e.g., one processing requests that are primarily queries that don't change the server state), the approach can be highly successful, offering scalable parallelism and fault tolerance for the same price.

15.4 Related Reading

On virtual synchrony: (see Birman and van Renesse [1994, 1996,] Powell [1996]); but see also Birman and Joseph (February 1987, November 1987), Birman and van Renesse (1996), Dolev and Malkhi, Schiper and Raynal.

On extended virtual synchrony: (see Malkhi); but see also Agarwal, Amir, Keidar and Dolev, Moser et al. (1996).

On uses of the virtual synchrony model: (see Birman and Joseph [November 1987], Birman and van Renesse [1994]).

On primary-backup schemes: (see Budhiraja et al.).

A discussion of other approaches to the same problems can be found in Cristian (1996).

CHAPTER 16 ✧ ✧ ✧ ✧

Consistency in Distributed Systems

CONTENTS

In the previous chapters, we examined options for implementing replicated data in various group membership models and looked at protocols for ordering conflicting actions under various ordering goals. We then showed how these protocols could be used as the basis of a computational model, virtual synchrony, in which members of distributed process groups see events that occur within those groups in consistent orders and with failure-atomicity guarantees and are consequently able to behave in consistent ways.

What is lacking, however, is a more general synthesis, which we seek to provide in this chapter. Key ideas underlying virtual synchrony are:

▶ Self-defining system and process group membership, in which processes are excluded from a system, if necessary, to permit continued progress.

▶ Tools for joining a group, state transfer, communication, and reporting new membership views.

▶ Depending on the model, a concept of primary components of the system.

▶ Algorithms that seek to achieve internal (as opposed to dynamically uniform) consistency.

▶ Distributed consistency achieved by ordering conflicting replicated events in consistent ways at the processes that observe those events.

The remainder of this chapter reviews these points relative to the alternatives we touched upon in developing our protocols and tools.

16.1 Consistency in the Static and Dynamic Membership Models

In the static model, the system is understood to be the set of places at which processes that act on behalf of the system execute. Here, the system is a relatively fixed collection of resources, which experience dynamic disruptions of communication connectivity, process failures, and re-starts. Obviously, a static system may not be static over very long periods of time, but the time scale on which membership of the full set of places where members run is understood to be long compared to the time scale at which these other events occur. The protocols for adding new members to the static set or dropping them are treated as being outside of the normal execution model. In cases where the system is symmetric, meaning that any correct execution of the system would also have been correct if the process identifiers were permuted, static systems rely on agreement protocols within which the majority of the statically defined composition of the full system must participate, directly or indirectly.

The dynamic model employs a concept of system membership that is self-defined and for this reason less difficult to support than the static one. Dynamic systems add and lose members on a very short time scale compared to static ones. In the case where the system is symmetric, the set of processes that must participate in decisions is based on a majority of a dynamically defined group; this is a weaker requirement than for the static model and hence permits progress under conditions when a static system would not make progress. (See Figure 16.1.)

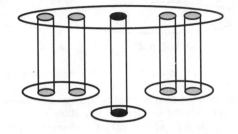

FIGURE 16.1. Static and dynamic views of a single set of sites. From a static perspective, the set has fixed membership but changing connectivity and availability properties—for example, the black nodes may be available and the gray ones treated as not available. Depending upon how such a system is implemented, it may be impossible to perform certain types of operations (notably, updates) unless a majority of the nodes are available. The dynamic perspective treats the system as if it were partitioned into a set of components whose membership is self-defined. Here, the black component might be the primary one and the gray components nonprimary. In contrast to the static approach, the primary component remains available, if primaryness can be deduced within the system. If communication is possible between two components, they are expected to merge their states in this model. Neither perspective is more correct than the other: The most appropriate way to view a system will typically depend upon the application, and different parts of the same application may sometimes require different approaches to membership. However, in the dynamic model, it is frequently important to track one of the components as being primary for the system, restricting certain classes of actions to occur only in this component (or not at all, if the primaryness attribute cannot be tracked after a complex series of failures).

These points are already significant when one considers what it means to say that a protocol is live in the two settings. However, before focusing on liveness, we review the question of consistency.

Consistency in a static model is typically defined with regard to an external observer, who may be capable of comparing the state and actions of a process that has become partitioned from the other processes in the system with the states and actions of the processes that remain connected. Such an external observer could be a disk that contains a database that will eventually have to be reintegrated and reconciled with other databases maintained by the processes remaining in the connected portion of the system, an external device or physical process with which the system processes interact, or some form of external communication technology that lacks the flexibility of message passing but may still transfer information in some way between system processes.

Consistency in a dynamic system is a much more internal concept (see Figure 16.2). In essence, a dynamic form of consistency requires that processes permitted to interact with one another will never observe contradictions in their states, which are detectable by comparing the contents of messages they exchange. Obviously, process states and the system state evolve through time, but the idea here is that if process p sends a message to process q that in some way reflects state information shared by them, process q should never conclude that the message sent by process p is impossible on the basis of what q itself has seen in regard to this shared state. If the

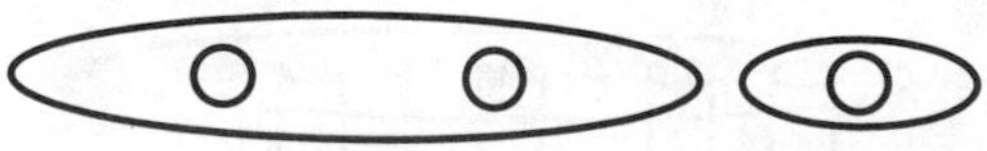

$x = 2, 4, 6, 8$ $x = 2, 4, 7, 9$

FIGURE 16.2. Dynamic (or interactive) consistency is the guarantee that the members of a given system component will maintain mutually consistent states (here, by agreeing upon the sequence of values that a variable, x, has taken). If a protocol is not dynamically uniform, it may allow a process that becomes partitioned from a component of the system to observe events in a way that is inconsistent with the event ordering observed within that component. Thus, in this example, the component on the right (consisting of a single process) observes x to take on the values 7 and 9, while the larger component on the left sees x pass through only even values. By pronouncing at most one of these components to be the primary one for the system, we can impose a sensible interpretation on this scenario. Alternatives are to use dynamically uniform protocols with external consistency guarantees. Such protocols can be supported both in the dynamic membership model and the static one, where this guarantee is almost always required. However, they are far more costly than protocols that do not provide dynamic uniformity.

state shared by p and q is a replicated variable, and q has observed that variable to increment only by 2s from 0 to its current value of 40, it would be inconsistent if p sent a message, ostensibly reflecting a past state, in which the variable's value was 7. For q such a state would not merely be stale, it would be impossible, since q believes itself to have seen the identical sequence of events, and the variable never had the value 7 in q's history.

Although this example is unrealistic, it corresponds to more realistic scenarios in which dynamic consistency is precisely what one wants—for example, when a set of processes divides the work of performing some operation using a coordinator-cohort rule, or by exploiting a mutually perceived ranking to partition a database, dynamic consistency is required for the partitioning to make sense. Dynamic consistency is also what one might desire from the Web proxies and servers that maintain copies of a document: They should agree on the version of the document that is the most current one and provide guarantees to the user that the most current document is returned in response to a request.

The significance of the specific example described above is thus not that applications often care about the past state of a replicated variable, but rather that cooperation or coordination or synchronization in distributed settings all involve cases in which a process, p, may need to reason about the state and actions of some other process, q. When this occurs, p can be understood to be using a form of replicated system state that it believes itself to share with q. Our shared variable has now become the shared concept of the state of a lock or the shared list of members and ranking of members for a process group to which both belong. Inconsistency in these cases means that the system is visibly misbehaving: Two processes both think they have locked the same variable, or each thinks the other holds the lock when neither in fact holds it. Perhaps both processes consider themselves primary for some request, or perhaps neither does. Both may search the first half of a database, each thinking the other is searching the second half. These same issues only get worse if we move to larger numbers of processes.

Of course, as the system evolves through time, it may be that p once held a lock but no longer does. So the issue is not so much one of being continuously consistent, but of seeing mutually consistent and mutually evolving histories of the system state. In effect, if the processes in a system see the same events in the same order, they can remain consistent with one another. This extremely general concept is at the heart of all forms of distributed consistency.

In the purest sense, the dynamic system model is entirely concerned with freedom from detectable inconsistencies in the logically derivable system state. This concept is well defined in part because of the following rule: When a dynamic system considers some process to have failed, communication to that process is permanently severed. Under such a rule, p cannot communicate to q unless both are still within the same component of the possibly partitioned system, and the protocols for dynamic systems operate in a manner that maintains consistency within subsets of processes residing in the same component. The system may allow a process to be inconsistent with the state of the system as a whole, but it does so only when that process is considered to have failed; it will never be allowed to rejoin the system until it has done something to correct its (presumably inconsistent) state.

The ability to take such an action permits dynamic systems to make progress when a static system might have to wait for a disconnected process to reconnect itself or for a failed process to be restarted. Thus, a process in the dynamic model can sometimes (often, in fact) make progress while a process in the static model would not be able to do so.

The static model, on the other hand, is in many ways a more intuitive and simpler one than the dynamic one. It is easy to draw an analogy between a static set of resources and a statically defined set of system processes, and external consistency constraints, being very strong, are also easy to understand. The dynamic model is in some sense superficially easy to understand, but much harder to understand upon close study. Suppose we are told that process p is a member of a dynamically defined system component and sets a replicated variable x to 7. In a static system we would have concluded that, if the system is guaranteeing the consistency of this action, p was safe in taking it. In a dynamic system, it may be that it is too early to know if p is a valid member of the system and that setting x to 7 is a safe action in the broader sense. The problem is that future events may cause the system to reconfigure itself in a way that excludes p and leads to an evolution of system state in which x never does take on the value 7. Moreover, the asynchronous nature of communication means that even if in real time p sets x to 7 before being excluded by the other system members as if it were faulty, in the logical system model, p's action occurs *after* it has been excluded from the system.

Where external actions are to be taken, the introduction of time offers us a way to work around this dilemma. Recall our air traffic control example (see Section 13.10.5). Sharing a clock with the remainder of the system, p can be warned with adequate time to avoid a situation where two processes ever own the air traffic space at the same time. Of course, this does not eliminate the problem that during the period after it became disconnected and before the remainder of the system took over, p may have initiated actions. We can resolve this issue by acknowledging that it is impossible to improve on the solution and by asking the application program to take an appropriate action. In this specific example, p would warn the air traffic controller that actions taken within the past δ seconds may not have been properly recorded by the main system, and connection to it has now been lost. With a person in the loop, such a solution would seem adequate. In fact, there is little choice, for no system that takes actions at multiple locations can ever be precisely sure of its state if a failure occurs while such an action is underway.

Faced with such seemingly troubling scenarios, one asks why we consider the dynamic model at all. Part of the answer is that the guarantees it offers are almost as strong as those for the static case, and yet it can often make progress when a static solution would be unable to do so. Moreover, the static model sometimes just doesn't fit a problem. Web proxies, for example, are a very dynamic and unpredictable set: The truth is out there, but a server will not be able to predict in advance just where copies of its documents may end up (imagine the case where one Web proxy obtains a copy of a document from some other Web proxy!). But perhaps the best answer is, as we saw in previous chapters, that the weaker model permits dramatically improved performance, perhaps by a factor of hundreds if our goal is to replicate data.

Both the static and dynamic system models offer a strong form of consistency whereby the state of the system is guaranteed to be consistent and coordinated over large numbers of components. But while taking an action in the static model can require a fairly slow, multiphase protocol, the dynamic system is often able to exploit asynchronous single-phase protocols, such as the nonuniform *fbcast* and *cbcast* primitives, for similar purposes. It is no exaggeration to say that these asynchronous protocols may result in levels of performance that are hundreds of times superior to those achievable when subjected to static consistency and membership constraints—for example, the Horus system is able to send nearly 85,000 small multicasts per second to update a variable replicated between two processes. This figure drops to about 50 updates per second when using a quorum-style replication scheme and to perhaps 1,500 per second when using an RPC scheme that is disconnected from any concept of consistency. The latency improvements can be even larger: In Horus, there are latency differences of as much as three orders of magnitude between typical figures for the dynamic case and typical protocols for taking actions in a static or dynamically uniform manner.

In practical work with dynamic system models, we typically need to assume that the system is basically well behaved, despite experiencing some infrequent rate of failures. Under such an assumption, the model is easy to work with and makes sense. If a system experiences frequent failures (relative to the time it takes to reconfigure itself or otherwise repair the failures), the static model becomes more and more appealing and the dynamic one less and less predictable. Fortunately, most real systems are built with extremely reliable components, which, indeed, experience infrequent failures. This pragmatic consideration explains why dynamically consistent distributed systems have become popular: The model behaves reasonably well in real environments, and its performance is superior compared to what can be achieved in the static model.

Indeed, one way to understand the performance advantage of the dynamic model is that by precomputing membership information, the dynamic algorithms represent optimizations of the static algorithms. As one looks closely at the algorithms, they seem more and more similar in a basic way, and perhaps this explains why that should be the case. In effect, the static and dynamic models are very similar, but the static algorithms (such as quorum data replication) tend to compute the membership information they need on each operation, while the dynamic ones precompute this information and are built using a much simpler fail-stop model. However, although this perspective is intuitively appealing, I have never seen it elaborated in any sort of detailed formal treatment.

Moreover, it is important to realize that the external concept of consistency associated with static models is in some ways much stronger, and consequently more restrictive, than is necessary for realistic applications. This can translate to periods of mandatory unavailability, where a static system model forces us to stop and wait while a dynamic consistency model permits reconfiguration and progress. Many distributed systems contain services of various kinds that have small server states (which can therefore be transferred to a new server when it joins the system) and that are only of interest when they are operational and connected to the system as a whole. Mutual consistency between the servers and the states of the applications using them is all that one needs in such internal uses of a consistency-preserving technology. If a dynamic approach is dramatically faster than a static one, so much the better for the dynamic approach!

These comments should not be taken to suggest that a dynamic system can *always* make progress even when a static one must wait. In recent work, Chandra and his colleagues have established that a result similar to the FLP result holds for group membership protocols (see Chandra et al. [1996])—hence, there are conditions under which an asynchronous system can be prevented from reaching consensus upon its own membership and therefore prevented from making progress. Other researchers (myself included) have pinned down precise conditions (in various models) under which dynamic membership consensus protocols are guaranteed to make progress (see Babaoglu et al. [1995], Friedman et al., Guerraoui and Schiper, Neiger), and the good news is that for most practical settings the answer is that such protocols make progress with overwhelmingly high probability if the probability of failures and message loss is uniform and independent over the processes and messages sent in the system. In effect, only partitioning failures or a very intelligent adversary (one that in practice could never be implemented) can prevent these systems from making progress.

Thus, we know that *all* of these models face conditions under which progress is not possible. Research is still underway on pinning down the precise conditions when progress *is* possible in each approach: the maximum rates of failures that dynamic systems can sustain. But as a practical matter, the evidence is that all of these models are perfectly reasonable for building reliable distributed systems. The theoretical impossibility results do not appear to represent practical impediments to implementing reliable distributed software; they simply tell us that there will be conditions that these reliability approaches cannot overcome. The choice, in a practical sense, is to match the performance and consistency properties of the solution to the performance and consistency requirements of the application. The weaker the requirements, the better the performance we can achieve.

Our study also revealed two other issues that deserve comment: the need, or lack thereof, for a *primary component* in a partitioned membership model and the broader but related question of how consistency is tied to ordering properties in distributed environments.

The question of a primary component is readily understood in terms of the air traffic control example we looked at earlier. In that example, there was a need to take authoritative action within a service on behalf of the system as a whole. In effect, a representative of a service needed to be sure that it could safely allow an air traffic controller to take a certain action, meaning that it ran no risk of being contradicted by any other process (or, in the case of a possible partitioning failure, that before any other process could start taking potentially conflicting actions, a timeout would elapse and the air traffic controller would be warned that this representative of the service was now out of touch with the primary partition).

In the static system model, there is only a single concept of the system as a whole, and actions are taken upon the authority of the full system membership. Naturally, it can take time to obtain majority acquiescence in an action (see Keidar and Dolev)—hence, this is a model in which some actions may be delayed for a considerable period of time. However, when an action is actually taken, it is taken on behalf of the full system.

In the dynamic model we lose this guarantee and face the prospect that our concept of consistency can become trivial because of system partitioning failures—a dynamic system could partition arbitrarily, with each component having its own concept of authoritative action. For purely *internal* purposes, such a concept of consistency may be adequate, in the sense that it still permits work to be shared among the processes that compose the system, and, as noted above, it may be sufficient to avoid the risk that the states of processes will be directly inconsistent in a way that is readily detectable. The state merge problem (see Babaoglu et al. [1996], Malkhi), which occurs when two components of a partitioned system reestablish communication connectivity and must reconcile their states, is where such problems are normally resolved (and the normal resolution is to simply take the state of one partition as being the official system state, abandoning the other). As noted in Chapter 13, this challenge has led researchers working on the Relacs system in Bologna to propose a set of tools, combined with a set of guarantees that relate to view installation, which simplify the development of applications that can operate in this manner (see Babaoglu et al. [1996]).

The downside of allowing simultaneous progress in multiple components of a partitioned dynamic system, however, is that there is no meaningful form of consistency that can be guaranteed between the components, unless one is prepared to pay the high cost of using only dynamically uniform message-delivery protocols. In particular, the impossibility of guaranteeing progress among the participants in a consensus protocol implies that when a system partitions, there will be situations in which we can define the membership of both components but cannot decide how to terminate protocols that were underway at the time of the partitioning event. Consequences of this observation include the implication that when nonuniform protocols are employed, it will be impossible to ensure that the components have consistent histories (in terms of the events that occurred and the ordering of events) for their past prior to the partitioning event. In practice, one component, or both, may be irreconcilably inconsistent with the other.

There is no obvious way to merge states in such a situation: The only real option is to arbitrarily pick one component's state as the official one and replace the other component's state with this state, perhaps reapplying any updates that occurred in the unofficial partition. Such an approach, however, can be understood as one in which the primary component is simply selected when the network partition is corrected rather than when it forms. If there is a reasonable basis on which to make the decision, why delay it?

As we saw in the previous chapter, there are two broad ways to deal with this problem. The one I favor is to define a concept of a *primary component* of a partitioned system and to track primaryness when the partitioning event first occurs. The system can then enforce the rule that nonprimary components must not trust their own histories of the past state of the system and certainly should not undertake authoritative actions on behalf of the system as a whole. A nonprimary component may, for example, continue to operate a device that it owns, but that may not be reliable for use in instructing an air traffic controller about the status of airspace sectors or other global forms of state-sensitive data unless it was updated using dynamically uniform protocols.

Of course, a dynamic distributed system can lose its primary component, and, making matters still more difficult, there may be patterns of partial communication connectivity within which a static distributed system model can make progress but no primary partition can be formed—hence, a dynamic model must block. Suppose, for example, that a system partitions so that all of its members are disconnected from one another. Now we can selectively reenable connections so that over time a majority of a static system membership set are able to vote in favor of an action. Such a pattern of communication could allow progress—for example, there is the protocol of Keidar and Dolev, cited several times previously, in which an action can be terminated entirely on the basis of point-to-point connections. However, as we commented, this protocol delays actions until a majority of the processes in the entire system know about them, which will often take a very long time.

My work has not needed to directly engage these issues because of the underlying assumption that rates of failure are relatively low and that partitioning failures are infrequent and rapidly repaired. Such assumptions let us conclude that these types of partitioning scenarios just don't occur in typical local area networks and typical distributed systems.

On the other hand, frequent periods of partitioned operation *could* occur in very mobile situations, such as when units are active on a battlefield. They are simply less likely to occur in applications such as air traffic control systems or other conventional distributed environments. Thus, there are probably systems that should use a static model with partial communication connectivity as their basic model, systems that should use a primary component consistency model, and perhaps still other systems for which a virtual synchrony model that doesn't track primaryness would suffice. These represent successively higher levels of availability, and even the lowest level retains a meaningful concept of distributed consistency. At the same time, they represent diminishing concepts of consistency in any absolute sense. This suggests that there are unavoidable tradeoffs in the design of reliable distributed systems for critical applications.

The two-tiered architecture discussed in the previous chapter can be recognized as a response to this impossibility result. Such an approach explicitly trades higher availability for weaker consistency in the LAN subsystems, while favoring strong consistency at the expense of reduced availability in the WAN layer (which might run a protocol based on the Chandra/Toueg consensus algorithm). The LAN level of a system might use nonuniform protocols for speed, while the WAN level uses tools and protocols similar to the ones proposed by the Transis effort or by Babaoglu's group in their work on Relacs.

We alluded briefly to the connection between consistency and order. This topic is perhaps an appropriate one on which to end our review of the models. Starting with Lamport's earliest work on distributed computing systems, it was already clear that consistency and the ordering of distributed events are closely linked. Over time, it has become apparent that distributed systems contain what are essentially two forms of knowledge or information. Static knowledge is information that is well known to all of the processes in the system at the outset—for example, the membership of a static system is a form of static knowledge. Being well known, it can be exploited in a decentralized but consistent manner. Other forms of static knowledge can include knowledge of the protocol that processes use, knowledge that some processes are more important than others, or knowledge that certain classes of events can only occur in certain places within the system as a whole. (See Figure 16.3.)

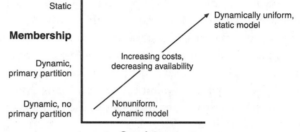

FIGURE 16.3. Conceptual options for the distributed system designer. Even when one seeks consistency, there are choices concerning how strong the consistency desired should be and which membership model to use. The least-costly and highest-availability solution for replicating data, for example, looks only for internal consistency within dynamically defined partitions of a system and does not limit progress to the primary partition. This model, as we have suggested, may be too weak for practical purposes. A slightly less available approach, which maintains the same high level of performance, allows progress only in the primary partition. As one introduces further constraints, such as dynamic uniformity or a static system model, costs rise and availability falls, but the system model becomes simpler and simpler to understand. The most costly and restrictive model sacrifices nearly three orders of magnitude of performance in some studies relative to the least-costly one. Within any given model, the degree of ordering required for multicasts introduces further fine-grained cost/benefit tradeoffs.

Dynamic knowledge is information that stems from unpredicted events occurring within the system—either as a consequence of nondeterminism of the members, failures or event orderings that are determined by external physical processes, or inputs from external users of the system. The events that occur within a distributed system are frequently associated with the need to update the system state in response to dynamic events. To the degree that system state is replicated, or is reflected in the states of multiple system processes, these dynamic updates of the state will need to occur at multiple places. In the work we have presented here, process groups are the places where such state resides, and multicasts are used to update such state.

Viewed from this perspective, it becomes apparent that *consistency is order,* in the sense that the distributed aspects of the system state are entirely defined by process groups and multicasts to those groups, and these abstractions, in turn, are defined entirely in terms of ordering and atomicity. Moreover, to the degree that the system membership is self-defined, as in the dynamic models, atomicity is also an order-based abstraction.

This reasoning leads to the conclusion that the deepest of the properties in a distributed system concerned with consistency may be the ordering in which distributed events are scheduled to occur. As we have seen, there are many ways to order events, but the schemes all depend upon either explicit participation by a majority of the system processes or upon dynamically changing membership, managed by a group membership protocol. These protocols, in turn, depend upon majority action (by a dynamically defined majority). Moreover, when examined closely, all the dynamic protocols depend upon some concept of token or special permission, which enables the process holding that permission to take actions on behalf of the system as a whole. One is strongly inclined to speculate that in this observation lies the grain of a general theory of distributed computing, in which all forms of consistency and all forms of progress could be related to membership and in which dynamic membership could be related to the liveness of token passing or leader election protocols. At the time of this writing, I am not aware of any clear presentation of this theory of all possible behaviors for asynchronous distributed systems, but perhaps it will emerge in the not-too-distant future.

Our goals in this book remain practical, however, and we now have powerful practical tools to bring to bear on the problems of reliability and robustness in critical applications. Even knowing that our solutions will not be able to guarantee progress under all possible asynchronous conditions, we have seen enough to know how to guarantee that *when* progress is made, consistency will be preserved. There are promising signs of emerging understanding of the conditions under which progress can be made, and the evidence is that the prognosis is really quite good: If a system rarely loses messages and rarely experiences real failures (or mistakenly detects failures), the system will be able to reconfigure itself dynamically and make progress while maintaining consistency.

As to the tradeoffs between the static and dynamic model, it may be that real applications should employ mixtures of the two. The static model is more costly in most settings (perhaps not in heavily partitioned ones), and it may be much more expensive if the goal is merely to update the state of a distributed server or a set of Web pages managed on a collection of Web proxies. The dynamic primary component model, while overcoming these problems, lacks external safety guarantees that may sometimes be needed. The nonprimary component model lacks consistency and the ability to initiate authoritative actions, but perhaps this ability is not always needed. Complex distributed systems of the future may well incorporate multiple levels of consistency, using the cheapest one that suffices for a given purpose.

16.2 General Remarks concerning Causal and Total Ordering

The entire concept of providing ordered message delivery has been a source of considerable controversy within the community that develops distributed software (see van Renesse [1993]). Causal ordering has been especially controversial, but even total ordering is opposed by some researchers (see Cheriton and Skeen), although others have been critical of the arguments advanced in this area (see Birman [1994], Cooper [1994], van Renesse [1994]). The CATOCS controversy came to a head in 1993, and although it seems no longer to interest the research community, it would also be hard to claim that there is a generally accepted resolution of the question.

Underlying the debate are tradeoffs between consistency, ordering, and cost. As we have seen, ordering is an important form of consistency. In the next chapter, we will develop a variety of powerful tools for exploiting ordering, especially for implementing replicated data efficiently. Thus, since the first work on consistency and replication with process groups, there has been an emphasis on ordering. Some systems, such as the Isis Toolkit, which I developed in the mid-1980s, made extensive use of causal ordering because of its relatively high performance and low latency. Isis, in fact, enforces causally delivered ordering as a systemwide default, although, as we saw in Chapter 14, such a design point is in some ways risky. The Isis approach makes certain types of asynchronous algorithms very easy to implement, but it has important cost implications; developers of sophisticated Isis applications sometimes need to disable the causal ordering mechanism to avoid these costs. Other systems, such as Amoeba, looked at the same issues but concluded that causal ordering is rarely needed if total ordering can be made fast enough.

We have now seen a sampling of the sorts of uses to which ordered group communication can be put. Moreover, earlier sections of this book have established the potential value of these sorts of solutions in settings such as the Web, financial trading systems, and highly available database or file servers.

Nonetheless, there is a third community of researchers (Cheriton and Skeen are best known within this group), who have concluded that ordered communication is almost never matched with the needs of the application. These researchers cite their success in developing distributed support for equity trading in financial settings and work in factory automation—both settings in which developers have reported good results using distributed message-bus technologies (TIB is the one used by Cheriton and Skeen), which offer little in the sense of distributed consistency or fault-tolerant guarantees. To the degree that the need occurs for consistency within these applications, Cheriton and Skeen have found ways to reduce the consistency requirements of the *application* rather than providing stronger consistency within a system to respond to a strong application-level consistency requirement (the NFS example from Chapter 7 comes to mind). Broadly, this led them to a mindset that favors the use of stateless architectures, nonreplicated data, and simple fault-tolerant solutions in which one restarts a failed server and leaves it to the clients to reconnect. Cheriton and Skeen suggest that such a point of view is the logical extension of the end-to-end argument (see Saltzer et al.), which they interpret as an argument that each application must take direct responsibility for guaranteeing its own behavior.

Cheriton and Skeen also make some very specific points. They are critical of system-level support for causal or total ordering guarantees. They argue that communication ordering properties are better left to customized application-level protocols, which can also incorporate other sorts of application-specific properties. In support of this view, they present applications that need stronger ordering guarantees and applications that need weaker ones, arguing that in the former case causal or total ordering will be inadequate, and in the latter case it will be overkill. Their analysis led them to conclude that in *almost all cases,* causal ordering is either more than the application needs (and more costly) or less than the application needs (in which case the application must add some higher-level ordering protocol of its own). They also produced similar results for total ordering (see Cheriton and Skeen).

Unfortunately, while making some good points, Cheriton and Skeen's paper also includes a number of questionable claims, including some outright errors, which have been refuted in other papers (see Birman [1994], Cooper [1994], van Renesse [1994]). Cheriton and Skeen claim that causal ordering algorithms have an overhead on messages that increases as n^2, where n is the number of processes in the system as a whole. Yet we have seen that causal ordering for group multicasts, which they claim to be discussing, can easily be provided with a vector clock, whose length is linear in the number of active senders in a group (rarely more than two or three processes). In more complex settings, compression techniques can often be used to bound the vector timestamp to a small size. This example is just one of several specific points about which Cheriton and Skeen make statements that could be disputed on technical grounds.

The entire approach to causal ordering adopted by Cheriton and Skeen is also unusual. In this chapter, we have seen that causal order is often needed when one seeks to *optimize* an algorithm expressed originally in terms of totally ordered communication and that total ordering is useful because, in a state machine style of distributed system, by presenting the same inputs to the various processes in a group in the same order, their states can be kept consistent. Cheriton and Skeen never address this use of ordering, focusing instead on causal and total order in the context of a publish/subscribe architecture in which a small number of data publishers send data that a large number of consumers receive and process and in which there are no consistency requirements that span the consumer processes. This example somewhat misses the point of the preceding chapters, where we made extensive use of total ordering primarily for consistent replication of data and of causal ordering as a relaxation of total ordering where the sender has some form of mutual exclusion within the group.

It is my feeling that Cheriton and Skeen's most effective argument is one based on the end-to-end philosophy. They suggest, in effect, that although many applications will benefit from properties such as fault tolerance, ordering, or other communication guarantees, no single primitive is capable of capturing all possible properties without imposing absurdly high costs for the applications that required weaker guarantees. Our observation about the cost of dynamically uniform strong ordering bears this out: Here we see a very strong property, but it is also thousands of times more costly than a rather similar but weaker property! If one makes the weaker version of a primitive the default, the application programmer will need to be careful not to be surprised by its nonuniform behavior; the stronger version may just be too costly for many applications.

Cheriton and Skeen generalize from similar observations based on their own examples and conclude that the application should implement its own ordering protocols.

However, we have seen that these protocols are not trivial and implementing them would not be an easy undertaking. It also seems unreasonable to expect the average application designer to implement a special-purpose, hand- crafted protocol for each specific need. In practice, if ordering and atomicity properties are not provided by the computing system, it seems unlikely that applications will be able to make any use of these concepts at all. Thus, even if one agrees with the end-to-end philosophy, one might disagree that it implies that each application programmer should implement nearly identical and rather complex ordering and consistency protocols, because no single protocol will suffice for all uses.

Current systems, including the Horus system, usually adopt a middle ground, in which the ordering and atomicity properties of the communication system are viewed as options that can be selectively enabled (Chapter 18). The designer can in this way match the ordering property of a communication primitive to the intended use. If Cheriton and Skeen were using Horus, their arguments would warn us not to enable such and such a property for a particular application because the application doesn't need the property and the property is costly. Other parts of their work would be seen to argue in favor of additional properties beyond the ones normally provided by Horus. As it happens, Horus is easily extended to accommodate such special needs. Thus, the reasoning of Cheriton and Skeen can be seen as critical of systems that adopt a single, all-or-nothing approach to ordering or atomicity, but not of systems such as Horus that seek to be more general and flexible.

The benefits of providing stronger communication tools in a system are that the resulting protocols can be highly optimized and refined, giving much better performance than could be achieved by a typical application developer working over a very general but very weak communication infrastructure. To the degree that Cheriton and Skeen are correct and application developers will need to implement special-purpose ordering properties, such a system can also provide powerful support for the necessary protocol development tasks. In either case, the effort required from the developer is reduced and the reliability and performance of the resulting applications improved.

We mentioned that the community has been particularly uncomfortable with the causal ordering property. Within a system such as Horus, causal order is normally used *as an optimization* of total order, in settings where the algorithm was designed to use a totally ordered communication primitive but exhibits a pattern communication for which the causal order is also a total one. We will return to this point, but we mention it now simply to stress that the explicit use of casually ordered communication, much criticized by Cheriton and Skeen, is actually quite uncommon. More typical is a process of refinement, whereby an application is gradually extended to use less and less costly communication primitives in order to optimize performance. The enforcement of causal ordering, systemwide, is not likely to become standard in future distributed systems. When *cbcast* is substituted for *abcast,* communication may cease to be totally ordered; but any situation in which messages arrive in different orders at different members will be due to events that commute. Thus, their *effect* on the group state will be as if the messages had been received in a total order—even if the actual sequence of events is different.

In contrast, much of the discussion and controversy surrounding causal order occurs when causal order is considered not as an optimization, but rather as an ordering property which one might employ by default—just as a stream provides FIFO ordering by default. Indeed, the analogy is a very good one, because causal ordering is an extension of FIFO ordering. Additionally, much of the argument over causal order uses examples in which point-to-point messages are sent asynchronously, using systemwide causal order to ensure that later messages arrive after earlier ones. There is some merit to this, because the assumption of systemwide causal ordering permits some very asynchronous algorithms to be expressed elegantly and simply. It would be a shame to lose the option of exploiting such algorithms. However, systemwide causal order is not really the main use of causal order, and one could easily live without such a guarantee. Point-to-point messages can also be sent using a fast RPC protocol, and saving a few hundred microseconds at the cost of a substantial systemwide overhead seems like a very questionable design choice; systems such as Horus obtain systemwide causality, if needed, by waiting for asynchronously transmitted messages to become stable.

On the other hand, when causal order is used as an optimization of atomic or total order, the performance benefits can be huge. So we face a performance argument, in fact, in which the rejection of causal order involves an acceptance of higher than necessary latencies, particularly for replicated data.

Notice that if asynchronous *cbcast* is only used to replace *abcast* in settings where the resulting delivery order will be unchanged, the associated process group can still be programmed under the assumption that all group members will see the same events in the same order. As it turns out, there are cases in which the handling of messages commute and the members may not even need to see messages in identical ordering in order to behave as if they did. There are major advantages to exploiting these cases: Doing so potentially reduces idle time (since the latency to message delivery is lower, a member can start work on a request sooner, if the *cbcast* encodes a request that will cause the recipient to perform a computation). Moreover, the risk that a Heisenbug will cause all group members to fail simultaneously is reduced, because the members do not process the requests in identical orders and Heisenbugs are likely to be very sensitive to the detailed ordering of events within a process. Yet one still presents the algorithm in the group and thinks of the group as if all the communication within it were totally ordered.

16.3 Summary and Conclusion

There has been a great deal of debate over the concepts of consistency and reliability in distributed systems (which are sometimes seen as violating end-to-end principles) and of causal or total ordering (which are sometimes too weak or too strong for the needs of a specific application that does need ordering). Finally, although we have not focused on this here, there is the criticism that technologies such as the ones we have reviewed do not fit with standard styles of distributed system development.

As to the first concern, the best argument for consistency and reliability is to simply exhibit classes of critical distributed computing systems that will not be sufficiently available unless data are replicated and will not be trustworthy unless these data are replicated consistently. One would not want to conclude that *most* distributed applications need these properties: Today, the ones that do remain a fairly small subset of the total. However, this subset is growing rapidly. Moreover, even if one believed that consistency and reliability are extremely important in a great many applications, one would not want to impose potentially costly communication properties systemwide, especially in applications with very large numbers of overlapping process groups. To do so is to invite poor performance, although there may be specific situations where the enforcement of strong properties within small sets of groups is desirable or necessary.

Turning to the second issue, it is clearly true that different applications have different ordering needs. The best solution to this problem is to offer systems that permit the ordering and consistency properties of a communication primitive or process group to be tailored to their needs. If the designer is concerned about paying the minimum price for the properties an application really requires, such a system can then be configured to only offer the requested properties. Later in the book, we will see that the Horus system implements just such an approach.

Finally, as to the last issue, it is true that we have presented a distributed computing model that, so far, may not seem very closely tied to the software engineering tools normally used to implement distributed systems. In the next chapter we study this practical issue, looking at how group communication tools and virtual synchrony can be applied to real systems that may have been implemented using other technologies.

16.4 Related Reading

On concepts of consistency in distributed systems: (see Birman and van Renesse [1994, 1996]); in the case of partitionable systems: (see Amir, Keidar and Dolev, Malkhi, Moser et al. [1996]).

On the causal controversy: (see van Renesse [1993]).

On the dispute over CATOCS: (see Cheriton and Skeen); but see also Birman (1994), Cooper (1994), van Renesse (1994) for responses.

The end-to-end argument was first put forward in Saltzer et al.

Regarding recent theoretical work on tradeoffs between consistency and availability: (see Babaoglu et al. [1995], Chandra et al. [1996], Fisher et al., Friedman et al.).

CHAPTER 17 ✧ ✧ ✧ ✧

Retrofitting Reliability into Complex Systems

CONTENTS

This chapter deals with options for presenting group computing tools to the application developer. Two broad approaches are considered: those involving wrappers, which encapsulate an existing piece of software in an environment that transparently extends its properties—for example, by introducing fault tolerance through replication or security—and those based upon toolkits, which provide explicit procedure-call interfaces. We will not examine specific examples of such systems now, but will instead focus on the advantages and disadvantages of each approach and on their limitations. In subsequent chapters, however, we discuss a real system, which I have worked on, and present it in substantial detail; in Chapter 26 we review a number of other systems in the same area.

17.1   Wrappers and Toolkits

The introduction of reliability technologies into a complex application raises two sorts of issues. One is that many applications contain substantial amounts of preexisting software or make use of off-the-shelf components (the military and government favor the acronym COTS for this, meaning "components off the shelf"; presumably because OTSC is hard to pronounce!). In these cases, the developer is extremely limited in terms of the ways that the old technology can be modified. A *wrapper* is a technology that overcomes this problem by intercepting events at some interface between the unmodifiable technology and the external environment (see Jones), replacing the original behavior of that interface with an extended behavior, which confers a desired property on the wrapped component, extends the interface itself with new functionality, or otherwise offers a virtualized environment within which the old component executes. Wrapping is a powerful technical option for hardening existing software, although it also has some practical limitations. In this section, we'll review a number of approaches to performing the wrapping operation itself, as well as a number of types of interventions that wrappers can enable.

An alternative to wrapping is to explicitly develop a new application program designed from the outset with the reliability technology in mind—for example, we might set out to build an authentication service for a distributed environment that implements a particular encryption technology and that uses replication to avoid denial of service when some of its server processes fail. Such a program would be said to use a *toolkit* style of distributed computing, in which the sorts of algorithms developed in the previous chapter are explicitly invoked to accomplish a desired task. A toolkit approach packages potentially complex mechanisms, such as replicated data with locking, behind easy-to-use interfaces (in the case of replicated data, *LOCK, READ,* and *UPDATE* operations). The disadvantage of such an approach is that it can be hard to glue a reliability tool into an arbitrary piece of code, and the tools themselves will often reflect design tradeoffs that limit generality. Thus, toolkits can be very powerful but are in some sense inflexible: They adopt a programming paradigm, and, having done so, it is potentially difficult to use the functionality encapsulated within the toolkit in a setting other than the one envisioned by the tool designer.

Toolkits can also take other forms—for example, one could view a firewall, which filters messages entering and exiting a distributed application, as a tool for enforcing a limited security policy. When one uses this broader interpretation of the term, toolkits include quite a variety of presentations of reliability technologies. In addition to the case of firewalls, a toolkit could package a reliable communication technology as a message bus, a system monitoring and management technology, a fault-tolerant file system or database system, or a wide area name service (Table 17.1). Moreover, one can view a programming language that offers primitives for reliable computing as a form of toolkit.

TABLE 17.1. Types of Toolkits Useful in Building or Hardening Distributed Systems

Toolkit	Description
Server replication	Tools and techniques for replicating data to achieve high availability, load-balancing, scalable parallelism, very large memory-mapped caches, and so forth. Cluster APIs for management and exploitation of clusters.
Video server	Technologies for striping video data across multiple servers, isochronous replay, and single replay when multiple clients request the same data.
WAN replication	Technologies for data diffusion among servers that make up a corporate network.
Client groupware	Integration of group conferencing and cooperative work tools into Java agents, TCL/TK, or other GUI builders and client-side applications.
Client reliability	Mechanisms for transparently fault-tolerant RPC to servers, consistent data subscription for sets of clients that monitor the same data source, and so forth.
System management	Tools for instrumenting a distributed system and performing reactive control. Different solutions might be needed when instrumenting the network itself, cluster-style servers, and user-developed applications.
Firewalls and containment tools	Tools for restricting the behavior of an application or for protecting it against a potentially hostile environment—for example, such a toolkit might provide a bank with a way to install a partially trusted client/server application in order to permit its normal operations while preventing unauthorized ones.

As we see in Table 17.1, each toolkit would address a set of application-specific problems, presenting an API specialized to the programming language or environment within which the toolkit will be used and to the task at hand. While it is also possible to develop extremely general toolkits, which seek to address a great variety of possible types of users, doing so can result in a presentation of the technology that is architecturally weak and doesn't guide users to the best system structure for solving their problems. In contrast, application-oriented toolkits often reflect strong structural assumptions, which are known to result in solutions that perform well and achieve high reliability.

In practice, many realistic distributed applications require a mixture of toolkit solutions and wrappers. To the degree that a system has new functionality, which can be developed with a reliability technology in mind, the designer is afforded a great deal of flexibility and power through the execution model supported (e.g., transactional serializability or virtual synchrony) and may be able to provide sophisticated functionality that would not otherwise be feasible. On the other hand, in any system that reuses large amounts of old code, wrappers can be invaluable by shielding the previously developed functionality from the programming model and assumptions of the toolkit.

17.1.1 Wrapper Technologies

In our usage, a wrapper is any technology that intercepts an existing execution path in a manner transparent to the wrapped application or component. By wrapping a component, the developer is able to virtualize the wrapped interface, introducing an extended version with new functionality or other desirable properties. In particular, wrappers can be used to introduce various robustness mechanisms, such as replication for fault tolerance or message encryption for security.

Wrapping at Object Interfaces

Object-oriented interfaces are the best example of a wrapping technology (Figure 17.1), and systems built using CORBA or OLE-2 are, in effect, prewrapped in a manner that makes it easy to introduce new technologies or to substitute a hardened implementation of a service for a nonrobust one. Suppose, for example, that a CORBA implementation of a client/server system turns out to be unavailable because the server has sometimes crashed. Earlier, when discussing CORBA, we pointed out that the CORBA architectural features in support of dynamic reconfiguration or fail-over are difficult to use. If, however, a CORBA service could be replaced with a process group (object group) implementing the same functionality, the problem becomes trivial. Technologies such as Orbix+Isis and Electra, described in Chapter 18, provide precisely this ability. In effect, the CORBA interface wraps the service in such a manner that any other service providing a compatible interface can be substituted for the original one transparently.

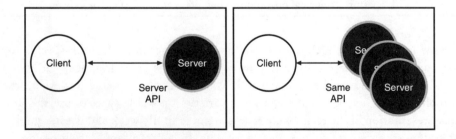

FIGURE 17.1. Object-oriented interfaces permit the easy substitution of a reliable service for a less-reliable one. They represent a simple example of a wrapper technology. However, one can often wrap a system component even if it were not built using object-oriented tools.

Wrapping by Library Replacement

Even when we lack an object-oriented architecture, similar ideas can often be employed to achieve these sorts of objectives. As an example, one can potentially wrap a program by relinking it with a modified version of a library procedure that it calls. In the relinked program, the code will still issue the same procedure calls as it did in the past. But control will now pass to the wrapper procedures, which can take actions other than those taken by the original versions.

In practice, this specific wrapping method would only work on older operating systems, because of the way that libraries are implemented on typical modern operating systems. Until fairly recently, it was typical for linkers to operate by making a single pass over the application program, building a *symbol table* and a list of *unresolved external references*. The linker would then make a single pass over the library (which would typically be represented as a directory containing object files or as an archive of object files), examining the symbol table for each contained object and linking it to the application program if the symbols it declares include any of the remaining unresolved external references. This process causes the size of the program object to grow, and it results in extensions both to the symbol table and, potentially, to the list of unresolved external references. As the linking process continues, these references will in turn be resolved, until there are no remaining external references. At that point, the linker assigns addresses to the various object modules and builds a single program file, which it writes out. In some systems, the actual object files are not copied into the program, but are instead loaded dynamically when first referenced at run time. (See Figure 17.2.)

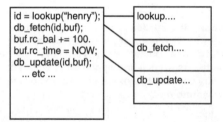

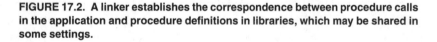

FIGURE 17.2. A linker establishes the correspondence between procedure calls in the application and procedure definitions in libraries, which may be shared in some settings.

Operating systems and linkers have evolved, however, in response to pressure for more efficient use of computer memory. Most modern operating systems support some form of shared libraries. In the shared library schemes, it would be impossible to replace just one procedure in the shared library. Any wrapper technology for a shared library environment would then involve reimplementing all the procedures defined by the shared library—a daunting prospect.

Wrapping by Object Code Editing

Object code editing is an example of a recent wrapping technology that has been exploited in a number of research and commercial application settings. The approach was originally developed by

Wahbe, Lucco, Anderson, and Graham and involves analysis of the object code files before or during the linking process. A variety of object code transformations are possible. Lucco, for example, uses object code editing to enforce type safety and to eliminate the risk of address boundary violations in modules that will run without memory protection—a software fault isolation technique.

For purposes of wrapping, object code editing would permit the selective remapping of certain procedure calls into calls to wrapper functions, which could then issue calls to the original procedures if desired (see Figure 17.3). In this manner, an application that uses the UNIX *sendto* system call to transmit a message could be transformed into one that calls *filter_sendto* (perhaps even passing additional arguments). This procedure, presumably after filtering outgoing messages, could then call *sendto* if a message survives its output filtering criteria. Notice that an approximation to this result can be obtained by simply reading in the symbol table of the application's object file and modifying entries prior to the linking stage.

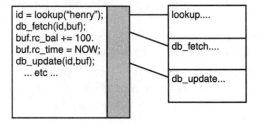

FIGURE 17.3. A wrapper (gray) intercepts selected procedure calls or interface invocations, permitting the introduction of new functionality transparently to the application or library. The wrapper may itself forward the calls to the library, but it can also perform other operations. Wrappers are an important option for introducing reliability into an existing application, which may be too complex to rewrite or to modify easily with explicit procedure calls to a reliability toolkit or some other new technology.

One important application of object code editing, discussed earlier, involves importing untrustworthy code into a client's Web browser. When we discussed this option in Section 10.9, we described it simply as a security enhancement tool. Clearly, however, the same idea could be useful in many other settings. Thus, it makes sense to understand object code editing as a wrapping technology and how specific use of it in Web browser applications might permit us to increase our level of trust in applications that would otherwise represent a serious security threat.

Wrapping with Interposition Agents and Buddy Processes

Until now, we have focused on wrappers that operate directly upon the application process and that live in its address space. However, wrappers need not be so intrusive.

Interposition involves placing some sort of object or process in between an existing object or process and its users. An interposition architecture based on what are called "coprocesses" or "buddy processes" is a simple way to implement this approach, particularly for developers familiar with UNIX pipes (Figure 17.4). Such an architecture involves replacing the connections from an existing process to the outside world with an interface to a buddy process that has a much more sophisticated view of the external environment—for example, perhaps the existing program is basically designed to process a pipeline of data, record by record, or to process batch-style files containing large numbers of records. The buddy process might employ a pipe or file system interface to the original application, which will often continue to execute as if it were still reading batch files or commands typed by a user at a terminal; therefore, it may not need to be modified. To the outside world, however, the interface seen is the one presented by the buddy process, which may now exploit sophisticated technologies such as CORBA, DCE, the Isis Toolkit or Horus, a message bus, and so forth. (One can also imagine embedding the buddy process directly into the address space of the original application, coroutine style, but this is likely to be much more complex and the benefit may be small unless the connection from the buddy process to the older application is known to represent a bottleneck.) The pair of processes would be treated as a single entity for purposes of system management and reliability: They would run on the same platform and be set up so that if one fails, the other automatically fails too.

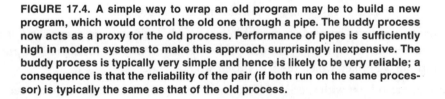

FIGURE 17.4. A simple way to wrap an old program may be to build a new program, which would control the old one through a pipe. The buddy process now acts as a proxy for the old process. Performance of pipes is sufficiently high in modern systems to make this approach surprisingly inexpensive. The buddy process is typically very simple and hence is likely to be very reliable; a consequence is that the reliability of the pair (if both run on the same processor) is typically the same as that of the old process.

Interposition wrappers may also be supported by the operating system. Many operating systems provide some form of packet filter capability, which would permit a user-supplied procedure to examine incoming or outgoing messages, selectively operating on them in various ways. Clearly, a packet filter can implement wrapping. The stream communication abstraction in UNIX, discussed in Chapter 5, supports a related form of wrapping, in which stream modules are pushed and popped from a protocol stack. Pushing a stream module onto the stack is a way of wrapping the stream with some new functionality implemented in the module. The stream still looks the same to its users, but its behavior changes.

Interposition wrappers have been elevated to a real art form in the Chorus operating system (see Rozier et al. [Fall 1988, December 1988]), which is object oriented and uses object invocation for procedure and system calls. In Chorus, an object invocation is done by specifying a procedure to invoke and providing a handle referencing the target object. If a different handle is specified for the original one, and the object referenced has the same or a superset of the interface of the original object, the same call will pass control to a new object. This object now represents a wrapper. Chorus uses this technique extensively for a great variety of purposes, including the sorts of security and reliability objectives discussed above.

Wrapping Communication Infrastructures: Virtual Private Networks

Sometime in the near future, it may become possible to wrap an application by replacing the communication infrastructure it uses with a virtual infrastructure. A great deal of work on the Internet and on telecommunication information architectures is concerned with developing a technology base that can support virtual private networks, having special security or quality-of-service guarantees. A virtual network could also wrap an application—for example, by imposing a firewall interface between certain classes of components or by encrypting data so that intruders can be prevented from eavesdropping.

The concept of a virtual private network runs along the following lines. In Section 10.8 we saw how agent languages such as Java permit a server to download special-purpose display software into a client's browser. One could also imagine doing this in the network communication infrastructure itself, so that the network routing and switching nodes would be in a position to provide customized behavior on behalf of specialized applications needing particular, nonstandard communication features. We call the resulting structure a virtual private network because, from the perspective of each individual user, the network seems to be a dedicated one with precisely the properties needed by the application. This is a virtual behavior, however, in the sense that it is superimposed on a physical network of a more general nature. Uses to which a virtual private network (VPN) could be put include the following:

▶ Support for a security infrastructure within which only legitimate users can send or receive messages. This behavior might be accomplished by requiring that messages be signed using some form of VPN key, which the VPN itself would validate.

▶ Communication links with special video transmission properties, such as guarantees of limited loss rate or real-time delivery (so-called "isochronous" communication).

▶ Tools for stepping down data rates when a slow participant conferences to individuals who all share higher-speed video systems. Here, the VPN would filter the video data, sending through only a small percentage of the frames to reduce load on the slow link.

▶ Concealing link-level redundancy from the user. In current networks, although it is possible to build a redundant communication infrastructure that will remain connected even if a link fails, one often must assign two IP addresses to each process in the network, and the application itself must sense that problems have developed and switch from one to the other explicitly. A VPN could hide this mechanism, providing protection against link failures in a manner transparent to the user.

Wrappers: Some Final Thoughts

Wrappers will be familiar to the system engineering community, which has long employed these sorts of hacks to attach an old piece of code to a new system component. By giving the approach an appealing name, we are not trying to suggest that it represents a breakthrough in technology. On the contrary, the point is simply that there can be many ways to introduce new technologies into a distributed system and not all of them require that the system be rebuilt from scratch.

Given the option, it is certainly desirable to build with the robustness goals and tools that will be used in mind. But lacking that option, one is not necessarily forced to abandon the use of a robustness-enhancing tool. There are often back-door mechanisms by which such tools can be slipped under the covers or otherwise introduced in a largely transparent, nonintrusive manner. Doing so will preserve the large investment an organization may have made in its existing infrastructure and applications and should be viewed as a positive option—not a setback for the developer who seeks to harden a system. Preservation of the existing technology base must be given a high priority in any distributed system development effort, and wrappers represent an important tool in trying to accomplish this goal.

17.1.2 Introducing Robustness in Wrapped Applications

Our purpose in this book is to understand how reliability can be enhanced through the appropriate use of distributed computing technologies. How do wrappers help in this undertaking? Examples of robustness properties that wrappers can use in an application include the following:

▶ *Fault tolerance:* Here, the role of the wrapper is to replace the existing I/O interface between an application and its external environment with one that replicates inputs so that each of a set of replicas of the application will see the same inputs. The wrapper also plays a role in collating the outputs, so that a replicated application will appear to produce a single output, albeit more reliably than if it were not replicated. To my knowledge, the first such use was in a protocol proposed by Anita Borg as part of a system called Aurogen (see Borg et al. [1983, 1985]), and the approach was later generalized by Eric Cooper in his work at Berkeley on a system called Circus (see Cooper [1985]), and in the Isis system, which I developed at Cornell University (see Birman and Joseph [November 1987]). Generally, these techniques assume that the wrapped application is completely deterministic, although later we will see an example in which a wrapper can deal with nondeterminism by carefully tracing the nondeterministic actions of a primary process and then replaying those actions in a replica.

▶ *Caching:* Many applications use remote services in a client/server manner, through some form of RPC interface. Such interfaces can potentially be wrapped to extend their functionality— for example, a database system might evolve over time to support caching of data within its clients in order to take advantage of patterns of repeated access to the same data items, which are common in most distributed applications. To avoid changing the client programs, the database system could wrap an existing interface with a wrapper that manages the cached data, satisfying requests out of the cache when possible and otherwise forwarding them to the server. Notice that the set of clients managing the same cached data item represents a form of process group, within which the cached data can be viewed as a form of replicated data.

▶ *Security and authentication:* A wrapper that intercepts incoming and outgoing messages can secure communication by, for example, encrypting those messages or adding a signature field as they depart and decrypting incoming messages or validating the signature field. Invalid messages can either be discarded silently, or some form of I/O failure can be reported to the application program. This type of wrapper needs access to a cryptographic subsystem for performing encryption or generating signatures. Notice that in this case, a single application may constitute a form of *security enclave,* having the property that all components of the application share certain classes of cryptographic secrets. It follows that the set of wrappers associated with the application can be considered as a form of process group, despite the fact that it may not be necessary to explicitly represent that group at run time or communicate to it as a group.

▶ *Firewall protection:* A wrapper can perform the same sort of actions as a firewall, intercepting incoming or outgoing messages and applying some form of filtering to them—passing only those messages that satisfy the filtering criteria. Such a wrapper would be placed at each of the I/O boundaries between the application and its external environment. As in the case of the security enclave just mentioned, a firewall can be viewed as a set of processes surrounding a protected application or encircling an application to protect the remainder of the system from its potentially unauthorized behavior. If the ring contains multiple members—multiple firewall processes—the structure of a process group is again present, even if the group is not explicitly represented by the system—for example, all firewall processes need to use consistent filtering policies if a firewall is to behave correctly in a distributed setting.

▶ *Monitoring and tracing or logging:* A wrapper can monitor the use of a specific interface or set of interfaces and can trigger certain actions under conditions that depend on the flow of data through those interfaces. A wrapper could be used, for example, to log the actions of an application for purposes of tracing the overall performance and efficiency of a system, or, in a more active role, it could be used to enforce a security policy under which an application has an associated behavioral profile and in which deviation from that profile of expected behavior potentially triggers interventions by an oversight mechanism. Such a security policy would be called an *in-depth security mechanism,* meaning that, unlike a security policy applied merely at the perimeter of the system, it would continue to be applied in an active way throughout the lifetime of an application or its access to the system.

▶ *Quality-of-service negotiation:* A wrapper could be placed around a communication connection for which the application has implicit behavioral requirements, such as minimum performance, throughput, loss rate requirements, or maximum latency limits. The wrapper could then play a role either in negotiation with the underlying network infrastructure to ensure that the required quality of service is provided or in triggering reconfiguration of an application if the necessary quality of service cannot be obtained. Since many applications are built with *implicit* requirements of this sort, such a wrapper would really play the role of making *explicit* an existing (but not expressed) aspect of the application. One reason such a wrapper might make sense would be that future networks may be able to offer guarantees of quality of service even when current networks do not. Thus, an existing application might in the future be wrapped to take advantage of those new properties with little or no change to the underlying application software itself.

▶ *Language-level wrappers:* Wrappers can also operate at the level of a programming language or an interpreted run-time environment. In Chapter 18, for example, we will describe a case in which the TCL/TK programming language was extended to introduce fault tolerance by wrapping some of its standard interfaces with extended ones. Similarly, we will see that fault tolerance and load-balancing can often be introduced into object-oriented programming languages, such as C++, Ada, or SmallTalk, by introducing new object classes that are transparently replicated or that use other transparent extensions of their normal functionality. An existing application can then benefit from replication by simply using these objects in place of the ones previously used.

The above list is at best very partial. What it illustrates is that given the idea of using wrappers to reach into a system and manage or modify it, one can imagine a great variety of possible interventions that would have the effect of introducing fault tolerance or other forms of robustness, such as security, system management, or an explicit declaration of requirements that the application places on its environment.

These examples also illustrate another point: When wrappers are used to introduce a robustness property, it is often the case that some form of distributed process group structure will be present in the resulting system. As previously noted, the system may not need to actually represent such a structure and may not try to take advantage of it per se. However, it is also clear that the ability to represent such structures and to program using them explicitly could confer important benefits on a distributed environment. The wrappers could, for example, use consistently replicated and dynamically updated data to vary some sort of security policy. Thus, a firewall could be made dynamic, capable of varying its filtering behavior in response to changing requirements on the part of the application or environment. A monitoring mechanism could communicate information among its representatives in an attempt to detect correlated behaviors or attacks on a system. A caching mechanism can ensure the consistency of its cached data by updating these data dynamically.

Wrappers do not always require process group support, but the two technologies are well matched to one another. Where a process group technology is available, the developer of a wrapper can potentially benefit from it to provide sophisticated functionality, which would otherwise be difficult to implement. Moreover, some types of wrappers are only meaningful if process group communication is available.

17.1.3 Toolkit Technologies

In the introduction to this chapter, we noted that wrappers will often have limitations—for example, although it is fairly easy to use wrappers to replicate a completely deterministic application to make it fault tolerant, it is much harder to do so if an application is not deterministic. And, unfortunately, many applications are nondeterministic for obvious reasons—for example, an application that is sensitive to time (e.g., timestamps on files or messages, clock values, timeouts) will be nondeterministic to the degree that it is difficult to guarantee that the behavior of a replica will be the same without ensuring that the replica sees the same time values and receives timer interrupts at the same point in its execution. The UNIX *select* system call is a source of nondeterminism, as are interactions with devices. Any time an application uses *ftell* to measure the amount of data available in an incoming communication connection, this introduces a form of nondeterminism. Asynchronous I/O mechanisms, common in many systems, are also potentially nondeterministic. Parallel or preemptive multithreaded applications are potentially the most nondeterministic of all.

In cases such as these, there may be no obvious way that a wrapper could be introduced to transparently confer some desired reliability property. Alternatively, it may be possible to do so but impractical in terms of cost or complexity. In such cases, it is sometimes hard to avoid building a new version of the application in question, in which explicit use is made of the desired reliability technology. Generally, such approaches involve what is called a *toolkit* methodology.

In a toolkit, the desired technology is prepackaged, usually in the form of procedure calls (Table 17.2). These provide the functionality needed by the application, but without requiring that the user understand the reasoning that led the toolkit developer to decide that in one situation *cbcast* was a good choice of communication primitive, but that in another *abcast* was a better option. A toolkit for managing replicated data might offer an abstract data type called a replicated data item, perhaps with some form of name and some sort of representation, such as a vector or an *n*-dimensional array. Operations appropriate to the data type would then be offered: *UPDATE, READ,* and *LOCK* being the obvious ones for a replicated data item (in addition to any other operations that might be needed to initialize the object, detach from it when no longer using it, etc.). Other examples of typical toolkit functionality might include transactional interfaces, mechanisms for performing distributed load-balancing or fault-tolerant request execution, tools for publish/subscribe styles of communication, tuple-space tools for implementing an abstraction similar to the one in the Linda tuple-oriented parallel programming environment, and so forth. The potential list of tools is really unlimited, particularly if such issues as distributed system security are also considered.

Toolkits often include other elements of a distributed environment, such as a name space for managing names of objects, a concept of a communication end point object, process group communication support, message data structures and message manipulation functionality, lightweight threads or other event notification interfaces, and so forth. Alternatively, a toolkit may assume that the user is already working with a distributed computing environment, such as the DCE environment or Sun Microsystem's ONC environment. The advantage of such an assumption is that it reduces the scope of the toolkit itself to those issues explicitly associated with its model; the disadvantage is that it compels the toolkit user to use the environment in question, thus reducing portability.

TABLE 17.2. Typical Interfaces in Toolkits for Process Group Computing*

Tool	Description
Load-balancing	Provides mechanisms for building a load-balanced server, which can handle more work as the number of group members increases.
Guaranteed execution	Provides fault tolerance in RPC-style request execution, normally in a manner that is transparent to the client.
Locking	Provides synchronization or some form of token passing.
Replicated data	Provides for data replication, with interfaces to read and write data, as well as selectable properties, such as data persistence, dynamic uniformity, and the type of data integrity guarantees supported.
Logging	Maintains logs and checkpoints and provides playback.
Wide area spooling	Provides tools for integrating LAN systems into a WAN solution.
Membership ranking	Within a process group, provides a ranking on the members that can be used to subdivide tasks or load-balance work.
Monitoring and control	Provides interfaces for instrumenting communication into and out of a group and for controlling some aspects of communication.
State transfer	Supports the transfer of group state to a joining process.
Bulk transfer	Supports out-of-band transfer of very large blocks of data.
Shared memory	Tools for managing shared memory regions within a process group. The members can then use these tools for communication that is difficult or expensive to represent in terms of message passing.

*In typical practice, a set of toolkits would be needed, each aimed at a different class of problems. The interfaces listed above would be typical for a server replication toolkit, but might not be appropriate for building a cluster-style multimedia video server or a caching Web proxy with dynamic update and document consistency guarantees.

17.1.4 Distributed Programming Languages

The reader may recall the discussion of agent programming languages and other *fourth-generation languages* (4GLs), which package powerful computing tools in the form of special-purpose programming environments. Java is the best-known example of such a language, albeit aimed at a setting in which reliability is taken primarily to mean security of the user's system against viruses, worms, and other forms of intrusion. PowerBuilder and Visual BASIC will soon emerge as important alternatives to Java. Other types of agent-oriented programming languages include TCL/TK (see Ousterhout [1994]) and TACOMA (see Johansen et al. [June 1995]).

Although existing distributed programming languages lack group communication features and few make provisions for reliability or fault tolerance, one can extend many such languages without difficult. The resulting enhanced language can be viewed as a form of distributed computing toolkit in which the tools are tightly integrated with the language. For example, in Chapter 18, we will see how the TCL/TK GUI development environment was converted into a distributed groupware system by integrating it with Horus. The resulting system is a powerful prototyping tool, but in fact could actually support production applications as well; Brian Smith at Cornell University is using this infrastructure in support of a new video conferencing system, and it could also be employed as a groupware- and computer-supported cooperative work (CSCW) programming tool.

Similarly, one can integrate a technology such as Horus into a Web browser such as the HotJava browser—in this way providing the option of group communication support directly to Java applets and applications. We'll discuss this type of functionality and the opportunities it might create in Section 17.4.

17.2 Wrapping a Simple RPC Server

To illustrate the idea of wrapping for reliability, consider a simple RPC server designed for a financial setting. A common problem that occurs in banking is to compute the theoretical price for a bond; this involves a calculation that potentially reflects current and projected interest rates, market conditions and volatility (expected price fluctuations), dependency of the priced bond on other securities, and myriad other factors. Typically, the necessary model and input data are represented in the form of a server, which clients access using RPC requests. Each RPC can be reissued as often as necessary: The results may not be identical (because the server is continuously updating the parameters to its model), but any particular result should be valid for at least a brief period of time.

Now, suppose that we have developed such a server, but only after putting it into operation do we begin to be concerned about its availability. A typical scenario might be that the server has evolved over time, so that although it was really quite simple and easy to restart after crashes when first introduced, it can now require an hour or more to restart itself after failures. The result is that if the server does fail, the disruption could be extremely costly.

An analysis of the causes of failure is likely to reveal that the server itself is fairly stable, although a low residual rate of crashes is observed. Perhaps there is a lingering suspicion that some changes recently introduced to handle the possible unification of European currencies after 1997 are buggy and are causing crashes. The development team is working on this problem and expects to have a new version in a few months, but management, being pragmatic, doubts that this will be the end of the software-reliability issues for this server. Meanwhile, however, routine maintenance and communication link problems are believed to be at least as serious a source of downtime. Finally, although the server hardware is relatively robust, it has definitely caused at least two major outages during the past year, and loss of power associated with a fire triggered additional downtime recently.

In such a situation, it may be extremely important to take steps to improve server reliability. But clearly, rebuilding the server from scratch would be an impractical step given the evolutionary nature of the software that it uses. Such an effort could take months or years, and when traders perceive a problem, they are rarely prepared to wait years for a solution.

The introduction of reliable hardware and networks could improve matters substantially. A dual network connection to the server, for example, would permit messages to route around problematic network components such as faulty routers or damaged bridges. But the software and management failures would remain an issue. Upgrading to a fault-tolerant hardware platform on which to run the server would clearly improve reliability, but only to a degree. If the software is in fact responsible for many of the failures that are being observed, all of these steps will only eliminate some fraction of the outages.

An approach that replicates the server using wrappers, however, might be very appealing in this setting. As stated, the server state seems to be dependent on pricing inputs to it, but not on queries. Thus, a solution such as the one illustrated in Figure 17.5 can be considered. Here, the inputs that determine server behavior are replicated using broadcasts to a process group. The queries are load-balanced by directing the queries for any given client to one or another member of the server process group. The architecture has substantial design flexibility in this regard: The clients can be managed as a group, with their queries carefully programmed to match each client to a different, optimally selected server. Alternatively, the clients can use a random policy to issue requests to the servers. If a server is unreasonably slow to respond, or has clearly failed, the same request could be reissued to some other server (or, if the request itself may have caused the failure, a slightly modified version of the request could be issued to some other server). Moreover, the use of wrappers makes it easy to see how such an approach can be introduced transparently (without changing existing server or client code). Perhaps the only really difficult problem would be to restart a server while the system is already active.

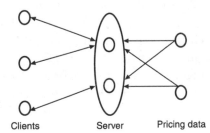

FIGURE 17.5. A client/server application can be wrapped to introduce fault tolerance and load-balancing with few or no changes to the existing code.

In fact, even this problem may not be so difficult to solve. The same wrappers that are used to replace the connection from the data sources to the server with a broadcast to the replicated server group can potentially be set up to log input to the server group members in the order that they are delivered. To start a new server, this information can be transferred to it using a state transfer from the old members, after which any new inputs can be delivered. When the new server is fully initialized, a message can then be sent to the client wrappers informing them that the new server is able to accept requests. To optimize this process, it may be possible to launch the server using a checkpoint, replaying only those logged events that changed the server state after the checkpoint was created. These steps would have the effect of minimizing the impact of the slow server restart on perceived system performance.

This discussion is not entirely hypothetical. I am aware of a number of settings in which problems such as this were solved precisely in this manner. The use of wrappers is clearly an effective way to introduce reliability or other properties (such as load-balancing) transparently, or nearly so, in complex settings characterized by substantial preexisting applications.

17.3   Wrapping a Web Server

The techniques of the preceding section could also be used to develop a fault-tolerant version of a Web server. However, whereas the example presented above concerned a database server used only for queries, many Web servers also offer applications that become active in response to data submitted by the user through a form-fill or similar interface. To wrap such a server for fault tolerance, one would need to first confirm that its implementation is deterministic if these sorts of operations are invoked in the same order at the replicas. Given such information, the *abcast* protocol could be used to ensure that the replicas all see the same inputs in the same order. Since the replicas would now take the same actions against the same state, the first response received could be passed back to the user; subsequent duplicate responses can be ignored.

A slightly more elaborate approach is commonly used to introduce load-balancing within a set of replicated Web servers for query accesses, while fully replicating update accesses to keep the copies in consistent states. The HTTP protocol is sufficiently sophisticated to make this an easy task: For each retrieval (*get*) request received, a front-end Web server simply returns a different server's address from which that retrieval request should be satisfied, using a temporary redirection error code. This requires no changes to the HTTP protocol, Web browsers, or Web servers, and although purists might consider it to be a form of hack, the benefits of introducing load-balancing without having to redesign HTTP are so substantial that within the Web development community, the approach is viewed as an important design paradigm. In the terminology of this chapter, the front-end server wraps the cluster of back-end machines.

17.4   Hardening Other Aspects of the Web

A wrapped Web server just hints at the potential that group communication tools may have in future enterprise uses of the Web. As seen in Table 17.3 and Figures 17.6 and 17.7, the expansion of the Web into groupware applications and environments, computer-supported cooperative work (CSCW), and dynamic information publication applications creates challenges, which the tools we developed in Chapters 13 though 16 could be used to solve.

TABLE 17.3. Potential Uses of Groups in Internet Systems

Application domain	Uses of process groups
Server replication	• High availability, fault tolerance • State transfer to restarted process • Scalable parallelism and automatic load-balancing • Coherent caching for local data access • Database replication for high availability
Data dissemination	• Dynamic update of documents in the Web or of fields in documents • Video data transmission to group conference browsers with video viewers • Updates to parameters of a parallel program • Updates to spreadsheet values displayed to browsers showing financial data • Database updates to database GUI viewers • Publish/subscribe applications
System management	• Propagate management information base (MIB) updates to visualization systems • Propagate knowledge of the set of servers that compose a service • Rank the members of a server set for subdividing the work • Detect failures and recoveries and trigger consistent, coordinated action • Coordinate actions when multiple processes can all handle some event • Rebalance load when a server becomes overloaded, fails, or recovers
Security applications	• Dynamically updating firewall profiles • Updating security keys and authorization information • Replicating authorization servers or directories for high availability • Splitting secrets to raise the barrier faced by potential intruders • Wrapping components to enforce behavior limitations (a form of firewall that is placed close to the component and monitors the behavior of the application as a whole)

Today, a typical enterprise that makes use of a number of Web servers treats each server as an independently managed platform and has little control over the cache coherency policies of the Web proxy servers residing between the end user and the Web servers. With group replication and load-balancing, we could transform these Web servers into fault-tolerant, parallel processing systems. Such a step would bring benefits such as high availability and scalable performance, enabling the enterprise to reduce the risk of server overload when a popular document is under heavy demand. Web servers will increasingly be used as video servers, capturing video input (such as conferences and short presentations by company experts on topics of near-term interest, news stories off the wire, etc.), in which case such scalable parallelism may be critical to both data archiving (which often involves computationally costly techniques such as compression) and playback.

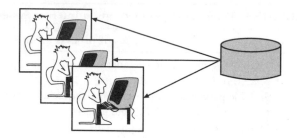

FIGURE 17.6. Web server transmits continuous updates to documents or video feeds to a group of users. Depending upon the properties of the group communication technology employed, the users may be guaranteed of seeing identical sequences of input, of seeing data synchronously, of security from external intrusion or interference, and so forth. Such a capability is most conveniently packaged by integrating group communication directly into a Web agent language such as Java or Visual BASIC—for example, by extending the HotJava browser with group communication protocols that could then be used through a groupware API.

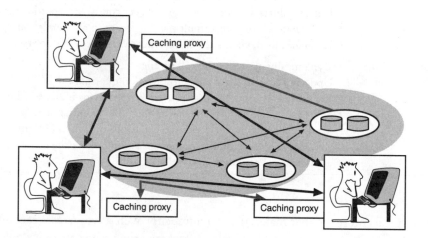

FIGURE 17.7. Potential group communication uses in Web applications occur at several levels. Web servers can be replicated for fault tolerance and load-balancing or integrated into wide area structures, which might span large corporations with many sites. Caching Web proxies could be fixed to provide guarantees of data consistency, and digital encryption or signatures could be used to protect the overall enterprise against intrusion or attack. Moreover, one can foresee integrating group communication directly into agent languages such as Java, thereby creating a natural tool for building cooperative groupware applications. A key to successfully realizing this vision will be to design wrappers or toolkit APIs that are both natural and easy to use for the different levels of abstraction and purposes seen here: Clearly, the tools one would want to use in building an interactive multimedia groupware object would be very different from those one would use to replicate a Web server.

RETROFITTING RELIABILITY INTO COMPLEX SYSTEMS   Chapter 17

Wide area group tools could also be used to integrate these servers into a wide area architecture, which would be seamless, presenting users with the abstraction of a single, highly consistent, high-availability Web service—yet internally self-managed and structured. Such a multiserver system might implement data migration policies, moving data to keep them close to the users who demand these data most often, and wide area replication of frequently requested critical information, while also providing guarantees of rapid update and consistency. Later, we will be looking at security technologies that could also be provided through such an enterprise architecture, permitting a company to limit access to its critical data to just those users who have been authorized—for example, through provision of a Fortezza card (see Section 19.3.4).

Turning to the caching Web proxies, group communication tools would permit us to replace the standard caching policy with a stateful coherent caching mechanism. In contrast with the typical situation today, where a Web page may be stale, such an approach would allow a server to reliably send out a message that would invalidate or refresh any cached data that have changed since these data were copied. Moreover, by drawing on CORBA functionality, one could begin to deal with document groups (sets of documents with hyperlinks to one another) and multidocument structures in a more sophisticated manner.

Group communication tools can also play a role in the delivery of data to end users. Consider, for example, the idea of treating a message within a group as a Java-style self-displaying object, a topic we touched upon earlier. In effect, the server could manufacture and broadcast to a set of users an actively self-constructed entity. Now, if group tools are available within the browsers themselves, these applets could cooperate with one another to animate a scene in a way that all participants in the group conferencing session can observe or to mediate among a set of concurrent actions initiated by different users. Users would download the current state of such an applet and then receive (or generate) updates, observing these in a consistent order with respect to other concurrent users. Indeed, the applet itself could be made self-modifying—for example, by sending out new code if actions taken by the users demand it (zooming for higher resolution, for example, might cause an applet to replace itself with one suited for accurate display of fine-grained detail).

Thus, one could imagine a world of active multidocuments in which the objects retrieved by different users would be mutually consistent, dynamically updated, and able to communicate with one another, and in which updates originating on the Web servers would be automatically and rapidly propagated to the documents themselves. Such a technology would permit a major step forward in conferencing tools and is likely to be needed in some settings, such as telemedicine (remote surgery or consultations), military strategic/tactical analysis, and remote teleoperation of devices. It would enable a new generation of interactive multiparticipant network games or simulations, and it could support the sort of cooperation needed in commercial or financial transactions requiring simultaneous actions in multiple markets or multiple countries. The potential seems nearly unlimited. Moreover, all of these are applications that would appear very difficult to realize in the absence of a consistent group communication architecture and that demand a high level of reliability in order to be useful within the intended community.

Obviously, our wrapped Web server represents just the tip of a potentially large application domain. While it is difficult to say with any certainty that this type of system will ever be of commercial importance, or to predict the time frame in which it might become operational, it seems plausible that the pressures that today are pushing more and more organizations and corporations onto the Web will tomorrow translate into pressure for consistent, predictable, and rapidly updated groupware tools and objects. The match of the technologies we have presented with this likely need is good, although the *packaging* of group communication tools to work naturally and easily within such applications will certainly demand additional research and development. In particular, notice that the tools and APIs one might desire at the level of a replicated Web server will look completely different from those that would make sense in a multimedia groupware conferencing system. This is one reason that systems such as Horus need flexibility, both at the level of how they behave and how they look. Nonetheless, the development of appropriate APIs ultimately seems like a small obstacle. I am confident that group communication tools will come to play a large role in the enterprise Web computing systems of the coming decades.

17.5 Unbreakable Stream Connections

Motivated by Section 17.4, we now consider a more complex example. In Chapter 5, we discuss unreliability issues associated with stream-style communication. In this chapter, we discuss extensions to Web servers that might make them reliable. However, consider the client browser: It will typically connect to such a server through a stream (a TCP connection, to be specific). Thus, it makes sense to ask how group communication tools can help us overcome some of the problems we noted in our original discussion of streams and their behavior when failures occur. After all, if we want our Web technology to be *completely* reliable and to handle failures in a completely transparent manner, we will need to solve this problem.

Our analysis will lead to a mixed conclusion, and indeed one reason for including this section in the book is to illustrate the challenges created by real-world considerations and the sort of tradeoffs that result. A constraint underlying the discussion will be the assumption that we are concerned with a client and a server, and that the server (but not the client) is to be replicated for increased availability. The client, on the other hand, uses a completely standard and unmodified implementation of some stream-style reliable protocol. In the following text, we will use TCP as our example for such a protocol, although the same discussion would make sense for other stream-style protocols. This constraint prevents us from using a solution such as the protocol discussed in Section 13.13.

Notice, however, that these constraints are somewhat arbitrary. While there may be important benefits in avoiding modification of the client systems, these benefits are unlikely to appeal to a developer who will need to pay a high cost, in complexity or performance, for the transparency afforded by such a solution. Moreover, in a world where servers can download agents to the client, it may be quite simple to download a special-purpose applet, which causes the client's system to simply talk to the server through some new, special-purpose protocol. This alternative will underlie much of the discussion of this chapter. We will see that under certain conditions, a very transparent stream protocol from the client to the server can be made reliable at low cost, and this class of solutions will be discussed in some detail. Under other conditions, we will encounter dead ends in which either the complexity or performance overheads exceed the likely threshold of pain at which the nonmember to group protocols would make more sense. Such solutions are consequently of limited practical interest, and we will discuss them only superficially.

17.5.1 Reliability Options for Stream Communication

What would it mean to say that a stream connection is "more reliable" than the ones considered in Chapter 5? Two types of answers make sense. A sensible starting point would be to overcome the failure-reporting problems of stream connections by rewiring the failure mechanisms of some standard stream protocol to the GMS input and output. More precisely, we can introduce wrappers for this purpose. Depending on how the stream package was implemented, this could be very easy (i.e., if the stream module is implemented using source code available to the developer and can easily be modified), but it could more often represent a tremendously difficult undertaking. The problem is that standard computer systems generally place such code inside the O/S kernel and protect it against modification by users.

In light of our constraint that the client be unmodified, this rewiring will only occur within the server. Nonetheless, it can have the effect of avoiding inconsistent failure scenarios, if a client is connected to multiple servers. In such cases, we will now be sure that if one server concludes that a client has failed, all servers will react consistently.

The use of wrappers to provide consistent failure reporting requires that code be added to intercept failure detections in the stream package, modifying the reporting of such events so that they become upcalls to the GMS service. To do this, one would first modify the channel protocol so that each process using the protocol registers itself with the GMS and so that any process p connected to some other process q asks the GMS to monitor q and to report failures. Next, suppose that the original code implementing the connection had a procedure called *break_connection*, which gets called when the number of retransmission attempts for some packet exceeds a threshold. The developer would modify these parts of the code to instead issue upcalls to the GMS service, informing it that the end point has apparently failed. This will cause the GMS to run a protocol excluding the end point and eventually to issue downcalls to all the processes monitoring the end point that has been excluded. When the GMS reports that the end point of the connection has failed, the old code associated with *break_connection* would be executed. Notice, however, that in the original implementation, each process independently detects (apparent) failures and immediately executes *break_connection*. With this change, each process continues to independently detect failures, but *all* processes execute *break_connection* if any one does. This interaction is illustrated in Figure 17.8. Moreover, the example isn't completely hypothetical: There are public-domain implementations of the TCP protocol stack that run in user space. My students have carried out this transformation successfully and demonstrated that the resulting technology indeed exhibits consistent failure-reporting semantics. On the other hand, one could question whether the benefits of this change justify the effort.

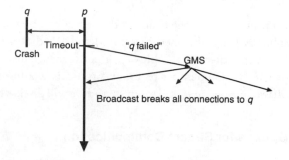

FIGURE 17.8. Modified stream protocol reports detected failures to the GMS, which breaks all connections to a failed or excluded process simultaneously. To introduce this behavior, the original interfaces by which the protocol detects and reports failures within itself would be wrapped to interconnect the detection mechanism with the GMS and to connect the GMS output back to the stream protocol.

A more ambitious goal would be to support a stream connection to a group of processes having the property that the members can now emulate the behavior of a single, very reliable, nonreplicated service. Solving this problem potentially involves much more effort than for our initial intervention. Here, we need a way to ensure that the stream connection (say, TCP) survives the failure of subsets of the members and that the members can stay in consistent states even when failures occur. We will want our solution to be easy to use (in particular, it would be best if the clients of such a connection could employ standard versions of the stream protocol, with all the changes being made on the server-side). Also, we will want the solution to be as efficient as possible, so that the cost of using a reliable service through such a connection is as close as possible to the cost of using an unreliable service through a conventional stream connection. Such an arrangement is illustrated in Figures 17.9 and 17.10; the former figure shows how this might work, while the latter shows how the resulting structure appears to the client system.

Recall that this problem would be straightforward to solve if we had the freedom to modify the application on the client-side of the connection. In that case, it would suffice to implement an interface that looks like the standard stream interface for the client computing platform, but operates using one of the client-to-group protocols developed in Section 13.13. Thus, in the specific case of a Java-enabled Web browser, where there is a realistic option for downloading an agent that can use such a nonstandard protocol to talk to the server, there is a relatively simple solution to this problem.

The same problem would become more difficult if our goal were to fool a standard stream protocol such as TCP into believing that it is communicating with a single, nonfaulty process using that protocol, when in fact the destination is group. As we will see, a completely general solution may be so costly that implementation of a direct client-to-group broadcast might be highly advantageous. However, for a somewhat constrained class of stream protocols and applications, a very transparent, very general solution is achievable.

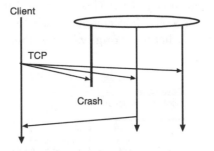

FIGURE 17.9. A more elaborate solution to the reliable stream problem. In the desired protocol, the client uses a completely standard stream protocol, such as TCP, but the messages are delivered as reliable broadcasts within the group. Unless all members fail, the client sees an unbroken TCP connection to what appears to be a continuously available server. Only if all members fail does the channel break from the perspective of the client.

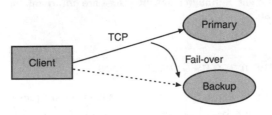

FIGURE 17.10. A successful implementation of the reliable stream protocol would provide clients with the illusion of a completely transparent fail-over, under which their TCP connection to a primary server would automatically and seamlessly switch from a primary server to the backup if a failure occurs.

17.5.2 An Unbreakable Stream That Mimics TCP

To address this issue, we will need to assume that there is a version of the stream protocol that has been isolated in the form of a protocol module with a well-defined interface. To simplify the discussion, assume that we are talking about a TCP protocol (other stream protocols could be treated the same way; only the details would change). At the bottom, the protocol accepts incoming IP packets from the network and sends back IP packets containing acknowledgments, retransmission requests, and outgoing TCP data (outgoing segments). Internally, the module has an interface to the timer subsystem of the machine on which it is running, using this to read the time and to schedule timer interrupts. Fortunately, not many protocol implementations of this sort make use of threaded concurrency, but if the module in question does so, the interface from it to the subsystem implementing lightweight threads would also have to be considered as part of its interface to the outside world. Finally, there is the interface by which the module interacts with the application process: This consists of its read and write interface and perhaps a control interface (this would implement the UNIX *ioctl, ftell,* and *select* interfaces or the equivalent operations on other operating systems). This environment is illustrated in Figure 17.11.

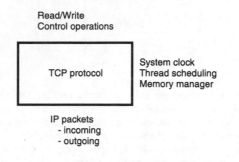

FIGURE 17.11. The TCP protocol can be viewed as a black box with interfaces to its environment. Although these interfaces are nontrivial, they are not over-whelmingly so.

17.5.3 Nondeterminism and Its Consequences

To use the load-balancing or primary-backup replication techniques presented earlier, together with a fault-tolerant scheme based on wrappers, we need to understand how to control any nondeterminism associated with this protocol. Specifically, let's assume that we intercept incoming events by replacing the various interfaces that connect the TCP protocol to the outside world with modified interfaces that will try to keep a set of backup processes in sync with a primary. How hard would it be to make such a solution work?

Given an accurately modeled TCP protocol, this problem is not as difficult as one might expect. Our enumeration of interfaces has reduced the TCP protocol itself to a state machine, which can be thought of as receiving incoming events from its varied interfaces, computing, and then performing output events. Even access by the protocol to the clock can be thought of as an output event (sent to the clock) followed by an input event (a response from the clock to the protocol). It follows that we can arrange for a primary copy of the TCP protocol to broadcast a script of the full set of its interactions with the outside world. Such a script would list the events that occurred to the protocol and its actions: First, it received an IP data packet containing the following byte sequence; then it issued a request to read the local clock; next, a read request was received from the application; 18 bytes of data were returned.

If a copy of the TCP protocol module has access to such a script—earlier we called this a "trace"—it can precisely emulate the actions of the primary merely by replaying the same input events in the same order. If our interface specification were complete and accurate, the backup will faithfully perform the exact same actions in the same order. This approach can be extended to encapsulate the application process as well: Given a complete characterization of the application process's interface to the external environment, the actions taken by the primary copy can be traced in such a manner that the actions of the replicas will emulate it in an accurate manner. An analysis of the protocol itself will generally be needed to convince ourselves that we know precisely how it can be nondeterministic and that the required information can be encoded into trace messages.

Obviously, there are sources of nondeterminism that can be very hard to deal with. Interrupt-driven behavior and thread scheduling are two instances of such problems, and any sort of direct access by the driver to hardware properties of the computer or attached peripherals runs the risk of introducing similar problems. One can imagine noting the time at which an interrupt occurs and forcing the backup to replay interrupts at the right time—noting the time at which thread scheduling actions occur and replaying these in the same manner. Other kinds of nondeterminism, on the other hand, may be much easier to deal with. If the application program itself is deterministic and will write back the identical data if given identical inputs, the main source of nondeterminism seen by the protocol stack may be that associated with the relative ordering of timeouts and *write* operations. This ordering information, and the associated clock values when timeouts occur, can be encoded very concisely. Notice, however, that even the knowledge that no timeout occurred before the application sent a message to the client may be significant to the state of the protocol.

17.5.4 Dealing with Arbitrary Nondeterminism

A paper by Bressoud and Schneider recently suggested a way to extend this trace-driven approach to software fault tolerance to make entire machines fault tolerant, including the operating system and all the applications running on the machine (see Bressoud and Schneider). They do this using the special hardware properties of certain classes of modern microprocessors. Their work operates at the level of the CPU itself and involves noting the time at which interrupts occur. Specifically, the method requires a special hardware register, which measures time in machine cycles and is saved as part of the interrupt sequence.

The same register can also be set to a value, in which case an interrupt is generated at the desired time. Using this feature on the backup, the Bressoud and Schneider solution operates by repeatedly setting the cycle counter to the time at which the next interrupt should occur. All other types of interrupts are disabled, and the machine is allowed to execute up to the point when the counter fires. Then, an interrupt indistinguishable from the one that occurred on the primary is generated. The method is most readily applicable to machines with very few I/O connections: ideally, just a communication interface and an interface to the clock. Unfortunately, the hardware required is available only on HP's PA-RISC microprocessors.

Returning to our problem, it is easy to see that the key factor limiting a solution will be the degree of nondeterminism present in the TCP protocol. Motivated by Bressoud and Schneider's work, it may sometimes be possible to modify a TCP protocol that includes nondeterminism (such as concurrent threads or interrupts) into a protocol that is deterministic and hence describable by a trace (e.g., by replacing the threads and interrupts with a nonthreaded polling method). As noted in the introduction to this section, some forms of design complexity are best viewed as an argument for the nonmember to a group protocols discussed in Section 13.13, and any substantial change to the TCP protocol itself to eliminate nondeterminism probably falls into this class of complex interventions, which should be viewed with skepticism. In the remainder of this section, we will assume that it is reasonably easy to trace the actions of our TCP protocol and that the volume of trace information is reasonably low; otherwise, the method simply should not be used.

17.5.5 Replicating the IP Address

Our transformation leaves two questions open. First, we need to resolve a simple matter, which is to ensure that the backup will actually receive incoming IP packets after taking over from the primary in the event of a failure. The specific issue is as follows. Normally, the receive side of a stream connection is identifiable by the address (the IP address) to which packets are sent by TCP. This address would normally consist of the IP address of the machine itself and is locked into the TCP protocol of the client system. As a consequence, IP packets sent by the client are only received at one site, which represents a single point of failure for the protocol. We need a way to shift the address to a different location to enable a backup to take over after a crash.

This problem can be solved by manufacturing virtual IP addresses, which don't correspond to any real machine on the network. It turns out that the IP address of a machine is assigned during the boot sequence and that there is typically some form of protected system call by which a new address can be assigned. Indeed, if a machine resides on multiple networks, it may have multiple IP addresses, since the IP address is typically used to index routing tables. Thus, it is perfectly practical to assign a single machine a true address and one or more virtual addresses. We can use this feature to assign the TCP end point such an address and to reassign that address to a backup after a failure. In UNIX, this is done using the *ifconfig* system command.

17.5.6 Maximizing Concurrency by Relaxing Multicast Ordering

The other lingering problem is concerned with maximizing performance. Had we not intervened to replicate the TCP state machine, it would reside on the critical path that determines I/O latency and throughput from client to server and back. Suddenly, we have modified this path to insert what may turn out to be a large number of multicasts to the replicas. How costly will these be?

Specifically, we need to understand the conditions under which a replicated TCP stack can perform as well, or nearly as well, as a nonreplicated one. The rationale is similar to the one we encountered in discussing sources of nondeterminism: If the performance hit associated with a reliable stream is high, it makes more sense to simply modify the client to use a protocol knowledgeable about the presence of a group.

The cheapest case for our protocol occurs when the stream connection is used as a pipe, with unidirectional communication from the client to the server. In this case, notice that all the multicasts are initiated by the primary copy of the TCP protocol stack until a failure occurs.

Only then will multicasts begin to be initiated by a backup process, namely the new primary. For this purpose, a sender-ordered protocol, the primitive we called *fbcast,* would be sufficient. Now, the costs of *fbcast* occur in several ways. There is the fixed overhead of creating the message and passing it to the local multicast subsystem, which may be as small as a few tens of instructions for a very small message and an efficient implementation of *fbcast.* There is a background cost associated with the protocol, but this would normally not impact the latency seen in the primary server, which is the one measurable by the client. Next, there is a bandwidth cost: Every byte that reaches the primary will need to be forwarded to the clients; in some cases this may represent a problem, although it will often be of minor importance because most communication devices are capable of sustaining much higher loads than the TCP protocol itself can produce. Finally, however, there is an issue of waiting for *fbcast* stability, which we will discuss momentarily.

To the degree that bandwidth proves to be a problem, one can imagine developing a protocol in which the full group of servers would present the same IP address to the network, much as in the IP multicast protocol previously discussed. However, with such an approach, there is the risk that some data segments may not reach some of the clients, and there will be a need to retransmit data that any client misses. This starts to sound like a complex and costly undertaking, so in keeping with our initial constraints, we will assume that bandwidth is not a problem. If it is expected to be a problem, a protocol knowledgeable about the presence of a group should be used.

The stability issue to which we alluded is the following. Consider a TCP-level acknowledgment or some other message sent by the TCP protocol from the primary server to the client. When such a message is received, the client's outgoing window will be updated, clearing frames associated with any data that were acknowledged. If the primary now crashes, there will be no possibility of reconstructing the data that were garbage collected. Thus, we see that before sending any TCP-level message from the primary to the client, the primary should wait until any causally prior messages have reached the backups. This constraint applies both to incoming TCP-level messages from the client to the server and to trace messages that the primary may have generated in the course of handling incoming data. In the terminology of Chapter 13, we need to be sure that these causally prior messages are stable at the backup.

For one-directional streams, this problem does not represent a serious source of delay, because the only messages affected are acknowledgments generated within TCP, and the use of a larger TCP window suffices to hide the higher latency. In effect, there is no situation in which the *application* would be forced to wait for stability of the *fbcast* that conveys trace information to the backup processes. In pipelines with multiple stages, each successive write may need to delay for stability of the prior *fbcast*, but such delays are likely to be hidden by concurrency. Moreover, one-directional streams are easy to detect, because the protocol can simply assume itself to be in a unidirectional mode until the server attempts to send information back to the client.

In the more general case, however, such as an RPC or object invocation that runs over a stream, it is likely that a single, very small *fbcast* will need to be sent from the primary server to its backups immediately after each *write* operation by the application to the TCP stream. This *fbcast* becomes stable when it has reached its destinations; round-trip times in typical modern multicast systems, such as Horus, are in the range of .7 ms to 1.4 ms for such events. Thus, responses from the server to the client may be delayed by about 1 ms to achieve fault tolerance. Such a cost may seem small to some users and large to others. It can be viewed as the price of transparency, since comparable delays would not have occurred in applications where complete transparency on the client-side was not an objective. This is illustrated in Figure 17.12.

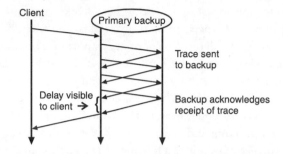

FIGURE 17.12. The latency introduced by replication is largely invisible to the client. As seen here, most trace information reaches the backup while the primary is still computing. Only the last trace message, sent after the primary issues its reply and before that reply can be sent from primary to client, introduces noticeable latency in the critical path that limits RPC round-trip times over the replicated TCP channel. The problem is that if the trace information causally prior to the reply were lost in a crash, the primary's state could not be reproduced by the backup. Thus, this information must be stable at the backup before the reply is transmitted. Unfortunately, the use of timeouts in the TCP protocol stack implies that such information will be generated. One can imagine other protocols, however, or optimizations to the TCP protocol, in which the primary would have extremely little or no trace information to send to the backup; replication of such protocols for fault tolerance would introduce minimal additional latency.

Thus, we find ourselves back at the same limitation cited earlier for the TCP protocol itself. In the simple case of a largely one-way communication channel, and to the degree that the protocol and the application are deterministic, the replication method will have minimal impact on system performance. As we move away from this simple case into more complicated ones, the protocol becomes much more complicated and imposes increasingly visible overheads, which would not have been incurred if the client were simply modified to use a protocol knowledgeable about the presence of a group of servers.

When our modifications are feasible, notice also that the role of the virtual synchrony model is fairly limited. The model lets us overlook issues of agreement on membership in the server group and lets us implement an *fbcast* protocol, which will be delivered atomically before failure notifications occur for the primary server, if it fails. These guarantees greatly simplify the protocol, which isn't all that simple in any case. One could argue that without them, the solution would be impractically complicated. The model does not, however, introduce any particularly complex reasoning of its own.

17.5.7   State Transfer Issues

Our discussion overlooked the issues relating to launching of new servers and of transferring state information to bring them up to date. For applications and protocols in which state is easily represented, the solution presented here can easily be modified to accommodate joins with state transfer to the joining process. Otherwise, it may be best to launch a sufficient set of replicas at the time the connection is first made, so that even if failures occur, the group will still have an adequate number of servers to continue providing response.

17.5.8   Discussion

Although we presented the unbreakable stream problem as a hypothetical problem, I have supervised several research projects that pursued precisely such an analysis and ultimately implemented unbreakable TCP connections for various purposes. One of these projects focused on the case of TCP channels to mobile users, whose handheld computers might need to connect to a succession of base stations as the computer was moved around (see Cho and Birman), while another looked at TCP in a more standard LAN setting, where the focus was on transparent fail-over (see Wong). Thus, with modest effort, the problem can be solved as described in the preceding text.

Our analysis suggests that in situations where we are not limited by bandwidth and where the stream protocol to be modified is available to the developer and has modest nondeterminism, a reliability transformation that uses wrappers to introduce fault tolerance through replication might be worthwhile. The impact on performance and complexity would be reasonably low, and the performance costs may actually be hidden by concurrency. However, if any of these conditions does not hold, the introduced complexity and performance overhead may begin to seem excessive for the degree of transparency such a reliable protocol can afford. Finally, we have observed that there will probably be a small latency impact associated with our transformation in RPC-style interactions, but that this cost would probably be hidden in pipeline-style uses of streams—again, because of the concurrency achieved between the protocol used to transmit trace information to a replica and the computation occurring in the application itself.

In the specific case of a Web browser connected to an enhanced Web server, one might well look at these tradeoffs and costs and conclude that the benefit of providing reliability of this sort would not be worth the additional complexity and development effort. First, the possibility of downloading a Java applet containing a protocol such as the ones developed for nonmember to group communication in Chapter 13 may represent the easiest path to a solution. Moreover, even in situations where downloading such an applet is unrealistic, the decision to use an unreliable stream might have relatively minor consequences. Such a decision would mean that the Web server group would remain available even if some of its members were to fail, but a client actually using a Web server at the instant it crashes might see the failure as just another case of Web access failure. However, the actual frequency of such events will surely be very low, and perhaps the impact when they do occur will be sufficiently minor to make the design point an acceptable one. This is especially likely to be the case if some of the other reasons for a Web operation failure remain in our hardened design—for example, a DNS timeout. After all, if the hardened system can still fail from time to time (although, one hopes, infrequently!), making a large investment to eliminate what may be a statistically small percentage of the remaining failure cases might not make a lot of sense.

These sorts of tradeoffs are inevitable in complex distributed systems, and it is important that the developer keep one eye on the balance. It is very appealing to imagine a technology that would let us replicate a server, making it fault tolerant, in a manner that would be completely transparent to its clients. An unbreakable TCP stream connecting the client to the server seems like a natural and inevitably desirable feature. Yet the alternative of building a protocol whereby the client would know it was communicating to the group, or that would conceal such interactions beneath a layer of software presenting a stream-like interface, must be weighed against the presumed disadvantages of needing to modify (or at least recompile) the client program. (See Figure 17.13.)

Pro	• Totally transparent fail-over • Uninterrupted service to client • Client system not changed at all
Con	• With Java, might have a reasonably easy way to modify the client system • Solution is complex • Performance penalty may be substantial for some patterns of use • Doesn't address other causes of failure, such as the ones discussed in Part II of the book

FIGURE 17.13. Tradeoffs to be considered when looking at the decision to implement a transparently fault-tolerant TCP stream protocol. In many settings, the arguments *against* doing so would dominate.

In my experience, one often encounters such tradeoffs between performance and transparency or complexity and transparency. Transparency is a good thing, and the use of wrappers can provide a route for very transparent hardening of a distributed system. However, transparency should generally not be elevated to the level of a religion. In effect, a technology should be as transparent as possible, consistent with the need to keep the software used simple and the overheads associated with it low. When these properties fall into question, a less-transparent solution should be seriously considered.

17.6 Building a Replicated TCP Protocol Using a Toolkit

The above analysis may leave the reader with the sense that even if one can wrap a TCP protocol for fault tolerance, the benefits of doing so are outweighed by the complexity of dealing with black-box nondeterminism. But it is important to keep in mind that the limitations associated with the solution we developed stemmed specifically from the attempt to use a wrapper to avoid modifying the TCP protocol itself. If, in contrast, we were in a position to implement a TCP protocol of our own, the same issues could be circumvented.

In particular, the question of nondeterminism underlies most of the performance concerns raised above. Were we to design a TCP implementation specifically for the purpose of replicating it, we could probably eliminate most or all of this nondeterminism through a design that cleverly hides nondeterministic events behind other sorts of communication. Our TCP protocol could be designed, for example, to check for timeouts and to send acknowledgment messages using timestamps placed on messages by the primary copy of the protocol stack. Each time the primary process receives a message or any other form of input, it could timestamp the outgoing copies of that message with the time at which it saw the event. If the protocol used these inputs to trigger timeout-related events, we could avoid the need to send much of the trace information discussed in the preceding sections.

Conversely, whereas our wrapper would be forced to trace interrupt events and thread-switching events in order to overcome nondeterminism, an explicitly replicated approach might use concurrent threads only for logically independent tasks, ensuring that in any situation where there is a sensitivity to thread-scheduling order, that order is deterministically fixed by the external sequence of events received by the TCP protocol stack.

Another concern of ours was that the application program might sometimes request a full buffer of data and yet be passed a partial buffer by the TCP implementation. Knowing that such a situation creates an overhead in a replicated TCP stream, one might implement the protocol to never return a partial result from a *read* operation unless the remote end of the stream has been closed. Knowing that all reads block until the stream closes or 8 KB are available for the reader, the nondeterminism associated with read requests can be greatly reduced or even eliminated.

These observations having been made, however, it should also be mentioned that our analysis in this section has been superficial. Moreover, if a toolkit approach is used within the TCP stack on the server-side, it may be reasonable to extend the approach to encompass the client-side as well. Such an effort clearly represents a potential research (or product) opportunity and goes beyond any investigation of this topic of which I am aware. In the interest of brevity, we will not develop this discussion at the present time.

17.7 Reliable Distributed Shared Memory

Distributed shared memories are a "hot topic" in the distributed system research community. In this section, we look at the idea of implementing a wrapper for the UNIX *mmap* (or *shrmem*) functions, which are normally used to map files and memory regions into the address space of user applications and are shared between concurrently executing processes. The extension we consider here provides for the sharing of memory-mapped objects over a virtually synchronous communication architecture running on a high-speed communication network. One might use such a system as a repository for rapidly changing visual information in the form of Web pages: The provider of the information would update a local mapped copy directly in memory, while the subscribers could map the region directly into the memory of a display device and in this way obtain a direct I/O path between the data source and the remote display. Other uses might include parallel scientific computations, in which the shared memory represents the shared state of the parallel computation; a collaborative workplace or virtual reality environment shared between a number of users; a simulation of a conference or meeting room populated by the participants in a teleconference; or some other abstraction.

In studying this problem, we should comment at the outset that the topic is an area of active research by several operating system groups worldwide (see Ahamad et al., Carter, Feeley et al., Felton and Zahorjan, Gharachorloo et al., Johnson et al., Li and Hudak), but I am not aware of any effort that has looked at the implementation of a *reliable* shared memory using process group technology. To the degree that there has been work on this subject, the emphasis has tended to be on settings in which reliability issues are secondary to questions of functionality and performance. Our goal in this section, then, is to look at another nontrivial example of how group communication might be used to solve a challenging contemporary problem, but not to claim that our solution is a real one with known performance and latency properties.

17.7.1 The Shared Memory Wrapper Abstraction

As for the case of the unbreakable TCP connection, our solution will start with an appropriate wrapper technology. In many UNIX-like operating systems there is a mechanism available for mapping a file into the memory of a process, sharing memory between concurrently executing processes or doing both at the same time. The UNIX system calls supporting this functionality are called *shrmem* or *mmap*, depending on the version of UNIX one is using; a related interface called *semctl* provides access to a semaphore-based mutual-exclusion mechanism. By wrapping these interfaces (e.g., by intercepting calls to them, checking the arguments and special-casing certain calls using new code, and passing other calls to the operating system itself), the functionality of the shared memory subsystem can potentially be extended. Our design makes use of such a wrapper.

In particular, if we assume that there will be a *distributed shared memory daemon* process (DSMD) running on each node where our extended memory-mapping functionality will be used, we can adopt an approach whereby certain memory-mapped operations are recognized as being operations on the DSM and are handled through cooperation with the DSMD. The recognition that an operation is remote can be supported in either of two ways. One simple option is to introduce a new file system object called a DSM object, which is recognizable through a special file type, file name extension (such as .dsm), or some other attribute. The file contents can then be treated as a handle on the DSM object itself by the DSM subsystem. A second option is to extend the options field supported by the existing shared memory system calls with extra bits, one of which could indicate that the request refers to a region of the DSM. In a similar manner, we can extend the concept of semaphore names (which are normally positive integers in UNIX) to include a DSM semaphore name space for which operations are recognizable as being distributed synchronization requests.

Having identified a DSM request, that request can then be handled through a protocol with the DSMD process. In particular, we can adopt the rule that all distributed shared memory is implemented as locally shared memory between the application process and the DSMD process, which the DSMD process arranges to maintain in a coherent manner with regard to other processes mapping the same region of memory. The DSMD process thus functions as a type of server, handling requests associated with semaphore operations or events that involve the mapped memory and managing the mapped regions themselves as parts of its own address space. It will be the role of the DSMD servers as a group to cooperate to implement the DSM abstractions in a correct manner; the system call wrappers are thereby kept extremely small and simple, functioning mainly by passing requests through to the DSMD or to the local copy of the operating system, depending on the nature of the system call that was intercepted. This is illustrated in Figure 17.14.

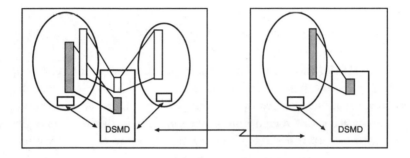

FIGURE 17.14. Two machines share memory through the intermediary of a distributed shared memory daemon that runs on each. A wrapper (shown as small boxes) intercepts memory mapping and semaphore system calls, redirecting DSM operations to the DSMD. The DMSD processes sharing a given region of memory belong to a process group and cooperate to provide coherent, fault-tolerant behavior. The best implementation of the abstraction depends upon the expected pattern of sharing and the origin of updates.

For design simplicity, it will be helpful to consider the DSM architecture as being volatile: DSM regions exist only while one or more processes are mapping them, and there is no persistent disk storage associated with them, except perhaps for purposes of paging if the region is too large to maintain in memory. We can view the DSM as a whole as being a collection of objects or *regions*, each having a base address within the DSM, a size, and perhaps access restrictions and security properties. A region might be associated with a file system name, or it could be allocated using some form of DSM region manager server; we will not address this issue here.

Notice that our design has reduced the issue to one of maintaining replicated data and performing synchronization with a collection of superimposed process groups (one on behalf of each shared memory region). The DMSD processes that map a given region would also belong to the corresponding process group. The properties of that process group and the algorithms used to maintain the data in it can now be tuned to match well with the patterns of access expected from the application processes using it.

17.7.2 Memory Coherency Options for Distributed Shared Memory

In any distributed memory architecture, memory coherence is one of the hardest issues to address. Abstractly, the coherence properties of a memory characterize the degree to which that memory is guaranteed to behave like a single, nonshared memory that handles every memory access directly. Because our memory is not resident at any single location, but is shared among the processes that happen to be mapping it at a given time, there are a number of options in regard to the degree to which these copies should be coherent. The major choices correspond to the options for shared memory on parallel processors and consist of the following:

1 *Strong consistency:* In this model, the DSM is expected to behave precisely as a single nonreplicated memory might have behaved. In effect, there is a single global serialization order for all read and write operations.

2 *Weak consistency:* In this model, the DSM can be highly inconsistent. Updates propagate after an unspecified and possibly long delay, and copies of the mapped region may differ significantly for this reason.

3 *Release consistency (DASH project):* This model assumes that conflicting read or update accesses to memory are always serialized (protected) using mutual-exclusion locks, such as the semaphore system calls intercepted by our wrapper. The model requires that if process p obtains a lock associated with a region from process q, then p will also observe the results of any update that q has performed. However, if p tries to access the DSM without properly locking the memory, the outcome can be unpredictable.

4 *Causal consistency (Neiger and Hutto):* In this model, the causal relationship between reads and updates is tracked; the memory must provide the property that if access b occurs after access a in a causal sense, then b will observe the results of access a.

There are additional models, but these four already represent a sufficient variety of options to present us with some reasonable design choices. To implement strong consistency, it will be necessary to order all update operations, raising the question of how this can be accomplished. The memory protection mechanisms of a virtual memory system offer the needed flexibility: If we imagine all the DSMD processes for a given region to have locked that region as read-only, then the memory will refuse updates and will trivially achieve the strong consistency property. Suppose now that each time an update occurs, we intercept the resulting page fault in our wrapper and request that the local DSMD process enable the memory region for write access. The DSMD process can do this by obtaining a token, perhaps using the *cbcast*-based token-passing algorithm we developed earlier. It can then unlock the region for local writes and permit the local process to continue. After a suitable period of time, or when it next learns of an update access attempt by some other process, it can relock the local copy of the region, *cbcast* the changed portions, and then pass the token. Notice that although *cbcast* is used to implement this policy, the desired behavior could also have been obtained using *abcast* for all the operations or even using *fbcast* and sequencing the update and token-passing messages at the sender (the latter would have the advantage of requiring just a single field to represent the sequence number). With all these approaches, the resulting behavior is that of the strongly consistent memory. The strongly consistent memory will also be causally consistent in the case of the implementation that uses *cbcast*; for the alternative implementations this will depend upon the details of the scheme being used.

The release consistency model can be implemented in a similar manner, except that in this case, the token passing is associated with semaphore operations, and there is no need to communicate changes to a page until the corresponding semaphore is released. Of course, there may be performance reasons that would favor transmitting updates before the semaphore is released, but the release consistency model itself does not require us to do so.

Consider now the degree of match between these design options and the expected patterns of use for a DSM. It is likely that a DSM will either be updated primarily from one source at a time or in a random way by the processes that use it, simply because this is the pattern seen for other types of distributed applications that maintain replicated data. For the case where there is a primary data source, both the strong and release consistency models will work equally well: The update lock will tend to remain at the site where the updates are done, and other copies of the DSM will passively receive incoming updates. If the update source moves around, however, there may be advantages to the release consistency implementation: Although the programmer is compelled to include extra code (to lock objects in a way that guarantees determinism), these locks may be obtained more efficiently than in the case of strong consistency, where the implementation we proposed might move the update lock around more frequently than necessary, incurring a high overhead in the process. Further, the release consistency implementation avoids the need to trap page faults in the application, and in this manner avoids a potentially high overhead for updates. (See Figure 17.15.)

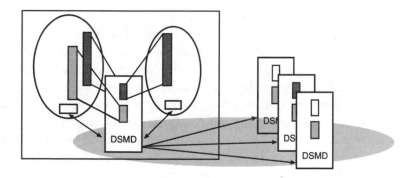

FIGURE 17.15. The proposed solution maps the DSM problem to a more familiar one: replicated data with locking within a virtually synchronous process group. Only one of several overlapped groups is shown; another group would be used for the dark gray memory region, another for the white one, and so forth. Virtual synchrony provides us with simple solutions for what would otherwise be tricky problems, such as ensuring the coherence of the distributed memory, handling failures and dynamic join events, and dealing with protection. Actually implementing such an architecture using the Horus system over an ATM network would be an interesting research project and would raise interesting performance challenges, but the basic problem is clearly very closely matched to the model for which virtual synchrony and Horus were developed.

These considerations make release consistency an appealing model for our DSM, despite its dependence on the use of semaphore-style locking. Of course, should an application desire a weak consistency model or need strong consistency, we now know how both models can be implemented.

However, there are also issues that the consistency model overlooks and that could be quite important in a practical DSM. Many applications that operate on shared memory will be sensitive to the latency with which updates are propagated, and there will be a subset in which other communication patterns and properties are needed—for example, video algorithms will want to send a full frame at a time and will need guarantees of throughput and latency from the underlying communication architecture. Accordingly, our design should include one additional interface by which a knowledgeable application can specify the desired update properties to the DSM. This *dsmctl* system call would be used to specify both the pattern of updates that the application will generate (random, page based, isochronous) and also the maximum latency and other special requirements for acceptable performance. The DSMD can then use this information to schedule its communication appropriately. If available, the *page dirty bit* provided by the virtual memory hardware can be checked periodically by the DSMD; if not available, shared regions that are mapped for update can be transmitted in their entirety at the frequency requested by the user.

17.7.3 False Sharing

False sharing is a phenomenon seen on parallel shared memory machines that corresponds to thrashing in a virtual memory architecture. The problem arises when multiple logically unrelated objects are mapped to the same shared memory region or page by an accident of storage allocation. When these objects are updated in parallel, the memory subsystem is unable to detect that the updates are independent ones and treats the situation as one in which the processes doing the updates are contending for the same object. In our implementation of strong consistency, the update token would bounce around in this case, resulting in a huge overhead for token passing and page fault handing on the client systems. Yet the problem also points to an issue in our proposed release consistency scheme—the *granularity of locking.* In particular, it becomes clear that the semaphores used for locking must have the same granularity as the objects the DSMD transmits for updates—most likely a page. Otherwise, because the DSMD lacks a fine-grained concept of data access, when an object is updated on a page and the semaphore locking that object is released, the entire page will be transmitted to other processes mapping the page, potentially overwriting parts of the page that the semaphore was not considered to lock and which are in fact not even up-to-date on the node that held the lock.

Our DSM architecture can only work if the granularity of locking is at the page level or region level, and, in either case, false sharing could now occur as a visible problem for the developer. Rather than trying to overcome this problem, it may be best to simply caution the user: The DSM architecture we have proposed here will perform poorly if an application is subject to false sharing; hence, such applications may need to be redesigned to arrange for concurrently updated but logically unrelated objects to reside in different regions or at least on different pages, and, in the case of release consistency, they must be locked by separate semaphores.

17.7.4 Demand Paging and Intelligent Prefetching

We cited the case of frequent and time-critical updates, but there is another style of DSM use that will require more or less the opposite treatment. Suppose that the DSM region is extremely large and most applications access it in a sparse manner. Then, even if a region is mapped by some process, it may not be necessary or even desirable to actively update that region each time some process updates some part of the data area. In such cases, a demand paging model, whereby a portion of the DSM is maintained as current only if the process holding that region is actually accessing it, makes more sense.

Although we will not tackle the problem here, for reasons of brevity, it would be desirable for such large regions to be managed as multiple subregions, shrinking the process group for a given subregion to include only those processes that are actively updating it or reading it. With such an approach, one arrives at a form of *demand paging*, in which a process, upon attempting to access a subregion not currently mapped into its address space, experiences a page fault. To resolve the fault the DSMD would join the process group for that subregion, transferring the current state of the subregion (or just those updates that have occurred since the process was last a memory) and then enabling read or update access to the subregion and resuming local computation.

Notice that the virtual synchrony properties of the state transfer make it easy to describe a solution to what would otherwise be a tricky synchronization problem! Lacking the virtual synchrony model, it would not be at all simple to coordinate the addition of a new memory to a subregion group and to integrate the state transfer operation with updates that may be occurring dynamically. The model makes it easy to do so and still be able to guarantee that release consistency or strong consistency will be observed by the DSM user. On the other hand, recall that virtual synchrony comes with no guarantees of real-time performance (a topic to which we will return in Chapter 20), and hence support for dynamically adjusting the members of a process group that maps a given region or subregion may be incompatible with providing real-time performance and latency guarantees. For situations in which such guarantees are desired, it may be wise to disable this form of dynamicism unless the requirements are fairly weak ones.

Demand paging systems perform best if the relatively costly operations involved in fetching a page are performed shortly before the page fault actually takes place, in order to overlap useful computation with the paging-in activity and to minimize the delay associated with actually servicing the page fault when it occurs. Accordingly, it would be advisable to implement some form of prefetching policy, whereby the DSMD, recognizing a pattern of access (such as sequential access to a series of subregions), would assume that this pattern will continue into the future and would join subregion groups in anticipation of the future need. Our architecture creates a convenient context within which to implement such a policy.

17.7.5 Fault Tolerance Issues

Our DSM will have a natural form of fault tolerance, which arises directly from the fault tolerance of the virtual synchrony model used by the DSMD processes to form process groups and propagate updates. The issues that occur are primarily ones associated with the possibility of a failure by a process while it is doing an update. Such an event might leave the DSM corrupted and a semaphore in the locked state (the token for the group would be at the process that failed).

A good way to solve this problem would be to introduce a new kind of page fault exception into the DSM model; this could be called a *page corruption* exception. In such an approach, when a process holding an update lock or semaphore for a page or region fails, any subsequent access by some other process mapping that region would result in a corruption trap. The handler for such a trap would be granted the update lock or semaphore and would be required to restore the page to a consistent state. The next update would be understood to clear the corruption bit, so that processes not attempting to access the page during the period of corruption would be completely unaware that a problem had occurred.

17.7.6 Security and Protection Considerations

The reliability of a DSM should extend beyond issues of fault tolerance and detecting potential corruption to also include guarantees of protection and security or privacy if desired. We have not yet treated security issues in this book and defer discussion of the options until Chapter 19. In brief, one could arrange for the data on the wire to be encrypted so that eavesdroppers lacking an appropriate key would be unable to map a protected segment and unable to make sense of any intercepted updates. Depending on the degree to which the system implementing virtual synchrony is trusted, weaker security options might include some form of user-ID–based access control in which unauthorized users are prevented from joining the group. Because the DSMD must join a process group to gain access to a DSM segment, the group join operation can include authorization keys for use in determining whether or not access should be granted. Alternatively, if the DSMD process itself can be trusted, it can perform a mapping from local user-IDs on the host machine where it is running to global user-IDs in a protection domain associated with the DSM, permitting access under UNIX-style restrictions.

17.7.7 Summary and Discussion

The previous examples in this chapter illustrated some of the challenges that can be encountered when attempting to exploit group structures in implementing a distributed system. In contrast, the DSM example shows how simple and elegant solutions can be (and how easy it can be to understand them) when the match of problem and tool turns out to be close. Our architecture would, in principle, be a highly efficient one: The costs and overheads are predominately those of the virtual synchrony communication architecture, and, in the case of the release consistency model, no additional overhead beyond this is imposed. As we will see, the model can perform extremely well over an appropriate software implementation architecture and with high-speed hardware. The video-mapped shared memory suggested at the start of this section is not an unreasonable prospect. Moreover, by wrapping the standard shared memory mechanisms for a setting, the DSM abstraction can be made extremely transparent.

Of course, the feasibility of this architecture depends upon having a suitable shared memory subsystem available for use between the DSMD and its clients; our solutions have required that we be able to manipulate memory protection bits from the DSMD, trap page faults by the client processes, restart them after servicing, and sense the state of a page (dirty or clean) to avoid undesired excess communication. Some operating systems, such as Mach or the commercial OSF/1 system, provide interfaces by which this would be possible; others do not or offer only part of the support that might be needed.

17.8 Related Reading

On wrappers and technologies that can support them: (see Jones, Rozier et al. [Fall 1988, December 1988], Wahbe et al.).

On the Isis Toolkit: (see Birman and Joseph [November 1987], Birman and van Renesse [1994]). (Information on the most current APIs should be obtained directly from the company that markets the Isis product line; their Web page is http://www.isis.com.)

On agents: (see Gosling and McGilton [1995a, 1995b], Johansen et al. [June 1995], Ousterhout [1994]).

On virtual fault tolerance: (see Bressoud and Schneider).

On shared memory: (see Ahamad et al., Carter, Feeley et al., Felton and Zahorjan, Gharachorloo et al., Johnson et al., Li and Hudak). Tanenbaum also discusses shared memory: (see Tanenbaum), and Coulouris treats the topic as well: (see Coulouris et al.).

CHAPTER 18 ✧ ✧ ✧ ✧

Reliable Distributed Computing Systems

CONTENTS

The purpose of this chapter is to shift our attention away from protocol issues to architectural considerations associated with the implementation of process group computing solutions. Although there has been a great deal of work in this area, we focus on the Horus system,[1] because that system is well matched to the presentation of this book. Horus is available for researchers in academic or industrial settings (at no fee) and may be used in conjunction with this book as a platform on which to base experiments and to gain some hands-on experience with reliable distributed computing.

18.1 Architectural Considerations in Reliable Systems

The reader may feel that Part II of this book and the first chapters of Part III have lost one of the important themes of Part I—namely, the growing importance of architectural structure and modularity in reliable distributed systems and, indeed, in structuring distributed systems of all types. Our goal in this chapter, in part, is to reestablish some of these principles in the context of the group computing constructs introduced in Chapters 13 through 17. Specifically, we will explore the embedding of group communication support into a modular system architecture.

Historically, group computing and data replication tools have tended to overlook the importance of architectural structure. These technologies have traditionally been presented in what might be called a flat architecture: one in which the APIs provided by the system are fixed, correspond closely to the group construct and associated communication primitives, and are more or less uniformly accessible from any application making use of the group communication environment anywhere in the system.

In practice, however, the use of group communication will vary considerably depending upon what one is attempting to do. Consider the examples that occurred in Chapter 17, when we discussed group computing in the context of enterprise Web applications:

▶ Groups used to replicate a Web server for load-balancing, fault tolerance, or scalable performance through parallelism.

▶ Groups used to interconnect a set of Web servers in order to create the illusion of a single, corporate-wide server within which objects might migrate or be replicated to varying degrees, depending on usage patterns.

▶ Groups corresponding to the set of Web proxy servers that cache a given data item and are used to invalidate those cached copies or to refresh them when they change.

---

[1] The ancient Egyptian religion teaches that after the world was created, the gods Osiris and Seth engaged in an epic battle for control of the earth. Osiris, who is associated with good, was defeated and hacked to pieces by the evil Seth, and his body was scattered over the Nile Delta. The goddess Isis gathered the fragments and magically restored Osiris to life. He descended to rule the Underworld and, with Isis, fathered a child, Horus, who went on to defeat Seth. When we developed the Isis Toolkit, the image of a system that puts the pieces together after a failure appealed to us, and we named the system accordingly, although the failures that the toolkit can handle are a little less extreme than the one that Osiris experienced! Later, when we developed Horus, it seemed appropriate to again allude to the Egyptian myth. However, it may take some time to determine whether the Horus system will go on to banish unreliability and inconsistency from the Information Superhighway.

- Groups used to distribute Java applets to users cooperating in conferencing applications or other groupware applications (we gave a number of examples in Chapter 17 and won't repeat them here).

- Groups used to distribute updates to documents, or other forms of updates, to Java applets running close to the client browsers.

- Groups formed among the set of Java applets, running on behalf of clients, for the purpose of multicasting updates or other changes to the state of the group session among the participants.

- Groups associated with security keys employed in a virtual private network.

Clearly, these uses correspond to applications that would be implemented at very different levels of programming abstraction and for which the most appropriate presentation of the group technology would vary dramatically. Several of these represent potential uses of wrappers, but others would match better with toolkit interfaces and still others with special-purpose, high-level programming languages. Even within those subclasses, one would expect considerable variation in terms of what is wrapped, the context in which those tools or languages are provided, and the nature of the tools themselves. No single solution could possibly satisfy all these potential types of developers and users. On the contrary, any system that offers just a single interface to all of its users is likely to confuse its users and to be perceived as complex and difficult to learn. Returning to our historical observation, the tendency to offer group communication tools through a flat interface (one that looks the same to all applications and that offers identical capabilities no matter where it is used in the system) has proved to be an obstacle to the adoption of these technologies, because the resulting tools tend to be conceptually mismatched with the developer's goals and mindset.

Indeed, the lesson goes further than this. Although we have presented group communication as an obvious and elegant step, the experience of programming with groups can be quite a bit more challenging than the developer might expect. Obtaining good performance is not always an easy thing, and the challenge of doing so increases greatly if groups are deployed in an unstructured way, creating complex patterns of overlap within which the loads placed on individual group members may vary widely from process to process. Thus, what may seem obvious and elegant to the reader, can start to seem clumsy and complex to the developer, who is struggling to obtain predictable performance and graceful scalability.

These observations argue for a more structured presentation of group computing technologies: one in which the tools and APIs provided are aimed at a specific class of users and will guide those users to a harmonious and simple solution to the problems anticipated for that class of users. If the same technology will also support some other community of users, a second set of tools and APIs should be offered to them. Thus, the tools provided for Web server replication might look very different from those available to the developer of a Java display applet, even if both the applet and the Web server turn out to offer functionality occurring in a group communication subsystem. I believe that far too little attention has been given to this issue up to the present and that this has emerged as a significant obstacle to the widespread use of reliability technologies.

At a minimum, focusing only on issues associated with replication (as opposed to security, system management, or real time), it would appear that three layers of APIs are needed (Figure 18.1). The lowest layer is the one aimed at uses within servers, the middle layer focuses on interconnection and management of servers within a WAN setting, and the third layer focuses on client-side issues and interfaces. Such layers may be further subdivided: Perhaps the client layer offers a collection of transactional database tools and a collection of Java groupware interfaces, while the server layer offers tools for multimedia data transmission, consistent replication and coordinated control, and fault tolerance through active replication. This view of the issues now places unusual demands on the underlying communication system: not only must it potentially look different for different classes of users, but it may also need to offer very different properties for different classes of users. Security and management subsystems would introduce additional APIs, which may well be further structured. Real-time subsystems are likely to require still further structure and interfaces.

Functionality of a client-level API:
- Fault-tolerant remote procedure call
- Reliable, unbreakable streams to servers
- Consistent or reliable subscriptions to data published by servers
- Tools for forming groupware sessions involving other client systems

Functionality of a WAN server API:
- Tools for consistently replicating data within wide area or corporate networks
- Technology for updating global state and for merging after a partitioning failure is corrected
- Security tools for creating virtual private networks
- Management tools for control and supervision

Functionality of a cluster-server API:
- Tools for building fault-tolerant servers (ideally, as transparently as possible)
- Load-balancing and scalable parallelism support
- Management tools for system servicing and automatic reconfiguration
- Facilities for on-line upgrade

Other cases that may require specialized APIs:
- Multimedia data transport protocols (special quality-of-service or real-time properties)
- Security (key management and authentication APIs)
- Debugging and instrumentation
- Very large scale data diffusion

FIGURE 18.1. Different levels of a system may require different styles of group computing support. A simple client/server architecture gives rise to three levels of API. Further structure might be introduced in a multimedia setting (where special protocols may be needed for video data movement or to provide time-synchronous functionality), in a transactional database setting (where clients may expect an SQL-oriented interface), or in a security setting (where APIs will focus on authentication and key management).

18.2 Horus: A Flexible Group Communication System

The observations in the preceding section may seem to yield an ambiguous situation. On the one hand, we have seen in previous chapters that process group environments for distributed computing represent a promising step toward robustness for mission-critical distributed applications. Process groups have a natural correspondence with data or services that have been replicated for availability or as part of a coherent cache, such as might be used to ensure the consistency of documents managed by a set of Web proxies. They can be used to support highly available security domains. Also, group mechanisms fit well with an emerging generation of intelligent network and collaborative work applications.

Yet we have also seen that there are many options concerning how process groups should look and behave. The requirements that applications place on a group infrastructure can vary tremendously, and there may be fundamental tradeoffs between semantics and performance. Even the most appropriate way to present the group abstraction to the application depends on the setting.

The Horus system responds to this observation by providing an unusually flexible group communication model to application developers. This flexibility extends to system interfaces; the properties provided by a protocol stack; and even the configuration of Horus itself, which can run in user space, in an operating system kernel or microkernel, or be split between them. Horus can be used through any of several application interfaces. These include toolkit-style interfaces and wrappers, which hide group functionality behind UNIX communication system calls, the TCL/TK programming language, and other distributed computing constructs. The intent is that it should be possible to slide Horus beneath an existing system as transparently as possible—for example, to introduce fault tolerance or security without requiring substantial changes to the system being hardened (see Bressoud and Schneider).

A basic goal of Horus is to provide efficient support for the virtually synchronous execution model. However, although often desirable, properties such as virtual synchrony may sometimes be unwanted, introduce unnecessary overheads, or conflict with other objectives such as real-time guarantees. Moreover, the optimal implementation of a desired group communication property sometimes depends on the run-time environment. In an insecure environment, one might accept the overhead of data encryption but wish to avoid this cost when running inside a firewall. On a platform such as the IBM SP2, which has reliable message transmission, protocols for message retransmission would be superfluous.

Accordingly, Horus provides an architecture whereby the protocol supporting a group can be varied, at run time, to match the specific requirements of its application and environment. Virtual synchrony is only one of the options available, and, even when it is selected, the specific ordering properties that messages will respect, the flow-control policies used, and other details can be fine-tuned. Horus obtains this flexibility by using a structured framework for protocol composition, which incorporates ideas from systems such as the UNIX stream framework and the x-Kernel, but replaces point-to-point communication with group communication as the fundamental abstraction. In Horus, group communication support is provided by stacking protocol modules having a regular architecture, where each module has a separate responsibility. A process group can be optimized by dynamically including or excluding particular modules from its protocol stack.

18.2.1 A Layered Process Group Architecture

It is useful to think of Horus's central protocol abstraction as resembling a Lego block; the Horus system is thus similar to a box of Lego blocks. Each type of block implements a micro-protocol, which provides a different communication feature. To promote the combination of these blocks into macroprotocols with desired properties, the blocks have standardized top and bottom interfaces, which allow them to be stacked on top of each other at run time in a variety of ways (see Figure 18.2). Obviously, not every sort of protocol block makes sense above or below every other sort. But the conceptual value of the architecture is that where it makes sense to create a new protocol by restacking existing blocks in a new way, doing so is straightforward.

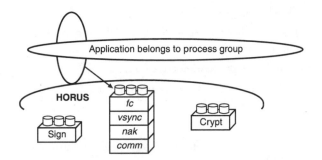

FIGURE 18.2. Group protocol layers can be stacked at run time like Lego blocks and support applications through one of several application programmer interfaces. Shown is an application program belonging to a single process group, supported by a Horus protocol stack of four layers: "fc," the flow-control layer; "vsync," the layer implementing virtually synchronous process group views; "nak," a layer using negative acknowledgments to overcome communication failures; and "comm," which interfaces Horus to a network. The application would often use Horus through a wrapper, which might conceal this group functionality, but it can also do so using a toolkit. The layers illustrated here are imaginary; some real layers are shown in Table 18.1. Horus supports many layers, but not all need be used in any particular stack: Shown here are two security layers (one for signing messages and one for encrypting their contents), which were not used for this particular application.

Technically, each Horus protocol block is a software module with a set of entry points for downcall and upcall procedures—for example, there is a downcall to send a message and an upcall to receive a message. Each layer is identified by an ASCII name and registers its upcall and downcall handlers at initialization time. There is a strong similarity between Horus protocol blocks and object classes in an object-oriented inheritance scheme, and readers may wish to think of protocol blocks as members of a class hierarchy.

TABLE 18.1. Microprotocols Available in Horus

Layer	Description
COM	The COM layer provides the Horus group interface to such low-level protocols as IP, UDP, and some ATM interfaces.
NAK	This layer implements a negative acknowledgment-based message retransmission protocol.
CYCLE	Multimedia message dissemination using Smith's cyclic UDP protocol
PARCLD	Hierarchical message dissemination (parent-child layer)
FRAG	Fragmentation and reassembly of large messages
MBRSHIP	This layer provides each member with a list of end points believed to be accessible. It runs a group membership consensus protocol to provide its users with a virtually synchronous execution model.
FC	Flow-control layer
TOTAL	Totally ordered message delivery
STABLE	This layer detects when a message has been delivered to all destination end points and can consequently be garbage collected.
CRYPT	Encryption and decryption of message body
MERGE	Location and merging of multiple group instances

To see how this works, consider the Horus *message_send* operation. It looks up the message send entry in the topmost block and invokes that function. This function may add a header to the message and will then typically invoke *message_send* again. This time, control passes to the message send function in the layer below it. This repeats itself recursively until the bottommost block is reached and invokes a driver to actually send the message.

The specific layers currently supported by Horus solve such problems as interfacing the system to varied communication transport mechanisms, overcoming lost packets, encryption and decryption, maintaining group membership, helping a process that joins a group obtain the state of the group, merging a group that has partitioned, flow control, and so forth. Horus also includes tools to assist in the development and debugging of new layers.

Each stack of blocks is carefully shielded from other stacks. It has its own prioritized threads and has controlled access to available memory through a mechanism called *memory channels*. Horus has a memory scheduler, which dynamically assigns the rate at which each stack can allocate memory, depending on availability and priority, so that no stack can monopolize the available memory. This is particularly important inside a kernel or if one of the stacks has soft real-time requirements.

Besides threads and memory channels, each stack deals with three other types of objects: end points, groups, and messages. The end-point object models the communicating entity. Depending on the application, it may correspond to a machine, a process, a thread, a socket, a port, and so forth. An end point has an address and can send and receive messages. However, as we will see later, messages are not addressed to end points, but to groups. The end-point address is used for membership purposes.

A *group object* is used to maintain the local protocol state on an end point. Associated with each group object is the *group address*, to which messages are sent, and a *view*: a list of destination end-point addresses believed to be accessible group members. Since a group object is purely local, Horus technically allows different end points to have different views of the same group. An end point may have multiple group objects, allowing it to communicate with different groups and views. A user can install new views when processes crash or recover and can use one of several membership protocols to reach some form of agreement on views between multiple group objects in the same group.

The message object is a local storage structure. Its interface includes operations to push and pop protocol headers. Messages are passed from layer to layer by passing a pointer and never need to be copied.

A thread at the bottommost layer waits for messages arriving on the network interface. When a message arrives, the bottommost layer (typically COM) pops off its header and passes the message on to the layer above it. This repeats itself recursively. If necessary, a layer may drop a message or buffer it for delayed delivery. When multiple messages arrive simultaneously, it may be important to enforce an order on the delivery of the messages. However, since each message is delivered using its own thread, this ordering may be lost, depending on the scheduling policies used by the thread scheduler. Therefore, Horus numbers the messages and uses *event count* synchronization variables (see Reed and Kanodia) to reconstruct the order where necessary.

18.3 Protocol Stacks

The microprotocol architecture of Horus would not be of great value unless the various classes of process group protocols we might wish to support could be simplified, by being expressed as stacks of layers; perform well; and share significant functionality. The experience with Horus in this regard has been very positive.

The stacks shown in Figure 18.3 all implement virtually synchronous process groups. The leftmost stack provides totally ordered, flow-controlled communication over the group membership abstraction. The layers FRAG, NAK, and COM, respectively, break large messages into smaller ones, overcome packet loss using negative acknowledgments, and interface Horus to the underlying transport protocols. The adjacent stack is similar, but provides weaker ordering and includes a layer supporting state transfer to a process joining a group or when groups merge after a network partition. To the right is a stack that supports scaling through a hierarchical structure, in which each parent process is responsible for a set of child processes. The dual stack illustrated in this case represents a feature whereby a message can be routed down one of several stacks, depending on the type of processing required. Additional protocol blocks provide functionality such as data encryption, packing small messages for efficient communication, isochronous communication (useful in multimedia systems), and so forth.

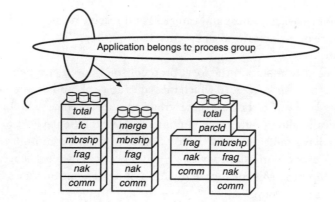

FIGURE 18.3. The Horus stacks are shielded from each other and have their own threads and memory, each of which is provided through a scheduler. Each stack can be thought of as a small program executing inside Horus. Although this feature is not shown, a stack can be split between the user's address space and the kernel, permitting the user to add customized features to a stack while benefiting from the performance of a kernel-based protocol implementation.

In order for Horus layers to fit like Lego blocks, they each must provide the same downcall and upcall interfaces. A lesson learned from the x-Kernel is that if the interface is not rich enough, extensive use will be made of general-purpose control operations (similar to *ioctl*), which reduce configuration flexibility. (Since the control operations are unique to a layer, the Lego blocks would not fit as easily.) The *Horus Common Protocol Interface* (HCPI), therefore supports an extensive interface, which supports all common operations in group communication systems, going beyond the functionality of earlier layered systems such as the x-Kernel. Furthermore, the HCPI is designed for multiprocessing and is completely asynchronous and reentrant.

Broadly, the HCPI interfaces fall into two categories. Those in the first group are concerned with sending and receiving messages and the stability of messages.[2] The second category of Horus operations is concerned with membership. In the down direction, it lets an application or layer control the group membership used by layers below it. As upcalls, these report membership changes, communication problems, and other related events to the application.

While supporting the same HCPI, each Horus layer runs a different protocol—each implementing a different property. Although Horus allows layers to be stacked in any order (and even multiple times), most layers require certain semantics from layers below them, imposing a partial order on the stacking. These constraints have been tabulated. Given information about the

---

[2]It is common to say that a message is *stable* when processing has completed and associated information can be garbage collected. Horus standardizes the handling of stability information, but leaves the actual semantics of stability to the user. Thus, an application for which stability means "logged to disk" can share this Horus functionality with an application for which stability means "displayed on the screen."

properties of the network transport service, and the properties provided by the application, it is often possible to automatically generate a minimal protocol stack to achieve a desired property.

Layered protocol architectures sometimes perform poorly. Traditional layered systems impose an order on which protocols process messages, limiting opportunities for optimization and imposing excessive overhead. Clark and Tennenhouse have suggested that the key to good performance rests in *Integrated Layer Processing* (ILP) (see Abbott and Peterson, Braun and Diot, Clark and Tennenhouse, Karamcheti and Chien, Kay and Pasquale). Systems based on the ILP principle avoid interlayer ordering constraints and can perform as well as monolithically structured systems. Horus is consistent with ILP: There are no intrinsic ordering constraints on processing, so unnecessary synchronization delays are avoided. Moreover, as we will see, Horus supports an optional protocol accelerator, which greatly improves the performance of the layered protocols making use of it.

18.4 Using Horus to Build a Robust Groupware Application

Earlier, we commented that Horus can be hidden behind standard application programmer interfaces. A good illustration of how this is done occurred when we interfaced the TCL/TK graphical programming language to Horus. A challenge posed by running systems such as Horus side by side with a package such as X Windows or TCL/TK is that such packages are rarely designed with threads or Horus communication stacks in mind. To avoid a complex integration task, we therefore chose to run TCL/TK as a separate thread in an address space shared with Horus. Horus intercepts certain system calls issued by TCL/TK, such as the UNIX *open* and *socket* system calls. We call this resulting mechanism an *intercept proxy*; it is a special type of wrapper oriented toward intercepting this type of system call. The proxy redirects the system calls, invoking Horus functions, which will create Horus process groups and register appropriate protocol stacks at run time. Subsequent I/O operations on these group I/O sockets are mapped to Horus communication functions.

To make Horus accessible within TCL applications, two new functions were registered with the TCL interpreter. One creates end-point objects, and the other creates group addresses. The end-point object itself can create a group object using a group address. Group objects are used to send and receive messages. Received messages result in calls to TCL code that typically interpret the message as a TCL command. This yields a powerful framework: a distributed, fault-tolerant, whiteboard application can be built using only eight short lines of TCL code over a Horus stack of seven protocols.

To validate our approach, we ported a sophisticated TCL/TK application to Horus. The Continuous Media Toolkit (CMT) (see Rowe and Smith) is a TCL/TK extension providing objects that read or output audio and video data. These objects can be linked together in pipelines and are synchronized by a *logical timestamp* object. This object may be set to run slower or faster than the real clock or even backwards. This allows stop, slow motion, fast forward, and rewind functions to be implemented.

Architecturally, CMT consists of a multimedia server process, which multicasts video and audio to a set of clients. We decided to replicate the server using a primary-backup approach, where the backup servers stand by to back up failed or slow primaries.

The original CMT implementation depends on extensions to TCL/TK. These implement a master-slave relationship between the machines, provide for a form of logical timestamp synchronization between them, and support a real-time communication protocol called Cyclic UDP. The Cyclic UDP implementation consists of two halves: a sink object, which accepts multimedia data from another CMT object, and a source object, which produces multimedia data and passes it on to another CMT object (see Figure 18.4a). The resulting system is distributed but intolerant of failures and does not allow for multicast.

By using Horus, it was straightforward to extend CMT with fault tolerance and multicast capabilities. Five Horus stacks were required. One of these is hidden from the application and implements a clock synchronization protocol (see Cristian [1989]). It uses a Horus layer called

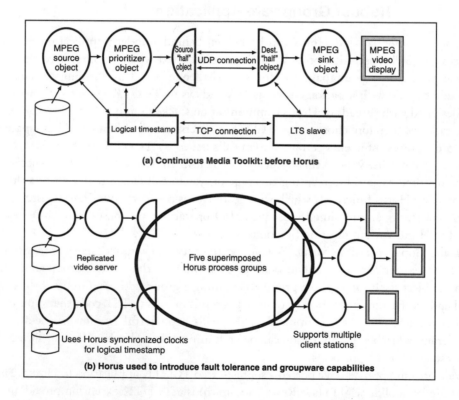

(a) Continuous Media Toolkit: before Horus

(b) Horus used to introduce fault tolerance and groupware capabilities

FIGURE 18.4. This illustrates an example of a video service implemented using the Continuous Media Toolkit. MPEG is a video compression standard. In (a), a standard, fault-intolerant setup is depicted. In (b), Horus was used to implement a fault-tolerant version that is also able to multicast to a set of clients.

MERGE to ensure that the different machines will find each other automatically (even after network partitions), and it employs the virtual synchrony property to rank the processes, assigning the lowest-ranked machine to maintain a master clock on behalf of the others. The second stack synchronizes the speeds and offsets with respect to real time of the logical timestamp objects. To keep these values consistent, it is necessary that they be updated in the same order. Therefore, this stack is similar to the previous one, but includes a Horus protocol block, which places a total order on multicast messages delivered within the group.[3] The third stack tracks the list of servers and clients. Using a deterministic rule based on the process ranking maintained by the virtual synchrony layer, one server decides to multicast the video, and one server, usually the same, decides to multicast the audio. This setup is shown in Figure 18.4b.

To disseminate the multimedia data, we used two identical stacks—one for audio and one for video. The key component in these is a protocol block, which implements a multimedia generalization of the cyclic UDP protocol. The algorithm is similar to FRAG, but it will reassemble messages arriving out of order and drop messages with missing fragments.

One might expect that a huge amount of recoding would have been required to accomplish these changes. However, all the necessary work was completed using 42 lines of TCL code. An additional 160 lines of C code support the CMT frame buffers in Horus. Two new Horus layers were needed, but were developed by adapting existing layers; they consist of 1,800 lines of C code and 300 lines of TCL code, respectively (ignoring the comments and lines common to all layers). Thus, with relatively little effort and little code, a complex application written with no expectation that process group computing might later be valuable was modified to exploit Horus functionality.

18.5   Using Horus to Harden CORBA Applications

The introduction of process groups into CMT required sophistication with Horus and its intercept proxies. Many potential users would lack the sophistication and knowledge of Horus required to do this; hence, we recognized a need for a way to introduce Horus functionality in a more transparent way. This goal evokes an image of plug-and-play robustness; it leads one to think in terms of an object-oriented approach to group computing.

Early in this book, we looked at CORBA, noting that object-oriented distributed applications that comply with the CORBA ORB specification and support the IOP protocol can invoke one another's methods with relative ease. Our work resulted in a CORBA-compliant interface to Horus, which we call Electra (see Maffeis). Electra can be used without Horus, and vice versa, but the combination represents a more complete system.

---

[3]This protocol differs from the *Total* protocol in the Trans/Total (see Moser et al. [1996]) project in that the Horus protocol only rotates the token among the current set of senders, while the Trans/Total protocol rotates the token among all members.

In Electra, applications are provided with ways to build Horus process groups and to directly exploit the virtual synchrony model. Moreover, Electra objects can be aggregated to form object groups, and object references can be bound to both singleton objects and object groups. An implication of the interoperability of CORBA implementations is that Electra object groups can be invoked from *any* CORBA-compliant distributed application, regardless of the CORBA platform on which it is running, without special provisions for group communication. This means that a service can be made fault tolerant without changing its clients.

When a method invocation occurs within Electra, object-group references are detected and transformed into multicasts to the member objects (see Figure 18.5). Requests can be issued either in transparent mode, where only the first arriving member reply is returned to the client application, or in nontransparent mode, permitting the client to access the full set of responses from individual group members. The transparent mode is used by clients to communicate with replicated CORBA objects, while the nontransparent mode is employed with object groups whose members perform different tasks. Clients submit a request either in a synchronous, asynchronous, or deferred-synchronous way.

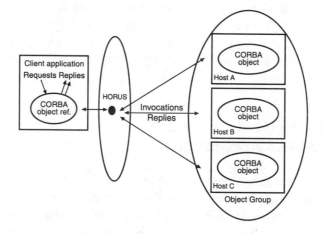

FIGURE 18.5. Object-group communication in Electra, a CORBA-compliant ORB, which uses Horus to implement group multicast. The invocation method can be changed depending on the intended use. Orbix+Isis and the COOL-ORB are examples of commercial products that support object groups.

The integration of Horus into Electra shows that group programming can be provided in a natural, transparent way with popular programming methodologies. The resulting technology permits the user to plug in group communication tools anywhere that a CORBA application has a suitable interface. To the degree that process group computing interfaces and abstractions represent an impediment to their use in commercial software, technologies such as Electra suggest a possible middle ground, in which fault tolerance, security, and other group-based mechanisms can be introduced late in the design cycle of a sophisticated distributed application.

18.6 Basic Performance of Horus

A major concern of the Horus architecture is the overhead of layering—hence, we now focus on this issue. This section presents the overall performance of Horus on a system of Sun SPARC10 workstations running SunOS 4.1.3, communicating through a loaded Ethernet. We used two network transport protocols: normal UDP and UDP with the Deering IP multicast extensions (see Deering [1988]) (shown as "Deering").

To highlight some of the performance numbers: Horus achieves a one-way latency of 1.2 ms over an unordered virtual synchrony stack (over ATM, it is currently 0.7 ms) and, using a totally ordered layer over the same stack, 7,500 one-byte messages per second. Given an application that can accept lists of messages in a single receive operation, we can drive up the total number of messages per second to over 75,000 using the FC flow-control layer, which buffers heavily using the message list capabilities of Horus (see Friedman and van Renesse [July 1995]). Horus easily reached the Ethernet 1,007 KB/sec maximum bandwidth with a message size smaller than 1 KB.

The performance test program has each member do exactly the same thing: Send k messages and wait for $k * (n - 1)$ messages of size s, where s is the number of members. This way we simulate an application that imposes a high load on the system while occasionally synchronizing on intermediate results.

Figure 18.6 depicts the one-way communication latency of one-byte Horus messages. As can be seen, hardware multicast is a big win, especially when the message size goes up. In this figure, we compare FIFO to totally ordered communication. For small messages we get a FIFO one-way latency of about 1.5 ms and a totally ordered one-way latency of about 6.7 ms. A problem with the totally ordered layer is that it can be inefficient when senders send single messages at random, with a high degree of concurrent sending by different group members. With just one sender, the one-way latency drops to 1.6 ms.

Figure 18.7 shows the number of one-byte messages per second that can be achieved for three cases. For normal UDP and Deering UDP the throughput is fairly constant. For totally ordered communication we see that the throughput becomes better if we send more messages per round (because of increased concurrency). Perhaps surprisingly, the throughput also becomes better as the number of members in the group goes up. The reason for this is threefold. First, with more members there are more senders. Second, with more members it takes longer to order messages, and thus more messages can be packed together and sent out in single network packets. Third, the ordering protocol allows only one sender on the network at a time, thus introducing flow control and reducing collisions.

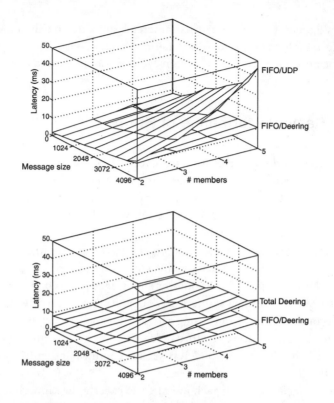

FIGURE 18.6. The top graph of this figure compares the one-way latency of one-byte FIFO Horus messages over straight UDP and UDP with the Deering IP multicast extensions. The bottom graph of the figure compares the performance of total and FIFO order of Horus, both over UDP multicast.

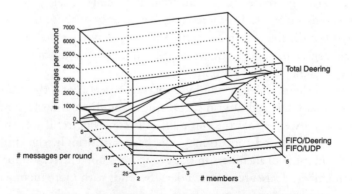

FIGURE 18.7. This graph depicts the message throughput for virtually synchronous, FIFO-ordered communication over normal UDP and Deering UDP, as well as for totally ordering communication over Deering UDP.

18.7 Masking the Overhead of Protocol Layering

Although layering of protocols can be advocated as a way of dealing with the complexity of computer communication, it is also criticized for its performance overhead. Recent work by van Renesse has yielded considerable insight regarding the design of protocols, which van Renesse uses to mask the overhead of layering in Horus. The fundamental idea is very similar to client caching in a file system. With these new techniques, he achieves an order of magnitude improvement in end-to-end message latency in the Horus communication framework, compared to the best latency possible using Horus without these optimizations. Over an ATM network, the approach permits applications to send and deliver messages of varying levels of semantics in about 85μs, using a protocol stack written in ML, an interpreted functional language. In contrast, the performance figures given in the previous section were for a version of Horus coded in C—carefully optimized by hand but without use of the protocol accelerator.

Having presented this material in seminars, I have noticed that the systems community seems to respond to the very mention of the ML language with skepticism, and it is perhaps appropriate to comment on this before continuing. First, the reader should keep in mind that a technology such as Horus is simply a tool used to harden a system. It makes little difference whether such a tool is internally coded in C, assembly language, LISP, or ML if it works well for the desired purpose. The decision to work with a version of Horus coded in ML is not one that would impact the *use* of Horus in applications that work with the technology through wrappers or toolkit interfaces. However, as we will see here and in Chapter 25, it does bring some important benefits to Horus itself, notably the potential for us to harden the system using formal software analysis tools. Moreover, although ML is often viewed as obscure and of academic interest only, the version of ML used in our work on Horus is not really so different from LISP or C++ once one becomes accustomed to the syntax. Finally, as we will see here, the performance of Horus coded in ML is actually better than that of Horus coded in C, at least for certain patterns of communication. Thus, we would hope that the reader will recognize that the work reported here is in fact very practical.

As we saw in earlier chapters, modern network technology allows for very low latency communication—for example, the U-Net (see von Eicken et al. [1995]) interface to ATM achieves 75 μs round-trip communication as long as the message is 40 bytes or smaller. On the other hand, if a message is larger, it will not fit in a single ATM cell, significantly increasing the latency. This points to two basic concerns: first, that systems such as Horus need to be designed to take full advantage of the potential performance of current communication technology, and, second, that to do so, it will be important that Horus protocols use small headers and introduce minimal processing overhead.

Unfortunately, these properties are not typical of the protocol layers needed to implement virtual synchrony. Many of these protocols are complex, and layering introduces additional overhead of its own. One source of overhead is interfacing: crossing a layer costs some CPU cycles. The other is header overhead. Each layer uses its own header, which is prepended to every message and usually padded so that each header is aligned on a four- or eight-byte boundary. Combining this with a trend to very large addresses (of which at least two per message are needed), it is impossible to have the total amount of header space be less than 40 bytes.

The Horus Protocol Accelerator (Horus PA) eliminates these overheads almost entirely and offers the potential of one to three orders of magnitude of latency improvement over the protocol implementations described in the previous section—for example, we looked at the impact of the Horus PA on an ML (see Milner et al.) implementation of a protocol stack with five layers. The ML code is interpreted (although in the future it will be compiled) and is therefore relatively slow compared to compiled C code. Nevertheless, between two SunOS user processes on two SPARC20s connected by a 155 MB/sec ATM network, the Horus PA permits these layers to achieve a round-trip latency of 175 μs, down from about 1.5 ms in the original Horus system (written in C).

The Horus PA achieves its results using three techniques. First, message header fields that never change are only sent once. Second, the rest of the header information is carefully packed, ignoring layer boundaries, typically leading to headers that are much smaller than 40 bytes and thus leaving room to fit a small message within a single U-Net packet. Third, a semiautomatic transformation is done on the send and delivery operations, splitting them into two parts: one that updates or checks the header but not the protocol state, and the other vice versa. The first part is then executed by a special packet filter (both in the send and the delivery path) to circumvent the actual protocol layers whenever possible. The second part is executed, as much as possible, when the application is idle or blocked.

18.7.1 Reducing Header Overhead

In traditional layered protocol systems, each protocol layer designs its own header data structure. The headers are concatenated and prepended to each user message. For convenience, each header is aligned to a four- or eight-byte boundary to allow easy access. In systems such as the x-Kernel or Horus, where many simple protocols may be stacked on top of each other, this may lead to extensive padding overhead.

Some fields in the headers, such as the source and destination addresses, never change from message to message. Yet, instead of agreeing on these values, they are frequently included in every message and used as the identifier of the connection to the peer. Since addresses tend to be large (and they are getting larger to deal with the rapid growth of the Internet), this results in significant use of space for what are essentially constants of the connection. Moreover, notice that the connection itself may already be identifiable from other information. On an ATM network, connections are named by a small four-byte VPI/VCI pair, and every packet carries this information. Thus, constants such as sender and destination addresses are implied by the connection identifier and including them in the header is superfluous.

The Horus PA exploits these observations to reduce header sizes to a bare minimum. The approach starts by dividing header fields into four *classes*:

▶ *Connection identification:* Fields that never change during the period of a connection, such as sender and destination.

- *Protocol-specific information:* Fields that are important for the correct delivery of the particular message frame. Examples are the sequence number of a message or the message type (Horus messages have types, such as "data," "ack," or "nack"). These fields must be deterministically implied by the protocol state—not on the message contents or the time at which it was sent.

- *Message-specific information:* Fields that need to accompany the message, such as the message length and checksum or a timestamp. Typically, such information depends only on the message—not on the protocol state.

- *Gossip:* Fields that technically do not need to accompany the message but are included for efficiency.

Each layer is expected to declare the header fields that it will use during initialization, and it subsequently accesses fields using a collection of highly optimized functions implemented by the Horus PA. These functions extract values directly from headers, if they are present, or otherwise compute the appropriate field value and return that instead. This permits the Horus PA to precompute header templates that have optimized layouts, with a minimum of wasted space.

Horus includes the protocol-specific and message-specific information in every message. Currently, although not technically necessary, gossip information is also included, since it is usually small. However, since the connection identification fields never change, they are only included occasionally, since they tend to be large.

A 64-bit miniheader is placed on each message to indicate which headers it actually includes. Two bits of this are used to indicate whether or not the connection identification is present in the message and to designate the byte ordering for bytes in the message. The remaining 62 bits are a *connection cookie*, which is a magic number established in the connection identification header, selected randomly, to identify the connection.

The idea is that the first message sent over a connection will be a connection identifier, specifying the cookie to use and providing an initial copy of the connection identification fields. Subsequent messages need only contain the identification field if it has changed. Since the connection identification fields tend to include very large identifiers, this mechanism reduces the amount of header space in the normal case significantly—for example, in the version of Horus that van Renesse used in his tests, the connection identification typically occupies about 76 bytes.

18.7.2 Eliminating Layered Protocol Processing Overhead

In most protocol implementations, layered or not, a great deal of processing must be done between the application's send operation and the time that the message is actually sent out onto the network. The same is true between the arrival of a message and the delivery to the application. The Horus PA reduces the length of the critical path by updating the protocol state only after a message has been sent or delivered and by precomputing any statically predictable protocol-specific header fields, so that the necessary values will be known *before* the application generates the next message (Figure 18.8). These methods work because the protocol-specific information for most messages can be predicted (calculated) before the message is sent or delivered. (Recall that, as noted above, such information must not depend on the message contents or the time on which it was sent.) Each connection maintains a predicted protocol-specific header for the next send operation and another for the next delivery (much like a read-ahead strategy in a file system). For sending, the gossip information can be predicted as well, since this does not depend on the message contents. The idea is a bit like that of prefetching in a file system.

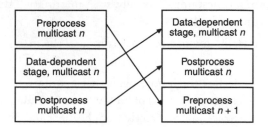

FIGURE 18.8. Restructuring a protocol layer to reduce the critical path. By moving data-dependent code to the front, delays for sending the next message are minimized. Postprocessing of the current multicast and preprocessing of the next multicast (all computation that can be done before seeing the actual contents of the message) are shifted to occur after the current multicast has been sent and hence concurrently with application-level computing.

Thus, when a message is actually sent, only the message-specific header will need to be generated. This is done using a *packet filter* (see Mogul et al.), which is constructed at the time of layer initialization. Packet filters are programmed using a simple programming language (a dialect of ML), and they operate by extracting from the message information needed to form the message-specific header. A filter can also hand-off a message to the associated layer for special handling—for example, if a message fails to satisfy some assumption that was used in predicting the protocol-specific header. In the usual case, the message-specific header will be computed, other headers prepended from the precomputed versions, and the message transmitted with no additional delay. Because the header fields have fixed and precomputed sizes, a header template can be filled in with no copying, and scatter-send/scatter-gather hardware can be used to transmit the header and message as a single packet without copying them first to a single place. This reduces the computational cost of sending or delivering a message to a bare minimum, although it leaves some background costs in the form of prediction code, which must be executed before the next message is sent or delivered.

18.7.3 Message Packing

The Horus PA as described so far will reduce the latency of individual messages significantly, but only if they are spaced out far enough to allow time for postprocessing. If not, messages will have to wait until the postprocessing of every previous message completes (somewhat like a process that reads file system records faster than they can be prefetched). To reduce this overhead, the Horus PA uses *message packing* (see Friedman and van Renesse [July 1995]) to deal with backlogs. The idea is a very simple one. After the postprocessing of a send operation completes, the PA checks to see if there are messages waiting. If there are more than one, the PA will pack these messages together into a single message. The single message is now processed in the usual way, which takes only one preprocessing and postprocessing phase. When the packed message is ready for delivery, it is unpacked and the messages are individually delivered to the application.

Returning to our file system analogy, the approach is similar to one in which the application could indicate that it plans to read three 1 KB data blocks. Rather than fetching them one by one, the file system can now fetch them all at the same time. Doing so amortizes the overhead associated with fetching the blocks, permitting better utilization of network bandwidth.

18.7.4 Performance of Horus with the Protocol Accelerator

The Horus PA dramatically improved the performance of the system over the base figures described earlier (which were themselves comparable to the best performance figures cited for other systems). With the accelerator, one-way latencies dropped to as little as 85µs (compared to 35µs for the U-Net implementation over which the accelerator was tested). As many as 85,000 one-byte messages could be sent and delivered per second over a protocol stack of five layers implementing the virtual synchrony model within a group of two members. For RPC-style interactions, 2,600 round trips per second were achieved. These latency figures, however, represent a best-case scenario in which the frequency of messages was low enough to permit the predictive mechanisms to operate; when they become overloaded, latency increases to about 425µs for the same test pattern. This points to a strong dependency of the method on the speed of the code used to implement layers.

van Renesse's work on the Horus PA made use of a version of the ML programming language which was interpreted, not compiled. ML turns out to be a very useful language for specifying Horus layers: It lends itself to formal analysis and permits packet filters to actually be constructed at run time; moreover, the programming model is well matched to the functional style of programming used to implement Horus layers. ML compiler technology is rapidly evolving, and when the Horus PA is moved to a compiled version of ML the sustainable load should rise and these maximum latency figures drop.

The Horus PA does suffer from some limitations. Message fragmentation and reassembly is not supported by the PA—hence, the preprocessing of large messages must be handled explicitly by the protocol stack. Some technical complications result from this design decision, but it reduces the complexity of the PA and improves the maximum performance achievable using it. A second limitation is that the PA must be used by all parties to a communication stack. However, this is not an unreasonable restriction, since Horus has the same sort of limitation with regard to the stacks themselves (all members of a group must use identical or at least compatible protocol stacks).

18.8 Scalability

Up to the present, this book has largely overlooked issues associated with protocol scalability. Although a serious treatment of scalability in the general sense might require a whole book in itself, the purpose of this section is to set out some general remarks on the subject, as we have approached it in the Horus project. It is perhaps worthwhile to comment that, overall, surprisingly little is known about scaling reliable distributed systems.

If one looks at the scalability of Horus protocols, as we did earlier in presenting some basic Horus performance figures, it is clear that Horus performs well for groups with small numbers of members and for moderately large groups when IP multicast is available as a hardware tool to reduce the cost of moving large volumes of data to large numbers of destinations. Yet although these graphs are correct, they may be misleading. In fact, as systems such as Horus are scaled to larger and larger numbers of participating processes, they experience steadily growing overheads in the form of acknowledgments and negative acknowledgments from the recipient processes to the senders. A consequence is that if these systems are used with very large numbers of participating processes, the backflow associated with these types of messages and with flow control becomes a serious problem.

A simple thought experiment suffices to illustrate that there are probably fundamental limits on reliability in very large networks. Suppose that a communication network is extremely reliable, but that the processes using it are designed to distrust that network and to assume that it may actually malfunction by losing messages. Moreover, assume that these processes are in fact closely rate-matched (the consumers of data keep up with the producers), but again that the system is designed to deal with individual processes that lag far behind. Now, were it not for the backflow of messages to the senders, this hypothetical system might perform very well near the limits of the hardware. It could potentially be scaled just by adding new recipient processes and, with no changes at all, continue to provide a high level of reliability.

However, the backflow messages will substantially impact this simple and rosy scenario. They represent a source of overhead, and, in the case of flow-control messages, if they are not received, the sender may be forced to stop and wait for them. Now, the performance of the sender side is coupled to the timely and reliable reception of backflow messages, and, as we scale the number of recipients connected to the system, we can anticipate a traffic jam phenomenon at the sender's interface (protocol designers call this an acknowledgment "implosion"), which will cause traffic to get increasingly bursty and performance to drop. In effect, the attempt to protect against the mere risk of data loss or flow-control mismatches is likely to slash the maximum achievable performance of the system. Now, obtaining a stable delivery of data near the limits of our technology will become a tremendously difficult juggling problem, in which the protocol developer must trade the transmission of backflow messages against their performance impact.

Graduate students Guerney Hunt and Michael Kalantar have studied aspects of this problem in their Ph.D. dissertations at Cornell University—both using special-purpose experimental tools (i.e., neither actually experimented on Horus or a similar system; Kalantar, in fact, worked mostly with a simulator). Hunt's work was on flow control in a very large scale system. He concluded that most forms of backflow were unworkable on a large scale, and he ultimately proposed a rate-based flow-control scheme in which the sender limits the transmission rate for data to match what the receivers can accommodate (see Hunt). Kalantar looked at the impact of multicast ordering on latency, asking how frequently an ordering property such as causal or total ordering would significantly impact the latency of message delivery (see Kalantar). He found that although ordering had a fairly small impact on latency, there were other, much more important, phenomena that represented serious potential concerns.

In particular, Kalantar discovered that as he scaled the size of his simulation, message latencies tended to become unstable and bursty. He hypothesized that in large-scale protocols, the domain of stable performance becomes smaller and smaller. In such situations, a slight perturbation of the overall system—for example, because of a lost message—could cause much of the remainder of the system to block due to reliability or ordering constraints. Now, the system would shift into what is sometimes called a *convoy* behavior, in which long message backlogs build up and are never really eliminated; they may shift from place to place, but stable, smooth delivery is generally not restored. In effect, a bursty scheduling behavior represents a more stable configuration of the overall system than one in which message delivery is extremely regular and smooth, at least if the number of recipients is large and the presented load is a substantial percentage of the maximum achievable (so that there is little slack bandwidth with which the system can catch up after an overload develops).

Hunt and Kalantar's observations are not really surprising ones. It makes sense that it should be easy to provide reliability or ordering when far from the saturation point of the hardware and much harder to do so as the communication or processor speed limits are approached.

Over many years of working with Isis and Horus, I have gained considerable experience with these sorts of scaling and flow-control problems. Realistically, the conclusion can only be called a mixed one. On the positive side, it seems that one can easily build a reliable system if the communication load is not expected to exceed, say, 20 percent of the capacity of the hardware. With a little luck, one can even push this to as high as perhaps 40 percent of the hardware. (Happily, hardware is becoming so fast that this may still represent a very satisfactory level of performance far into the future!)

However, as the load presented to the system rises beyond this threshold, or if the number of destinations for a typical message becomes very large (hundreds), it becomes increasingly difficult to guarantee reliability and flow control. A fundamental tradeoff seems to be present: One can send data and hope that these data will arrive, and, by doing so, one may be able to operate quite reliably near the limits of the hardware. But, of course, if a process falls behind, it may lose large numbers of messages before it recovers, and no mechanism is provided to let it recover these messages from any form of backup storage. On the other hand, one can operate in a less-demanding performance range and in this case provide reliability, ordering, and performance guarantees. In between the two, however, lies a domain that is extremely difficult in an engineering sense and often requires a very high level of software complexity, which will necessarily reduce reliability. Moreover, one can raise serious questions about the stability of message-passing systems that operate in this intermediate domain, where the load presented is near the limits of what can be accomplished. The typical experience with such systems is that they perform well, most of the time, but once something fails, the system falls so far behind that it can never again catch up—in effect, any perturbation can shift such a system into the domain of overloads and hopeless backlogs.

Where does Horus position itself in this spectrum? Although the performance data shown earlier may suggest that the system seeks to provide scalable reliability, it is more likely that successful Horus applications will seek one property or the other, but not both at once or at least not both when performance is demanding. In Horus, this is done by using multiple protocol stacks, in which the protocol stacks providing strong properties are used much less frequently, while the protocol stacks providing weaker reliability properties may be used for high-volume communication.

As an example, suppose that Horus were to be used to build a stock trading system. It might be very important to ensure that certain classes of trading information will reach all clients, and, for this sort of information, a stack with strong reliability properties could be used. But as a general rule, the majority of communication in such systems will be in the form of bid/offered pricing, which may not need to be delivered quite so reliably: If a price quote is dropped, the loss won't be serious as long as the next quote has a good probability of getting through. Thus, one can visualize such a system as having two superimposed architectures: one that has much less traffic and much stronger reliability requirements and a second one with much greater traffic but weaker properties. We saw a similar structure in the Horus application to the CMT system: Here, the stronger logical properties were reserved for coordination, timestamp generation, and agreement on such data as system membership. The actual flow of video data was through a protocol stack with very different properties: stronger temporal guarantees, but weaker reliability properties. In building scalable reliable systems, such tradeoffs may be intrinsic.

In general, this leads to a number of interesting problems having to do with the synchronization and ordering of data when multiple communication streams are involved. Researchers at the Hebrew University in Jerusalem, working with a system similar to Horus called Transis (and with Horus itself), have begun to investigate this issue. Their work on providing strong communication semantics in applications that mix multiple quality-of-service properties at the transport level promises to make such multiprotocol systems more and more manageable and controlled (see Chockler et al.).

More broadly, it seems likely that one could develop a theoretical argument to the effect that reliability properties are fundamentally at odds with high performance. While one can scale reliable systems, they appear to be intrinsically unstable if the result of the scaling is to push the overall system anywhere close to the maximum performance of the technology used. Perhaps some future effort to model these classes of systems will reveal the basic reasons for this relationship and point to classes of protocols that degrade gracefully while remaining stable under steadily increasing scale and load. Until then, however, the heuristic I recommend is to scale systems, by all means, but to be extremely careful not to expect the highest levels of reliability, performance, and scale simultaneously. To do so is to move beyond the limits of problems we know how to solve and perhaps to expect the impossible. Instead, the most demanding systems must somehow be split into subsystems that demand high performance but can manage with weaker reliability properties and subsystems that need reliability but will not be subjected to extreme performance demands.

18.9   Related Reading

Chapter 26 includes a review of related research activities, which we will not duplicate here.

On the Horus system: (see Birman and van Renesse [1996], Friedman and van Renesse [July 1995], van Renesse et al.).

On Horus used in a real-time telephone switching application: (see Friedman and Birman).

On virtual fault tolerance: (see Bressoud and Schneider).

On layered protocols: (see Abbott and Peterson, Braun and Diot, Clark and Tennenhouse, Karamcheti and Chien, Kay and Pasquale).

On event counters: (see Reed and Kanodia).

On the Continuous Media Toolkit: (see Rowe and Smith).

On U-Net (see von Eicken et al. [1995]).

On packet filters (in Mach): (see Mogul et al.).

Chapter 25 discusses verification of the Horus protocols in more detail; this chapter focuses on the same ML implementation of Horus to which the Protocol Accelerator was applied.

CHAPTER 19 ✧ ✧ ✧ ✧

Security Options for Distributed Settings

CONTENTS

19.1   Security Options for Distributed Settings

The use of distributed computing systems for storage of sensitive data and in commercial applications has created significant pressure to improve the security options available to software developers. Yet distributed system security has many possible interpretations, corresponding to very different forms of guarantees, and even the contemporary distributed systems that claim to be secure often suffer from basic security weaknesses. In this chapter we will review some of the major security technologies, look at the nature of their guarantees and their limitations and discuss some of the issues raised when we require that a security system also guarantee high availability.

The technologies we consider here span a range of approaches. At the low end of the spectrum are firewall technologies and other *perimeter defense mechanisms*, which operate by restricting access or communication across specified system boundaries. These technologies are extremely popular but very limited in their capabilities. In particular, once an intruder has found a way to work around the firewall or log into the system, the protection benefit is lost.

Internal to a distributed system one typically finds *access control mechanisms*, which are often based on the UNIX model of user and group IDs employed to limit access to shared resources such as file systems. When these are used in stateless settings, serious problems occur, which we will discuss here and will return to later in Chapter 23. Access control mechanisms rarely extend to communication, and this is perhaps their most serious security exposure. In fact, many communication systems are open to attack by a clever intruder able to guess what port numbers will be used by the protocols within the system: Secrecy of port numbers is a common security dependency in modern distributed software.

Stateful protection mechanisms operate by maintaining strong concepts of session and channel state and authenticating use at the time that communication sessions are established. These schemes adopt the approach that after a user has been validated, the difficulty of breaking into the user's session will represent an obstacle to intrusion.

Authentication-based security systems employ some scheme to authenticate the user running each application; the method may be highly reliable or less so, depending on the setting (see Denning, Needham and Schroeder). Individual communication sessions are then protected using some form of key, which is negotiated using a trusted agent. Messages may be encrypted or signed in this approach, resulting in very strong security guarantees. However, the costs of the overall approach can also be high, because of the intrinsically high costs of data encryption and signature schemes. Moreover, such methods may involve nontrivial modifications of the application programs being used, and they may be unsuitable for embedded settings in which no user would be available to periodically enter passwords or other authentication data. The best-known system of this sort is Kerberos, developed by MIT's Project Athena, and our review will focus on the approaches used in that system (see Schiller, Steiner et al.). (See Figure 19.1.)

Multilevel distributed system security architectures are based on a government security standard developed in the mid-1980s. This security model is very strong, but it has proven to be difficult to implement and it requires extensive effort on the part of application developers. Perhaps for

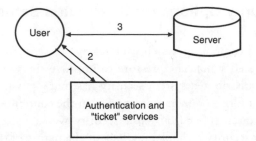

FIGURE 19.1. MIT's Project Athena developed the Kerberos security architecture. Kerberos or a similar mechanism is found at the core of many distributed system security technologies today. In this approach, an authentication service is used as a trusted intermediary to create secure channels, using DES encryption for security. During step 1, the user employs a password as a DES key to request that a connection be established to the remote server. The authentication server, which knows the user's password, constructs a session key, which is sent back in duplicated form—one copy by the user and one encrypted with the server's secret key (step 2). The session key is now used between the user and server (step 3), providing the server with trusted information about user identification. In practice, Kerberos avoids the need to keep user passwords around by trading the user's password for a session to the "ticket granting service," which then acts as the user's proxy in establishing connections to necessary servers, but the idea is unchanged. Kerberos session keys expire and must be periodically renewed—hence, even if an intruder gains physical access to the user's machine, the period during which illicit actions are possible is limited.

these reasons, this approach has not been widely successful. Moreover, the pressure to use off-the-shelf technologies has made it difficult even for the government to build systems that enforce multilevel security.

Traditional security technologies have not considered availability when failures occur, creating exposure to attacks whereby critical system components are shut down, overloaded, or partitioned away from application programs that depend upon them. Recent research has begun to address these concerns, resulting in a new generation of highly available security technologies. However, when one considers failures in the context of a security subsystem, the benign failure models of earlier chapters must be called into question. Thus, work in this area has included a reexamination of Byzantine failure models, questioning whether extremely robust authentication servers can be built that will remain available even if Byzantine failures occur. Progress in this direction has been encouraging, as has work on using process groups to provide security guarantees that go beyond those available in a single server.

In the future, technologies supporting digital cash and digital commerce are likely to be of increasing importance and will often depend upon the use of trusted banking agents and strong forms of encryption, such as the RSA or DES standards (see Desmedt, Diffie and Hellman, Rivest et al.). Progress in this area has been very rapid and we will review some of the major approaches.

Yet, if the progress in distributed system security has been impressive, the limitations on such systems remain quite serious. On the whole, it remains difficult to secure a distributed system and very hard to add security to a technology that already exists and must be treated as a form of black box. The best-known technologies, such as Kerberos, are still used only sporadically. This makes it hard to implement customized security mechanisms and leaves the average distributed system quite open to attack. Break-ins and security violations are extremely common in the most standard distributed computing environments, and there seems to be at best a shallow commitment by the major software vendors to improve the security of their basic product lines. These observations raise troubling questions about the security to be expected from the emerging generation of extremely critical distributed systems, many of which will be implemented using standard software solutions on standard platforms. Until distributed system security is difficult to *disable*, as opposed to being difficult to enable, we may continue to read about intrusions of increasingly serious nature, and we will continue to be at risk for serious intrusions into our personal medical records, banking and financial systems, and personal computing environments.

19.2 Perimeter Defense Technologies

It is common to protect a distributed system by erecting barriers around it. Examples include the password control associated with dial-in ports; dial-back mechanisms, which some systems use to restrict access to a set of predesignated telephone numbers; and firewalls through which incoming and outgoing messages must pass. Each of these technologies has important limitations.

Password control systems are subject to attack by password-guessing mechanisms and by intruders who find ways to capture packets containing passwords as they are transmitted over the Internet or some other external networking technology. So-called password "sniffers" became a serious threat to system security in the mid-1990s and illustrate that the general Internet is not the benign environment it was in the early days of distributed computing, when most Internet users knew each other by name. Typical sniffers operate by exhibiting an IP address for some other legitimate machine on the network or by placing their network interfaces into a special mode, in which all passing packets will be accepted. They then scan the traffic captured for packets that might have originated in a log-in sequence. With a bit of knowledge about how such packets normally look, it is not hard to reliably capture passwords as they are routed through the Internet. Sniffers have also been used to capture credit card information and to break into e-mail correspondence.

Dial-up systems are often perceived as being more secure than direct network connections, but this is not necessarily the case. The major problem is that many systems use their dial-up connections for data and file transfer and as a sending and receiving point for fax communications—hence, the corresponding telephone numbers are stored in various standard data files, often with connection information. An intruder who breaks into one system may in this manner learn dial-up numbers for other systems and may even find log ins and passwords, which will make it easy to break in to them, as well. Moreover, the telephone system itself is increasingly complex and, as an unavoidable side-effect, increasingly vulnerable to intrusions. This creates the threat that a telephone connection over which communication protocols are running may be increasingly open to attack by a clever hacker, who breaks into the telephone system itself.

Dial-back mechanisms, whereby the system calls the user back, clearly increase the hurdle an intruder must cross to penetrate a system relative to one in which the caller is assumed to be a potentially legitimate user. However, such systems depend for their security upon the integrity of the telephone system, which, as we have noted, can be subverted. In particular, the emergence of mobile telephones and the introduction of mobility mechanisms into telephone switching systems create a path by which an intruder can potentially redirect a telephone dial-back to a telephone number other than the intended one. Such a mechanism is a good example of a security technology that can protect against benign attacks but would be considerably more exposed to well-organized malicious ones.

Firewalls have become popular as a form of protection against communication-level attacks on distributed systems. Many of these technologies operate using *packet filters* and must be instantiated at all the access points to a distributed network. Each copy of the firewall will have a *filtering control policy* in the form of a set of rules for deciding which packets to reject and which to pass through; although firewalls that can check packet content have been proposed, typical filtering is on the basis of protocol type, sender and destination addresses, and port numbers. Thus, for example, packets can be allowed through if they are addressed to the e-mail or FTP server on a particular node; otherwise they are rejected. Often, firewalls are combined with *proxy* mechanisms, which permit file transfer and remote log in through an intermediary system enforcing further restrictions. The use of proxies for the transfer of public Web pages and FTP areas has also become common: In these cases, the proxy is configured as a mirror of some protected internal file system area, copying changed files to the less-secure external area periodically.

Other technologies commonly used to implement firewalls include application-level proxies and routers. With these approaches, small fragments of user-supplied code (or programs obtained from the firewall vendor) are permitted to examine the incoming and outgoing packet streams. These programs run in a loop, waiting for the next incoming or outgoing message, performing an acceptance test upon it, and then either discarding the message or permitting it to continue. The possibility of logging the message and maintaining additional statistics on traffic is also commonly supported.

The major problem associated with firewall technologies is that they represent a single point of failure: If the firewall is breached, the intruder essentially could gain free run of the enclosed system. Intruders may know of ways to attack specific firewalls—perhaps learned through study of the code used to implement the firewall, secret backdoor mechanisms included by the original fire wall developers for reasons of their own, or by compromising some of the software components included in the application itself. Having broken in, it may be possible to establish connections to servers that will be fooled into trusting the intruder or to otherwise act to attack the system from within. Reiterating the point made above, an increasingly serious exposure is created by the explosive growth of telecommunications. In the past, a dedicated leased line could safely be treated as an internal technology linking components of a distributed system within its firewall. Now, however, such a line must be viewed as a potential point of intrusion.

These considerations are increasingly leading corporations to implement what are called *virtual private networks* (see Figure 19.2) in which communication is authenticated (typically using a hardware signature scheme) so that all messages originating outside of the legitimately accepted sources will be rejected. In settings where security is vital, these sorts of measures are likely to considerably increase the robustness of the network to attack. However, the cost remains high, and as a consequence it seems unlikely that the average network will offer this sort of cryptographic protection in the near future. Thus, while the prospects for strong security may be promising in certain settings, such as military systems or electronic banking systems, the more routine computing environments, on which the great majority of sensitive applications run, remain open to a great variety of attacks and are likely to continue to have such exposure well into the next decade.

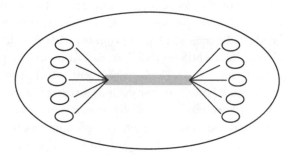

FIGURE 19.2. A long-haul connection internal to a distributed system (gray) represents a potential point of attack. Developers often protect systems with firewalls on the periphery but overlook the risk that the communication infrastructure itself may be compromised, offering the intruder a backdoor approach into the protected environment. Although some corporations are protecting themselves against such threats by using encryption techniques to create virtual private networks, most mundane communication systems are increasingly at risk.

This situation may seem pessimistic; however, in many respects, the story is far from over. Although it may seem extremely negative to think in such terms, it is probable that future information terrorists and warfare tactics will include some of these forms of attack and perhaps others that are hard to anticipate until they have first been experienced. Short of a major shift in mindset on the part of vendors, the situation is not likely to improve anytime soon. We may need to wait until a generation of new technologies has displaced the majority of the existing infrastructure—a process that could take some 10 to 15 years. Thus, information security is likely to remain a serious problem—at least until the year 2010 or later.

Although we will now move on to other topics in security, we note that defensive management techniques can be coupled with security-oriented wrappers to raise the barriers in systems that use firewall technologies for protection. We will return to this subject in Chapter 23.

19.3 Access Control Technologies

Access control techniques operate by restricting use of system resources on the basis of user or group identifiers, which are typically fixed at log-in time—for example, by validation of a password. It is typical that these policies trust the operating system, its key services, and the network. In particular, the log-in program is trusted to obtain the password and correctly check it against the database of system passwords, granting the user permission to work under the user-ID or group-ID only if a match is detected. The log-in system trusts the file server or Network Information Server to respond correctly with database entries that can be used safely in this authentication process, and the resource manager (typically, an NFS server or database server) trusts the ensemble, believing that all packets presented to it as "valid NFS packets" or "valid XYZbase requests" originated at a trusted source.[1]

These dependencies are only rarely enforced rigorously. Thus, one could potentially attack an access control system by taking over a computer, rebooting it as the root or superuser, directing the system to change the user-ID to any desired value, and then starting work as the specified user. An intruder could replace the standard log-in program with a modified one—introducing a false NIS, which would emulate the NIS protocol but substitute invalid password records. One could even code one's own version of the NFS client protocol, which, operating from user space as a normal RPC application, could misrepresent itself as a trusted source of NFS requests. All these attacks on the NFS have been used successfully at one time or another, and many of the loopholes have been closed by one or more of the major vendors. Yet the fact remains that file and database servers continue to be largely trusting of the major operating system components on the nodes where they run and where their clients run.

Perhaps the most serious limitation associated with access control mechanisms is that they generally do not extend to the communication subsystem: typically, any process can issue an RPC message to any address it wishes to place in a message and can attempt to connect to any stream end point for which it possesses an address. In practice, these exposures are hard to exploit, because a process that undertakes to do so will need to guess the addresses being used by the applications it attacks. Precisely to reduce this risk, many applications exploit *randomly generated* end-point addresses, so that an intruder would be forced to guess a pseudorandom number to break into a critical server. However, pseudorandom numbers may be less random than intended, particularly if an intruder has access to the pseudorandom number-generation scheme as well as samples of the values recently produced.

Such break-ins are more common than one might expect—for example, in 1994 an attack on X11 servers was discovered in which an intruder found a way to deduce the connection port number that would be used. Sending a message that would cause the X11 server to prepare to

---

[1]Not all file systems are exposed to such problems—for example, the AFS file system has a sophisticated stateful client/server architecture, which is also much more robust to attack. AFS has become popular, but it is less widely used than NFS.

accept a new connection to a shell command window, the intruder managed to connect to the server and to send a few commands to it. Not surprisingly, this proved sufficient to open the door to a full-fledged penetration. Moreover, the attack was orchestrated in such a manner as to trick typical firewalls into forwarding these poisoned messages, even through the normal firewall protection policy should have required that they be rejected. Until the nature of the attack was understood, the approach permitted intrusion into a wide variety of firewall-protected systems.

To give a sense of how exposed typical distributed systems currently are, Table 19.1 presents some of the assumptions made by the NFS file server technology when it is run without the security technology available from some vendors (in practice, NFS security is rarely enabled in systems that are protected by firewalls; the security mechanisms are hard to administer in heterogeneous environments and can slow down the NFS system significantly). We have listed typical assumptions of the NFS, the normal reason that this assumption holds, and one or more attacks that operate by emulation of the normal NFS environment in a way that the server is unable to detect. The statelessness of the NFS server makes it particularly easy to attack, but most client/server systems have similar dependencies and are similarly exposed.

TABLE 19.1. NFS Security Assumptions

NFS assumption	Dependent on . . .
O/S integrity	NFS protocol messages originate only in trusted subsystems or the kernel
	Attacks: Introduce a computer running an open operating system; modify the NFS subsystem. Develop a user-level program to implement the NFS client protocol; use it to emulate a legitimate NFS client issuing requests under any desired user-ID.
Authentication	Assumes that user- and group-ID information is valid
	Attacks: Spoof the Network Information Server or NFS response packets so that authentication will be done against a falsified password database. Compromise the log-in program. Reboot the system or log-in using the root or superuser account; then change the user-ID or group-ID to the desired one and issue NFS requests.
Network integrity	Assumes that communication over the network is secure
	Attacks: Intercept network packets, reading file system data and modifying data written. Replay NFS commands, perhaps with modifications.

One can only feel serious concern when these security exposures are contemplated against the backdrop of increasingly critical applications that trust client/server technologies such as NFS—for example, it is very common to store sensitive files on unprotected NFS servers. As we noted, there is an NFS security standard, but it is vendor-specific and may be impractical to use in heterogeneous environments. A hospital system, for example, is necessarily heterogeneous: The workstations used in such systems must interoperate with a great variety of special-purpose devices and peripherals, produced by many vendors. Thus, in precisely the setting one might hope would use strong data protection, one typically finds proprietary solutions or unprotected use of standard file servers! Indeed, many hospitals might be prevented from using a strong security policy, because, since so many individuals need access to a patient's record, any form of restriction would effectively be nullified.

Thus, in a setting where protection of data is not just important but is actually legally mandated, it may be very easy for an intruder to break in. While such an individual might find it hard to walk up to a typical hospital computing station and break through its password protection, by connecting a portable laptop computer to the hospital Ethernet (potentially a much easier task), it would be easy to gain access to the protected files stored on the hospital's servers. Such security exposures are already a potentially serious issue, and the problem will only grow more serious with time.

When we first discussed the NFS security issues, we pointed out that there are other file systems that do quite a bit better in this regard, such as the AFS system developed originally at Carnegie Mellon University and now commercialized by Transarc. AFS, however, is not considered to be standard and many vendors provide NFS as part of their basic product line, while AFS is a commercial product from a third party. Thus, the emergence of more secure file system technologies faces formidable practical barriers. It is unfortunate but entirely likely that the same is true for other reliability and security technologies.

19.4 Authentication Schemes and Kerberos

The weak points of typical computing environments are readily seen to be their authentication mechanisms and their blind trust in the security of the communication subsystem. Best known among the technologies that respond to these issues is MIT's Kerberos system, developed as part of Project Athena.

Kerberos makes use of encryption; therefore, it will be useful to start by reviewing the existing encryption technologies and their limitations. Although a number of encryption schemes have been proposed, the most popular ones at the time of this writing are the RSA public key algorithms and the DES encryption standard.

19.4.1 RSA and DES

RSA (see Rivest et al.) is an implementation of a public key cryptosystem (see Diffie and Hellman), which exploits properties of modular exponentiation. In practice, the method operates by generating pairs of *keys*, which are distributed to the users and programs within a distributed system. One key within each pair is the *private* key and is kept secret. The other key is *public*, as is an encryption function, *crypt(key, object)*. The encryption function has a number of useful properties. Suppose that we denote the public key of some user as K and the private key of that user as K^{-1}. Then *crypt*(K,*crypt*(K^{-1}, M)) = *crypt*(K^{-1},*crypt*(K, M)) = M—that is, encryption by the public key will decrypt an object encrypted previously with the private key and vice versa. Moreover, even if keys A and B are unrelated, encryption is commutative: $crypt(A,crypt(B, M)) = crypt(B,crypt(A, M))$.

In typical use, public keys are published in some form of trusted directory service (see Birrell, *Fortezza Application Developers Guide*). If process A wants to send a secure message to process B (this message could only have originated in process A and can only be read by process B), A sends $crypt(A^{-1},crypt(B, M))$ to B, and B computes $crypt(B^{-1},crypt(A, M))$ to extract the message. Here, we have used A and A^{-1} as shorthand for the public and private keys of processes A and B. A can send a message that only B can read by computing the simpler $crypt(B, M)$ and can sign a message to prove that the message was seen by A by attaching $crypt(A^{-1}, digest(M))$ to the message, where $digest(M)$ is a function that computes some sort of small number reflecting the contents of M, perhaps using an error-correcting code for this purpose. Upon reception, process B can compute the digest of the received message and compare this with the result of decrypting the signature sent by A using A's public key. The message can be validated by verifying that these values match (see Denning).

A process can also be asked to encrypt or sign a blinded message when using the RSA scheme. To solve the former problem, process A is presented with $M' = crypt(B, M)$. If A computes $M'' = crypt(A^{-1}, M')$, then $crypt(B^{-1}, M'')$ will yield $crypt(A^{-1}, M)$ without A having ever seen M. Given an appropriate message digest function, the same approach also allows a process to sign a message without being able to read that message.

In contrast, the DES standard (see *Data Encryption Standard*, Diffie and Hellman) is based on shared secret keys, in which two users or processes exchanging a message will both have a copy of the key for messages sent between them. Separate functions are provided for encryption and decryption of a message. Similar to the RSA scheme, DES can also be used to encrypt a digest of a message as proof that the message has not been tampered with. Blinding mechanisms for DES are, however, not available at the present time.

DES is the basis of a government standard, which specifies a standard key size and can be implemented in hardware. Although the standard key size is large enough to provide security for most applications, the key is still small enough to permit it to be broken using a supercomputing system or a large number of powerful workstations in a distributed environment. This is viewed by the government as a virtue of the scheme, because it provides the possibility of decrypting messages for purposes of criminal investigation or national security. When using DES, it is possible to convert plain text (such as a password) into a DES key; in effect, a password can be used to encrypt information so that it can only be decrypted by a process that also has a copy of that password. As will be seen, this is the central feature that makes possible DES-based authentication architectures such as Kerberos (see Schiller, Steiner et al.).

More recently, a security standard has been proposed for use in telecommunication environments. This standard, Capstone, was designed for telephone communication but is not specific to telephony; it involves a form of key for each user and supports what is called *key escrow*, whereby the government is able to reconstruct the key by combining two portions of it, which are stored in secure and independent locations (see Denning and Branstad). The objective of this work is to permit secure and private use of telephones while preserving the government's right to wiretap with appropriate court orders. The Clipper chip, which implements Capstone in hardware, is also used in the Fortezza PCMCIA card, described further in Section 19.3.4.

Both DES and the Capstone security standard are the subjects of vigorous debate. On the one hand, such methods limit privacy and personal security, because the government is able to break both schemes and indeed may have taken steps to make them easier to break than is widely known. On the other hand, the growing use of information systems by criminal organizations clearly poses a serious threat to security and privacy as well, and it is obviously desirable for the government to be able to combat such organizations. Meanwhile, the fundamental security of methods such as RSA and DES is not known—for example, although it is conjectured that RSA is very difficult to break, in 1995 it was shown that in some cases, information about the amount of time needed to compute the *crypt* function could provide data that substantially reduce the difficulty of breaking the encryption scheme. Meanwhile, clever uses of large numbers of computers have made it possible to break DES encryption. These ongoing tensions between social obligations of privacy and security and the obligation of the government to oppose criminality, as well as between the strength of cryptographic systems and the attacks upon them, can be expected to continue into the coming decades.

19.4.2 Kerberos

The Kerberos system is a widely used implementation of secure communication channels, based on the DES encryption scheme (see Schiller, Steiner et al.). Integrated into the DCE environment, Kerberos is currently a de facto standard in the UNIX community. The approach genuinely offers a major improvement in security over that which is traditionally available within UNIX. Its primary limitation is that applications using Kerberos must be modified to create communication channels using the Kerberos secure channel facilities. Although this may seem to be a minor point, it represents a surprisingly serious one for potential Kerberos users, since application software using Kerberos is not yet common. Nonetheless, Kerberos has had some important successes; one of these is its use in the AFS system, discussed earlier (see Satyanarayanan).

The basic Kerberos protocols revolve around the use of a trusted authentication server, which creates session keys between clients and servers upon demand. The basic scheme is as follows. At the time the user logs in, he or she presents a name and password to a log-in agent, which runs in a trusted mode on the user's machine. The user can now create sessions with the various servers that he or she accesses—for example, to communicate with an AFS server, the user requests that the authentication server create a new unique session key and send it back in two forms: one for use by the user's machine and one for use by the file server.

The authentication server, which has a copy of the user's password and also the secret key of the server itself, creates a new DES session key and encrypts it using the user's password. A copy of the session key encrypted with the server's secret key is also included. The resulting information is sent back to the user, where it is decrypted.

The user now sends a message to the remote server asking it to open a session. The server can easily validate that the session key is legitimate, since it has been encrypted with its own secret key, which could only have been done by the authentication server. The session key also contains trustworthy information concerning the user-ID, workstation-ID, and the expiration time of the key itself. Thus, the server knows with certainty who is using it, where the user is working, and how long the session can remain open without a refreshed session key.

It can be seen that there is a risk associated with this method, since it uses the user's password as an encryption key and hence must keep it in memory for a long period of time. Perhaps the user trusts the log-in agent, but does not wish to trust the entire run-time environment over long periods. A clever intruder might be able to simply walk up to a temporarily unused workstation and steal the key from it, reusing it later at will.

Accordingly, Kerberos actually works by exchanging the user's password for a type of one-time password, which has a limited lifetime and is stored only at a *ticket granting service* with which a session is established as soon as the user logs in. The user sends requests to make new connections to this ticket granting service instead of to the original authentication service during the normal course of work; it encrypts them not with the user's password, but with this one-time session key. The only threat now is that an intruder might somehow manage to execute commands while the user is logged in (e.g., by sitting down at a machine while the normal user is getting a cup of coffee). This threat is a real one, but it is minor compared to the others that concern us. Moreover, since all the keys actually stored on the system have limited validity, even if one is stolen, it can only be used briefly before it expires. In particular, if the session key to the ticket granting service expires, the user is required to type in his or her password again, and an intruder would have no way of obtaining the password in this model other than grabbing it during the initial protocol to create a session with the ticket granting service or by breaking into the authentication server itself.

Once a session exists, communication to and from the file server can be done in the clear, in which case the file server can use the user-ID information established during the connection setup to authenticate file access, or it can be signed, giving a somewhat stronger guarantee that the channel protocol has not been compromised in any way, or even encrypted, in which case data exchanged are only accessible by the user and the server. In practice, the initial channel authentication, which also provides strong authentication guarantees for the user-ID and group-ID information to be employed in restricting file access, suffices for most purposes. An overview of the protocol is seen in Figure 19.1.

The Kerberos protocol has been proven secure against most forms of attack (see Lampson et al.); one of its few dependencies is its trust in the system time servers, which are used to detect expiration of session keys (see Gong). Moreover, the technology has been shown to scale to large installations using an approach whereby authentication servers for multiple protection domains can be linked to create session keys spanning wide areas. Perhaps the most serious exposure of the technology is that associated with partitioned operation. If a portion of the network is cut off from the authentication server for its part of the network, Kerberos session keys will begin to expire, and it will be impossible to refresh them with new keys. Gradually, such a component of the network will lose the ability to operate, even between applications and servers residing entirely within the partitioned component. In future applications requiring support for mobility, with links forming and being cut very dynamically, the Kerberos design would require additional development.

A less-obvious exposure to the Kerberos approach is that associated with active attacks on its authentication and ticket granting server. The server is a software system operating on standard computing platforms, and those platforms are often subject to attack over the network. A knowledgeable user might be able to concoct a poison pill by building a message, which will look sufficiently legitimate, to be passed to a standard service on the node; this message will then provoke the node into crashing by exploiting some known intolerance to incorrect input. The fragility of contemporary systems to this sort of attack is well known to protocol developers, many of whom have the experience of repeatedly crashing the machines with which they work during the debugging stages of a development effort. Thus, one could imagine an attack on Kerberos or a similar system aimed not at breaking through its security architecture, but rather at repeatedly crashing the authentication server, with the effect of denying service to legitimate users.

Kerberos supports the ability to prefabricate and cache session keys (tickets) for current users, and this mechanism would offer a period of respite to a system subjected to a denial of service attack. However, after a sufficient period of time, such an attack would effectively shut down the system.

Within military circles, there is an old story (perhaps not true) about an admiral who used a new generation of information-based battle management system in a training exercise. Unfortunately, the story goes, the system had an absolute requirement that all accesses to sensitive data be logged on an audit trail, which for that system was printed on a protected line printer. At some point during the exercise the line printer jammed or ran low on paper, and the audit capability shut down. The system, now unable to record the required audit records, therefore denied the admiral access to his databases of troop movements and enemy positions. Moreover, the same problem rippled through the system, preventing all forms of legitimate but sensitive data access.

The developer of a secure system often thinks of his or her task as being that of protecting critical data from the "bad guys." But any distributed system has a more immediate obligation, which is to make data and critical services available to the "good guys." Denial of service in the name of security may be as serious a problem as providing service to an unauthorized user. Indeed, the admiral in the story is now said to have a profound distrust of computing systems. Having no choice but to use computers, in his command the security mechanisms are disabled. (The military phrase is, "He runs all his computers at system high.") This illustrates a fundamental point, which is overlooked by most security technologies today: Security cannot be treated independent of other aspects of reliability.

19.4.3 ONC Security and NFS

Sun Microsystems, Inc., has developed an RPC standard, which it calls Open Network Computing (ONC), around the protocols used to communicate with NFS servers and similar systems.

ONC includes an authentication technology, which can protect against most of the spoofing attacks previously described. Similar to a Kerberos system, this technology operates by obtaining unforgeable authorization information at the time a user logs into a network. The NFS is able to use this information to validate accesses as being from legitimate workstations and to strengthen its access control policies. If desired, the technology can also encrypt data to protect against network intruders who monitor passing messages.

Much like Kerberos, the NFS security technology is considered by many users to have limitations and to be subject to indirect forms of attack. Perhaps the most serious limitations are those associated with export of the technology: Companies such as Sun export their products, and U.S. government restrictions prevent the export of encryption technologies. As a result, it is impractical for Sun to enable the NFS protection mechanisms by default or to envision an open standard, allowing complete interoperability between client and server systems from multiple vendors (the major benefit of NFS), which, at the same time, would be secure. The problem here is the obvious one: Not all client and server systems are manufactured in the United States!

Beyond the heterogeneity issue is the problem of management of a security technology in complex settings. Although ONC security works well for NFS systems in fairly simple systems based entirely on Sun products, serious management challenges occur in complex system configurations, where users are spread over a large physical area, or in systems using heterogeneous hardware and software sources. With security disabled, these problems vanish. Finally, the same availability issues raised in our discussion of Kerberos pose a potential problem for ONC security. Thus, it is perhaps not surprising that these technologies have not been adopted on a widespread basis. Such considerations raise the question of how one might wrap a technology such as NFS, which was not developed with security in mind, so that security can be superimposed without changing the underlying software. One can also ask about monitoring a system to detect intrusions as a proactive alternative to hardening a system against intrusions and then betting that the security scheme will in fact provide the desired protection. We discuss these issues further in Chapter 23.

19.4.4 Fortezza

Fortezza is a recently introduced hardware-based security technology oriented toward users of portable computers and other PC-compatible computing systems (see *Fortezza Application Developers Guide*, Denning and Branstad). Fortezza can be understood both as an architecture and as an implementation of that architecture. In this section, we briefly describe both perspectives.

Viewed as an architecture, Fortezza represents a standard way to attach a public key cryptographic protocol to a computer system. Fortezza consists of a set of software interfaces, which standardize the interface to its cryptographic engine, which is itself implemented as a hardware device that plugs into the PCMCIA slot of a standard personal computer. The idea is that a variety of hardware devices might eventually exist that are compatible with this standard. Some, such as military security technology, might be highly restricted and not suitable for export; others, such as an internally accepted security standard for commercial transactions, might be less restricted and safe for export. By designing software systems to use the Fortezza interfaces, the distributed application becomes independent of its security technology and very general. Depending upon the Fortezza card that is actually used in a given setting, the security properties of the resulting system may be strengthened or weakened. When no security is desired at all, the Fortezza functions become "no-ops": Calls to them take no action and are extremely inexpensive.

Viewed as an implementation, Fortezza is an initial version of a credit card-sized PCMCIA card compatible with the standard and the associated software interfaces implementing the architecture. The initial Fortezza cards use the Clipper chip, which implements a cryptographic protocol called Capstone—for example, the interfaces define a function, *CI_Encrypt*, and another function, *CI_Decrypt*, which, respectively, convert a data record provided by the user into and out of its encrypted form. The initial version of the card implements the Capstone cryptographic integrated circuit. It stores the private key information needed for each of its possible users and the public keys needed for cryptography. The card performs the digital signature and hash functions needed to sign messages, provides public and private key functions, and supports block data encryption and decryption at high speeds. Other cards could be produced that would implement other encryption technologies using the same interfaces but different methods.

Although we will not discuss this point in detail, readers should be aware that Fortezza supports what is called *key escrow* (see Denning and Branstad), meaning that the underlying technology permits a third party to assemble the private key of a Fortezza user from information stored at one or more trusted locations (two, in the specific case of the Capstone protocol). Key escrow is controversial because of public concerns about the degree to which the law enforcement authorities who maintain these locations can themselves be trusted and about the security of the escrow databases. On one hand, it can be argued that in the absence of such an escrow mechanism, it will be easy for criminals to exploit secure communications for illegal purposes, such as money laundering and drug transactions. Key escrow permits law enforcement organizations to wiretap such communication. But on the other side of the coin, one can argue that freedom of speech should extend to the freedom to encrypt data for privacy. The issue is an active topic of public debate.

Many authentication schemes are secure either because of something the user "knows," which is used to establish authorization, or something the user "has." Fortezza is designed to have both properties: Each user is expected to remember a personal identification code (PIN), and the card cannot be used unless the PIN has been entered recently. At the same time, the card itself is required to perform secure functions, and it stores the user's private keys in a trustworthy manner. When a user correctly enters his or her PIN, Fortezza behaves according to a standard public key encryption scheme, as described earlier. (As an aside, it should be noted that the Clipper-based Fortezza PCMCIA card does not implement this PIN functionality.)

In order to authenticate a message as coming from user A, such a scheme requires a way to determine the public key associated with user A. For this purpose, Fortezza uses a secured X.500-compatible directory, in which user identifications are saved with what are called "certificates." A certificate consists of a version number, a serial number, the issuer's signature algorithm, the issuer's distinguished name validity period (after which the name is considered to have expired), the subject's distinguished name, the subject's public key, and the issuer's signature for the certificate as a whole. The issuer of a certificate will typically be an X.500 server administered by a trusted agency or entity on behalf of the Fortezza authentication domain.

In a typical use, Fortezza is designed with built-in knowledge of the public keys associated with the trusted directory services that are appropriate for use in a given domain. A standard protocol is supported by which these keys can be refreshed prior to the expiration of the distinguished name on behalf of which they were issued. In this manner, the card itself knows whether or not it can trust a given X.500 directory agent, because the certificates issued by that agent are either correctly, and hence securely, signed, or they are not and hence are invalid. Thus, although an intruder could potentially masquerade as an X.500 directory server, without the private key information of the server it will be impossible to issue valid certificates and forge public key information. Short of breaking the cryptographic system itself, the intruder's only option is to seek to deny service by somehow preventing the Fortezza user from obtaining needed public keys. If successful, such an attack could in principle last long enough for the names involved to expire, at which point the card must be reprogrammed or replaced. However, secured information will never be revealed, even if the system is attacked in this manner, and incorrect authentication will never occur.

Although Fortezza is designed as a PCMCIA card, the same technology could be implemented in a true credit card with a microprocessor embedded in it. Such a system would then be a very suitable basis for commercial transactions over the Internet. The primary risk would be one in which the computer itself becomes compromised and takes advantage of the user's card and PIN during the period when both are present and valid to perform undesired actions on behalf of that user. Such a risk is essentially unavoidable, however, in any system using software as an intermediary between the user and the services that he or she requests. With Fortezza or a similar technology, the period of vulnerability is kept to a minimum: It holds only for as long as the card is in the machine, the PIN has been entered, and the associated timeout has not yet occurred. Although this still represents an exposure, it is difficult to see how the risk could be further reduced.

19.5 Availability and Security

Recent research on the introduction of availability into Kerberos-like architectures has revealed considerable potential for overcoming the availability limitations of the basic Kerberos approach. As we have seen, Kerberos is dependent on the availability of its authentication server for the generation of new protection keys. Should the server fail or become partitioned away from the applications depending on it, the establishment of new channels and the renewal of keys for old channels will cease to be possible, eventually shutting down the system.

In a Ph.D. dissertation based on an early version of the Horus system, Reiter showed that process groups could be used to build highly available authentication servers (see Reiter [1993, 1994], Reiter et al. [1992, 1995]). His work included a secure join protocol for adding new processes to such a group; methods for securely replicating data and for securing the ordering properties of a group communication primitive (including the causal property); and an analysis of availability issues, which occur in key distribution when such a server is employed. Interestingly, Reiter's approach does not require that the time service used in a system such as Kerberos be replicated: His techniques have a very weak dependency on time.

Process group technologies permit Reiter to propose a number of exotic new security options as well. Still working with Horus, he explored the use of split secret mechanisms to ensure that in a group of n processes (see Desmedt, Desmedt et al., Frankel, Frankel and Desmedt, Herlihy and Tygar, Laih and Harn), the availability of any $n - k$ members would suffice to maintain secure and available access to that group. In this work, Reiter uses a state machine approach: The individual members have identical states and respond to incoming requests in an identical manner. Accordingly, his focus was on implementing state machines in environments with intruders and on signing responses in such a way that $n - k$ signatures by members would be recognizable as a group signature carrying the authority of the group as a whole.

A related approach can be developed in which the servers split a secret in such a manner that none of the servers in the group has access to the full data, and yet clients can reconstruct these data provided that $n - k$ or more of the servers are correct. Such a split secret scheme might be useful if the group needs to maintain a secret that none of its individual members can be trusted to manage appropriately.

Techniques such as these can be carried in many directions. Reiter, after leaving the Horus project, started work on a system called Rampart at AT&T (see Reiter [1996]). Rampart provides secure group functionality under assumptions of Byzantine failures and is used to build extremely secure group-based mechanisms for use by less stringently secured applications in a more general setting—for example, Rampart could be the basis of an authentication service, a service used to maintain billing information in a shared environment, a digital cash technology, or a strongly secured firewall technology.

Cooper, also working with Horus, has explored the use of process groups as a blinding mechanism. The concept here originated with work by Chaum, who showed how privacy can be enforced in distributed systems by mixing information from many sources in a manner that prevents an intruder from matching an individual data item to its source or tracing a data item from source to destination (see Chaum). Cooper's work shows how a replicated service can actually mix up the contents of messages from multiple sources to create a private and secure e-mail repository (see Cooper [1994]). In his approach, the process group-based mail repository service stores mail on behalf of many users. A protocol is given for placing mail into the service, retrieving mail from it, and for dealing with vacations; the scheme offers privacy (intruders cannot determine sources and destinations of messages) and security (intruders cannot see the contents of messages) under a variety of attacks and can also be made fault tolerant through replication.

Intended for large-scale mobile applications, Cooper's work would permit exchanging messages between processes in a large office complex or a city without revealing the physical location of the principals—however, this type of communication is notoriously insecure. The emergence of digital commerce may expose technology users to very serious intrusions on their privacy and finances. Work such as that done by Reiter, Chaum, and Cooper suggests that security and privacy should be possible even with the levels of availability that will be needed when initiating commercial transactions from mobile devices.

19.6 Related Reading

On Kerberos: (see Schiller, Steiner et al.).

On associated theory: (see Bellovin and Merritt, Lampson et al.).

On RSA and DES: (see Denning, Desmedt, Diffie and Hellman, Rivest et al.).

On Fortezza: Most information is on-line, but Denning and Branstad includes a brief review.

On Rampart: (see Reiter [1993, 1994], Reiter et al. [1992, 1995]).

On split-key cryptographic techniques and associated theory: (see Desmedt, Desmedt et al., Frankel, Frankel and Desmedt, Herlihy and Tygar, Laih and Harn).

On mixing techniques: (see Chaum, Cho and Birman, Cooper [1994]).

CHAPTER 20

Clock Synchronization and Synchronous Systems

CONTENTS

P revious chapters of this book have made a number of uses of clocks or time in distributed protocols. In this chapter, we look more closely at the underlying issues. Our focus is on aspects of real-time computing that are specific to distributed protocols and systems.

20.1 Clock Synchronization

Clock synchronization is an example of a topic that until the recent past represented an important area for distributed system research (see Clegg and Marzullo, Cristian [1989], Cristian and Fetzer, Kopetz and Ochsenreiter, Lamport [1984], Lamport and Melliar-Smith, Marzullo [1984], Srikanth and Toueg, Verissimo); overviews of the field can be found in Liskov, Simons et al. The introduction of the global positioning system, in the early 1990s, greatly changed the situation. As recently as five years ago, a book such as this would have treated the problem in considerable detail, to the benefit of the reader, because the topic is an elegant one and the clock-based protocols that have been proposed are interesting to read and analyze. Today, however, it seems more appropriate to touch only briefly on the subject.

The general problem of clock synchronization occurs because the computers in a distributed system typically use internal clocks as their primary time source. On most systems, these clocks are accurate to within a few seconds per day, but there can be surprising exceptions to the rule. PCs, for example, may operate in power-saving modes, in which even the clock is slowed down or stopped, making it impossible for the system to gauge real time reliably. At the other end of the spectrum, the global positioning system (GPS) has introduced an inexpensive way to obtain accurate timing information using a radio receiver; time obtained in this manner is accurate to within a few milliseconds unless the GPS signal itself is distorted by unusual atmospheric conditions or problems with the antenna used to receive the signal. Break-out 20.1 discusses the MARS system, which uses clock synchronization for real-time control.

Traditionally, clock synchronization was treated in the context of a group of peers, each possessing an equivalent local clock, with known accuracy and drift properties. The goal in such a system was typically to design an agreement protocol by which the clocks could be kept as close as possible to real time and with which the tendency of individual clocks to drift (from one another and to do so with respect to real time) could be controlled. To accomplish this, processes would periodically exchange time readings, running a protocol by which a software clock could be constructed having substantially better properties than that of any of the individual participating programs—with the potential to overcome outright failures whereby a clock might drift at an excessive rate or return completely erroneous values.

Key parameters to such a protocol are the expected and maximum communication latencies of the system. It can be shown that these values limit the quality of clock synchronization achievable in a system by introducing uncertainty in the values exchanged between processes—for example, if the latency of the communication system between p and q is known to vary in the range $[0, \varepsilon]$, any clock reading that p sends to q will potentially be aged by ε time units by the time q receives it. When latency is also bounded below, a method developed by Verissimo (briefly

MARS: A Distributed System for Real-Time Control

The MARS system uses clock synchronization as the basis of an efficient fault tolerance method, implemented using pairs of processing components interconnected by redundant communication links. The basic approach is as follows (see Damm et al., Kopetz and Ochsenreiter, Kopetz and Verissimo).

A very high quality of clock synchronization is achieved using a synchronization method that resides close to the hardware (a broadcast-style bus). Implemented in part using a special-purpose device controller, clocks can be synchronized to well under a millisecond and, if a source of accurate timing information is available, can be both precise and accurate to within this degree of precision.

Applications of MARS consist of directly controlled hardware, such as robotic units or components of a vehicle. Each processor is duplicated, as is the program that runs on it, and each action is taken redundantly. Normally, every message will be sent four times: once by each processor on each message bus. The architecture is completely deterministic in the sense that all processes see the same events in the same order and base actions on synchronized temporal information in such a way that even clock readings will be identical when identical tasks are performed. Software tools

for scheduling periodic actions and for performing actions after a timer expires are provided by the MARS operating system, which is a very simple execution environment concerned primarily with scheduling and message passing.

MARS is designed for very simple control programs and assumes that these programs fail by halting (the programs are expected to self-check their actions for sanity and shut down if an error is detected). In the event that a component does fail, this can be detected by the absence of messages from it or by their late arrival. Such a failed component is taken off-line for replacement and reintegrated into the system the next time it is restarted from scratch. These assumptions are typical of in-flight systems for aircraft and factory-floor process control systems.

Although MARS is not a particularly elaborate or general technology, it is extremely effective within its domain of intended use. The assumptions made are felt to be reasonable ones for this class of application, and although there are limitations on the classes of failures that MARS can tolerate, the system is also remarkably simple and modular, benefiting from precisely those limitations and assumptions. The performance of the system is extremely good for the same reasons.

presented here) can achieve clock precisions bounded by the *variation* in latency. In light of the high speed of modern communication systems, these limits represent a remarkably high degree of synchronization: It is rarely necessary to time events to within accuracies of a millisecond or less, and these limits tell us that it should be possible to synchronize clocks to that degree if desired.

Modern computing systems face a form of clock synchronization problem that is easier to solve than the most general version of the problem. If such systems make use of time at all, it is common to introduce two or more GPS receivers—in this manner creating a number of system time sources. Devices consisting of nothing more than a GPS receiver and a network interface can, for example, be placed directly on a shared communication bus. The machines sharing that bus will now receive time packets at some frequency, observing identical values at nearly identical time. (See Figure 20.1.)

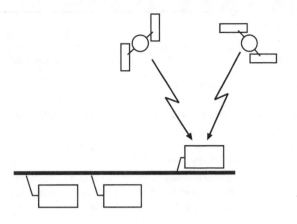

FIGURE 20.1. The global positioning system is a satellite network, which broadcasts highly accurate time values worldwide. Although intended for accurate position location, GPS systems are also making accurate real-time information available at low cost.

If the device driver associated with the network device is able to identify these incoming time packets, it can be used to set the local clock of the host machine to extremely high precision; if not, an application should be able to do so with reasonable accuracy. Given data for the average access and propagation delays for packets sent over the communication hardware, the associated latency can be added to the incoming time value, producing an even more accurate result. In such a manner, systems in which real time is important can synchronize processor clocks to within milliseconds, obviating the need for any sophisticated application-level synchronization algorithm. After all, the delays associated with passing a message through an operating system up to the application, scheduling the application process if it were in a blocked state, and paging in the event of a possible page fault are substantial compared with the clock accuracy achievable in this manner. Moreover, it is very unlikely that a GPS time source would fail other than by crashing. If noncrash failures are a concern, a simple solution is to collect sets of readings from three GPS sources, exclude the outlying values, and take the remaining value as the correct one. (See Figure 20.2 for definitions of some useful terms.)

In light of this development, it has become desirable to consider distributed computing systems as falling into two classes. Systems in which time is important for reliability can readily include accurate time sources and should do so. Systems in which time is not important for reliability should be designed to avoid all use of workstation clock values, using elapsed time on a local clock to trigger timer-based events such as retransmission of messages or timeout, but not exchanging time values between processes or making spurious use of time. For the purposes of such elapsed timers, the clocks on typical processors are more than adequate: A clock that is accurate to a few seconds per day will measure a 100 ms timeout with impressive accuracy.

Where clocks are known to drift, Verissimo and Rodrigues have suggested an elegant method for maintaining very precise clocks (see Verissimo and Rodrigues); see also Clegg and Marzullo.

Accuracy is a characterization of the degree to which a correct clock can differ from an external clock that gives the true time. A clock synchronization protocol guaranteeing highly accurate clocks thus provides the assurance that a correct clock will return a value within some known maximum error of the value that the external clock would return. In some settings, accuracy is expressed as an absolute bound; in others, accuracy is expressed as a maximum rate of drift—in this case, the accuracy of the clock at a given time is a function of how long the clock has been free-running since the last round of the synchronization protocol.

Skew is a measure of the difference between clock readings for a pair of processes whose clocks have been sampled at the same instant in real time.

Precision is a characterization of the degree to which any pair of correct clocks can differ over time. As with accuracy, a precision may be given as a constant upper bound on the skew or as a maximum rate of drift of the skews for pairs of correct clocks.

FIGURE 20.2. Definitions of accuracy, skew, and precision for synchronized clocks in distributed settings.

This protocol, called *a-posteriori clock synchronization*, operates roughly as follows. A process other than the GPS receiver initiates clock synchronization periodically (for fault tolerance, two or more processes can run the algorithm concurrently). Upon deciding to synchronize clocks, this process sends out a *resynchronize* message, including its own clock value in the message and setting this value as close as possible to when the message is transmitted on the wire—for example, the device driver can set the clock field in the header of an outgoing message just before setting up the DMA transfer to the network.

Upon arrival in destination machines, each recipient notes its local clock value, again doing this as close as possible to the wire. The recipients send back messages containing their clock values at the time of the receipt. The difference between these measured clock values and that of the initiator will be latency from the initiator to the receivers plus the drift of the recipient's clock relative to the clock of the initiator. Thus, if the initiator believes it to be three o'clock and the latency of communication is 1 ms, the value, –31 ms, would correspond to a recipient whose clock was showing 30 ms before three o'clock when the initiator sent the message, while the value 121 ms would be returned by a process whose clock read three o'clock plus 122 ms at the time the initiator sent the message. Variations in latency will cause these values to be slightly higher or lower than in this example: perhaps –31.050 ms or 120.980 ms.

The synchronization algorithm now selects one of the participants as the official clock of the system. It does so either by selecting a value returned from a process with a GPS receiver, if one is included, or by sorting the returned differences and selecting the median. It subtracts this value from the other differences. The vector will now have small numbers in it if, as assumed, the latency from initiator to participants is fairly constant over the set. The values in the vector will represent the distance that the corresponding participant's clock has drifted with respect to the reference clock. Given an estimate of the message latency between the reference process and the initiator, the initiator can also compute the drift of its own clock—for example, a process may learn that its clock has drifted by −32 ms since the last synchronization event. Any sort of reliable multicast protocol can be used to return the correction factors to the participants.

To actually correct a clock that has drifted, it is common to use an idea introduced by Srikanth and Toueg. The approach involves gradually compensating for the drift under the assumption that the rate of drift is constant. Thus, if a process has drifted 120 ms over a one-minute period, the clock might be modified in software to introduce a compensating drift rate of −240 ms over the next minute, in this manner correcting the original 120 ms and overcoming the continuing 120 ms drift of its own clock during the period. Such an adjustment occurs gradually, avoiding noticeable jumps in the clock value that might confuse an application program.

The above discussion has oversimplified the protocol: The method is actually more complicated because it needs to account for a variety of possible failure modes; this is done by running several rounds of the protocol and selecting, from among the candidate clocks appearing best in each round, that round and clock for which the overall expected precision and accuracy is likely to be best.

Verissimo and Rodrigues's algorithm is optimally precise but not necessarily the best for obtaining optimal accuracy: The best-known solution to that problem is the protocol of Srikanth and Toueg mentioned above. However, when a GPS receiver is present in a distributed system having a standard broadcast-style LAN architecture, the a-posteriori method will be optimal in both respects—accuracy and precision—with clock accuracies comparable in magnitude to the variation in message latencies from initiator to recipients. These variations can be extremely small: Numbers in the tens of microseconds are typical. Thus, in a worldwide environment with GPS receivers, one can imagine an inexpensive software and hardware combination permitting processes anywhere in the world to measure time accurately to a few tens of microseconds. Accuracies such as this are adequate for even the most demanding real-time uses.

Unfortunately, neither of these methods is actually employed by typical commercial computing systems. At the time of this writing, the situation is best characterized as a transitional one. There are well-known and relatively standard software clock synchronization solutions available for most networks, but the standards rarely span multiple vendor systems. Heterogeneous networks are thus likely to exhibit considerable time drift from processor to processor. On the other hand, most workstations have time sources available that can be marginally trusted, at a resolution of seconds or tens of seconds. This is fortunate, because the use of time is growing in applications such as security systems, where session keys and other generated secrets are normally viewed as having limited lifetimes. Thus, security systems may depend upon their time sources for correct behavior. Meanwhile, vendors seem to be closer and closer to including GPS receivers as a standard component of network servers, making cheap and accurate time more and more standard. The a-posteriori protocol, if used widely, could result in a situation where all computers worldwide would share clocks synchronized to within a few tens of milliseconds—a development that would facilitate major advances in communication support for video on demand and group conferencing systems.

20.2 Timed-Asynchronous Protocols

Given a network of computers that share an accurate time source, it is possible to design broadcast protocols to guarantee real-time properties as well as other properties, such as failure-atomicity or totally ordered delivery. The best-known work in this area is that of Cristian, Aghili, Strong, and Dolev and is widely cited as the CASD protocol suite or the Δ-T atomic broadcast protocols (see Cristian et al. [1985, 1990]). These protocols are designed for a static membership model, although Cristian later extended the network model to dynamically track the formation and merging of components in the event of network partitioning failures, again with real-time guarantees on the resulting protocols. In the remainder of this section, we present these protocols in the simple case where processes fail only by crashing or by having clocks that lie outside of the acceptable range for correct clocks—where messages are lost but not corrupted. The protocols have often been called synchronous, but Cristian favors the term "timed asynchronous" (see Cristian and Schmuck), and this is the one we use here.

The CASD protocols seek to guarantee that in a time period during which a set of processes is *continuously operational* and *connected*, this set will deliver the same messages at the same time and in the same order. There is a subtlety here, to which we will return: A process may not be able to detect that it has been nonoperational for a period of time and that therefore it may not have the guarantee of expecting correct behavior from the protocols. But we start by considering the simple scenario of a network consisting of a collection of n processes, k of which may be faulty. Moreover, "same time" must be understood to be limited by the clock skew: Because processor clocks may differ by as much as ε, two correct processors undertaking to perform the same action at the same time may in fact do so as much as ε time units apart.

The CASD protocol is designed for a network in which packets must be routed; the network diameter, d, is the maximum number of hops a packet may have to take to reach a destination node from a source node. It is understood that failures will not cause the network to become disconnected. Although individual packets can be lost in the network, it is assumed that there is a known limit on the number of packets that will actually be lost in any single run of the protocol. Finally, multicast networks are not modeled as such: An Ethernet or FDDI is treated as a set of point-to-point links.

The CASD protocol operates as follows. A process (which may itself be faulty) creates a message and labels it with a timestamp, t (from its local clock), and its process identifier. It then forwards the message to all processors reachable over communication links directly connected to it. These processes accept incoming messages. A message is *discarded* if it is a duplicate of a message that has been seen previously or if the timestamp on the message falls outside a range of currently feasible valid timestamps. Otherwise, the incoming message is *relayed* over all communication links except the one on which it was received. This results in the exchange of $O(n^2)$ messages, as illustrated in Figure 20.3.

A process holding a message waits until time $t + \Delta$ on its local clock (here, t is the time when the message was sent) and then delivers it in the order determined by the sender's timestamp, breaking ties using the processor ID of the sender. For suitable validity limits and Δ, this protocol can be shown to overcome crash failures, limited numbers of communication failures, and incorrect clock values on the part of the sender or intermediary relay processes.

The calculation of this parameter is based on the following reasoning: For the range of behaviors possible in the system, there is a corresponding maximum latency after which a message that originates at a faulty process and that has been forwarded only by faulty processes finally reaches a correct process and is accepted as valid. From this point forward, there is an additional maximum latency before the message has reached all correct processes, limited by the maximum number of network packet losses that can occur. Finally, any specific recipient may consider itself to be the earliest of the correct processes to have received the message and will assume that other correct processes will be the last to receive a copy. From this reasoning, a value can be assigned to Δ such that at time $t + \Delta$, every correct process will have a copy of the message and will know that all other correct processes also have a copy. It is therefore safe to deliver the message at time $t + \Delta$: The other processes will do so as well, within a time skew of ε, corresponding to the maximum difference in clock values for any two correct processes. This is illustrated in Figure 20.3, where time $t + b$ corresponds to $t + \Delta - \varepsilon/2$ and $t + c$ corresponds to $t + \Delta + \varepsilon/2$.

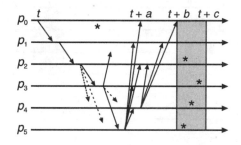

FIGURE 20.3. In the CASD protocol, messages are delivered with real-time guarantees despite a variety of possible failures. In this example for a fully connected network ($d = 1$), processes p_0 and p_1 are faulty and send the message only to one destination each. p_2 and p_3 are correct but experience communication failures, which prevent the message from being forwarded to the full set of correct processors. Eventually, however, the full set of possible failures has been exhausted and the message reaches all correct destinations even if the execution is a worst-case one. In this example, the message finally reaches its last destination at time $t + a$. The processors now delay delivery of the message under a best-case/worst-case analysis, whereby each process reasons that it may have received the message in the minimum possible time but that others may receive it after the maximum possible time and yet assume that they too had received the message after a minimal delay. When this delay has elapsed, all correct processes know that all other correct processes have the message and are prepared to deliver it; delivery then takes place during a period bounded above and below by the clock synchronization constant ε (shown as [$t + b$, $t + c$] in the figure). Incorrect processes may fail to deliver the message, as in the case of p_1; may deliver outside of the window, as does p_0; or may deliver messages rejected by all correct processes.

Although we will not develop the actual formulas here, because the analysis would be fairly long, it is not hard to develop a basic intuition into the reasoning behind this protocol. If we are safe in assuming that there are at most f faulty processes in the network and that the network itself loses no more than k packets during a run of the protocol, it must follow that a broadcast will reach at least one operational process, which will forward it successfully to every other operational process within $f + k$ rounds. A process using the protocol simply waits long enough to be able to deduce that every other process must have a copy of the message, after which it delivers the message in timestamp order.

Because all the operational processes will have received the same messages and use the same timestamp values when ordering them for delivery, the delivered messages are the same and in the same order at all correct processes. However, this may not be the case at *incorrect* processes—namely, those for which the various temporal limits and constants of the analysis do not hold or those that failed to send or receive messages the protocol requires them to send or receive. (See Figure 20.4.)

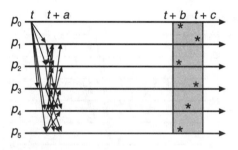

FIGURE 20.4. A run of the CASD protocol in which no failures occur. After a flurry of message exchanges during which $O(n^2)$ messages are sent and received, the protocol lies quiescent until delivery occurs. The delay to delivery is unaffected by the good fortune of the protocol in having reached all the participants so rapidly. Notice that as normally presented, the protocol makes no use of broadcast hardware.

Clearly, when a protocol such as this one is used in a practical setting, it will be advantageous to reduce the value of Δ as much as possible, since Δ is essentially a minimum latency for the protocol. For this reason, the CASD protocol is usually considered in a broadcast network for which the network diameter, d, is 1; processes and communication are assumed to be quite reliable (hence, these failure limits are reduced to numbers such as 1); and clocks are assumed to be very closely synchronized for the operational processes in the network. With these sorts of assumptions, Δ, which would have a value of about three seconds in the local area network used by the Computer Science Department at Cornell, can be reduced into the range of 100–150 ms. Such a squeezing of the protocol leads to runs such as the one shown in Figure 20.5.

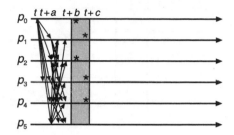

FIGURE 20.5. More aggressive parameter settings and assumptions can substantially reduce the delay before delivery occurs.

We noted that there is a subtle issue associated with the definition of "operational" in the goals of the CASD protocol. The problem occurs when we consider a process that is technically faulty because its clock has drifted outside the limits assumed for a correct process; with the clock synchronization methods reviewed above, this is an unavoidable risk, which grows as the assumed limits become tighter. This is also true when using Cristian's recommended clock synchronization protocol (see Cristian [1989])—that is, the same actions that we took to reduce Δ also have the side-effect of making it more likely that a process will be considered faulty.

Such a process is only faulty in a technical sense. Viewed from above, we can see that its clock is slightly too fast or too slow, perhaps only five or ten milliseconds from the admissible range. Internally, the process considers itself quite operational and would be unable to detect this type of fault even if it tried to do so. Yet, because it is faulty in the formal sense of violating our conditions on correct processes, the guarantees of the protocol may no longer hold for such a process: It may deliver messages that no other process delivered, fail to deliver messages that every other process delivered successfully, or deliver messages outside the normal time range within which delivery should have occurred. Even worse, the process may then drift back into the range considered normal and hence recover to an operational state immediately after this condition occurs. The outcome might be a run more like the one shown in Figure 20.6.

Thus, although the CASD protocol offers strong temporal and fault-tolerant properties to correct processes, the guarantees of these protocols may appear weaker to a process using them, because such a process has no way to know, or to learn, whether or not it is one of the correct ones. In some sense, the protocol has a concept of system membership built into it, but this information is not available to the processes in the system. The effect is to relax all the properties of the protocol suite, which is perhaps best understood as being probabilistically reliable for this reason.

A stronger statement could be made if failures were detectable so that such a process could later learn that its state was potentially inconsistent with that of other processes. There has been some encouraging work on strengthening the properties of this protocol by layering additional mechanisms over it. Gopal et al., for example, have shown how the CASD protocols can be extended to guarantee causal ordering and to overcome some forms of inconsistency (see Gopal

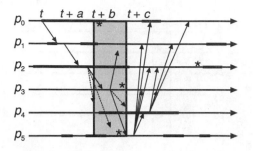

FIGURE 20.6. In this case, overly aggressive parameter settings have caused many processes to be incorrect in the eyes of the protocol, illustrated by bold intervals on the process timelines (each process is considered incorrect during a bold interval—for example, because its clock has drifted too far from the global mean). The real-time and atomicity properties are considerably weakened; moreover, participating processes have no way to determine if they were correct or incorrect on a given run of the protocol. Here, the messages that arrive prior to time *t + c* are considered as valid by the protocol; the others arrive too late and are ignored by correct processes.

et al.). In Chapter 22 we will see how a CASD-like protocol can be made to offer stronger guarantees under a slightly different system model, in which probabilistic information is available in regard to the properties of communication channels.

In the Portuguese NavTech project, Almeida and Verissimo have explored a class of protocols that superimpose a background state exchange mechanism on a CASD-like protocol structure (see Figure 20.7). In this approach, processes within the system periodically send snapshots of aspects of their state to one another using unreliable all-to-all message exchanges over dedicated but low bandwidth links. The resulting n^2 message exchange leaves the correct processes with accurate information about one another's states prior to the last message exchange and with partially accurate information as of the current exchange (the limitation is due to the possibility that messages may be lost by the communication subsystem). In particular, the sender of a CASD-style broadcast may now learn that it has reached all its destinations. During the subsequent exchange of messages, information gained in the previous exchange can be exploited— for example, to initiate an early delivery of a timed broadcast protocol. Unfortunately, however, the mechanism does not offer an obvious way to assist the correct processes in maintaining mutually consistent knowledge concerning which processes are correct and which are not: To accomplish that goal, one would need to go further by implementing a process group membership service superimposed on the real-time processes in the system. This limitation is apparent when one looks at possible uses for information that can be gathered through such a message exchange: It can be used to adjust protocol parameters in limited ways, but generally cannot be used to solve problems in which the correct processes must have mutually consistent views of shared parameters or other forms of replicated state.

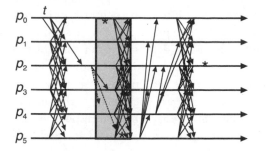

FIGURE 20.7. In the NavTech protocol suite developed by Almeida and Verissimo, periodic background exchanges of state (dark intervals) cut through the normal message traffic, permitting such optimizations as early message delivery and offering information for use in overcoming inconsistency. However, short of running a group membership protocol in the background communication channel, there are limits to the forms of inconsistency that this method can actually detect and correct.

It would be interesting to explore an architecture in which real-time protocols are knowingly superimposed on virtually synchronous process groups, using a high-priority background channel such as the one introduced in Almeida's work to support the virtually synchronous group. With such a hybrid approach, it would be possible to exclude faulty processes from a system within a known delay after the fault occurs; adjust protocol parameters such as the delay to delivery by correct processes, so that the system will adaptively seek out the best possible delay for a given configuration; or combine the use of coherently replicated data and state with real-time updates to other forms of data and state. An approach that uses reserved-capacity, high-priority channels, such as the ones introduced by Almeida, could be used to support such a solution. At the time of this writing, however, I am not aware of any project that has implemented such an architecture.

This brings us back to the normal implementation of the CASD protocol suite. The user of such a protocol must expect that the distributed system as a whole may contain processes that are contaminated by having updated their states on the basis of messages that were handled differently than those at the correct processes. Such processes are not in any way prevented from initiating new messages, which will be received by all processes and which will presumably reflect this inconsistent state in direct or subtle ways. Indeed, over time, almost any process may be viewed as incorrect for one or another run of the protocol; hence, such contamination is likely to be pervasive and is capable of spreading. Mechanisms for ensuring that such a system will converge back into a mutually consistent state should a divergence of states occur are needed when these protocols are used. Alternatively, one can restrict the use of the protocols to forms of information that need not be absolutely correct, or one can use them only as input to algorithms that are tolerant of a certain level of noise in their inputs. One should not, for example, use them as the basis for a coherently replicated cache or as the basis of a safety critical decision that must be made consistently at multiple locations in a system.

The CASD protocols represent an interesting contrast with the virtual synchrony protocols we discussed earlier in this book. Those protocols tolerate similar types of failures, but lack any concept of time and offer no temporal delivery guarantees. On the other hand, they do offer strong logical guarantees: the consistency properties stressed at the time we discussed them. CASD, as we have now seen, lacks this concept of consistency, but has a very strong temporal guarantee when used by processes that are operational within its model. CASD is weakened by the sorts of optimizations that improve its temporal responsiveness, but it would be very unlikely to misbehave if a large value of Δ were considered acceptable. Thus, we have what appears to be a basic tradeoff between logical guarantees and temporal ones. It is intriguing to speculate that such tradeoffs may be fundamental ones.

The tradeoff is also noticeable in the delay of the protocol. For large values of Δ the CASD protocol provides very strong guarantees, but also has a very large latency to delivery. This is the converse of the situation for the virtually synchronous *fbcast* or *cbcast* protocol, which has a very low latency to delivery in the usual case and very strong *logical* guarantees, but no meaningful real-time guarantees. Indeed, the installation of new views of a process group can delay a *cbcast* for a period of time, which grows with the size of the group and which also will be affected by the level of communication traffic in the group at the time of the view installation. Thus, *cbcast* can be understood as rushing to deliver its messages and often doing so in far less time and with far fewer messages than a protocol such as CASD. However, if correct temporal behavior by the correct processes is critical to the application, *cbcast* is not able to offer this guarantee in better than a probabilistic sense (and even a probabilistic argument would involve an analysis of *cbcast* protocol results).

One might characterize the basic difference here as one of pessimism versus optimism. The *cbcast* style of protocols is generally optimistic in its expectations from the system: It is expected that failures will be relatively uncommon events and will be optimized for the earliest possible delivery if a failure does occur. These protocols can give extremely low latency (two or more orders of magnitude better than the CASD style of protocol) and can be extremely predictable in their behavior provided that the network load is light, paging and other delays do not occur, and failures are genuinely infrequent. Indeed, if one could be *certain* that these conditions held, a protocol such as *cbcast* could be the basis of a real-time system, and it would perform perhaps thousands of times better than the timed-asynchronous style of system. But hoping that a condition holds and proving that it holds are two different matters.

The CASD suite of protocols and other work by Cristian's group on the timed-asynchronous model can be viewed as relatively pessimistic, in the sense that for a given set of assumptions, these protocols are designed to expect and to overcome a worst-case execution. If CASD is used in a setting where it is known that the number of failures will be low, the protocol can be optimized to benefit from this. As we have seen, however, the protocol will only work to the degree that the assumptions are valid and that most operational processes will be considered as correct. When this ceases to be the case, the CASD protocols break down and will appear to behave incorrectly from the point of view of processes that, in the eyes of the system model, are now considered to flicker in and out of the zone of correct behavior. But the merit of this protocol suite is that if the assumptions are valid ones, the protocols are *guaranteed* to satisfy their real-time properties.

As noted above, Cristian has also worked on group membership in the timed-asynchronous model. Researchers in the Delta-4 project in Europe have also proposed integrated models in which temporal guarantees and logical guarantees were integrated into a single protocol suite (see Powell [1991], Rodrigues and Verissimo [1989], Rodrigues et al., Verissimo [1993, 1994]). For brevity, however, we will not present these protocols here.

20.3 Adapting Virtual Synchrony for Real-Time Settings

Friedman has developed a real-time protocol suite for Horus, which works by trying to improve the expected behavior of the virtual synchrony group protocols rather than by starting with temporal assumptions and deriving provable protocol behaviors as in the case of CASD (see Friedman and van Renesse [August 1995]). Although in a preliminary state, this work has yielded some interesting results. Among these, Friedman has developed a view installation and message-delivery architecture for Horus, which draws on the Transis idea of distinguishing safe from unsafe message delivery states. In Freidman's protocols, safe states are those for which the virtual synchrony properties hold, while unsafe ones are states for which real-time guarantees can be offered but in which weaker properties than the usual virtual synchrony properties hold.

One way to understand Friedman's approach is to think of a system in which each message and view is delivered twice (he implements this behavior, however, with a more efficient upcall mechanism). The initial delivery occurs with real-time guarantees of bounded latency from sending to reception or bounded delay from when an event that will change the group view occurs to when that view is delivered. However, the initial delivery may occur before the virtually synchronous one. The second delivery has the virtual synchrony properties and may report a group view different from the initial one, albeit in limited ways (specifically, such a view can be smaller than the original one but never larger—processes can fail but not join). The idea is that the application can now select between virtual synchrony properties and real-time ones, using the real-time delivery event for time-critical tasks and the virtually synchronous event for tasks in which logical consistency of the actions by group members are critical. Notice that a similar behavior could be had by placing a Horus protocol stack running a real-time protocol side by side in the same processes with a Horus protocol stack supporting virtual synchrony and sending all events through both stacks. Friedman's scheme also guarantees that event orderings in the two stacks will be the same, unless the time constraints make this impossible; two side-by-side stacks might differ in their event orderings or other aspects of the execution. Break-out 20.2 discusses other research in this field.

In support of the effort to introduce real-time protocols into Horus, Vogels and Mosse have investigated the addition of real-time scheduling features to Horus, message and thread priorities, and preallocation mechanisms whereby resources needed for a computation can be pinned down in advance to avoid risk of delay if a needed resource is not available during a time-critical task.

Other Research Projects

A major goal of the research community is to develop a reliable distributed system for real-time applications. Such a system would have the ability to perform time-critical tasks with guarantees that deadlines will be respected and that other required levels of performance will be achieved, even if failures occur while the system is running.

A distributed real-time system would be built from real-time components: operating systems that support task priorities, preallocation of resources, special-purpose scheduling, and time-driven mechanisms permitting the user to implement real-time applications and to be sure that there is nothing in the operating system that can prevent critical tasks from occurring on time.

Over this, one would layer real-time communication protocols coupled to a failure-detection mechanism of known maximum delay. Since messages may need to be buffered while the failure-detection mechanism has yet to kick in, an analysis of worst-case buffering requirements would also be required, so that

if a failure does occur, the system will not become overloaded due to the accumulation of data being sent to the failed node before reconfiguring itself to exclude that node.

Finally, such a system would need to demonstrate a viable methodology for actually building real-time distributed applications that would demonstrably tolerate failures while continuing to preserve response guarantees within the envelope specified by the designer.

A system such as Horus, with real-time protocols and synchronized clocks, may be suitable for solving problems such as these if they are not excessively demanding. If such a system is used far from its maximum capacity and with very large amounts of memory, and if the real-time limits are relatively weak ones, Horus will predictably operate within the desired bounds. An open problem of considerable interest is to pin down just what those bounds might be, so that we can strengthen such a statement and say that Horus can *guarantee* real-time behavior while also providing fault tolerance and consistency.

An early application of this real-time, fault-tolerant technology is the problem of building a telecommunication switch in which a cluster of computers control the actions taken as telephone calls are received (Figure 20.8). Such an application has a very simple architecture: The switch itself (based on the SS7 architecture) sees the incoming call and recognizes the class of telephone numbers as one requiring special treatment, as in the case of an 800 or 900 number in the United States. The switch creates a small descriptive message, giving the caller's telephone number, the destination, billing information, and a call identification number, and forwards this to a what is called an *intelligent network coprocessor*, or IN coprocessor. The coprocessor (traditionally implemented using a fault-tolerant computer system) is expected to perform a database query based on the telephone numbers and to determine the appropriate routing for the call, responding within a limited amount of time (typically, 100 ms). Typically, the switch will need to handle as many as 10,000 to 20,000 calls per second, dropping no more than some small percentage, and do this randomly even during periods when a failure is being serviced. The switch must never be down for more than a few seconds per year, although individual calls may sometimes have a small chance of not going through and may need to be redialed.

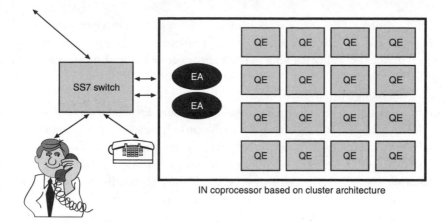

IN coprocessor based on cluster architecture

**FIGURE 20.8. Friedman has experimented with the use of a cluster of comput-
ing systems in support of a demanding real-time telecommunication application. On the left is a single switch, which handles telephone calls in the SS7
switching architecture. Somewhat simplifying the actual setup, we see local
telephones connected to the switch from below and lines connecting to other
switches above. SS7-compatible switches can be connected to adjunct processors, called IN coprocessors, which provide intelligent routing functionality and implement advanced services on behalf of the switch itself—for example,
if an 800-number call is received, the coprocessor would determine which line
to route the call on, and, if call forwarding were in use, the coprocessor would
reroute forwarded calls. Friedman's architecture uses Horus to support a cluster configuration within the IN coprocessor, an approach that provides very
large scalable memory for the query elements (which would typically map a
telephone directory into memory), load-balancing, and fault tolerance.**

The argument in favor of using a cluster of computers for this purpose is that such a system
potentially has greater computing power (and much aggregate main memory) than any single
processor could have. This may translate to the ability to keep a very large database in memory
for rapid access (spread among the nodes) or of executing a more sophisticated query strategy.
Moreover, whereas the upgrading of a fault-tolerant coprocessor may require that the switch be
shut down, one can potentially upgrade a cluster-style computer one node or one program at a time.

Without getting into the details, Friedman has demonstrated that systems such as Horus can
indeed be used to support such a model. He describes a system emulating this configuration of
telephone switch, servicing 22,000 calls per second while dropping no more than 1 to 3 percent even when a failure or recovery is actually being serviced. Friedman's design involves a pair
of external adapter nodes (EAs), which sense incoming calls and dispatch the corresponding
query onto pairs of query-processing nodes (QEs). Friedman batches requests and uses an innovative real-time, fault-tolerant protocol to optimize for the very high processing loads characterizing the application (see Friedman and Birman).

To solve this problem, Friedman's work combines the real-time mechanisms cited above with a number of other innovations, and it is fair to say that the application is not a straightforward one. However, the benefits of being able to use a cluster-style computing system in this manner could be dramatic: Such systems are quite inexpensive, and yet they may bring a great deal of performance and flexibility to the application, which would otherwise be very constrained by the physical limitations typical of any single-processor solution. Friedman's initial work focuses on memory scalability, but he is now extending the approach to seek a performance benefit from scale, as well as the possibility of benefit from memory mapping the database.

Although cast in the context of a telephone switching application, it should be noted that the type of real-time, client/server architecture being studied in Friedman's work is much more general. We have seen in earlier chapters of this book that the great majority of distributed systems have a client/server architecture, and this is also true for real-time systems, which typically look like client/server systems with time-critical response deadlines superimposed upon an otherwise conventional architecture. Thus, Friedman's work on telephone switching could also be applicable to process control systems, air traffic control systems, and other demanding applications that combine fault tolerance and real-time constraints.

Other work in this area includes Marzullo's research on the CORTO system, which includes such features as *periodic process groups*. These are process groups whose members periodically and within a bounded period of real time initiate synchronized actions. Marzullo has studied minimizing the communication overhead required in support of this periodic model, integrating real-time communication with other periodic or real-time actions, priority inversion in communication environments, and other topics in the area.

20.4   Related Reading

On clock synchronization, see the review in Simons et al.; other references include Cristian (1989), Kopetz and Ochsenreiter, Lamport (1984), Lamport and Melliar-Smith, Marzullo (1984), Srikanth and Toueg.

On the a-posteriori method: (see Clegg and Marzullo, Verissimo and Rodrigues).

On the CASD protocol: (see Cristian [1996], Cristian and Schmuck, Cristian et al. [1985, 1990], Gopal et al.).

On the MARS system: (see Damm et al., Kopetz and Ochsenreiter, Kopetz and Verissimo).

On Delta-4: (see Powell [1991, 1994], Rodrigues and Verissimo [1989], Rodrigues et al., Verissimo [1993, 1994]).

On real-time work with Horus: (see Friedman and Birman, Friedman and van Renesse [August 1995]).

CHAPTER 21 ✧ ✧ ✧ ✧

Transactional Systems

CONTENTS

21.1 Review of the Transactional Model

We first encountered the transactional execution model in Chapter 7, in conjunction with client/server architectures. As noted at that time, the model draws on a series of assumptions to arrive at a style of computing that is uniquely well suited to applications operating on databases. In this chapter we consider some of the details that Chapter 7 did not cover: notably the issues involved in implementing transactional storage mechanisms and the problems that occur when transactional architectures are extended to encompass transactional access to distributed objects in a reliable distributed system.

Without repeating the material covered earlier, it may be useful to start by reviewing the transactional model in light of what we have subsequently learned about other styles of distributed computing and distributed state. Notice first that the assumptions underlying the transactional approach are quite different from those underlying the virtual synchrony model. Transactional applications are expected to be structured in terms of the basic transactional constructs: *begin*, *read*, *update*, and *commit* or *abort*. They are assumed to have been written in isolation, so that they will operate correctly when applied to an idle database system in an initially consistent state. Each transaction, in effect, is a function transforming the database from a consistent state into a new consistent state. The database, for its part, is a well-defined entity: It manages data objects, has a limited interface by which transactions operate on it, and manages information using operations with well-understood semantics.

General-purpose distributed systems, and many client/server applications, match such a model only to a limited degree. The computations performed may or may not act upon saved data in a database, and even when they do, it will be difficult to isolate data access operations from other types of message-based interactions and operations.

Additionally, the basic reliability goals of the transactional model are tied closely to its programming model. The transactional reliability guarantees are basically this: If a server or client crashes, prior to the commit point of a transaction, a complete rollback of the server state will occur—it is as if the transaction had never been executed. There is a strong emphasis on recoverability of the database contents after a crash: Any committed transaction will have effects that survive repeated server crashes and restarts. This strong separation of computation from data, coupled with an emphasis on recoverability (as opposed, for example, to continuous availability), distinguishes the transactional approach from the process group replication schemes we have studied in the preceding chapters of this book.

One could ask whether general-purpose distributed programs couldn't be considered as transactional programs, in this manner mapping the general case to the transactional one. This turns out to be very hard to do. General purpose distributed programs lack a well-defined *begin* or *commit* point, and it would not always be practical to introduce such a structure—sometimes one could do so, but often it would be difficult. These programs lack a well-defined separation of program (transactional client) from persistent state (database); again, some applications could be represented this way, but many could not. Indeed, it is not unreasonable to remark that because of the powerful support that exists for database programming on modern computer systems, most database applications are in fact implemented using database systems. The applications that are left over are the ones where a database model either seems unnatural, fails to match some sort of external constraint, or would lead to extremely inefficient execution. This perspective argues that the distributed applications of interest to us will probably split into the transactional ones and others, which are unlikely to match the transactional model even if one tries to force them into it.

Nonetheless, the virtual synchrony model shares some elements of the transactional one: The serialization ordering of the transactional model is similar to the view-synchronous addressing and ordered delivery properties of a multicast to a process group.[1] Virtual synchrony can be considered as having substituted the concept of a multicast for the concept of the transaction itself: In virtual synchrony one talks about a single operation that affects multiple processes, while in transaction systems one talks about a sequence of *read* and *update* operations that are treated as a single atomic unit. The big difference is that whereas explicit data semantics are natural in the context of a database, they are absent in the communication-oriented world we considered when studying the virtual synchrony protocols.

As we examine the transactional approach in more detail, it is important to keep these similarities and differences in mind. One could imagine using process groups and group multicast to implement replicated databases, and in fact there have been several research projects that have done just this. A great many distributed systems combine transactional aspects with nontransactional ones, using transactions where a database or persistent data structure is present and using virtual synchrony to maintain consistently replicated in-memory structures to coordinate the actions of groups of processes and so forth. The models are different in their assumptions and goals, but are not incompatible. Indeed, there has been work on merging the execution models themselves, although we will not discuss this here.

Perhaps the most important point is the one stated at the start of this chapter: Transactions focus primarily on recoverability and serializability, while virtual synchrony focuses primarily on order-based consistency guarantees. This shift in emphasis has pervasive implications, and even if one could somehow merge the models, it is likely that they would still be used in different ways. Indeed, it is not uncommon for distributed system engineers to try to simplify their lives by using transactions throughout a complex distributed system as its sole source of reliability or by using virtual synchrony throughout, exploiting dynamically uniform protocols as the sole source of external consistency. Such approaches are rarely successful.

---

[1]One can imagine doing a multicast by *reading the view of the group and then writing to the group members* and updating the view of the group by *writing to the group view*. Such a transactional implementation of virtual synchrony would address some aspects of the model, such as view synchronous addressing, although it would not deal with others, such as the ordered gap-freedom requirement. More to the point, it would result in an extremely inefficient style of distributed computing, because every multicast to a process group would now require a database update. The analogy, then, is useful because it suggests that the fundamental approaches are closely related and differ more at the level of how one engineers such systems to maximize performance than in any more basic way. However, it is not an architecture one would want to implement!

It can be difficult to use transactions as the basic model for communication-based systems because the interactions between transactional applications are necessarily structured into invocations of operations and manipulation of persistent data. To send a message, a process will potentially need to write the message into a database, from which the destination process is expected to read it. Moreover, there is the need to arrange for the transaction to begin, to ensure that the operations issued are identifiable as part of the same transaction (they need some form of unique transaction ID), and to ascertain that the transaction is committed in a well-defined way. Finally, the application itself needs a sensible way to deal with abort.

While this matches well with database access, it is hard to map such a model to an air traffic application in which a controller works with a continuously updated screen, interacting with pilots, other controllers, and various services. These applications will typically involve running more than one program at a time; logically, some of the actions taken are part of the same transaction, but for long-running programs it may not be clear which actions belong to which transaction or when the transaction should begin or end. It would make sense to have the operations associated with different flights treated as different transactions, but it is not clear how the database system (if one is present) would recognize this distinction, unless the operations were entirely structured as transactional accesses to a shared database server.

Transactions pay a high price to guarantee the recoverability of persistent data. But in a distributed setting, which replicates critical data, recovering a very old and stale copy of the state of some server may be pointless; only very large databases would be worth the trouble of recovering into a known state. The idea that transactions are coded to execute against an idle system and to run in isolation is fundamentally at odds with the concept of a system in which long-running services cooperate explicitly, both to load-balance work internally and to coordinate actions with one another. Also, there is the issue of aborts: precisely what should an air traffic controller do if a necessary action is aborted for reasons internal to the execution model? Worse, if an abort occurs deep within a system, how should the program recover? Many systems have no obvious concept of an operator and must automatically recover from all conditions occurring during execution.

Similar objections can be raised for the case where virtual synchrony is applied to what are basically transactional databases. Here, the issue is perhaps less that of a constraining programming style than that the specific programming tools we have discussed seem disconnected from the specific needs of a distributed database system. In fact, there has been considerable interest in applying process group concepts to replication and other aspects of distributed data management, notably through the Newtop protocol of the Arjuna system (see Ezhilhelvan et al.), and we will point to some of the special requirements occurring from such explorations. To the application developer, however, the point is that if an application contains what are logically databases, transactional access to them may be highly appropriate, while if the application contains groups of dynamically adaptive, highly available processes and servers, those may be more appropriately treated using virtual synchrony. Virtual synchrony is also a more natural way to *manage* complex distributed systems; transactions seem poorly matched to this goal. Very elaborate reliable distributed systems will probably need both technologies.

In the remainder of this chapter, we move beyond these philosophical issues to more detailed technical ones concerning the implementation of transactional systems for distributed environments.

21.2 Implementation of a Transactional Storage System

In this section we briefly review some of the more important techniques used in implementing transactional storage systems. Our purpose is not to be exhaustive or even try to present the best techniques known, since this is such an important issue that to cover it in detail is beyond the scope of this book. However, there are several excellent books available on this subject (see Bernstein et al., Gray [1979], Gray and Reuter). Rather, we focus on basic techniques with the purpose of building insight into the reliability mechanisms needed when implementing transactional systems.

21.2.1 Write-Ahead Logging

A *write-ahead log* is a data structure used by a transactional system as a form of backup for the basic data structures that compose the database itself. Transactional systems *append* to the log by writing *log records* to it. These records can record the operations that were performed on the database, their outcome (commit or abort), and can include before or after images of data updated by an operation. The specific content of the log will depend upon the transactional system itself.

We say that a log satisfies a *write-ahead property* if there is a mechanism by which records associated with a particular transaction can be safely and persistently flushed to disk before (ahead of) updates to data records being done by that transaction. In a typical use of this property, the log will record before images (old values) of the records a transaction updates and commit records for that transaction. When the transaction does an update, the database system will first log the old value of the record being updated and then update the database record itself on disk. Provided that the write-ahead property is respected, the actual order of I/O operations done can potentially be changed to optimize use of the disk. Should the server crash, it can recover by reviewing the uncommitted transactions in the log and reinstalling the original values of any data records these had modified. The transactions themselves will now be forced to abort, if they have not already done so. Such a process rolls back the transactions that have not committed, leaving the committed ones in place. Later, the log can be garbage collected by cleaning out records for committed transactions (which will never need to be rolled back) and those for uncommitted transactions that have been successfully aborted (and hence need not be rolled back again). (See Figure 21.1.)

Although a write-ahead log is traditionally managed on the disk itself, there has been recent research on the use of nonvolatile RAM memory or active replication techniques to replace the log with some form of less-expensive structure (see Liskov et al.). Such trends are likely to

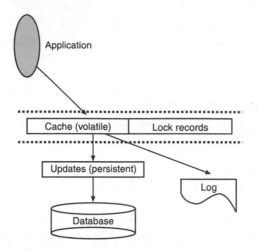

FIGURE 21.1. Overview of a transactional database server. Volatile data are used to maintain a high-speed cache of database records and for storage of lock records for uncommitted transactions. An updates list and the database itself store the data, while a write-ahead log is used to enable transactional rollback if an abort occurs and to ensure that updates done by committed transactions will be atomic and persistent. The log saves before or after images of updated data and lock records associated with a transaction running its commit protocol. Log records can be garbage collected after a transaction commits or aborts and the necessary updates to the database have been applied or rolled out.

continue as the relative performance gap between disks (which seems to have reached a performance limit of approximately 10 ms per disk access for a fast disk and as much as 40 to 50 ms per access for a slow one) and communication continue to grow.

21.2.2 Persistent Data Seen through an Updates List

Not all transactional systems perform updates to the persistent database at the time they are first issued. The decision to do updates directly depends on several factors; among these are the frequency with which transactions are expected to abort and the likelihood that the transaction will rewrite the same record repeatedly. The major alternative to performing direct updates on the database itself is to maintain some form of *updates list* in which database records that have been updated are saved. Each access to the database is first filtered through this updates storage object, and if the record being accessed has changed, the changed version is returned. The database itself is only accessed if the updates list does not contain the desired item, and any update made to the database is instead applied to this updates list.

The advantage of such a structure is that the database itself can be maintained in a very efficient search and access structure without requiring costly structural updates as each operation occurs. Periodically, the database can be updated to merge the committed updates from the updates list into the persistent part of the database, but this does not need to be done until there is a convenient time, perhaps while the database as a whole is under very light load. Moreover, as we will see shortly, the updates list can be generalized to deal with the nested transactions that occur when transactional databases are constructed using abstract data types.

The updates list data structure, if present, should not be confused with a cache or buffer pool. A database cache is a volatile data structure used to accelerate access to frequently used data items by maintaining them in high-speed memory. The updates list is a persistent data structure, which is logically part of the database itself. Its role is to provide the database system with a way of doing database updates without reorganizing the secondary index and other access structures needed to rapidly access items in the main portion of the database.

21.2.3 Nondistributed Commit Actions

To commit a transaction, it is necessary to ensure that its effects will be atomic even if the database server or client program fails during the commit procedure. In the nondistributed case, the required actions are as follows. First, all log records associated with updates done by the transaction are forced to the disk, as are *lock records* recording the locks currently held by the transaction. Once these actions are taken, the transaction is *prepared to commit*. A log record containing the *commit bit* is now written to disk; once it is recorded in a persistent manner in the log, the transaction is said to have *committed*.

Next, updates done by the transaction are applied to the updates list or database. In many transactional systems, this updating is done while the transaction is running, in which case this step (and the forcing of log records to disk) may have already occurred before the transaction reached the commit point.

Finally, when the updates have all been performed, the locks associated with the transaction are released and any log records associated with the transaction are freed for reuse by other transactions. The transaction is now said to be *stable*.

To abort a transaction, the log records associated with it are scanned and used to roll back any updates that may have been performed. All locks associated with the transaction are released, and the log records for the transaction are freed.

In the event that the client process should crash before requesting that the transaction commit or abort, the database server may *unilaterally abort* the transaction. This is done by executing the abort algorithm and later, if the client ever presents additional requests to the server, refusing them and returning an *already aborted* exception code.

Finally, in the event that the database server should crash, when it recovers it must execute a log-recovery procedure before reenabling access to the database. During this process, any transactions that are not shown as committed are aborted, and any updates that may have been done are backed out. Notice that if the log stored before images, backing out updates can be done by simply reinstalling the previous values of any records that were written by the transaction; this operation can be done as many times as necessary if the database server crashes repeatedly before recovering (i.e., the recovery operation is *idempotent*, meaning that it can be performed repeatedly with the same effect as if it had been performed only once).

For transactions shown as committed in the log, the database server recovers by completing the commit procedure and then freeing the log records. Abstractly, the database server can be thought of as recovering in a state where the committed transactions continue to hold any locks that they held at the time of the commit; this will be useful in the case of a distributed transaction on multiple databases.

21.3 Distributed Transactions and Multiphase Commit

When a transaction operates on multiple databases, it is said to be a *distributed transaction*. The commit problem now becomes the multiphase commit problem we discussed in Section 13.6.1. To commit, each participating database server is first asked to *prepare to commit*. If the server is unable to enter this state, it votes for abort; otherwise, it flushes log records and agrees that it is prepared. The transaction commits only if all the participating servers are prepared to commit; otherwise, it aborts. For this purpose, the transactional commit protocols presented earlier can be used without any modifications at all.

In the case of a database server recovery to the prepared state of a transaction, it is important for the server to act as if that transaction continues to hold any locks it held at the time it first became prepared to commit (including read locks, even if the transaction were a read-only one from the perspective of the database server in question). These locks should continue to be held until the outcome of the commit protocol is known and the transaction can complete by committing or aborting. When a transaction has read data at a server that subsequently crashes and recovers, having lost its read locks, before the transaction is prepared to commit, it must be aborted; otherwise, these broken read locks can permit some other transaction to acquire update locks, which should have been serialized after the transaction that first held the read locks, and to commit. Such a behavior is easily seen to result in nonserializable executions. Thus, a distributed transaction must include all database servers it has accessed in its commit protocol, not just the ones at which it performed updates.

21.4 Transactions on Replicated Data

A transactional system can replicate data by applying updates to all copies of the database, while load-balancing queries across the available copies (in a way that will not change the update serialization order being used). In the most standard approach, each database server is treated as a separate database, and each update is performed by updating at least a quorum of replicas. The transaction aborts if fewer than a quorum of replicas are operational. It should be noted that this method of replication, although much better known than other methods, performs poorly in comparison with the more-sophisticated method described in Section 21.7.

The reality is that few existing database servers make use of replication for high availability; therefore, the topic is primarily of academic interest. Transactional systems that are concerned with availability more often use primary-backup schemes in which a backup server periodically is passed a log of committed actions performed on a primary server. Such a scheme is faster (because the backup is not included in the commit protocol), but it also has a window during which updates by committed transactions can be temporarily lost (e.g., if the log records for a committed transaction have not yet reached the backup when the primary crashes). When this occurs, the lost updates are rediscovered later, after the primary recovers, and are either merged into the database or, if this would be inconsistent with the database state, user intervention is requested.

Another option is to use a spare computer connected by a dual-ported disk controller to a highly reliable RAID-style disk subsystem. If the primary computer on which the database is running fails, it can be restarted on the backup computer with little delay. The RAID disk system provides a degree of protection against hardware failures of the stored database in this case.

Although database replication for availability remains uncommon, there is a small but growing commercial market for systems that support distributed transactions. The limiting factor for widespread acceptance of these technologies remains performance. Whereas a nonreplicated, nondistributed transactional system may be able to achieve thousands or tens of thousands of short update and read transactions per second, distributed transactional protocols and replication slow such systems to perhaps hundreds of updates per second. Although such performance levels are adequate to sustain a moderately large market of customers, who value high availability or distributed consistency more than performance, the majority of the database marketplace remains focused on scalable, high-performance systems. Such customers are apparently prepared to accept the risk of downtime because of hardware or software crashes to gain an extra factor of 10 to 100 in performance. However, it should again be noted that process group technology may offer a compromise: combining high performance with replication for increased availability or scalable parallelism. We will return to this issue in Section 21.6.5.

21.5 Nested Transactions

Recall that at the beginning of this book, we suggested that object-oriented distributed system architectures are a natural match with client/server distributed system structures. This raises the question of how transactional reliability can be adapted to object-oriented distributed systems.

As we saw in Chapter 6, object-oriented distributed systems are typically treated as being composed of *active objects*, which invoke operations on *passive objects*. To some degree, of course, the distinction is an artificial one, because some passive objects have active computations associated with them—for example, to rearrange a data structure for better access behavior. However, to keep this section simple, we will accept the division. We can now ask if the active objects should be treated as transactions and the passive objects as small database servers.

Such a step leads to what are called *nested transactions* (see Moss). The sense in which these transactions are nested is that when an active object invokes an operation on an abstract object stored within an object-oriented database, that object may implement the operation by performing a series of operations on some other, more primitive, database object. An operation that inserts a name into a list of names maintained in a name server, for example, may be implemented by performing a series of updates on a file server in which the name list and associated values are actually stored. One now will have a tree-structured perspective on the transactions themselves, in which each level of object performs a transaction on the objects below it.

In such a tree, only the topmost level corresponds to an active object or program in the conventional sense. The intermediate levels of code correspond to the execution of methods (procedures) defined by the passive objects in the database. For these passive objects, transactions begin with the operation invocation by the invoking object and end when a result is returned—that is, procedure executions (operation invocations) are treated as starting with an implicit *begin* and ending with an implicit *commit* in the normal return case. Error conditions can be mapped to an *abort* outcome. The active object at the very top of the tree, in contrast, is said to *begin* a *top-level transaction* when it is started and to *commit* when it terminates normally. A nested transaction is shown in Figure 21.2.

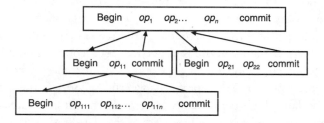
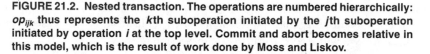

FIGURE 21.2. Nested transaction. The operations are numbered hierarchically: op_{ijk} thus represents the kth suboperation initiated by the jth suboperation initiated by operation i at the top level. Commit and abort becomes relative in this model, which is the result of work done by Moss and Liskov.

The nested transaction model can be used for objects that are colocated on a single object repository or for objects distributed among multiple repositories. In both cases, the basic elements of the resulting system architecture are similar to the approach used for a single-level transaction system. The details differ, however, because of the need to extend the concurrency control mechanisms to deal with nesting.

The easiest way to understand nested transactions is to view each subtransaction as a transaction that runs in a context created by its parent transaction and any committed sibling subtransactions the parent executed prior to it. Thus, operation op_{21} in Figure 21.2 should see a database state that corresponds to having executed the subtransaction below op_1 and committing it, even though the effects of that subtransaction will not become permanent and globally visible until the main transaction commits. This approach can be extended to deal with internal concurrency—for example, if op_1 were executed in parallel with op_2.

Moss proposed a concept of lock and data version inheritance to accomplish this goal. In his approach, each subtransaction operates by creating new versions of data items and acquiring locks, which are *inherited* by the subtransaction's immediate parent when the subtransaction commits or which return to the state prior to when the subtransaction began if it aborts. These inherited locks and data values are accessible to other subtransactions of the parent that now retains them, but they remain inaccessible to transactions outside of its scope. Moss's Ph.D. dissertation includes proof that this approach yields a nested version of two-phase locking, which guarantees serializable executions.

To implement a nested transaction system, it is usual to start by extending the updates list and locking subsystems of the database so that it will know about transactional nesting. Abstracting, the resulting architecture is one in which each lock and each data item are represented as a *stack* of locks or data items. When a new subtransaction is spawned, the abstract effect is to push a new copy of each lock or data item onto the top of the stack. Later, as the subtransaction acquires locks or updates these data items, the copy at the top of the stack is changed. Finally, when the subtransaction aborts, the topmost stack element is discarded; if it commits, the topmost stack item is popped, as well as the one below it, and then the topmost item is pushed back onto the stack. In a similar manner, the stack of lock records is maintained; the one difference is that if a subtransaction obtains a different class of lock than that held by the parent transaction, the lock is left in the more restrictive of the lock modes.

In practice, nested transactional systems are designed to be lazy, so the creation of new versions of data items or new lock records is delayed until absolutely necessary. Thus, the stack of data items and lock records is not actually generated unless it is needed to perform operations.

A similar abstraction is used to handle the commit and abort mechanisms. Abstractly, as a nested transaction executes, each level of the transaction tracks the data servers it visits, maintaining a list of *commit participants*. In order to commit or abort, the transaction will interact with the servers on this list. In practice, however, such an approach would require repeated execution of the multiphase commit protocols, which will have to run once for each internal node in the transaction tree and one more time for the root! Clearly, this would be prohibitively expensive.

To avoid this problem, Liskov's ARGUS group proposed an approach in which commit decisions are *deferred*, so that only the top-level commit protocol is actually executed as a multiphase protocol (see Ladin et al. [1990], Liskov and Scheifler, Liskov et al. [1987]). Intermediate commits are optimistically assumed successful, while aborts are executed directly by informing the commit participants of the outcome. Now, the issue occurs of how to handle an access by a subtransaction to a lock held by a sibling subtransaction or to a data item updated by a sibling. When this occurs, a protocol is executed by which the server tracks down a mutual parent and interrogates it about the outcomes, commit or abort, of the full transaction stack separating the two subtransactions. It then updates the stacks of data items and locks accordingly and allows the operation to proceed. In the case where a transaction rarely revisits data items, such a strategy reduces the cost of the nest transactional abstraction to the cost of a flat one-level transaction; the benefit is smaller as the degree of interference increases.

The reader may recall that Liskov's group also pioneered in the use of optimistic (or lazy) concurrency control schemes. Such approaches, which can be recognized as analogous to the use of asynchronous communication in a process group environment, allow a system to achieve high levels of internal concurrency, improving performance and processor utilization time by eliminating unneeded wait states—much as an asynchronous multicast eliminates delay when a multicast is sent in favor of later delays if a message arrives out of order at some destination. At the limit, these approaches converge upon one in which transactions on nonreplicated objects incur little overhead beyond that of the commit protocol run at the end of the top-level transaction, while transactions on replicated objects can be done largely asynchronously but with a similar overhead when the commit point is reached. These costs are low enough to be tolerable in many distributed settings, and it is likely that at some future time, a commercially viable, high-performance, object-oriented transaction technology will emerge as a serious design option for reliable data storage in distributed computing systems.

21.5.1 Comments on the Nested Transaction Model

Nested transactions were first introduced in the ARGUS project at MIT (see Moss) and were rapidly adopted by several other research projects, such as CLOUDS at the Georgia Institute of Technology and CMU's TABS and CAMELOT systems (see Spector) (predecessors of ENCINA, the commercial product marketed by Transarc). The model proved elegant but also difficult to implement efficiently and sometimes quirky. The current view of this technology is that it works best on object-oriented databases, which reside mostly on a single-storage server, but that it is less effective for general-purpose computing in which objects may be widely distributed and in which the distinction between active and passing objects can become blurred.

It is worthy of note that the same conclusions have been reached about database systems. During the mid-1980s, there was a push to develop database operating systems in which the database would take responsibility for more and more of the tasks traditionally handled by a general-purpose operating system. This trend culminated in systems such as IBM's AS/400 database server products, which achieve an extremely high level of integration between database and operation system functionality. Yet there are many communication applications that suffer a heavy performance penalty in these architectures, because direct point-to-point messages must be largely replaced by database updates followed by a read. While commercial products that take this approach offer optimizations capable of achieving the performance of general-purpose operating systems, users may require special training to understand how and when to exploit them. The trend at the time of this writing seems to be to integrate database servers into general-purpose distributed systems by including them on the network, but running nondatabase operating systems on the general-purpose computing nodes that support application programs. This is sometimes called the "open systems" approach to networked computing.

The following example illustrates the sort of problems that can occur when transactions are applied to objects that fit poorly with the database computing model. Consider a file system directory service implemented as an object-oriented data structure: In such an approach, the

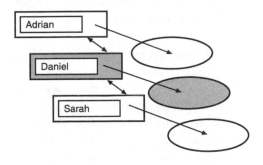

FIGURE 21.3. While a directory is being updated (in this case, the entry corresponding to "Daniel"), other transactions may be prevented from scanning the associated directory node by locks upon it, even if they are searching for some other record, such as the one corresponding to "Sarah" or "Adrian." Although a number of schemes can be used to work around such problems, they require sophistication by the developer, who must consider cases that can occur because of concurrency and arrange for concurrent transactions to cooperate implicitly to avoid inefficient patterns of execution. Such metadesign considerations run counter to the principle of independent design on which transactions are based and make the overall approach hard to use in general-purpose operating system settings.

directory would be a linked list of named objects, associating a name with some sort of abstract object corresponding to what would be a file in a conventional file system. Operations on a directory include searching it, scanning it sequentially, deleting and inserting entries, and updating the object nodes. Such a structure is illustrated in Figure 21.3.

A typical transaction in such a system might be a program that displays a graphical interface by which the user enters a name and then looks up the corresponding object. The contents of the object could then be displayed for the user to edit, and the changes, if any, saved into the object when the user finishes. Interfaces such as this are common in modern operating systems, such as Microsoft's Windows 95 or some of the more advanced versions of UNIX.

Viewed as an instance of a nested transaction, this program begins a transaction and then reads a series of directory records looking for the one that matches the name the user entered. The corresponding node would then be locked for update while the user scrutinizes its contents and updates it. The transaction commit would occur when the record is saved in its changed state. An example of such a locked record is highlighted in gray in Figure 21.3.

But now consider the situation if the system has any concurrency at all. While this process is occurring, the entire data structure may potentially be locked against operations by other transactions, even if they are not interested in the same record as the user is preparing to update! The problem is that any simplistic application of the nested transaction concurrency control rules will leave the top-level records that bind names to objects locked for either read or update and will leave all the directory records scanned while searching for the name entered by the user locked for reads. Other transactions will be unable to acquire conflicting forms of locks on these records and may thus be delayed until the user (who is perhaps heading down the hall for a cup of coffee!) terminates the interaction.

Many extensions to the nested transaction model have been proposed to cope with this sort of problem. ARGUS, for example, offers a way to perform operations outside the scope of a transaction and includes a way for a transaction to spawn new top-level transactions from deep within a nested execution. Weihl argues for a relaxation of the semantics of objects such as directory servers: In his view, overspecification of the interface of the directory service is the cause of this sort of problem, and he suggests extensions, such as unordered queues and nondeterministic interfaces, which correspond to implementations that give better performance. In his approach one would declare the directory to be an unordered semiqueue (an unordered set) and would implement a nontransactional search mechanism in which the search order is nondeterministic and does not need to involve an access to the locked record until all other records have been scanned. Shasha has developed families of concurrent data structures, in which semantic information is exploited to obtain highly concurrent transactional implementations of operations specific to the data type. Still other researchers have proposed that such problems be addressed by mixing transactional and nontransactional objects and have offered various rules to adapt the ACID properties to such an environment.

The example we gave above occurs in a data structure of unsurpassed simplicity. Similar issues would also be encountered in other data structures, such as doubly linked lists where order *does* matter, trees, hash tables, stacks, and so forth. In each case, a separate set of optimizations is needed to achieve optimal levels of concurrency.

Those who have worked with transactions have concluded that although the model works very well for databases, there are problems for which the transaction model is poorly matched. The argument is basically this: Although the various solutions suggested in the literature do work, they have complicated side-effects (interested readers may want to track down the literature concerned with terminating what are called "orphans of an aborted nested transaction," a problem that occurs when a nested transaction having active subtransactions aborts, eliminating the database state in which those subtransactions were spawned and exposing them to various forms of inconsistency). The resulting mechanisms are complex to work with, and many users might have problems using them correctly; some developers of nested transaction systems have suggested that only experts would be likely to actually build transactional objects, while most real users would work with libraries of preconstructed objects. Thus, even if mechanisms for overcoming these issues do exist, it seems clear that nested transactions do not represent an appropriate general-purpose reliability solution for nondatabase applications.

The commercial marketplace seems to have reached a similar decision. Transactional systems consist largely of relational databases (which may be used to store abstract data types, but in which the relationships between the objects are represented in the transactional tables) or transactional file-structured systems. Although many distributed, object-oriented, transactional systems have been developed, few seem to have made the transition from research prototype to commercial use.

In particular, many of the problems that are most easily solved using process groups are quite hard to solve using transactional solutions. The isolation property of transactions runs counter to the idea of load-balancing in a service replicated at several nodes or of passing a token within a group of cooperating processes. Conversely, however, transactional mechanisms bring a considerable infrastructure to the problem of implementing the ACID properties for applications that act upon persistent data stored in complex data structures, and this infrastructure is utterly lacking in the virtual synchrony model.

The implication is that while both models introduce reliability into distributed systems, they deal with very different reliability goals: recoverability on the one hand and availability on the other. While the models can be integrated so that one could use transactions within a virtual synchrony context and vice versa, there seems to be little hope that they could be merged into a single model that would provide all forms of reliability in a single, highly transparent environment. Integration and coexistence are, therefore, a more promising goal, which seems to be the one favored by industry and research groups.

21.6 Weak Consistency Models

There are some applications in which one requires most aspects of the transactional model, but where serializability in the strict sense is not practical to implement. Important among these are distributed systems in which a database must be accessed from a remote node, which is sometimes partitioned away from the system. In this situation, even if the remote node has a full copy of the database, it is potentially limited to read-only access. Even worse, the impossibility of building a nonblocking commit protocol for partitioned settings potentially prevents these read-only transactions from executing on the most current state of the database, since a network partitioning failure can leave a commit protocol in the prepared state at the remote site.

In practice, many distributed systems treat remote copies of databases as a form of second-class citizen. Such databases are often updated by periodic transfer of the log of recently committed transactions and are used only for read-only queries. Update transactions execute on a *primary copy* of the database. This approach avoids the need for a multiphase commit but has limited opportunity to benefit from the parallelism inherent in a distributed architecture. Moreover, the delay before updates reach the remote copies may be substantial, so that remote transactions will often execute against a stale copy of the database, with outcomes that may be inconsistent with the external environment—for example, a remote banking system may fail to reflect a recent deposit for hours or days.

In the following text, we briefly present some of the mechanisms that have been proposed as extensions to the transactional model to improve its usefulness in settings such as these.

21.6.1 Epsilon Serializability

Originally proposed by Pu, this is a model in which a preagreed strategy is used to limit the possible divergence between a primary database and its remote replicas (see Pu). Epsilon refers to the case where the database contains numeric data, and it is agreed that any value read by a transaction is within ε of the correct one.

Suppose, for example, that a remote transaction is executed to determine the current value of a bank balance, and the result obtained is \$500. If $\varepsilon = \$100$, we can conclude that the actual balance in the database (in the primary version) is no less than \$400 and no more than \$600. The benefit of this approach is that it relaxes the need to run costly synchronization protocols between remote copies of a database and the primary: Such protocols are only needed if an update might violate the constraint.

Continuing with our example, suppose we know that there are two replicas and one primary copy of the database. We can now allocate ranges within which these copies can independently perform update operations without interacting with one another to confirm that it is safe to do so. Thus, the primary copy and each replica might be limited to a maximum cumulative update of \$50 (larger updates would require a standard locking protocol). Even if the primary and one replica perform maximum increments to the balance of \$50, respectively, the remaining replica would still see a value within \$100 of the true value, and this remains true for any update the third replica might undertake. In general, the rule for this model is: The minimum and maximum cumulative updates done by other copies must be bounded by ε to ensure that a given copy will see a value within ε of the true one.

21.6.2 Weak and Strong Consistency in Partitioned Database Systems

During periods when a database system may be completely disconnected from other replicas of the same database, we will in general be unable to determine a safe serialization order for transactions originating at that disconnected copy.

Suppose that we want to implement a database system for use by soldiers in the field, where communication may be severely disrupted. The database could be a map showing troop positions, depots, the state of roads and bridges, and major targets. In such a situation, one can imagine transactions of varying degrees of urgency. A fairly routine transaction might be to update the record showing where an enemy outpost is located, indicating that there has been no change in the status of the outpost. At the other extreme would be an emergency query seeking to locate the closest medic or supply depot capable of servicing a given vehicle.

Serializability considerations underlie the consistency and correctness of the real database, but one would not necessarily want to wait for serializability to be guaranteed before making an informed guess about the location of a medical team. Thus, even if a transactional system requires time to achieve a completely stable ordering on transactions, there may be cases in which one would want it to process at least certain classes of transactions against the information presently available to it.

In his Ph.D. dissertation, Amir addressed this problem using the Transis system as a framework within which he constructed a working solution (see Amir); see also Amir et al., Davidson et al., Terry et al. His basic approach was to consider only transactions that can be represented as a single multicast to the database, which is understood to be managed by a process group of servers. (This is a fairly common assumption in transactional systems, and in fact most transactional applications indeed originate with a single database operation, which can be represented in a multicast or remote procedure call.) Amir's approach was to use *abcast* (the dynamically uniform or safe form) to distribute update transactions among the servers, which were designed to use a serialization order deterministically related to the incoming *abcast* order. Queries were implemented as local transactions requiring no interaction with remote database servers.

As we saw earlier, dynamically uniform *abcast* protocols must wait during partitioning failures in all but the primary component of the partitioned system. Thus, Amir's approach is subject to blocking in a site that has become partitioned away from the main system. Such a site may, in the general case, have a queue of undeliverable and partially ordered *abcasts*, which are waiting either for a final determination of their relative ordering or for a guarantee that dynamic uniformity will be achieved. Each such *abcast* corresponds to an update transaction, which could change the database state, perhaps in an order-sensitive way, and which cannot be safely applied until this information is known.

What Amir does next depends on the type of request presented to the system. If a request is urgent, it can be executed either against the last known completely safe state (ignoring these incomplete transactions) or against an approximation to the correct and current state (by applying these transactions, evaluating the database query, and then aborting the entire transaction). Finally, a normal update can simply wait until the safe and global ordering for the corresponding transaction is known, which may not occur until communication has been reestablished with remote sites.

Amir's work is not the only effort to have arrived at this solution to the problem. Working independently, a group at Xerox PARC developed a very similar approach to disconnected availability in the Bayou system (see Terry et al.). Their work is not expressed in terms of process groups and totally ordered, dynamically uniform, multicast, but the key ideas are the same. In other ways, the Bayou system is more sophisticated than the Transis-based one: It includes a substantial amount of constraint checking and automatic correction of inconsistencies that can creep into a database if urgent updates are permitted in a disconnected mode. Bayou is designed to support distributed management of calendars and scheduling of meetings in large organizations: a time-consuming activity, which often requires approximate decision making because some participants may be on the road or otherwise unavailable at the time a meeting must be scheduled.

21.6.3 Transactions on Multidatabase Systems

The Phoenix system (see Malloth), developed by Malloth, Guerraoui, Raynal, Schiper, and Wilhelm, adopts a similar philosophy but considers a different aspect of the problem. Starting with the same model used in Amir's work and in Bayou, where each transaction is initiated from a single multicast to the database servers, which form a process group, this effort asked how transactions operating upon multiple objects could be accommodated. Such considerations led them to propose a generalized multigroup atomic broadcast, which is totally ordered, dynamically uniform, and failure-atomic over multiple process groups to which it is sent (see Schiper and Raynal). The point of using this approach is that if a database is represented in fragments managed by separate servers, each of which is implemented in a process group, a single multicast would not otherwise suffice to do the desired updates. The Phoenix protocol used for this purpose is similar to the extended three-phase commit developed by Keidar for the Transis system and is considerably more efficient than sending multiple concurrent and asynchronous multicasts to the process groups and then running a multiphase commit on the full set of participants. Moreover, whereas such as multistep protocols would leave serious unresolved questions insofar as the view-synchronous addressing aspects of the virtual synchrony model are considered, the Phoenix protocol can be proved to guarantee this property within all of the destination groups.

21.6.4 Linearizability

Herlihy and Wing studied consistency issues from a more theoretical perspective (see Herlihy and Wing). In a paper on the *linearizability* model of database consistency, they suggested that object-oriented systems may find the full nested serializability model overly constraining, but still benefit from some forms of ordering guarantee. A nested execution is linearizable if the invocations of each object, considered independently of other objects, leave that object in a state that could have been reached by some sequential execution of the same operations, in an order consistent with the causal ordering on the original invocation sequence. In other words, this model says that an object may reorder the operations upon it and interleave their execution provided that it behaves as if it had executed operations one by one, in an order consistent with the (causal) order in which the invocations were presented to it.

Linearizability may seem like a very simple and obvious idea, but there are many distributed systems in which servers might not be guaranteed to respect this property. Such servers can allow concurrent transactions to interfere with one another or may reorder operations in ways that violate intuition (e.g., by executing a read-only operation on a state that is sufficiently old to be lacking some updates issued before the read by the same source). At the same time, notice that traditional serializability can be viewed as an extension of linearizability (although serializability does not require that the causal order of invocations be respected, few database systems intentionally violate this property). Herlihy and Wing argue that if designers of concurrent objects at least prove them to achieve linearizability, the objects will behave in an intuitive and consistent way when used in a complex distributed system; should one then wish to go further and superimpose a transactional structure over such a system, doing so simply requires stronger concurrency control. I am inclined to agree: Linearizability seems like an appropriate weakest consistency guarantee for the objects used in a distributed environment. The Herlihy and Wing paper develops this idea by presenting proof rules for demonstrating that an object implementation achieves linearizability; however, we will not discuss this issue here.

21.6.5 Transactions in Real-Time Systems

The option of using transactional reliability in real-time systems has been considered by a number of researchers, but the resulting techniques have apparently seen relatively little use in commercial products. There are a number of approaches that can be taken to this problem. Davidson is known for work on transactional concurrency control subject to real-time constraints; her approach involves extending the scheduling mechanisms used in transactional systems (notably, timestamped transactional systems) to seek to satisfy the additional constraints associated with the need to perform operations before a deadline expires.

Broadly, however, the transactional model is fairly complex and consequently not suited for use in settings where the temporal constraints have fine granularity with regard to the time needed to execute a typical transaction. In environments where there is substantial breathing room, transactions may be a useful technique even if there are real-time constraints to take into account, but as the temporal demands on the system rise, more and more deviation from the pure serializability model is typically needed in order to continue to guarantee timely response.

21.7 Advanced Replication Techniques

One of the more exciting research directions of which I am aware involves the use of process groups as a form of coherently replicated cache to accelerate access to a database. The idea can be understood as a synthesis of Liskov's work on the Harp file system (see Liskov et al. [1991]), my work on Isis (see Birman and van Renesse [1994]), and research by Seltzer and others on log-structured database systems (see Seltzer). However, I am not aware of any publication in which the contributions of these disparate systems are unified.

To understand the motivation for this work, it may help to briefly review the normal approach to replication in database systems. As was noted earlier, one can replicate a data item by maintaining multiple copies of that item on servers that will fail independently and updating the item using a transaction that either writes all copies or at least writes to a majority of copies of the item. However, such transactions are slowed by the quorum read and commit operations: The former will now be a distributed operation and hence subject to high overhead, while the latter is a cost not paid in the nondistributed or nonreplicated case.

For this reason, most commercial database systems are operated in a nondistributed manner, even in the case of technologies such as Tuxedo or Encina, which were developed specifically to support distributed transactional applications. Moreover, many commercial database systems provide a weak form of replication for high-availability applications, in which the absolute guarantees of a traditional serializability model are reduced to improve performance. The specific approach is often as follows. The database system is replicated between a primary and backup server, whose roles will be interchanged if the primary fails and later is repaired and recovers. The primary server will, while running, maintain a log of committed transactions, periodically transmitting it to the backup, which applies the corresponding updates. (See Figure 21.4.)

Notice that this protocol has a window of vulnerability. If a primary server is performing transactions rapidly, perhaps hundreds of them per second, the backup may lag by hundreds or thousands of transactions because of the delay associated with preparing and sending the log records. Should the primary server now crash, these transactions will be trapped in the log records:

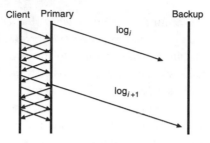

FIGURE 21.4. Many commercial database products achieve high availability using a weak replication policy, which can have a window of vulnerability. In this example, the last few transactions have not been logged to the backup and hence can be lost if the primary fails; the ones covered by log record log_{i+1}, on the other hand, are stable and will not be lost even if the primary fails. Although lost transactions will be recovered when the primary restarts, it may not be possible to reapply the updates automatically. A user intervenes in such cases.

They are committed and the client has potentially seen the result, but the backup will take over in a state that does not yet reflect the corresponding updates. Later, when the primary restarts, the lost transactions will be recovered and, hopefully, can still be applied without invalidating other actions that occurred in the interim; otherwise, a user is asked to intervene and correct the problem. The benefit of the architecture is that it gives higher availability without loss of performance; the hidden cost, however, is the risk that transactions will be rolled back by a failure, creating noticeable inconsistencies and a potentially difficult repair problem.

As it happens, we can do a less-costly job of replicating a database using process groups, and we may actually *gain* performance by doing so!

The idea is the following. Suppose that one were to consider a database as being represented by a checkpoint and a log of subsequent updates. At any point in time, the state of the database could be constructed by loading the checkpoint and then applying the updates to it; if the log were to grow too long, it could be truncated by forming a new checkpoint. This isn't an unusual way to actually view database systems: Seltzer's work on log-structured databases implemented a database this way and demonstrated some performance benefits by doing so. Liskov's research on Harp (a nontransactional file store, which was implemented using a log-based architecture) employed a similar idea, albeit in a system with nonvolatile RAM memory. Indeed, within the file system community, Rosenblum's work on LFS (a log-structured file system) revolutionized the architecture of many file system products (see Rosenblum and Ousterhout). So, it is entirely reasonable to adopt a similar approach to database systems.

Now, given a checkpoint and log representation of the database, a database server can be viewed as a process that caches the database contents in high-speed volatile memory. Each time the server is launched, it reconstructs this cached state from the most recent checkpoint and log of updates and subsequently transactions are executed out of the cache. To commit a transaction in this model, it suffices to force a description of the transaction to the log (perhaps as little as the

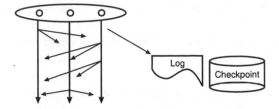

FIGURE 21.5. Future database systems may gain performance benefits by exploiting process groups as scalable parallel front ends, which cache the database in volatile memory and run transactions against this coherently cached state in a load-balanced manner. A persistent log is used to provide failure atomicity; because the log is a write-only structure, it can be optimized to give very high performance. The log would record a description of each update transaction and the serialization order that was used; only committed transactions need to be logged in this model.

transactional request itself and the serialization order that was used). The database state maintained in volatile memory by the server can safely be discarded after a failure; hence, the costly disk access associated with the standard database server architecture is avoided. Meanwhile, the log itself becomes an append-only structure, which is almost never reread, permitting a very efficient storage on disk. This is precisely the sort of logging studied by Rosenblum for file systems and by Seltzer for database systems; it is known to be very cost effective for small- to moderate-sized databases. Subsequent research has suggested that this approach can also be applied to very large databases. (See Figure 21.5.)

But now our process group technology offers a path to further performance improvements through parallelism. What we can do is use the lightweight replication methods in Chapter 15 to replicate the volatile, cached database state within a group of database servers, which can now use one of the load-balancing techniques from Section 15.3.3 to subdivide the work of performing transactions. Within this process group, there is *no need to run a multiphase commit protocol!* To see this, notice that just as the nonreplicated volatile server is merely a cache of the database log, so is the replicated group merely a volatile, cached database state.

When we claim that there is no need to run a multiphase commit protocol here, it may at first seem that such a claim is incorrect, since the log records associated with the transaction do need to be forced to disk (or to NVRAM if we use the Harp approach). If there is more than one log, there will need to be a coordination of this activity to ensure that either all logs reflect the committed transaction or that no log does. For availability, it may actually be necessary to replicate the log, and if this is done, a multiphase commit would be unavoidable. However, in many settings it might make sense to use just a single log server—for example, if the logging device is itself a RAID disk, then the intrinsic fault tolerance of the RAID technology could be adequate to provide the degree of availability required for our purposes. Thus, it may be better to say that there is no *inherent* reason for a multiphase commit protocol here, although in specific cases one may be needed.

The primary challenge associated with this approach is to implement a suitable concurrency control scheme in support of it. While optimistic methods are favored in the traditional work on distributed databases, it is not clear that they represent the best solution for this style of group-structured replication joined to a log-structured database architecture. In the case of pessimistic locking, a solution developed by Thomas Joseph and me in the mid-1980s is available. In this approach, data are replicated within a process group (see Joseph). Reads are done from any single local copy and writes are done by issuing an asynchronous *cbcast* to the entire group. Locking is done by obtaining local read locks and replicated write locks, the latter using a token-based scheme. The issue now occurs of read locks that can be broken by a failure; this is addressed by *reregistering* read locks at an operational database server if one of the group members fails. In the scheme Joseph explored, such reregistration occurs during the flush protocol used to reconfigure the group membership.

Next, Joseph introduced a rule whereby a write lock must be granted in the same process group view in which it was requested. If a write lock is requested in a group view where process p belongs to the group, and p fails before the lock is granted (perhaps because of a read lock some transaction held at process p), this forces the transaction requesting the write lock to release any locks it successfully acquired and to repeat its request. The repeated request will occur after the read lock has been reregistered, avoiding the need to abort a transaction because its read locks were broken by a failure. In such an approach, the need to support unilateral transaction abort is eliminated, because the log now provides persistency, and locks can never be lost within the process group (unless all its members fail, which is a special case). Transaction commit becomes an asynchronous *cbcast*, with the same low cost as the protocol used to do writes.

Readers familiar with transactional concurrency control may be puzzled by the similarity of this scheme to what is called the *available copies* replication method, an approach that is known to yield nonserializable executions (see Bernstein et al.). In fact, however, there is a subtle difference between Joseph's scheme and the available copies scheme—namely, that Joseph's approach depends on group membership changes to trigger lock reregistration, whereas the available copies scheme does not. Since group membership, in the virtual synchrony model, involves a consensus protocol, which provides consistent failure notification throughout the operational part of a system, the inconsistent failure detections that occur in the available copies approach do not occur. This somewhat obscure observation does not seem to be widely known within the database community.

Using Joseph's pessimistic locking scheme, a transaction that does not experience any failures will be able to do local reads at any copy of the replicated data objects on which it operates. The update and commit protocols both permit immediate local action at the group member where the transaction is active, together with an asynchronous *cbcast* to inform other members of the event. Only the acquisition of a write lock and the need to force the transaction description and commit record (including the serialization order that was used) involve a potential delay. This overhead, however, is counterbalanced by the performance benefits that come with scalable parallelism.

The result of this effort represents an interesting mixture of process group replication and database persistence properties. On one hand, we get the benefit of high-speed, memory-mapped database access, and we can use the very lightweight, nonuniform replication techniques that achieved such good performance in previous chapters. Moreover, we can potentially do load-balancing or other sorts of parallel processing within the group. Yet the logging method also gives us the persistence properties normally associated with transactions, and the concurrency control scheme provides for traditional transactional serializability. Moreover, this benefit is available without special hardware (such as NVRAM), although NVRAM would clearly be beneficial if one wanted to replicate the log itself for higher availability. I feel this approach offers the best of both worlds.

The integration of transactional constructs and process groups thus represents fertile territory for additional research, particularly of an experimental nature. As noted earlier, it is clear that developers of reliable distributed systems need group mechanisms for high availability and transactional mechanisms for persistence and recoverability of critical data. Integrated solutions offering both options in a clean way could lead to a much more complete and effective programming environment for developing the sorts of robust distributed applications needed in complex environments.

21.8 Related Reading

Chapter 26 includes a review of some of the major research projects in this area, which we will not attempt to duplicate here. For a general treatment of transactions: (see Bartlett et al., Gray and Reuter).

On the nested transaction model: (see Moss).

On disconnected operation in transactional systems: (see Amir, Amir et al., Davidson et al., Terry et al.).

On log-based transactional architectures: (see Birman and van Renesse [1994], Joseph, Liskov et al. [1991], Seltzer).

CHAPTER 22 ✧ ✧ ✧ ✧

Probabilistic Protocols

CONTENTS

The protocols considered in previous chapters of this book share certain basic assumptions concerning the way that a distributed behavior or a concept of distributed consistency is derived from the local behaviors of system components. Although we have explored a number of protocols, the general pattern involves reasoning about the possible system states observable by a correct process and generalizing from this to properties that are shared by sets of correct processes. This approach could be characterized as a deductive style of distributed computing, in which the causal history prior to an event is used to deduce system properties, and the possible deductions by different processes are shown to be consistent in the sense that, through exchanges of messages, they will not encounter direct contradictions in the deduced distributed state.

In support of this style of computing, we have reviewed a type of distributed system architecture that is hierarchical in structure or perhaps (as in Transis) composed of a set of hierarchical structures linked by some form of wide area protocol. There is little doubt that this leads to an effective technology for building very complex, highly reliable distributed systems. One might wonder, however, if there are *other* ways to achieve meaningful forms of consistent distributed behavior and, if so, whether the corresponding protocols might have advantages that would favor their use under conditions where the protocols we have seen up to now, for whatever reason, encounter limitations.

This line of reasoning has motivated some researchers to explore other styles of reliable distributed protocols, in which weaker assumptions are made about the behavior of the component programs but stronger ones are made about the network. Such an approach results in a form of protection against misbehavior, whereby a process fails to respect the rules of the protocol but is not detected as having failed. In this chapter we discuss the use of probabilistic techniques to implement reliable broadcast protocols and replicated data objects. Although we will see that there are important limitations on the resulting protocols, they also represent an interesting design point, which may be of practical value in important classes of distributed computing systems. Probabilistic protocols are not likely to replace the more traditional deductive protocols anytime soon, but they can be a useful addition to our repertoire of tools for constructing reliable distributed systems, particularly in settings where the load and timing properties of the system components are extremely predictable.

22.1 Designing Probabilistic Protocols

The protocols we will be looking at in this section are scalable and *probabilistically reliable*. Unlike the protocols presented previously, they are based on a probabilistic system model somewhat similar to the synchronous model, which we considered in our discussion of real- time protocols. In contrast to the asynchronous model, no mechanism for detecting failure is required.

These protocols are scalable in two senses. First, the message costs and latencies of the protocols grow slowly with the system size. Second, the reliability of the protocols, expressed in terms of the probability of a failed run of a protocol, approaches 0 at an exponential rate as the number of processes is increased. This scalable reliability is achieved through a form of gossip protocol, which is strongly self-stabilizing. Such a system has the following property: If it is disrupted into an inconsistent state, it will automatically restore itself to a consistent one given a sufficient period of time without failures. Our protocols (particularly for handling replicated data) also have this property.

The basic idea we will work with is illustrated in Figure 22.1, which shows a possible execution for a form of *gossip* protocol developed by Demers and others at Xerox PARC (see Demers et al.). In this example of a push gossip protocol, messages are diffused through a randomized flooding mechanism. The first time a process receives a message, it selects some fixed percentage of destinations from the set of processes that have not yet received it. The number of such destinations is said to be the *fanout* of the protocol, and the processes selected are picked randomly (a bit vector, carried on the messages, indicates which processes have received them). As these processes receive the message, they relay it in the same manner. Subsequently, if a process receives a duplicate copy of a message it has seen before, it discards the message silently.

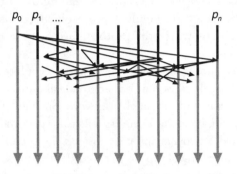

FIGURE 22.1. A push gossip protocol. Each process receiving a message picks some number of destinations (the fanout, 3 in the example shown) randomly among the processes known not to have received the message. Within a few rounds, the message has reached all destinations (in the figure, a process that has received a message is shown in gray). A pull protocol can be used to complement this behavior: A process periodically selects a few processes and solicits messages from them. Both approaches exhibit exponential convergence typical of epidemics in densely populated biological systems.

Gossip protocols will typically flood the network within a logarithmic number of rounds. This behavior is very similar to that of a biological epidemic; hence, such protocols are also known as *epidemic* ones (see Bailey). Notice that although each process may receive a message many times, the computational cost of detecting duplicates and discarding them is likely to be low. On the other hand, the cost of relaying them is a fixed function of the fanout regardless of the size of the network; this is cited as one of the benefits of the approach. The randomness of the protocols has the benefit of overcoming failures of individual processes, in contrast with protocols where each process has a specific role to play and must play it correctly, or fail detectably, for the protocol itself to terminate correctly. Figure 22.1 illustrates a push protocol, in the sense that processes with data push it to other processes lacking data by gossiping. A pull style of gossip can also be defined: In this approach, a process periodically solicits messages from some set of randomly selected processes. Moreover, the two schemes can be combined.

Demers and his colleagues have provided an analysis of the convergence and scaling properties of gossip protocols based on pushing, pulling, and combined mechanisms and have shown how these can overcome failures. They prove that both classes of protocols converge toward flooding at an exponential rate and demonstrate that they can be applied to real problems. The motivation for their work was a scaling problem, which occurred in the wide area mail system developed at PARC in the 1980s. As this system was used on a larger and larger scale, it began to exhibit consistency problems and had difficulties in accommodating mobile users. Demers and his colleagues showed that by reimplementing the e-mail system to use a gossip broadcast protocol they could overcome these problems, helping ensure timely and consistent e-mail services that were location independent and inexpensive.

22.2 Other Applications of Gossip Protocols

The protocol of Demers is not the first or the only to explore gossip-style information dissemination as a tool for communication in distributed systems. Other relevant work in this area includes that of Alon et al., who developed an information diffusion protocol using a technique similar to the one developed by Demers, and the protocol developed by Golding (1991) and Golding and Taylor, which uses gossip as a mechanism underlying a group membership algorithm for wide area applications.

22.3 Hayden's *pbcast* Primitive

In the style of protocol explored at Xerox, the actual rate with which messages will flood the network is not guaranteed, because of the risk of failures. Instead, these protocols guarantee that, given enough time, eventually either all or no correct processes will deliver a message. This property is called *eventual convergence*. Although eventual convergence is sufficient for many uses, the property is weaker than the guarantees of the protocols we used earlier to replicate data and perform synchronization, because eventual convergence does not provide bounds on message latency or ordering properties. Hayden has shown how gossip protocols can be extended to have these properties (see Hayden and Birman [1996]), and in this section we present the protocol he developed for this purpose. Hayden calls his protocol *pbcast*, characterizing it as a probabilistic analog of the *abcast* protocol for process groups.

The *pbcast* protocol is based on a number of assumptions about the environment, which may not hold in typical distributed systems. Thus, after presenting the protocol, we will need to ask ourselves when the protocol could appropriately be applied. If used in a setting where these assumptions are not valid, *pbcast* might not perform as well as the analysis would otherwise suggest.

Specifically, *pbcast* is designed for a static set of processes, which communicate synchronously over a fully connected, point-to-point network. The processes have unique, totally ordered identifiers and can toss weighted, independent random coins. Runs of the system proceed in a sequence of rounds in which messages sent in the current round are delivered in the next round.

There are two types of failures, both probabilistic in nature. The first is process failure. There is an independent, per-process probability of at most f_p that a process has a crash failure during the finite duration of a protocol. Such processes are called faulty. The second type of failure is message omission failure. There is an independent, per-message probability of at most f_m that a message between nonfaulty processes experiences a send omission failure. The union of all message omission failure events and process failure events is mutually independent. In this model, there are no malicious faults, spurious messages, or corruption of messages. We expect that both f_p and f_m are small probabilities (e.g., unless otherwise stated, the values used in the graphs in this chapter are $f_m = 0.05$ and $f_p = 0.001$).

The impact of the failure model can be visualized by thinking of the power that would be available to an adversary seeking to cause a run of the protocol to fail by manipulating the system within the bounds of the model. Such an adversary has these capabilities and restrictions:

▶ An adversary cannot use knowledge of future probabilistic outcomes, interfere with random coin tosses made by processes, cause correlated (nonindependent) failures to occur, or do anything not enumerated below.

▶ An adversary has complete knowledge of the history of the current run of the protocol.

▶ At the beginning of a run of the protocol, the adversary has the ability to individually set process failure rates, within the bounds $[0 \ldots f_p]$.

▶ For faulty processes, the adversary can choose an arbitrary point of failure.

▶ For messages, the adversary has the ability to individually set send omission failure probabilities within the bounds of $[0 \ldots f_m]$.

Note that although probabilities can be manipulated by the adversary, doing so can only make the system more reliable than the bounds, f_p and f_m.

The probabilistic analysis of the properties of the *pbcast* protocol is only valid in runs of the protocol in which the system obeys the model. In particular, the independence properties of the system model are quite strong and are not likely to be continuously realizable in an actual system—for example, partition failures in the sense of correlated communication failures do not occur in this model. Partitions can be simulated by the independent failures of several processes, but they are of low probability. However, the protocols we develop using *pbcast*, such as our replicated data protocol, remain safe even when the system degrades from the model. In addition, *pbcast*-based algorithms can be made self-healing—for instance, our replicated data protocol has guaranteed eventual convergence properties similar to normal gossip protocols: If the system recovers into a state that respects the model and remains in that state for sufficiently long, the protocol will eventually recover from the failure and reconverge to a consistent state.

22.3.1 Unordered *pbcast* Protocol

We begin with an unordered version of *pbcast* with static membership (see Figure 22.2). The protocol itself extends a basic gossip protocol with a quorum-based ordering algorithm inspired by the ordering scheme in CASD (see Cristian et al. [1985], Chandra and Toueg [1990]). What makes the protocol interesting is that it is tolerant of failures and, under the assumptions of the model, it can be analyzed formally.

```
(*State kept per pbcast: have I received a message regarding
  this pbcast yet?*)
let received_already = false

(*Initiate a pbcast.*)
to pbcast(msg):
  deliver_and_gossip(msg,k)

(*Handle message receipt.*)
on receive gossip(msg,round):
  deliver_and_gossip(msg,round)

(*Auxiliary function.*)
to deliver_and_gossip(msg,round):
  (*Do nothing if already received it.*)
  if received_already then return

(*Mark the message as being seen and deliver.*)
received_already := true
deliver(msg)

(*If last round, don't gossip.*)
if round = 0 then return

for each p in P:
  do with probability r:
    sendto p gossip(msg,round-1)
```

FIGURE 22.2. Unordered *pbcast* protocol. The function time() returns the current time expressed in rounds since the first round. Message receipt and *pbcast* are executed as atomic actions.

The protocol consists of a fixed number of rounds, in which each process participates in at most one round. A process initiates a *pbcast* by sending a message to a random set of other processes. When other processes receive a message for the first time, they gossip the message to some other randomly chosen members. Each process only gossips once: The first process does nothing after sending the initial messages and the other processes do nothing after sending their set of gossip messages. Processes choose the destinations for their gossip by tossing a weighted random coin for each other process to determine whether to send a gossip message to that process. Thus, the parameters of the protocol are:

- P: the set of processes in the system: $n = |P|$
- k: the number of rounds of gossip to run
- r: the probability that a process gossips to each other process (the weighting of the coin mentioned earlier)

The behavior of the gossip protocol mirrors a class of disease epidemics, which nearly always infects either almost all of a population or almost none of it. In the following text, we will show that *pbcast* has a bimodal delivery distribution, which stems from the epidemic behavior of the gossip protocol. The normal behavior of the protocol is for the gossip to flood the network in a random but exponential fashion. If r is sufficiently large, most processes will usually receive the gossip within a logarithmic number of rounds.

22.3.2 Adding Total Ordering

In the protocol shown in Figure 22.2, the *pbcast* messages are unordered. However, because the protocol runs in a fixed number of rounds of fixed length, it is trivial to extend it using the same method as was proposed in the CASD protocols (see Figure 22.3). By delaying the delivery of a message until it is known that all correct processes have a copy of that message, totally ordered delivery can be guaranteed. This yields a protocol similar to *abcast* in that it has totally ordered message delivery and reliability within the fixed membership of the process group invoking the primitive. It would not be difficult to introduce a further extension of the protocol for use in dynamic process groups, but we will not address that issue here.

22.3.3 Probabilistic Reliability and the Bimodal Delivery Distribution

Hayden has demonstrated that when the system respects the model, a *pbcast* is almost always delivered to most or to few processes and almost never to some processes. Such a delivery distribution is called a "bimodal" one and is depicted in Figure 22.4. The graphs show that varying numbers of processes will deliver a *pbcast*—for instance, the probability that 26 out of the 50 processes will deliver a *pbcast* is around 10^{-28}. Such a probabilistic guarantee is, for most practical purposes, a guarantee that the outcome cannot occur. This bimodal distribution property is presented here informally, but later we discuss the method used by Hayden to calculate the actual probability distributions for a particular configuration of *pbcast*.

A bimodal distribution is useful for voting-style protocols where, as an example, updates must be made at a majority of the processes to be valid; we saw examples of such protocols when we discussed quorum replication. Problems do occur in these sorts of protocols when failures cause a large number of processes, but not a majority, to carry out an update. *Pbcast* overcomes this difficulty through its bimodal delivery distribution by ensuring that votes will almost

```
(* Local state: message buffer and counter for generating
   unique identifiers.*)
let buffer = {}
let id_counter = 0

(* Initiate a pbcast.*)
to pbcast(msg):
  (* Create unique id for each message.*)
  let id = (my_id, id_counter)
  id_counter := id_counter + 1

  do_gossip(time(),id,msg,k)

(* Handle message receipt.*)
on receive gossip(timesent,id,msg,round):
  do_gossip(timesent,id,msg,round)

(* Handle timeouts.*)
on timeout(time):
  (*Check for messages ready for delivery. Assumes buffer is
   *scanned in lexicographic order of (sent,id).*)
  for each (sent,id,msg) in buffer:
    if sent + k + 1 = time then
      buffer := buffer\(sent,id,msg)
      deliver(msg)

(*Auxiliary function.*)
to do_gossip(timesent,id,msg,rnd):
  (*If have seen message already, do nothing.*)
  if (timesent,id,msg) in buffer then
    return

(*Buffer the message for later delivery, and then gossip.*)
buffer := buffer ∪(timesent,id,msg)
set_timer timesent + k + 1

(*If last round, do nothing more.*)
if rnd = 0 then return

for each p in P
  with probability r
    send p gossip(timesent,id,msg,rnd-1)
```

FIGURE 22.3. Ordered *pbcast* protocol, using the method of CASD.

always be weighted strongly for or against an update, and will very rarely be evenly divided. By counting votes, it can almost always be determined whether an update was valid or not, even in the presence of some failed processes.

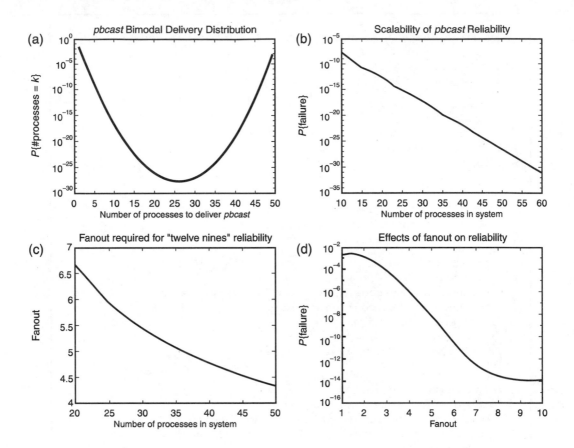

FIGURE 22.4. Graphs showing *pbcast* reliability, performance, and scaling.

With *pbcast*, the bad cases occur when "some" processes deliver the *pbcast*; these are the cases that *pbcast* makes unlikely to occur. We will call *pbcast*s delivered to "some" processes *failed pbcast*s and *pbcast*s delivered to "few" processes *invalid pbcast*s. The distinction anticipates the replicated data protocol presented in the following text, in which invalid *pbcast*s are inexpensive events and failed *pbcast*s are potentially costly.

To establish that *pbcast* does indeed have a bimodal delivery distribution, Hayden used a mixture of symbolic and computational methods. First, he computed a recurrence relation, which expresses the probability that a *pbcast* will be received by a processes at the end of round j, given that the message had been received by b processes at the end of round $j-1$, c of these for the first time. In the terminology of a biological infection, b denotes the number of processes that were infected during round $j-1$ and hence are infectious; the difference between a and b thus represents the number of *susceptible* processes that had not yet received a gossip message and that are successfully infected during round j.

The challenging aspect of this analysis is to deal with the impact of failures, which has the effect of making the variables in the recurrence relation random ones with binomial distributions. Hayden arrives at a recursive formula but not a closed form solution. However, such a formula is amenable to computational solutions and, by writing a program to calculate the various probabilities involved, he is able to arrive at the delivery distributions shown in Figure 22.4.

A potential risk in the analysis of *pbcast* is to assume, as may be done for many other protocols, that the worst case occurs when message loss is maximized. *Pbcast*'s failure mode occurs when there is a partial delivery of a *pbcast*. A pessimistic analysis must consider the case where local increases in the message-delivery probability decrease the reliability of the *overall pbcast protocol*. This makes the analysis quite a bit more difficult than the style of worst-case analysis used in protocols such as the CASD one, where the worst case is the one in which the maximum number of failures occurs.

22.3.4 An Extension to *pbcast*

When the process that initiates a *pbcast* is not faulty, it is possible to provide stronger guarantees for the *pbcast* delivery distribution. By having the process that starts a *pbcast* send more messages, an analysis can be given showing that if the sender is not faulty, the *pbcast* will almost always be delivered at most of the processes in the system. This is useful because an application can potentially take some actions knowing that its previous *pbcast* is almost certainly going to reach most of the processes in the system. The number of messages can be increased by having the process that initiates a *pbcast* use a higher value for *r* for just the first round of the *pbcast*. This extension is not used in the computations presented in the following text. Had the extension been included, the distributions would have favored bimodal delivery with even higher probabilities.

22.3.5 Evaluation and Scalability

The evaluation of *pbcast* is framed in the context of its scalability. As the number of processes increases, *pbcast* scales according to several metrics. First, the reliability of *pbcast* grows with system size. Second, the cost per participant, measured by number of messages sent or received, remains at or near constant as the system grows. Having made these claims, it must be said that the version of *pbcast* presented and analyzed for a network makes assumptions that become less and less realizable for large systems. In practice, this issue could be addressed with a more hierarchically structured protocol, but Hayden's analysis has not been extended to such a protocol. In this section, we will address the scaling characteristics according to the metrics previously listed and then discuss informally how *pbcast* can be adapted for large systems.

Reliability

Pbcast has the following property: As the number of processes participating in a *pbcast* grows, the protocol becomes more reliable. In order to demonstrate this, we present a graph, Figure 22.4(b), of *pbcast* reliability as the number of processes is varied between 10 and 60, fixing fanout and failure rates—for instance, the graph shows that with 20 processes the reliability is about

10^{-13}. The graph almost fits a straight line with slope $= -0.45$, thus the reliability of *pbcast* increases almost tenfold with every two processes added to the system.

Message Cost and Fanout

Although not immediately clear from the protocol, the message cost of the *pbcast* protocol is roughly a constant multiple of the number of processes in the system. In the worst cast, all processes can gossip to all other processes, causing $O(n^2)$ messages per *pbcast*. *r* will be set to cause some expected *fanout* of messages, so that on average a process should gossip to about *fanout* other processes, where *fanout* is some constant, in practice at most 10 (unless otherwise stated, *fanout* $= 7$ in the graphs presented in Figure 22.4). Figure 22.4(c) shows a graph of reliability versus *fanout* when the number of processes and other parameters is held constant—for instance, the graph shows that with a fanout of 7.0, *pbcast*'s reliability is about 10^{-13}. In general, the graph shows that the fanout can be increased to increase reliability, but eventually there are diminishing returns for the increased message cost.

On the other hand, *fanout* (and hence cost) can be decreased as the system grows, keeping the reliability at a fixed level. In Figure 22.4(d), reliability of at least "twelve nines" (i.e., the probability of a *failed pbcast* is less than or equal to 10^{-12}) is maintained, while the number of processes is increased. The graph shows that with 20 processes a *fanout* of 6.63 achieves "twelve nines" reliability, while with 50 processes a *fanout* of 4.32 is sufficient.

22.4 An Unscalable System Model

Although the protocol operating over the system model is scalable, the system model is not. The model assumes a flat network in which the cost of sending a message is the same between all pairs of processes. In reality, as a real system scales and the network loses the appearance of being flat, this assumption breaks down. There are two possible answers to this problem. The first is to consider *pbcast* suitable for scaling only to mid-sized systems (perhaps with as many as 100 processes). Certainly, on this size system, *pbcast* provides levels of reliability that are adequate for most applications. The second possible solution may be to structure *pbcast*'s message propagation hierarchically, so that a weaker system model, which scales to larger sizes, can be used. The structure of such a protocol, however, would likely complicate the analysis. Investigating the problem of scaling *pbcast* to be suitable for larger numbers of processes is an area of future work.

More broadly, the *pbcast* system model is one that would only be reasonable in certain settings. General-purpose distributed systems are not likely to guarantee clock synchronization and independence of failures and message delays, as assumed in the model. On the other hand, many dedicated networks would have the desired properties and hence could support *pbcast* if desired—for example, the networks that control telecommunication switches and satellite systems are often designed with guarantees of capacity and known maximum load; in settings such as these the assumptions required by *pbcast* would hold. An interesting issue concerns the use of *pbcast* for high-priority, infrequent tasks in a network that uses general-purpose computing technologies but supports a concept of message priority. In such settings the *pbcast* protocol messages might be infrequent enough to appear as independent events, permitting use of the protocol for special purposes but not for heavier loads or more frequent activities.

22.5 Replicated Data using *pbcast*

In presenting other reliable broadcast protocols, we used replicated data as our primary quality metric. How would a system that replicates data using *pbcast* be expected to behave, and how might such a replicated data object be used? In this section we briefly present a replication and synchronization protocol developed by Hayden and explore the associated issues.

22.5.1 Representation of Replicated Data

It is easiest to understand Hayden's scheme if the replicated data managed by the system are stored in the form of a history of values that the replicated data took on, linked by the updates that transformed the system data from each value to the successive one. In the following text, we will assume that the system contains a single replicated data object, although the generalization to a multiobject system is trivial.

22.5.2 Update Protocol

Hayden's scheme uses *pbcast* to transmit updates. Each process applies these updates to its local data copy as they are successfully delivered, in the order determined by the protocol. Were it not for the small but nonzero probability of a *failed pbcast*, this would represent an adequate solution to our problem. However, we know that *pbcast* is very sensitive to the assumptions made in developing the model—hence, there is some risk that if the system experiences a brief overload, or some other condition that pushes it away from its basic model, failed *pbcast*s may occur with unexpectedly high frequency, leaving the processes in the system with different update sequences. The updates will be ordered in the same way throughout but some may be missing updates and others may have an update that in fact reflects a failed *pbcast* and hence is not visible to many processes in the system.

To deal with this, we will use a *read* protocol, which can stabilize the system if it has returned to its normal operational mode (the resulting algorithm is said to be *self-stabilizing* for this reason). Specifically, associated with each update in the data queue, we will also have a distribution of the probability that the update, and all previous ones, is stable, meaning that the history of the queue prior to that update is believed to be complete and identical to the update queues maintained by other processes. For an incoming update, this distribution will be just the same as the basic *pbcast* reliability distribution; for older updates, it can change as the *read* algorithm is executed.

22.5.3 Read Protocol

Hayden's read algorithm distinguishes two types of *read* operation. A local read operation returns the current value of the data item, on the basis of the updates received up to the present. Such an operation has a probability of returning the correct value, which can be calculated using the reliability distributions of the updates. If each update rewrites the value of the data item, the probability will be just that of the last update; if updates are in some way state sensitive (such as an increment or decrement operation), this computation involves a recurrence relation.

Hayden also supports a *safe* read, which operates by using a gossip *pull* protocol to randomly compare the state of the process at which the read was performed with that of some number of other processes within the system. As the number of sampled states grows, it is possible to identify failed updates with higher and higher probability, permitting the processes that have used a safe read to clean these from their data histories. The result is that a read can be performed with any desired level of confidence at the cost of sampling a greater number of remote process states. Moreover, each time a *safe read* is performed, confidence in updates remaining on the queue will increase. In practice, Hayden finds that by sampling even a very small number of process states, a read can be done with what is effectively perfect safety: The probability of correctness soars to such a high level that it essentially converges to unity.

Conceptually, the application process that uses an unsafe local read should do so only under circumstances where the implications of an erroneous result are not all that serious for the end user. Perhaps these relate to current positions of aircraft remote from the one doing the read operations; these are of interest but are not a direct threat to safe navigation. In contrast, a process would use a safe read for operations having an external consistency requirement. A safe read is only risky if the network model is severely perturbed. Thus, a safe read might be used to obtain information about positions and trajectories of flights close to an aircraft of interest, in order to gain confidence that the resulting aircraft routing decisions are safe in a physical sense. However, the safe read may need to sample other process states and hence would be a slower operation.

One could question whether a probabilistic decision is ever safe. Hayden reasons that even a normal, nonprobabilistic distributed system is ultimately probabilistic because of the impracticality of proving complex systems correct. The many layers of software involved (compiler, operating system, protocols, applications), and the many services and servers involved, introduce a probabilistic element to any distributed or nondistributed computing application. In *pbcast* these probabilistics merely become part of the protocol properties and model, but this is not to say that they were not previously present—even if unknown to the developer.

22.5.4 Locking Protocol

Finally, Hayden presents a *pbcast*-based locking protocol, which is always safe and is probabilistically live. His protocol works by using *pbcast* to send out locking requests. A locking request that is known to have reached more than a majority of processes in the system is considered to be granted in the order given by the *pbcast* ordering algorithm. If the *pbcast* is unsuccessful or has an uncertain outcome, the lock is released (using any reliable point-to-point protocol or another *pbcast*) and then requested again (this may mean that a lock is successfully acquired but, because the requesting process is not able to confirm that a majority of processes granted the lock, is released without having been used). It is easy to see that this scheme will be safe, but also that it is only probabilistically live. After use of the lock, the process that requested it releases it. If desired, a timeout can be introduced to deal with the possibility that a lock will be requested by a process that subsequently fails.

22.6 Related Reading

Probabilistic protocols are an esoteric area of research, on which relatively little has been published.

On gossip protocols: (see Alon et al., Demers et al., Golding [1991], Golding and Taylor).

On the underlying theory: (see Bailey).

Hayden's work (see Hayden and Birman), draws on Chandra and Toueg (1990) and Cristian et al. (1985).

CHAPTER 23 ✦ ✦ ✦ ✦

Distributed System Management

CONTENTS

23.1 The Challenge of Distributed System Management

In distributed systems that are expected to overcome failures, reconfigure themselves to accommodate overloading or underloading, or modify their behavior in response to environmental phenomena, it is helpful to consider the system at two levels. At one level, the system is treated in terms of its behavior while the configuration is static. At the second level, issues of transition from configuration to configuration are addressed: These include sensing the conditions under which a reconfiguration is needed, taking the actions needed to reconfigure the physical layout of the system, and informing the components of the new structure.

Most of the material in this chapter is based loosely on work by Wood and Marzullo on a management and monitoring system called *Meta*, and on related work by Marzullo on issues of clock and sensor synchronization and fault tolerance (see Marzullo [1984, 1990], Marzullo et al., Wood [1991]). The Meta system is not the only one to have been developed for this purpose, and in fact it has been superseded by other systems and technologies since it was first introduced. However, Meta remains unusual for the elegance with which it treats the problem, and for that reason it is particularly well suited for presentation in this book. Moreover, Meta is one of the only systems to have dealt with reliability issues.

Marzullo and Wood treat the management issue as a form of programming problem, in which the inputs are events affecting the system configuration and the outputs are *actions* applied to the environment, sets of components, or individual components. *Metaprogramming* is the problem of developing the control rules used by the metasystem to manage the underlying controlled system.

In developing a system management structure, the following tasks must be performed:

▶ *Creating a system and environment model:* This establishes the conventions for naming the objects in the system, identifies the events that can occur for each type of object, and stipulates the actions that can be performed upon it.

▶ *Linking the model to the real system:* This is the process of instrumenting the managed system so that the various control points will be accessible to the metaprogram.

▶ *Developing the metaprograms:* This is the step during which control rules are specified, giving the conditions under which actions should be taken and the actions required.

▶ *Interpreting the control rules:* This step involves developing a metaprogram that acts upon the control rules, with the degree of reliability required by the application. A focus of this chapter will be on *fault tolerance of the interpretation mechanisms* and on *consistency of the actions taken.*

▶ *Visualizing the resulting metaenvironment:* A powerful benefit of using a metadescription and metacontrol language is that the controlled system can potentially be visualized using graphical tools, which show system states in an intuitive manner and permit operators to intervene when necessary.

The state of the art for network management and state visualization is extremely advanced, and tools for this purpose represent an important and growing software market. Less well understood is the problem of managing reliable applications in a consistent and fault-tolerant manner, and this is the issue on which our discussion will focus in the following sections.

23.2 A Relational System Model

Marzullo and Wood use a relational database both to model the system itself and the environment in which it runs. In this approach, the goal of the model is to establish the conventions by which the controlled entities and their environment can be referenced, to provide definitions of the relationships between them, and to provide definitions of the sensors and actuators associated with each type of component.

We assume that most readers are familiar with relational databases: They have been ubiquitous in settings ranging from personal finance to library computing systems. Such systems represent the basic entities of the database in the form of *tabular relations* whose entries are called *tuples*, each of which contains a unique identifier. Relationships between these entities are expressed by additional relations giving the identifiers for related entities—for example, suppose that we want to manage a system containing two types of servers: *file_servers* and *database_servers*. These servers execute on *server_nodes*. A varying number of *client_programs* execute on *client_nodes*. For simplicity, we will assume that there are only these three types of programs in the system and these two types of nodes. However, for reliability purposes, it may be that the file servers and database servers are replicated within process groups, and the collection of client programs may vary dynamically.

clid	uid	nid	sz	lreq
1	13/7	102	2201	READ
3	15/7	106	1840	READ
4	22/8	106	3103	WRITE

client_program

nid	load	memused	vmemused	memavail	IP addr.	protocol
102	3.5	4574	18544	642	128.13.71.2	SNMP
106	4.7	6620	24321	0	128.13.71.11	SNMP

client_nodes

fsid	load	nid	sz	uptime
13	12.2	67	1702	16:20:03
6	.30	33	620	12:22:11
27	3.5	25	980	1:02:19

file_servers

dbid	load	nid	sz	uptime
1	7.5	67	1888	16:21:02
2	6.2	25	9590	12:11:09
5	3.1	33	2890	1:21:02

database_servers

nid	load	memused	vmemused	memavail	IP addr.	protocol
67	18.1	6541	16187	6151	128.13.67.1	SNMP
25	9.6	6791	21981	6151	128.13.67.2	SNMP
33	10.7	5618	17566	4371	128.13.67.5	SNMP

server_nodes

FIGURE 23.1. Relational database used to represent system configuration.

Such a system can be represented by a collection of relations, or tables, whose contents change dynamically as the system configuration changes. If we temporarily defer dynamic aspects, such a system state may resemble the one shown in Figure 23.1. The relations that describe client systems are:

▶ *client_programs:* This relation has an entry (tuple) for each client program. The fields of the relation specify the unique identifier of the client (clid), its user-ID (uid), the last request issued by the client program (last_req), the current size of the client program process (sz), and so forth. The field called nid gives the client node on which the client is running—that is, this field relates the client_program entity to the client_node entity having that node ID.

▶ *client_nodes:* This relation has an entry for each node on which a client program might be running and, for that node, gives the current load on the node (load), physical memory in use (memused), virtual memory used (vmemused), and physical memory available (memavail).

▶ *file_servers:* This is a relation describing the file server processes and is similar to that for client processes.

▶ *database_servers:* This is a relation describing the database server processes and is similar to that for client processes.

▶ *server_nodes:* This relation describes the nodes on which file server and database server processes execute.

Notice that the dependency relationships between the entities are encoded directly into the tuples of the entity relations in this example. Thus, it is possible to query the load of the compute node on which a given server process is running in a simple way.

Additionally, it is useful to notice that there are natural process group relationships represented in Figure 23.1. Although we may not choose to represent the clients of a system as a process group in the explicit sense of our protocols from earlier in the book, such tables can encode groups in several ways. The entities shown can be treated as a group, as can the subsets of entities sharing some value in a field of their tuples, such as the processes residing on a given node. Marzullo and Wood use the term "aggregate" to describe these sorts of process groups, recalling similar use of this term in the field of database research.

23.3 Instrumentation Issues: Sensors, Actuators

The instrumentation problem involves obtaining values to fill in the fields of our modeled distributed system—for example, our server nodes are shown as having "loads," as are the servers themselves. One aspect of the instrumentation problem is to define a procedure for sampling these loads. A second consideration concerns the specific properties of each sensor. Notice that these different load sensors might not have the same units or be computed in the same way: Perhaps the load on a server is the average length of its queue of pending requests during a period of time, whereas a load on a server node is the average number of runnable programs on that node during some other period of time. Accordingly, we adopt this perspective: Values that can be obtained from a system are accessed through *sensors*, which have *types* and *properties*. Examples of sensor types include numeric sensors, sensors that return strings, and set-valued sensors. Properties include the continuity properties of a numeric sensor (e.g., whether or not it changes continuously), the precision of the sensor, and its possible range of values.

23.4 Management Information Bases: SNMP and CMIP

A management system will require a way to obtain sensor values from the instrumented entities. It is increasingly popular to do this using a standard called the Simple Network Management Protocol (SNMP), which defines procedure calls for accessing information in a Management Information Base, or MIB. SNMP is an IP-oriented protocol and uses a form of extended IP address to identify the values in the MIB: If a node has IP address 128.16.77.12, its load might, for example, be available as 128.16.77.12:45.71. A mapping from ASCII names to these IP address extensions is typically stored in the Domain Name Service (DNS) so that such a value can also be accessed as gunnlod.cs.cornell.edu:cpu/load. A trivial RPC protocol is used to query such values. Application programs on a node, with suitable permissions, use system calls to update the local MIB; the operating system is also instrumented and maintains the validity of standard system-level values.

The SNMP standard has become widely popular, but it is not the only such standard in current use. Common Management Information Protocol (CMIP) is a similar standard developed by the telecommunication industry; it differs in the details but is basically similar in its ability to represent values. SNMP and CMIP both standardize a great variety of sensors as well as the protocol used to access them: At the time of this writing, the SNMP standard included more than 4,000 sensor values that might be found in an MIB. However, any particular platform will only export some small subset of these sensors and any particular management application will only make use of a small collection of sensors, often permitting the user to reconfigure these aspects. Thus, of the 4,000 standard sensors, a typical system may in fact be instrumented using perhaps a dozen sensors, of which only two or three are critical to the management layer.

A monitoring system may also need a way to obtain sensor values directly from application processes, because both SNMP and CMIP have limitations on the nature of data they can represent, and both lack synchronization constructs, which may be necessary if a set of sensors must be updated atomically. In such cases, it is common to use RPC-oriented protocols to talk to special *monitoring interfaces* supported by the application itself; these interfaces could provide an SNMP-like behavior or a special-purpose solution. However, such approaches to monitoring are invasive and entail the development use of wrappers with monitoring interfaces or other modifications to the application. In light of such considerations, it is likely that SNMP and CMIP information bases will continue to be the more practical option for representing sensor values in distributed settings.

23.4.1 Sensors and Events

The ability to obtain a sensor's value is only a first step in dealing with dynamic sensors in a distributed system. The problem of computing with sensors also involves dealing with inaccuracy and possible sensor failures, developing a model for dealing with aggregates of sensors, and dealing with the issue of time and clock synchronization. Moreover, there are issues of dynamicism that occur if the group of instrumented entities changes over time. We need to understand how these issues can be addressed so that, given an instrumented system, we can define a meaningful concept of *events* that occur when a condition expressed over one or multiple sensors becomes true after having been false or becomes false after having been true.

Suppose that Figure 23.2 represents the loads on a group of database servers. We might wish to define an event called "database overloaded," which will occur if more than two-thirds of the servers in the group have loads in excess of 10. It can be seen that the servers in this group briefly satisfied this condition. Yet the sensor samples were taken in such a manner that this condition cannot be detected.

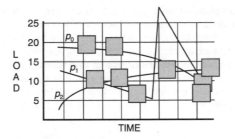

FIGURE 23.2. Imprecision in time and sensor values can confuse a monitoring system. Here, sensor readings (shaded boxes) are obtained from a distributed system. Not only is there inaccuracy in the values and time of the reading, but they are not sampled simultaneously. To ensure that its actions will be reasonable, a management system must address these issues.

Notice that the sensor readings are depicted as boxes. This is intended to reflect the concept of uncertainty in measurements: Temporal uncertainty yields a box with horizontal extend and value uncertainty yields a box with horizontal extent. Also visible here is the lack of temporal synchronization between the samples taken from different processes: Unless we have a real-time protocol and the associated infrastructure needed to sample a sensor accurately at a precise time, there is no obvious way to ensure that a set of data points represents a simultaneous system state. Simply sending messages to the database servers asking them to sample their states is a poor strategy for detecting overloads, since they may tend to process such requests primarily when lightly loaded (simply because a heavily loaded program may be too busy to update the database of sensor values or to notice incoming polling requests). Thus, we might obtain an artificially low measurement of load or one in which the servers are all sampled at different times and hence the different values cannot really be combined.

This point is illustrated in Figure 23.3, where we have postulated the use of a high-precision clock synchronization algorithm and some form of periodic process group mechanism that arranges for a high-priority load-checking procedure to be executed periodically. The sampling boxes are reduced in size and tend now to occur at the same point in time. But notice that some samples are missing. This illustrates yet another limitation associated with monitoring a complex system: Certain types of measurements may not always be possible—for example, if the

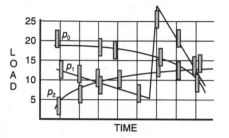

FIGURE 23.3. Sampling using a periodic process group. Here we assume that a wrapper, or some form of process group-oriented real-time mechanism, has been introduced to coordinate sampling times. Some samples are missing, corresponding to times at which the load for the corresponding process was not well defined or in which that process was unable to meet the deadlines associated with the real-time mechanism. In this example the samples are well synchronized but there are often missing values, raising the question of how a system can calculate an average given just two out of three or even one out of three values.

load on our servers is computed by calculating the length of a request queue data structure, there may be periods of time during which the queue is locked against access because it is being updated and is temporarily in an inconsistent state. If a sampling period happens to fall during such a lock-out period, we would be prevented from sampling the load for the corresponding server during that sampling period. Thus, we could improve on our situation by introducing a high-priority monitoring subsystem (perhaps in the form of a Horus-based wrapper, which could then take advantage of real-time protocols to coordinate and synchronize its sampling), but we would still be confronted with certain fundamental sources of uncertainty. In particular, we may now need to compute average load with one or even two missing values. As the desired accuracy of sampling rises, the probability that data will be missing will also rise; a similar phenomenon was observed when the CASD protocols were operated with smaller and smaller values of D, corresponding to stronger assumptions on the execution environment.

In the view of the monitoring subsystem, these factors blur the actual events that took place. As seen in Figure 23.4, the monitoring system is limited to an approximate concept of the range of values that the load sensor may have had, and this approximation may be quite poor (this figure is based on the samples from Figure 23.2). A higher sampling rate and more accurate sensors would improve upon the quality of this estimate, but missing values would creep in to limit the degree to which we can guarantee accuracy.

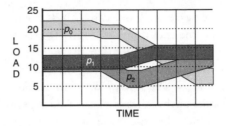

FIGURE 23.4. By interpolating between samples, a monitoring system can approximate the true behavior of a monitored application. But important detail can be lost if the sensor values are not sufficiently accurate and frequent. From this interpolated version of Figure 23.2 it is impossible to determine that the database system briefly became overloaded by having two servers that both exceeded loads of 15.

With these limitations in mind, we now move on to the question of events. To convert sensor values into events, a monitoring system will define a set of *event trigger conditions*—for example, a very simple overload condition might specify:

$$trigger\ overload\ when\ avg(s \in db\_servers: s.load) > 10$$

Here, we have used an informal notation to specify the aggregate consisting of all server processes and then computed the average values of the corresponding load fields. If this average exceeds 10, the overload event will occur; it will then remain disabled until the average load falls back below 10 and can then be triggered by the next increase beyond the threshold. Under what conditions should this event be triggered?

In our example, there was a brief period during which two-thirds of the database servers exceeded a load of 10 and during this period the true average load may well have also crossed the threshold. A system sampled as erratically as this one, however, would need to sustain an overload for a considerable period of time before such a condition would definitely be detectable. Moreover, even when the condition is detected, uncertainty in the sensor readings makes it difficult to know if the average actually exceeded the limit: One can in fact distinguish three cases—definitely below the limit, possibly above the limit, and definitely above the limit. Thus, there may be conditions under which the monitoring system will be uncertain of whether or not to trigger the overload event.

Circumstances may require that the interpretation of a condition be done in a particular manner. If an overload might trigger a catastrophic failure, the developer would probably prefer an aggressive solution: Should the load reach a point where the threshold might have been exceeded, the event should be raised. On the other hand, it may be that the more serious error would be to trigger the overload event when the load might actually be below the limits or might fall below soon. Such a scenario would argue for the more conservative approach.

To a limited degree, one could address such considerations by simply adjusting the limits. Thus, if we seek an aggressive solution but are working with a system that operates conservatively, we could reduce the threshold value by the expected imprecision in the sensors. By signaling an overload if the average definitely exceeds 7, one can address the possibility that the sensor readings were too low by 3 and that the true values averaged 10. However, a correct solution to this problem should also account for the possibility that the value might change more rapidly than the frequency of sampling, as in Figure 23.2. Knowing the maximum possible rate of change and assuming the worst, one might arrive at a system model more like the one shown in Figure 23.5. Here, the possible rate of change of the various sensor values permits the system to extrapolate the possible envelope within which the true value may lie. These curves are discontinuous because when a new reading is made, the resulting concrete data immediately narrow the envelope to the uncertainty built into the sensors themselves.

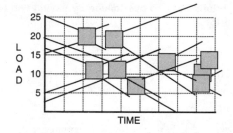

FIGURE 23.5. By factoring in the possible rate of change of a continuous sensor, the system can estimate possible sensor values under worst-case conditions. This permits a very conservative interpretation of trigger conditions.

Marzullo and Wood have developed a comprehensive theoretical treatment of these issues, dealing both with estimation of values and performing imprecise comparisons (see Wood [1993]). Their work results both in algorithms for combining and comparing sensor values and the suggestion that comparison operators be supported in two modes: one for the "possible" case and one for the "definite" one. They also provide some assistance with the problem of selecting an appropriate sampling rate to ensure that critical events will be detected correctly. Because some applications require rapid polling of sensors, they develop algorithms for transforming a distributed condition over a sensor aggregate into a set of local conditions that can be evaluated close to the monitored objects, where polling is comparatively inexpensive and can be done frequently. We will not cover their work in detail here, but interested readers will find discussion of these topics in Babaoglu and Marzullo, Marzullo (1990), Marzullo et al., Wood (1991).

23.4.2 Actuators

An actuator is the converse of a sensor: The management system assigns a value to it, and this causes some action to be taken. Actuators may be physical (e.g., a controller for a robot arm), logical (a parameter of a software system), or a connection to abstract actions (an actuator could cause a program to be executed on behalf of the management system). In the context of SNMP, an actuator can be approximated by having the external management program set a value in the

MIB that is periodically polled by the application program. More commonly, a monitoring program will place a remote agent at the locations where it may take actions and trigger those actions by RPC to it.

Thus, an actuator is the logical abstraction corresponding to any of the *actions* that a control policy can take. The policy will determine what action is desired, and then the action is performed by placing an appropriate value into the appropriate actuator or actuators. Actuators can be visualized as buttons, which can be pushed, and form-fill menus, which can be filled in and executed. Whereas a user might do these things through a GUI, a system control rule does these things by actuating an actuator or a set of actuators.

While the handling of faulty sensors is a fairly simple matter, dealing with potentially faulty actuators is quite a bit more complex. Marzullo and Wood studied this issue as part of a general treatment of aggregated actuators: groups of actuators having some type. One could, for example, define the group of *run a program* actuators associated with a set of computers (in practice, such an actuator would be a form of RPC interface to a remote execution facility; by placing a value into it, the remote execution facility could be asked to run the program corresponding to that value—e.g., the program with the same name that was written to the actuator or a program identified by an index into a table). One could then imagine a rule whereby, if one machine were unable to run the desired program, some other machine would be asked to do so: to run on any one of these machines, in effect. Rules for load-balanced execution could be superimposed on such an actuator aggregate. However, although the Meta system implemented some simple mechanisms along these lines, I am not aware of any use of the idea in commercial systems.

Marzullo and Wood have noted that at the limit, fault-tolerant actuator aggregates will need to run a protocol much like the Byzantine protocol—an analogy that brings to mind the copious research that has been conducted on what are called *embedded systems*, in which a control program is placed close to a hardware subsystem and used to manage or control some form of external process.

In most commercial monitoring and management systems, actuators are limited to very simple tasks, such as executing a program, changing a priority for a scheduling algorithm, and so forth. In effect, the actuator model is used to link the management policy to the external world, but not to superimpose any sort of more sophisticated abstractions upon it. The topic thus remains an intriguing area for future study.

23.5 Reactive Control in Distributed Settings

Having addressed the issues associated with modeling a distributed system and with interpreting its sensors, we now turn to the problem of *reactive control*, which occurs when triggered events are used to drive management policies.

Marzullo and Wood recommend that management policies be viewed as a database of control rules, which are bound to system components and become active when those components are active. Thus, a policy for managing a database server would be instantiated once for each database server in the system, and each rule would manage its own server as long as that server keeps running. A policy for the aggregate of database servers would be instantiated the first time a database server is started and would remain active as long as there are one or more servers in the system.

Each of these policies is described in the form of a script, giving the control rules to use for the corresponding component and the conditions under which those rules should apply. Policies are in this sense similar to a *state machine*. Each state defines a set of events for which it is monitoring, and if the event occurs, the machine transitions to a new state and takes some management action.

Thus, for our database server example, we might define a policy for managing the aggregate of database servers whereby an *overload* event, detected in the *normal* state, causes the system to add servers (up to a maximum of four), after which it might move to a *degraded operation* state, remaining in this state until a *load ok* event is detected (Figure 23.6). As it moves from state to state, such a policy might write values to actuators—for example, the rule shown here requires a means of launching and shutting down servers and for placing the group in a degraded mode. The actuator for the first of these cases would be an *execute process* actuator on the least-loaded machine, which can be picked by a simple operation on the *server_nodes* relation. To shut down a process, one would probably send that process a termination message or perhaps even send it a SIGTERM signal. To switch the servers into a *degraded query* mode, one would probably want an out-of-band signaling mechanism, since the message queues may be congested when this condition occurs. Thus, a degraded query actuator might simply be a bit, which can be set in the MIB associated with the database server; the server itself would check this bit periodically and switch in and out of degraded query mode accordingly. This illustrates that the concept of an actuator needs to be interpreted in a very flexible way.

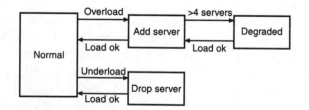

FIGURE 23.6. State machine for database server management. In the normal mode of operation, if overload is detected a server is added. However, if this would cause the server group to exceed four servers, the system degrades its quality of service until normal loading is restored. If the system is underloaded for a period of time, a server is dropped.

Approaches such as this raise a number of problems. One major issue is to pick an appropriate language in which to specify these rules. Marzullo and Wood proposed two such languages: a high-level language, which looked similar to a database query language, and a low-level one into which these high-level language rules could be compiled; the low-level language was similar to the PostScript language used for printers. Other work in this area has focused on popular command script languages, such as Perl. Questions about the appropriate execution environment for a rule and the methods by which a rule can interact with the system configuration are all potentially difficult. For the purposes of this chapter, however, I will not delve into the details of these language proposals, since I view the topic as one that remains open for further research.

23.6 Fault Tolerance by State Machine Replication

Having reduced our problem to one of interpreting a state machine, we find there are two remaining issues. The first is associated with the efficiency of our monitoring mechanism. If rules are evaluated at a computer that is remote from the place where the managed components are running, polling the sensors may become a costly source of overhead. Marzullo and Wood solve this problem in their Meta system by using an elegant technique: They compile conditions into local triggers, which can be evaluated entirely local to the monitored or managed component. Only potentially significant events need to be communicated to the state machine that is actually interpreting the rule. Thus, if we are concerned about load, we might find that the current average load is 7. If we ask the components to report their load to us if it rises by more than 1.333, we will certainly learn about any condition in which the load average has reached 10. Yet we can do so without reporting every intermediate value through which the servers pass, and indeed no communication may be needed at all in the normal case.

A second concern is that of ensuring the availability of the management environment. By using a process group (Meta was implemented over the Isis Toolkit), it was possible to implement this approach fault tolerantly. Events reported by a sensor to the policy were sent as *abcast* messages to a process group running replicas of the policy state machine. In this manner, the control policies needed to manage the system could remain operational even if some of the platforms on which the control software was running failed. Marzullo and Wood argue that availability is important in management systems: The control policies often tell a system how to reconfigure after a failure and hence must be running if the system itself is to reconfigure after a failure. One implication of this observation is that embedding monitoring and management functionality directly into the application itself may be a good idea, when it is practical to do so. In this manner, if any portion of the application survives a failure, so will the corresponding portion of the replicated management framework; one can then design the system so that if a sufficient amount of it remains operational, the management policies needed to recover will also be available and can be executed to reconfigure the system. Such a system can be said to be "self-managed."

23.7 Visualization of Distributed System States

A benefit of system management is that the existence of a system model and of a database of sensor values can support elegant visualization tools. Such tools go beyond the scope of this book; hence, we will not discuss the challenges of building them here. However, it is worthwhile to pause and stress the value of such a management interface. When building a reliable distributed system, one sets out to identify the potential causes of unreliability and to counter them systematically. A management infrastructure with suitable visualization tools will let an operator quickly and effectively understand the cause of such problems, which may slip through the original hardening process. These tools can be used to correct problems occurring at run time and can provide invaluable information concerning improvement of the system's self-management policies and technologies.

23.8 Correlated Events

It is common to assume that failures, recoveries, and other events requiring management actions are relatively infrequent and independent within a system. This avoids a number of thorny issues that would be raised by the potential for a sudden "storm" of events triggered when several things go wrong at the same time in a distributed system. We haven't talked about the concurrent execution of a set of rules (linearizability or serializability of the actions taken might be a sensible objective), and there is a broader issue of whether or not one even wants the same policy to be used while a system is operated in a routine manner as in the case where multiple events occur simultaneously.

Unfortunately, in a complex system, this assumption of independent failures and events is not likely to be satisfied. A power outage, for example, may cause half the nodes in a cluster computer to fail, and the management rules for the applications on that cluster will now be triggered simultaneously for many components of the surviving parts of the system. Moreover, if a system has complex dependencies between its components, the failure of a single component can cascade, resulting in a storm of secondary events and secondary failures, which are in fact symptoms of the original event. Attempting to correct these secondary problems will probably not be successful unless the core problem is identified and resolved.

Suppose, for example, that our database servers make use of the file servers. If a file server crashes, the database servers may hang until the file server subsystem has reconfigured itself to reallocate the workload for which the failed machine was originally responsible. During this period it is likely that we will detect an overload, but the overload is in fact purely a consequence of the file server reconfiguration. The correct management policy is to *inhibit* the overload condition for database servers while file server reconfiguration is underway. However, to understand this one needs to start by characterizing the dependencies of components upon one another and then pass from such a characterization to one giving dependent failure modes of the system as a whole. From this, it becomes clear that the correct way to deal with *independent* overload of the database server may be to add servers or to move to a degraded query mode. It also becomes clear that the best way to deal with *dependent* overload is entirely different; perhaps such a condition should be addressed by telling the client systems to stop submitting queries!

Traditional hardware engineering involves the development of *fault trees* by which such conditions can be characterized and by which appropriate solutions to occurring and/or intervening events can be charted and implemented. At the time of this writing, there has been little research on the use of analogous techniques in complex distributed systems, and the issues that might occur if one were to provide software support for a fault-tree analysis and problem resolution remain unknown. It is likely that the evolution of increasingly critical distributed systems will require development of increasingly sophisticated fault models, including techniques by which fault trees can be compiled into policies. The sorts of management policies discussed here would then be typical of those that might be used in a normal mode of operation; while other mechanisms, designed to deal with correlated failures or failures triggered by complex dependencies within the system, would be addressed through other policies, which might operate in very different ways.

23.9 Information Warfare and Defensive Tactics

In the introduction to this book we discussed the potential emergence of aggressive threats, which could include outright attack on the critical information assets on which the military or major segments of industry depend. One can confront such problems through security measures such as firewalls and access authorization. However, there is always the possibility that the security techniques themselves will depend upon an underlying technology, which could also be compromised—leaving the system defenseless.

To take the most obvious example, many distributed systems use firewalls for protection, mounting their primary defense against attack at the points where communication interfaces from the system to the outside world are placed. Yet such systems are rarely located at a single physical site, and they may have complex internal network topologies, including communication lines provided by telecommunication or Internet service providers. These lines are not perceived as connections to the external world and hence are often unprotected.

In practice, however, such a perspective places considerable trust in at least one external technology: that used to implement the dedicated lines themselves. If the Internet provider or telecommunication company is itself compromised, it may be that these dedicated lines can be tapped or even accessed freely by intruders. Such an intruder will now have circumvented the firewall protection and will be free to act within the protected system itself. The consequences of such a break-in can be very serious. We noted previously that the complexity of telecommunication systems and Internet architectures is increasing. Thus, the opportunities to compromise the underlying architectures used to support these dedicated lines are also increasing. Should the service provider itself be compromised, breaking into the network of a large organization may be much easier than doing so through the firewalls placed at its connections to the external world.

Recall the example of an NFS system used within a firewall (Section 19.1). When an NFS was operated within a firewall, we saw that it would be typical to disable its security mechanisms to improve performance. But an intruder would now be able to spoof, in a manner that would trick the NFS system into granting requests, by pretending to be a legitimate user on a legitimate machine. We saw that once through the firewall, there was little an intruder in an unprotected NFS environment would be prevented from doing: Files could potentially be read, rewritten, and dates and times reset with relative ease. Our challenge is to erect barriers that could convincingly detect and protect against such events.

The steps to repelling aggressive attacks, whether they threaten critical military assets or merely seek to falsify financial transfers from a large bank, will be to *detect* the intrusion, *quantify* the event by identifying the methods used and the system components under attack, and *respond* by modifying the behavior of the system, shutting down components, or disabling access points until the system itself can be made intolerant to the form of attack in question.

For purposes of *detection*, a DIW (defensive information warfare) monitoring system will typically rely on audit trails, which trace sensitive operations and permit comparison between patterns of access and authorized, or typical, patterns of activity. Such trails monitor classes of events that are of potential concern, such as file open operations, and are then filtered through programs maintaining historical information regarding permissible and typical patterns of access ("Vice-President Smith is permitted to access all account records, but in practice very rarely accesses any records except those for customers with whom she works directly," or "Admiral Walker normally reads and sends memos associated with naval operations in the southern Atlantic"). When violations of these normal patterns are detected, responses can be initiated, and if an intruder successfully penetrates a system, the audit trail will later permit the penetration to be localized and further problems to be prevented.

In the case of our NFS example, such a policy would involve intercepting file *open* requests somewhere on the path to the NFS server and comparing them with normal file access profiles. Suspicious patterns could be signaled to an operator, or simply discarded. Vice-President Smith would merely issue some sort of command to disable the protection system and then resume her unusual pattern of activity; an attacker would be much less likely to surmount this obstacle. These sorts of *application-specific* wrappers would substantially raise the barrier to an attack on the system, without necessarily paying the cost of general-purpose encryption or authentication, which can be high. (See Figures 23.7 and 23.8.)

Wrappers can be a valuable tool for compiling audit databases and for filtering messages in this manner. In general, such technologies are added *after the fact* to a system that was built *without knowledge that it would later be wrapped.* Thus, even if the system is compromised, it may not be able to anticipate and compensate for the protective mechanisms that will be used against it. The wrapper enjoys the advantage of stealth, provided of course that its performance impact is minimal. A wrapper can be used to control the actions of the wrapped component, to oversee those actions nonintrusively, or to build up profiles of typical behavior with which a new behavior

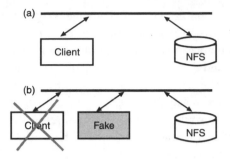

FIGURE 23.7. (a) The NFS protocol, when security is not enabled, involves trust in the IP address of client systems and the user/group IDs presented by clients as part of each request. (b) Should an intruder break into the network, it may be possible to masquerade as a legitimate client by killing some existing node with a poison pill or some other tactic and then sending fake NFS packets to the server. In this scenario, the NFS is unable to protect itself and may provide essentially unrestricted access to the files it manages. Wrappers around the client, the network, or the NFS unit could filter messages against a behavioral profile, which would permit such behavior to be detected or prevented.

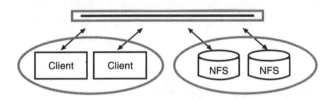

FIGURE 23.8. Application-specific wrappers surround various classes of system components in this hardened environment. The wrappers (gray) can intervene in lightweight ways, which are potentially less costly than strong security (encryption and authentication technologies) and yet can represent a significant barrier to intrusion. One can imagine a great variety of possible techniques that could be incorporated into such wrappers: For the NFS case they might maintain profiles of typical behavior, which would cause a database exploited by the wrapper group to detect unusual access patterns and to intervene. Here we have wrapped the clients, database servers, and network itself. Each wrapper might represent an intervention at a different place in the technology (the router of a network, the file I/O library on a client workstation, the network interface of an NFS server), and each might function in a different way.

can be compared. Moreover, a single system can support more than one form of wrapper for the same purpose—selecting the wrapper used randomly and arranging that a countermeasure to one wrapper will be detected as unusual behavior by another wrapper. Even an insider might have trouble overcoming such an approach. Work on wrappers for this purpose could be of great value in DIW management applications.

With regard to *quantifying attacks*, the clear challenge is to find ways of synthesizing patterns out of information gathered from physically dispersed components and to do so without excessively disrupting the performance or correctness of the monitored subsystems. One approach is to timestamp audit records and to send them to a central site for analysis; such a technology will depend upon bandwidth to the central site and heavy computing power to build up a centralized model of the decentralized behavior of the system from which patterns can be extracted. A second, distributed, approach would draw upon the natural process group structures present in reliable distributed systems and visible through the system model. In this approach the wrappers for a set of components would be programmed to detect unusual patterns of behavior in a distributed manner, operating close to the components themselves and basically compiling a pattern into its local subpattern. With the growing availability of high-performance process group support and clock synchronization technologies it may be increasingly practical to use such approaches, which have the benefit of imposing reduced loads on the communication system and exploiting decentralized and hence coarsely parallel computing power. Again, much research is needed in this area.

Turning to the problem of *response* when an attack has been detected, it is appealing to consider the introduction of wrappers for the purpose of placing selective firewalls close to groups of components. Suppose, for example, that every meaningful group of components was ringed by a firewall specifically tuned to the patterns of behavior seen in members of that group and capable of filtering incoming and outgoing messages under criteria associated with the message source and destination. Such a system could be visualized as having a potentially large number of superimposed firewalls intersecting in various ways.

Any given computation would now generate communication traversing one or many of these firewall boundaries, all implemented by wrappers in a lightweight manner. Now, if a threat were detected, the firewalls could be dynamically reprogrammed to inhibit communication to and from the compromised system components or to restrict such communication in a way believed sufficient to filter out the compromised traffic. Whereas the intruder in a conventional system needs only to violate a single firewall to gain free run of the system, the intruder in such a multilevel security containment environment would confront obstacle after obstacle, with each layer of firewalls implementing its own containment policy based on the normal patterns of communication for that subsystem or application. It seems likely that this would result in significant improvements in the robustness of critical systems against attacks.

Recalling our NFS example, one could imagine a solution in which the NFS traffic passes through filters, which do simple sanity checks on the origin and destination of packets and the degree to which these packets correspond to the expected network topology (Figure 23.8). By embedding such functionality into a packet router or the low levels of the network communication software used on the NFS itself, one could protect against many forms of NFS attacks even when the NFS security policy was not in place. Of course, protection against an attack requires that the attack be anticipated as a possible threat and that a suitable response be formulated, but at least the tools are available to potentially protect against such threats if the will exists to begin to exploit them. With few changes to the methodology by which distributed systems are typically constructed, it may be possible to introduce monitoring and protection mechanisms after the fact. This would make such systems far more robust against a great variety of possible threats and attacks. Moreover, by customizing the behavior of a technology, such as a firewall, to match it closely to the normal mode of operation of the protected system component or components, the complexity of breaking through that firewall can be greatly increased.

To summarize: If we face a growing threat of outright attacks on critical technologies and systems, the good news is that we also have a growing technology base from which we can obtain solutions to the problems of detecting those attacks (or explaining them in a postmortem analysis) and developing flexible and powerful responses. Wrapper technologies may permit the stealthy introduction of the resulting solutions into applications and subsystems not designed with such defensive abilities in mind and whose designers may not even be trustworthy: To the degree that the wrappers can be trusted, the resulting wrapped application may be trusted for specific purposes. While much work remains to be done before such a vision can be called a practical reality, it is clear that we have a growing number of powerful tools for use in the task. More research is needed, but we can already harden systems against a great number of simple attacks that many systems are unable to defend themselves against.

23.10 Related Reading

This chapter drew primarily from Babaoglu and Marzullo, Marzullo (1984, 1990), Marzullo et al., Wood (1991).

CHAPTER 24 ✧ ✧ ✧ ✧

Cluster Computer Architectures

CONTENTS

24.1 Introduction

A new generation of hardware is emerging in support of client/server applications: the so-called *cluster server* architectures, in which a collection of relatively standard compute nodes and storage nodes is interconnected using a high-speed communication device, often similar to a fast CSMA or ATM interconnect. Such cluster computers can be viewed as smaller cousins of the massively parallel supercomputers that emerged in the 1980s and now dominate the supercomputing market. However, they are more similar in many respects to distributed systems. In this chapter we look at the similarities and differences and discuss what impact these have on the solutions needed for reliability.

The specific definition of a cluster architecture remains elusive at the time of this writing: Many vendors offer computers that might be considered clusters, yet the term is sometimes applied to what might more properly be termed a multiprocessor, while some cluster computing systems are characterized by the press as coarse-grained parallel machines. Some vendors have applied the term cluster to primary-backup systems employing trivial (and potentially incorrect) fail-over technologies and lacking any tools for maintaining the consistency of the backup and the primary. Thus, the term is not merely disputed, but is also potentially misused.

For the purposes of this chapter, a cluster architecture is one based on:

▶ *Standard components:* A basic idea of cluster architectures is to take advantage of the price benefit associated with mass production of PCs and workstations—hence, clusters are typically built from completely standard components, albeit without displays and perhaps packaged in a nonstandard chassis. Typically, cluster architectures support at least two types of components: compute nodes and disk nodes. Some may also support network interface nodes and other special-purpose attachments.

▶ *High-speed interconnect:* The internal communication bus connecting the components is typically message-oriented and may be based on a standard ATM or high-speed CSMA connection device.

▶ *Cluster infrastructure:* This includes the chassis and cabling connecting components, the power supplies and communication adapters, and so forth.

▶ *Management subsystem:* A cluster is more than a rack-mounted pile of components, and the management subsystem plays what is often the dominant role in creating the abstraction of an integrated entity.

▶ *Cluster API:* This is a debatable term, but refers to the collection of system interfaces, available to program and system administrators, by which applications can take advantage of the clustered nature of the system. The API will typically include ways of determining the set of nodes on the cluster—monitoring their states, launching applications on nodes and monitoring them while they run, accessing the management functions of the cluster, and so forth.

Perhaps the most important attribute of cluster is that a separate copy of the operating system runs on each node. This is in contrast to more integrated multinode computers in which a single copy of the operating system controls many nodes at the same time: Such systems are best understood as true multiprocessors and will typically include special-purpose hardware supporting shared memory and permitting the operating system to control the multiple processors comprising the system. At the other end of the spectrum, a cluster is distinguished from a parallel computing system by having fewer nodes (normally 4–32, with configurations of 8 to 16 processors being common) and running a full operating system on each node, unlike the very stripped slave operating systems often used on nodes of parallel computers.

Our concern in this book is about reliability, and therefore it makes sense to focus on cluster computing systems intended for reliable or mission-critical applications. These include the Stratus RADIO computer (I played a role in developing that system and know it well); COMPAQ's Proliant line of cluster PC systems; the DECsafe Available Server Environment (see Digital Equipment Corporation); Sun Microsystem's Solaris MC system (see Khalidi et al.); other products from HP, Microsoft, and Tandem; and certain configurations of IBM's SP product line. Cluster computers that have a reliability concern must address many of the same replication and consistency issues we have considered in previous chapters; it makes sense to ask if we cannot adapt our solutions to a cluster computing environment. A more complete treatment of cluster computing can be found in Pfister.

24.2   Inside a High-Availability Cluster Product: The Stratus RADIO

The Stratus RADIO (Reliable Architecture for Distributed I/O) product is a good example of a cluster computer designed for high-availability applications. RADIO consists of a rack-mounted collection of between 6 and 24 compute and storage nodes. An individual rack of RADIO nodes can contain up to six of these nodes and will also have two communication nodes, its own power supplies, and so forth.[1]

The compute nodes are dual processors based on a standard PC architecture, with substantial on-board memory, small swap disks, and a fast clock cycle time. Each runs a standard PC operating system—UNIX or NT—and communicates with the other nodes over the internal communication network using completely standard communication protocols such as TCP/IP or the OSI protocol stack. Each node has its own IP address and runs its own copy of the operating system. These nodes are thus freestanding computers that share a hardware environment with one another but are capable of independent operation.

RADIO disk nodes are based on standard high-speed disk technology, but run a nonstandard operating system, which functions as a form of front end to the network. This permits RADIO nodes to share disks (much like *n-way* multiported disks on a more conventional computing

---

[1] I describe this system with some hesitation; as one of the developers of a commercial product, I am sensitive to the perception of a possible conflict of interest. To avoid such issues, the discussion is limited to the RADIO product on which I worked, and no attempt is made to compare this product with other products. Similar considerations entered into the review of distributed programming environments and transactional systems in previous chapters.

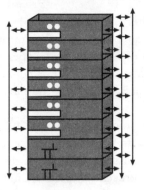

FIGURE 24.1. The RADIO cluster marketed by Stratus computer was designed for high availability. All components are duplicated for redundancy, and the compute and storage nodes are hot-pluggable without disrupting availability of critical applications. Illustrated are a six-node cluster of compute or storage nodes, the management network (on the left), and the dual internal networks (shown on the right). The bottom two nodes contain the network hardware, interfaces to external networks, power supplies, and other cluster management technology. Slots on the compute and storage node are for high-capacity floppy disks.

system), and software functionality for mirroring segments of disks is supported for high availability. The possibility of attaching other types of highly available disk technologies, such as RAID or SSA disk clusters, is also present. Any compute node can access any disk node within the cluster. (See Figure 24.1.)

RADIO communication is over an internal bus consisting of a fast Ethernet interface or an ATM switching interconnect. The network adapter nodes handle interconnection of multiple RADIO systems into a large, scalable system, as well as interconnection of the RADIO system with external networks connected to client systems. For availability reasons, the internal network is a dual one and the system can function transparently with either network out of service; similarly, dual connections to the client systems are assumed. RADIO also has a redundant power supply and can function normally with one or both power supplies in operation. It has an internal management network, which is used to monitor the status of the nodes and to provide the system administrator with control of the operator console interfaces to the PCs comprising the cluster. Such access can be from a local console or remotely over a communication interconnect from a central management site, which can be physically remote from the RADIO.

An interesting property of the RADIO management network is that it can detect a node going off-line (because of a self-diagnosed failure or because it was removed) or on-line within milliseconds. This is in striking contrast to the situation in a general distributed environment, where the same event may not be safely detectable until an extended period of pinging has occurred, because of the relatively high frequency of communication outages and minor partitioning failures. Thus, a type of failure detection that can take many seconds in a general setting is reduced to perhaps one hundredth of a second in this architecture.

All aspects of the RADIO architecture were designed with the avoidance of single points of failure in mind. Every component can be hot-swapped, meaning that each module is capable of being removed for service, without disrupting normal operation of the remainder of the cluster, and plugged back in after servicing or upgrading is complete. Moreover, the technology includes provisions for upgrades to new versions of the software or hardware, also without disrupting continuous operations.

The software used with a cluster product such as RADIO depends a great deal upon the goals of the application designer. RADIO itself is designed to be self-managed and includes a software infrastructure that exploits the Isis Distributed Computing Toolkit for this purpose. The application manager uses the general approach we discussed in Chapter 23 and operates by executing scripts written in the Perl command language using a fault-tolerant, state machine approach.

Applications can be run on the cluster without change (in which case they benefit from the hardware availability and management aspects of the platform but must be placed on the nodes by hand and will fail if the node on which they are running fails). Alternatively, an application can be managed by the RADIO application management technology, which involves developing management scripts indicating the aspects of the application state to monitor and what actions to take if a problem is detected. A RADIO system can also manage applications running external to it on client workstations or PCs.

A second way to benefit from a continuously available cluster is to use special-purpose applications designed specifically to take advantage of it. The RADIO system can run a variety of actively replicated software systems based upon the Isis Toolkit technology, including a continuously available database system, a continuously available version of the UNIX NFS server technology, and a message bus product supporting a publish/subscribe interface; all of these technologies are designed to gain performance as the size of the cluster is increased ("scalable parallelism").

Of course, there will always be applications that require the development of new continuously available servers. For this purpose, the RADIO user can draw upon the same active replication techniques presented earlier in this book. In particular, Orbix+Isis is available for the platform, offering a simple way to exploit replication and load-balancing in new applications, and Horus has been ported to the platform, providing a flexible and reconfigurable process group technology base for it.

I am less familiar with the details of some of the other cluster products available from other vendors and will not attempt to present any form of comparison of these products here. The experience of participating in the RADIO design, however, made it clear that there are many ways a cluster server can fail to be highly available, adequately manageable, or adequately functional. Striking the correct balance is a considerable challenge. There are no standard cluster architectures at this time, and existing cluster products employ varied hardware and software availability strategies and management technologies: Some are clearly deficient; others are very impressive but reflect different design tradeoffs than those made in developing RADIO. The developer of a critical application must therefore approach the selection of a cluster server architecture with considerable care, evaluating all aspects of the system design relative to the reliability goals of the application before selecting an appropriate server.

24.3 Reliability Goals for Cluster Servers

Cluster servers, if appropriately designed, have special properties that can be exploited in the communication and management software of the system. The motivation for these properties is best understood by looking at some of the goals typical application developers would be expected to have. These include the following:

▶ *Continuous availability:* Not all clusters are intended for critical applications, but in those that are, continuous availability is typically an important design goal. This objective has implications at all levels of the cluster architecture, including the hardware and interconnection technologies used, the cluster management technologies, the operating system functions offered to users through the cluster API, and the technologies used to implement applications on the cluster. Indeed, availability goals extend from the cluster into the surrounding network, because of considerations stemming from the environmental dependencies we identified in Figure 1.5.

▶ *Rapid failure detection:* This property is required if applications are to fail-over quickly when a node crashes or is removed for servicing.

▶ *Ease of management:* In critical settings, it is important to be able to rapidly localize sources of failures in order to fix the faulty system component or components and to have easy-to-use tools for upgrading system components and performing routine service. Cluster management environments must therefore include automated self-management solutions and must provide for remote management access to the cluster over a modem or computer network from which a remote operator is working.

▶ *Single-system image:* Although this goal is debatable, many clusters seek to provide the application developer with the illusion that the cluster is really a traditional single processor running a traditional operating system such as UNIX. The danger in such an approach is that unless some form of transparent availability technology is provided to the application designer, the application itself may not wish to view the cluster as a single system: Placement of replicas of critical services on failure-independent nodes, for example, requires that it be possible to distinguish the nodes in the clusters and separate them based on hardware properties of the machine. Yet at the same time, the application developer would clearly prefer not to deal with artificial barriers to programming the cluster as a single machine. Cluster designers struggle with the resulting dual goals: transparency and a single-system image on one hand, and ease of developing an extremely reliable application on the other hand. It sometimes comes as a surprise to the vendor that the first may not imply the second!

Some aspects of the single-system image concept considered in modern cluster designs are task migration and automated load-balancing, support for a coherent disk buffer cache on each node, direct paging and disk block access from the on-line memory of a different node, shared memory for processes running on different nodes, UNIX process ID and signal group functionality across nodes, and a single IP address for the cluster as a whole. Each of these can be beneficial in some settings and yet undesirable in others.

- *Load-balancing tools:* The major benefit of a cluster is the redundancy of hardware, permitting high availability, as well as the possibility of exploiting scalable parallelism. The application designer will be looking for tools to assist in this process by placing new tasks on lightly loaded nodes.

- *Task migration tools:* Another debatable aspect of the cluster debate: Some cluster technologies seek to automate the migration of application programs from node to node—a desirable property from the perspective of load-balancing and performance management, but potentially problematic if there are availability concerns governing task placement or if the functionality associated with this feature is extremely complex to support. Task migration support can vary from a process group mechanism, whereby the user can migrate a task (by adding a group member and doing a state transfer and then killing the old memory), to completely automated solutions, which transfer the virtual memory pages of a running application and in this way migrate its full state from node to node. I favor simpler solutions: Full-fledged transparent migration seems costly, nonstandard, and poses the risk that application programs intended to operate independently for increased reliability will instead find themselves on the same node and hence exposed to correlated failures.

- *Scalable parallelism tools:* Many applications will benefit from cluster architectures without requiring any changes to the code itself at all—for example, it may be that the application involves running large numbers of small processes (perhaps each command by a client on a workstation results in the execution of a process for that client on the server), and the placement policy for these processes is already sufficient to provide a benefit from the parallelism of the cluster. However, some servers will need to exploit parallelism in an explicit way, using tools such as the process group computing tools we discussed earlier. For users who need such tools, the cluster should provide well-integrated and highly tuned solutions, which will take advantage of any special properties of the hardware.

- *On-line upgrades:* Many clusters are designed to permit hot-plugging of hardware and software components, permitting repair of problems without disruption of the remainder of the cluster and offering a path whereby software revisions and bug fixes, or entirely new versions of the operating system or other major components, can be installed nondisruptively. Technology permitting such upgrades within software used by the application programs can be a difficult challenge.

- *Clock synchronization:* Many clusters have the capability of supporting highly synchronized clocks; not all take the step of actually providing this support.

- *Reliable communication:* Many clusters use communication hardware that can guarantee reliability if appropriately configured. Not all clusters offer this functionality to the application, however, even if the hardware is capable of supporting it: Modern operating systems are designed to discard messages when they become heavily loaded.

- *Security features:* Within the cluster, nodes may consider themselves to be in a shared security enclave, trusting one another in ways that would be unwise for applications on widely separated nodes in a general distributed environment.

▶ *No single point of failure:* Developers of reliable applications need to be sure that their system will not have any single point of failure. Cluster designers who develop highly available cluster products must adopt the same point of view to ensure that their products will be robust against all possible single failures and that this property extends from the lowest-level hardware components up to the software services that operate each node and provide management functionality.

These properties make cluster architectures extremely interesting for a number of applications. As an example, recall Friedman's work on cluster support for a telephone system using a centralized switch to handle some classes of calls, such as 800-number calls. Potentially, there may be as many as 10^7 telephone numbers in the associated database, of which each record will encode the correct action to take if the corresponding number is called. If these records require 100 bytes each to store, the database will contain perhaps 1 GB of data. To guarantee rapid response time for incoming calls (typically a switch must route a call in no more than 100 ms), one would prefer that these data reside in memory. However, the amount of memory here considerably exceeds what a typical computer can accommodate. A cluster architecture could easily provide memory-mapped access to a multigigabyte database, together with the fault-tolerant properties required for critical applications such as telephone switching.

This review has focused on positive aspects of the cluster approach, but there are also negative aspects. A cluster offers the *potential* for improved performance and availability, but actually achieving scalable parallelism in a specific application or implementing a highly available solution to a specific problem may be a very difficult undertaking. Moreover, a cluster will be physically located at a single site, and if the roof leaks or the power supply to the building is cut, the availability features of the cluster may be of little benefit. A distributed system replicating critical state between widely separated sites could well survive such disruptions with little or no interruptions in service. Thus, while it is safe to say that a cluster offers exciting options to the developer of a critical application, it is also clear that many issues must be addressed to achieve critical reliability, and clusters may or may not match the needs of a specific system.

24.4   Comparison with Fault-Tolerant Hardware

The reliability goals of a cluster computer are in many ways similar to those traditionally addressed through special-purpose, fault-tolerant hardware. In practice, however, the properties of cluster solutions are in some ways very different from what hardware fault tolerance typically provides. Although this text has not considered hardware fault tolerance up to now, it may be useful to point out some of these differences.

Hardware fault tolerance is usually achieved by building some form of self-checking logic into the basic hardware modules of a computer and off-lining a module that fails. A backup module continues operations on behalf of the failed one—for example, the same company that developed the RADIO system also manufactures a line of fault-tolerant computers called Continuum. The machines in this product line use a paired architecture in which each processor or memory component is duplicated twice. Pairs of CPUs and memories form a single module: By comparing the memory contents and CPU actions continuously, a module can be taken off-line instantly if any problem is detected. The remaining pair of processors remains operational until the damaged module is repaired, at which point the system is restored to its fully symmetric fault-tolerant mode.

Hardware fault tolerance of this sort offers protection against data integrity errors, such as undetectable multibit memory corruption, as well as protection against other forms of hardware failure. Application programs are continuously available without even a flicker when a failure does occur. But if a software failure occurs or a component of the system must be upgraded, there is no alternative but to take the system off-line to accomplish the repair. A cluster system will lack data integrity checks and may not detect hardware errors, permitting serious data corruption to occur and to spread. Failures that *are* detectable may not be detected very quickly: The software techniques used to detect some classes of failures and to reconfigure to recover from them may take many seconds to complete. However, cluster systems do offer a plausible response to the issues of software failure and upgrade through the use of process group technologies and state transfer. Moreover, in a redundant hardware system, one pays for the extra hardware needed to gain reliability, but gains no improvement in performance; a cluster will be less reliable but is also more cost-effective, since the processors can be kept busy doing nonidentical tasks while they also back up one another.

One sees from this that there is no single story concerning fault tolerance. Fault-tolerant hardware has found important roles in process control settings where a failure that results in system downtime will have immediate and costly implications. Telephone switching is a typical case and represents one of the largest applications of modern fault-tolerant computers; other examples occur in air traffic control, avionics, machine control, power systems, and so forth. These applications can't tolerate any downtime at all, even briefly. Clusters are a compromise offering scalability at a price: weaker fault-tolerance potential and slower reaction time. But they are also inexpensive and very flexible. It is likely that each class of computers will play a major role in future reliable systems, but that the associated markets will be different ones.

24.5 Protocol Optimizations

We noted earlier that it is appealing to consider the use of process group and transactional technologies on cluster computing systems. When this is done, a number of optimizations can potentially be introduced that greatly simplify these technologies and also hold the potential for improved performance.

In particular, a cluster system may have an *accurate means of sensing failures*—for example, the Stratus RADIO cluster has a management network that can rapidly sense hardware failures of nodes and can rapidly sense the introduction of a new node. A failure can thus be detected within a few milliseconds, and the necessary reconfiguration actions can be triggered almost immediately. The operating system itself will detect most forms of application failures. Thus, with the exception of failures that leave the operating system alive but unable to do useful work (hung), as well as similar failures in the application programs, a cluster system may be able to use a greatly simplified system membership management protocol. Even if the GMS protocol is used without change, its performance will improve because false failure detections are not a concern; the risk of partitioning is completely eliminated, permitting progress even when a small set of nodes are the only ones to survive a failure.

A cluster may have a *lossless message-passing capability*. If present, such a capability can eliminate the need for positive and negative acknowledgments. In an architecture such as that of Horus, this allows a major protocol layer to be omitted and offers potential performance benefits. Of course, the ability to exploit such a feature will depend upon the nature of flow control done in the application program: If data sources can outrun data sinks, message loss will be unavoidable.

A cluster may have *accurately synchronized clocks*. Such functionality would permit the introduction of real-time features and protocols and enable processes to cooperate in performing time-critical or periodic tasks.

A cluster will typically have *extremely uniform hardware properties*. These permit protocols to be tuned for better performance—for example, with knowledge of the true point-to-point latency, the decision to send acknowledgment and flow-control messages can be finely tuned, and timeouts can be closely matched to the true expected response time of a critical service. Notice, however, that the ability to exploit this feature will be limited by the properties of the operating system.

A cluster will have powerful *management capabilities*. These permit the development of rich and sophisticated software for system and application management, going well beyond what can be done in a distributed environment where the heterogeneity of platforms used forces a least-common-denominator approach in the management subsystem.

Thus, we see that although the same abstractions we considered in discussing reliability issues for general distributed computing systems may be extremely valuable in a cluster computing system, their implementation could be considerably affected by these properties. When using a layered protocol architecture, such as the one favored in Horus, the implications may be as minor as the need or lack of a particular Horus layer. More broadly, however, clusters offer the potential to exploit the uniformity of the system architecture to achieve better performance, improved management, and quicker response times. (See Figure 24.2.)

The bad news, of course, is that the cluster is not an isolated entity. Most cluster systems are integrated into heterogeneous distributed environments, and the same software that runs on the cluster will often include components residing outside of it, in client workstations or other components of the surrounding environment. We are again reminded of Figure 1.5, where some of the many technology dependencies present in a typical network were illustrated: The same dependencies will limit the effectiveness of our network as a whole, even if they have a somewhat reduced impact on the availability and reliability properties of the cluster itself. The designer who includes a cluster server into a large distributed system has not necessarily guaranteed high availability: A broader technology solution is needed. However, if a technology such as Horus can be used both on and off the cluster, a satisfactory response to this concern may emerge from the resulting software environment. The view will be one of software that is able to benefit from the local environment within the cluster, but that offers uniform features and a uniform programming model both on and off the cluster.

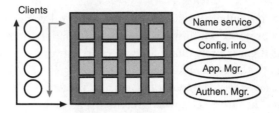

Clients

Name service
Config. info
App. Mgr.
Authen. Mgr.

FIGURE 24.2. When a cluster is used to implement a highly available client/server system, information will be needed regarding independent failure properties of the component nodes (white and gray here), connections to clients (black and gray), and dependencies the application may have on other services in the enclosing network environment. There may be substantial simplification possible by assuming that a technology such as Horus is available on all nodes in the cluster and the client nodes: The resulting uniformity of infrastructure could go a long way toward creating the context within which powerful applications can be supported. However, additional services would also be needed, such as services for determining the configuration of the network and cluster, programming management policies, and so forth. A cluster API would standardize these services and the interfaces by which they can be accessed, while also standardizing access to internal cluster communication and management mechanisms.

24.6 Cluster API Goals and Implementation

At the time of this writing (1996), there has been considerable interest in the formation of a standards organization to develop and prototype potential standards for a cluster API. Such an API would provide the developer with a predictable set of primitives, available on clusters from multiple vendors, which could be exploited in building highly portable cluster applications. Such a solution would free the developers of layered products from dependencies on any single cluster vendor.

The most appropriate content for a cluster API is clearly debatable. From the review of technologies we have undertaken in this book, one can quickly see that if the cluster includes availability and reliability as an important goal, the API could potentially include every technology we have discussed. At a minimum, a cluster API should address the following:

▶ It should provide software access to system configuration and management interfaces—that is, the API should permit an application to learn the topology of the cluster, failure and performance characteristics of components, and access to the various status monitoring features provided by the hardware. Standards should be specified for such properties as load and status (operational or failed) of cluster components, for naming the most standard components (compute and storage nodes, communication paths), and for specifying performance properties (speed, latency, etc.). Notice that some aspects of this could be addressed by mandating the use of SNMP or CMIP management information bases (MIBs) with standardized content; other aspects would not be resolved by such an approach.

- It should provide a simple architecture for exploiting the hardware features of the cluster—for example, it should address such issues as notification when system membership changes, communication among applications on nodes, and so forth.

- It should provide a uniform means of addressing applications. This is likely to require a two-level solution in which the cluster as a whole can be accessed using a single IP address, but its component systems will have secondary true IP addresses.

- It should provide O/S-level primitives for such aspects as launching an application on a node, monitoring the application as it runs, task migration (if supported), remote disk access (if supported), and so forth. The API should standardize these aspects even if they are considered optional, depending upon the particular architecture used by a given vendor. Thus, there might be an API for shared memory across cluster nodes, as well as a way to determine whether or not the vendor actually supports this feature.

- It should deal easily with off-cluster applications and not be restricted to issues internal to the cluster.

This brief review points to what is clearly a substantial topic. A system such as Horus might make a good starting point on a cluster API for dealing with system management and communication issues. However, doing so would still leave open important topics—for example, existing distributed systems lack any way to interrogate the network about its topology and performance properties. Without standards for this type of information, cluster applications would need to have their own proprietary configuration schemes, making them less easily portable and less standard. The problem is thus similar in many ways to the one tackled by the OMG group, which developed CORBA, or the OSF group, which developed DCE, and it may require a similar scale of effort to resolve it.

24.7 Related Reading

Most of the information in this chapter was drawn from materials available from Stratus Computer, Inc. For more general treatment of cluster computing, see Pfister.

Other clusters cited here include Digital Equipment Corporation, Khalidi, et al.

CHAPTER 25 ✧ ✧ ✧ ✧

Reasoning about Distributed Systems

CONTENTS

The preceding chapters have established a technology base with which existing distributed systems can be hardened, by selectively replacing components with versions that have been modified to withstand some class of anticipated threats, and with which new distributed systems can be constructed to be robust by design. Yet these tools are purely practical, in the sense that they represent software solutions to specific potential problems. What is lacking is a broader conceptual framework, which would let the designer characterize the behavior of a given system in a particular environment, reason about how that system might fail under various conditions, and predict the impact of a given robustness intervention on properties such as performance, security, and availability.

Early in the book we noted that it was not our intention to delve into the theory of distributed computing here, and this remains true now. Yet the ability to specify the behavior of a system and to reason about it, while representing a fairly formal activity, is not necessarily a theoretical one. In developing nondistributed systems, it is common practice to develop specifications characterizing the behavior of critical software modules and to use these specifications as an input to large-scale system design and visualization technologies.

In the remainder of this chapter, we review some of the options for describing a complex distributed system, and we specify the technologies it uses to achieve robust behavior. Such specifications necessarily extend to the environment (upon which the system may impose requirements) and the application (which may be assumed to respect various constraints or to behave in specified ways). We then look at the issue of exploiting these specifications to reason about the large-scale behavior of a distributed system by using its small-scale behavior as an input to an overall system model. We also explore some recent work on automatic synthesis of robust distributed systems by plugging application-specific data-handling methods into general-purpose frameworks; such steps can be viewed as providing methods for compiling robustness into a system that is specified using a high-level language. All of these areas are in need of additional study, and the conclusion of our review will indicate that the technology base for reasoning about distributed systems is very tentative at the time of this writing. However, the longer-term potential is considerable.

25.1 Dimensions of the System Validation Problem

The distributed system properties treated in this book can be understood as operating in a multi-dimensioned space of potential system properties. Viewing a system in this manner can be a useful step in sorting out the critical properties of a specific application and demonstrating that they hold.

Consider, for example, a process group system such as Horus. Horus operates over the lower levels of the OSI hierarchy and draws upon point-to-point properties of the routing subsystem and data-link layers of the transmission subsystem. The correct function of Horus ultimately depends upon the correct function of these lower layers: If packets cease to be routed to their destinations, or corrupted packets are passed up to the application layer without detection by the data-link layer, Horus will begin to malfunction. Two problem areas come to mind: the protocol used by UDP to reassemble fragmented packets into larger, arbitrary-sized packets and the methods used by the device driver and device controller to manage packet chains. These are nontrivial software procedures and are entirely capable of malfunctioning in subtle ways.

In my experience (just to emphasize that there is already a nontrivial problem here), protocols and systems have been observed to fail because of hardware that experienced an extremely high error rate for certain sizes of data packets (e.g., multiples of 204 bytes); an accidental configuration error that could cause a network to partition in a partial way; delivering only packets associated with certain protocols and losing others; and operating system interfaces that have failed—for example, by rejecting all UDP packets. (Since UDP is not guaranteed to be reliable, the vendor may have actually considered this behavior to be a correct one!) Other researchers have shown that network routing can malfunction so that process a can communicate normally with process b and b with both a and c, but so that a and c are incapable of direct communication. Such a behavior directly violates what we are told to assume about routing, and yet it occurs with disturbing frequency in the modern Internet! In one memorable event, the compiler malfunctioned in a way that caused a low-level Isis protocol to deadlock, despite the fact that the protocol as coded was actually correct. We are now faced with the following question: How can we possibly build reliable systems, if we can't trust our own assumptions?

One could argue that the point-to-point properties of the communication environment represent what is just a first dimension for potential analysis and formal verification. Moreover, actually writing down the properties of a communication environment will clearly represent an extremely difficult task. To start with, we expect a message-passing subsystem to route packets safely and correctly to their destinations (discarding corrupted ones), to be nongenerative (i.e., to deliver only packets that were legitimately produced by the application), and to be free of very delayed replays. In practice, we often make a further assumption that the network has short, bounded delivery delays. But to what degree are such assumptions actually valid? As with the routing assumption, it may be that these assumptions sometimes will break down.

Moving on, the point-to-point properties of a network will now be extended by higher-level protocols into end-to-end properties. For the OSI hierarchy, these correspond to stream protocols. Were we to verify such a protocol, we would need to first express the properties of the underlying system and then to characterize the desired protocol properties—for example, a stream protocol typically will detect and reject duplicated packets and place out-of-order packets back into the order of generation. Such protocols also handle retransmission, detection of failures, and may dynamically vary such parameters as window size or data-flow rate to match the properties of the sender, receiver, and network. Again, although the problem of actually doing this may seem simple at first glance, both the properties themselves and the task of validating them grow much more difficult when the true nature of the assumptions made and the guarantees offered are really pinned down. TCP, for example, uses a finite counter, which wraps when the field is full. If packet numbers can be reused, the protocol will not be correct in a network that may replay packets after long delays. The assumption that the network doesn't behave in this way is implicit in the typical TCP implementation, but would need to become explicit if we wanted to prove that TCP really works as desired. And then one would need to ask whether or not we really have grounds to trust this assumption—for example, what would happen if a router node somehow jammed up for a few hours but then suddenly resumed operation? Would it then replay packets hours after they were first sent?

In fact, there has been research on proving TCP correct under various assumptions, but to do so under realistic assumptions would be a daunting undertaking. This illustrates one of the points we will want to reconsider: Correctness analyses of protocols should probably not undertake to model the environment and protocol in excessive detail. Doing so is likely to load up the proof with a tremendous amount of detail and may ultimately yield only limited insight. Moreover, when the proof is finished, the developer may still be left facing a number of assumptions in which one can have no better than probabilistic confidence.

The problem grows still more complex as we consider group communication protocols. Not only do these protocols depend upon point-to-point and end-to-end assumptions at the level of communication channels over which they run, but they also depend upon the abstraction created by the group membership protocol (which may operate in a partitionable environment), and there may be further assumptions about the application itself. It is helpful to consider these protocols as having systemwide properties, such as the group abstraction itself or the dynamic uniformity guarantee of a protocol having this property—another dimension in the spectrum of properties. It turns out that even if we assume a very simple environmental model, capturing this systemwide perspective can be tremendously challenging. What does it mean to say that a group manages a replicated data object or that its members behave in a mutually consistent manner? Questions such as these remain active research topics today. Researchers work on them because there is considerable reason to hope that they can be solved, and because we gain considerable insight even from partial solutions. However, it may be some time before solutions begin to take on the character of useful tools that practitioners might exploit. We'll discuss this in more detail later in the chapter.

We observed that properties can be thought of as existing in a multidimensional space: multidimensional because there is a sense in which, for example, group properties seem to be orthogonal from the channel properties over which the group is implemented. Also, the real-time assumptions and guarantees of a real-time protocol suite represent a different dimension along which a system may seek to provide guarantees. This is similar to the security properties of the environment and system and their self-management abilities and guarantees.

If it is difficult to provide a formal proof that a TCP protocol guarantees the stream properties, it will be nearly impossible to do so for a complex system seeking to guarantee a range of reliability properties such as these. Of course, by isolating a question about a protocol, we can often provide a formal proof that under certain conditions, the protocol provides a desired property or that it will always make progress. When we talk about "reasoning about a distributed system," this is typically what we have in mind: the systematic analysis of some aspect of the system, under what may seem like extremely simple assumptions, to establish that when those assumptions hold, the system has the desired property.

Such a point of view can now lead us in any of several directions. We will review the state of affairs in some of the major research areas taking this sort of approach. Quite a bit is known, and the rate of progress is extremely encouraging. Over time, there is reason to expect that formal tools may emerge to assist the developer in putting together a system having a desired property or behavior, by using the sort of narrow and highly targeted analysis of properties emerging from this approach.

It should be stressed, however, that one can also adopt a more skeptical view of the whole subject. If it will ultimately be impossible to prove that systems have desired properties because of the complexity of the assumptions involved and the risk that these assumptions will not hold, we should seriously consider the possibility that reliable systems are fated to be unreliable, but that, hopefully, this will not occur frequently. In my experience, this is not an unrealistic perspective.

There is a community that takes this skepticism to an extreme degree: Feeling that reliability guarantees are ultimately meaningless, these developers make no effort to achieve reliability at all. More specifically, they resort to completely ad hoc methods, test their systems heavily, and accept as inevitable the remaining reliability and security loopholes. Such an approach seems to go too far: In effect, this community abandons all hope for any meaningful form of reliability at all because perfection is not attainable.

In particular, the techniques we have presented in the preceding chapters really do lead to systems that have useful properties and that can automatically adapt themselves to such accidents as infrequent failures and restarts, communication lines that break and later are reconnected, or limited security intrusions. They can guarantee real-time behavior, as long as the clocks are working well enough, the communication lines are responding the way they are expected to respond, and the applications themselves are not overloaded. And, most of the time, such assumptions are reasonable and the systems built this way can be considered as very reliable ones. It seems a bit extreme to discard these benefits simply because we doubt that the reliability of the resulting applications will achieve some absolute standard of perfection!

Sometimes, however, we must expect that our assumptions will prove to be incorrect and that the correctness of the system as a whole will indeed be compromised. Thus, while on the one hand it makes sense to invest using these tools to build systems and to invest effort in proving (to the extent that we can) that our techniques work, we should also design our systems to heal themselves when the unexpected occurs and the system state becomes corrupted as a result— that is, we can derive considerable benefit from reasoning about systems and using powerful tools, provided that we take the whole approach with a grain of salt.

Indeed, even if our assumptions were completely correct, we would still face this problem for other reasons. A system such as Horus is fairly complex: A typical protocol stack and its run-time support involve perhaps 25,000 lines of active code, not considering the code embedded into the operating system and compiler and the issues raised by communication with other machines, which may be running other systems or other revision levels of the same system. Thus, even if one were ready to "bet the bank" that a certain data replication protocol provides consistency and fault tolerance, it wouldn't hurt to design the system to either periodically reset the group members to have some known value for that replicated data or to at least periodically have the group check itself to detect errors.

Such a view argues, ultimately, that reliable distributed systems should be constructed to be as modular as possible: Not only will module boundaries serve as a natural place at which to specify behavior and undertake to prove that it will hold, but they will also serve as a natural firewall within which inconsistency or unexpected failures can be contained and repaired and within which self-checking mechanisms can operate. Recall that we discussed the controversy associated with extending group properties across group boundaries. Here, we now see a concrete reason not to do so casually: By attempting to extend properties across a broad system, we link together the correctness of the groups involved. In effect, we break down what might otherwise have been a form of protection boundary for properties. The same groups, with properties limited to communication that remains inside them, would be much more isolated from one another and much less likely to experience a systemwide failure if one of our assumptions turns out to have been invalid in some isolated corner of the system.

25.2 Process- and Message-Oriented Models

In discussing the protocols presented in previous chapters, we developed several execution models (the asynchronous model, the transactional model, the virtual synchrony model), which were expressed at the level of processes, messages, time, data objects, and relationships over these primitive elements. Today, such models are the ones most commonly used to formalize and reason about distributed systems. They are seen to be deficient, however, when one tries to apply them to a complex distributed application, composed of multiple subsystems having nontrivial properties and relationships among the components.

Most work on modeling distributed systems and protocols has focused on the use of very simplified formal descriptions of the network, the protocol itself, and the application, and using those descriptions to build up what might be called a "formal system" with provable properties. Although, as we commented earlier, it appears to be impractical to treat every aspect of a system in this manner, this type of formal system research can still increase the developer's confidence that a protocol does what it is supposed to do, does it when it is supposed to do it, and that it isn't subject to deadlock or other undesired problems in the normal course of events.

Broadly speaking, work in this area can be understood as falling into three categories. There is a community of researchers who are focused on developing distributed algorithms, which generally take the form of simple and very high level techniques for solving specific problems that are expressed in terms of a distributed computing model. We have touched upon some of this work, but have presented very little of what is a dynamic and extremely productive body of research. Problems that have been studied by this community include the detection of deadlock and other stable conditions in distributed systems, distributed discovery of graphs and computation of subgraphs (e.g., to identify optimal routes in a routed network), consensus and leader election, various forms of load-balanced execution, evaluation of predicates whose variables make reference to the states of processes in a distributed setting, and so forth. Recently, the group membership problem and virtual synchrony model have become a focus of attention for this community.

In building reliable distributed systems, one sometimes encounters problems such as these, and it can be very useful to gain background and insight by following the literature of the distributed algorithm community. Although we haven't covered much of this work here, there are other books available that do this in a very effective way. The reader might want to refer to Lynch's recent book (see Lynch), Andrews' book (see Andrews), and Schneider's forthcoming book (see Schneider [in press]). In addition, the interested reader should consider following academic conferences such as the ACM Proceedings on Distributed Systems (PODC) and the Workshop on Distributed Algorithms (WDAG). Many papers from these communities appear in Springer-Verlag's journal, *Distributed Computing*.

A second category of research in this area is concerned with the logical foundations of the systems and languages used to reason about distributed computing systems. The problem here is that we are only just beginning to find ways to express the properties of the sorts of reliability technologies this book has treated in some depth—for example, although the virtual synchrony model can be described in terms of a set of rules that such systems should follow, there is always a significant risk that such rules are somehow imprecise, leave open trivial solutions not having the properties desired in the formal specification, or are otherwise flawed. What one would ideally want is a highly expressive language in which such rules could be expressed rigorously— a logic for distributed computing systems.

In fact, there has been a great deal of research on logic used to reason about computing systems in general and distributed systems in particular. Leslie Lamport, for example, whose work was cited in connection with concepts of time and consistency, is in fact best known for his research on temporal logics, which he uses to express concurrent algorithms and to reason about complex distributed protocols and systems. Other researchers have looked at concepts of "knowledge" as they occur in distributed systems—for example, Halpern and Moses proposed what have become known as "knowledge logics" for distributed computing several years ago and showed that one can sometimes understand distributed protocols better by thinking in terms of facts that the system is able to deduce on the basis of messages that the processes making up the system have exchanged. Lynch has suggested that a type of logic automata might be used to understand the behavior of certain protocols and algorithms. And these are just a few instances from a very broad and vigorous field of study.

Overall, it seems clear that we have yet to see the definitive language or formal approach for describing properties of the sorts of reliable distributed systems treated in this book. Existing languages continue to be far from the actual written software, so protocols must be translated into the formalism (with the risk of mistakes or mistranslations), and it can be very difficult to reason about failures using formal tools. I have tried, for example, to formalize the properties of the group membership component of the virtual synchrony model with mixed results (see Chapters 13 and 14). To a large extent, the difficulty here seems to be that the temporal logic used to write down properties of the GMS has difficulty dealing with what might be called "branching" future executions.

Suppose that one were to write down the rule that a process communicating with the GMS sees the same sequence of events as do other processes in the system. This sounds simple until one considers that such a rule needs to be evaluated "at" a time and "on" a set of process states. But the purpose of the GMS is to maintain the set of processes for the system, and this set evolves dynamically through time. Moreover, since we lack accurate ways of detecting failures, any given process may suffer the misfortune of being partitioned away from the system and classified as having failed. The system as a whole may partition into two or more component subsystems, which experience different sequences of events. Thus, what looked like a very simple goal suddenly becomes extremely complex: Our rule must somehow be written in a way that can express all of these possible events. Worse still, at any given point in time for a process one may wish to say something about the state of that process; however, the actual status of that process relative to the other processes in the system may not be known until a complex consensus protocol has run long into the future. Perhaps, at time t, process p believes that replicated variable x has value 17. But at time $t + 10$ the system may conclude that process p was actually faulty at time t and that x never had this value.

In effect, the formal interpretation of an event that has already occurred in the past will not be decided until some sort of protocol runs. How should one deal with this form of uncertainty? One could write down a very complex property: Either p will turn out to be operational, in which case x is truly 17; or p will later prove to be partitioned away from the system, in which the official value of x may not be 17 after all. But this will clearly be awkward; moreover, once the fate of p is finally decided, the correct explanation of its behavior will suddenly become simplified. Existing systems for formalizing the behavior of distributed algorithms lack a good way to deal with this issue.

Thus, while research is making valuable inroads into the tools for formalizing the behavior of distributed systems, we still seem to be far from having the right set of tools for dealing with some of the more complex mechanisms considered in this book. A consequence is that those of us who work with reliability are forced to use relatively informal methods to specify our goals—at constant risk of mishap. On the one hand, it would be desirable to be more formal, but one finds that the languages and logics themselves are awkward for our purposes. On the other hand, one hesitates to abandon formality, because of the very big risk that mistakes will now enter our work.

Finally, there is a third major direction for research into distributed system specifications and formalisms. Were one to approach a logician with the comments discussed previously, the obvious first question to be posed would concern our ultimate goals: Do we actually know what sort of system we are trying to model? And, ironically, the answer would have to be negative. At the time of this writing, the distributed system community has yet to reach real agreement on the most appropriate system models to use. As we saw earlier, Cristian works in a timed asynchronous model (see Cristian [1996]), while Babaoglu favors a model based on reachability (see Babaoglu et al. [1994, 1995]). My own work has focused increasingly on Chandra and Toueg's approach of augmenting an asynchronous system with a failure detector (see Chandra and Toueg [1991], Friedman et al.). Clearly, the language won't do us a great deal of good unless we have some idea of what we want to do with it. The good news here is that all these approaches seem to work, at least in the sense that all three can be used to express the properties of the protocols and systems built and to show that they hold under various realistic conditions. But these models are also quite different from one another and, without some degree of consensus, we may not learn enough about any single model to begin to have an idea about what might be called a "distributed theory of everything." Indeed, the model we know the most about is the pure asynchronous model, and yet what we know is that this model is too weak to support the types of reliability that interest us here and that real systems are quite different from the ones described by the model.

This discussion has focused on logical properties of distributed systems, but could equally well have looked at security properties or temporal ones, with largely the same conclusions. In the area of security, for example, there has been some very important work on security logics (see Lampson et al.), which have been used to reason about the protocol used in Kerberos and other similar authentication subsystems. But this work doesn't consider issues such as failures and partitioning; it focuses on the simplest nonreplicated case of the authentication servers being considered. So while we do have some encouraging success, we face a major challenge in extending the same work to the more complex systems used in the real world.

For the case of systems with temporal properties, we have seen that the style of reasoning used in the CASD protocol was ultimately too conservative for its intended users (see Cristian et al. [1985]). Yet we know very little about this sort of protocol, in a formal sense, if we try to run it fast enough so that its properties start to become probabalistic ones, as discussed in Chapter 20. Friedman's work has shown how Horus can be used to obtain very predictable response time in a scalable, load-balanced server. But it seems impractical to *prove* that this approach should work in the manner that it is observed to work! By and large, the formal tools available for reasoning about temporal properties of fault-tolerant systems focus on systems that can be described conservatively. When we try to push predictability and performance close to their joint limits, we appear to enter a less-conservative world in which those formal tools no longer will serve us.

Traditionally, when formal models and tools fail us, computer scientists have turned to simulation studies. Yet reliable distributed systems can be too complex to model using simulations. Kalantar, whose work was reported in connection with scalability (see Kalantar), encountered just this problem in his doctoral research. Although he was able to gain significant insight into the fine-grained behavior of a complex protocol using a simulation study, he was not able to simplify the model enough to let him simulate the large-scale behavior of such systems: The simulation became swamped with detail and too slow. This is one reason why researchers such as Friedman have tended to build mock-ups of real applications and to study those mock-ups in great detail: Since they draw on real technology, one can be sure they are not simplistic. But one is also limited by such an approach, because it can be difficult to study the impact of specific parameters on real systems. How can one vary the latencies or bandwidths of a network used in a real system, for example, or arrange for a process to fail at a particularly awkward stage of a protocol? Real systems are notoriously difficult to test, and their use in this type of analysis represents a problem of at least comparable difficulty.

25.3 System Definition Languages

The first part of this book can be understood, in retrospect, as a study of client/server computing. Readers will recall that we started by looking at some of the underlying technical issues, such as the mechanisms by which messages can be addressed and transported, associated reliability issues, and computing models. But we then argued that the successful use of such technologies requires the development of computing environments within which the interfaces between the objects distributed over the network are published and standardized. This led to a review of CORBA, which has become a standard as the best available technology for documenting interfaces and managing systems that are modeled as collections of cooperating objects.

As we moved beyond these technologies into a broader collection of tools and protocols for replicating data, load-balancing, fault tolerance, and guaranteeing low latency and high data throughput, we lost the connection to specifications. It is clear how one can specify that a service manages two-dimensional tables—but it is not at all clear how to specify that the service is replicated on three sites, which must be selected to fail independently; accessible over two communication paths, which similarly fail independently; load-balanced; and capable of providing 100 ms response times to queries. This form of information, although potentially necessary for the purpose of describing a system correctly, is not represented in a typical system.

To a limited degree, we saw that one can model a distributed system using a form of relational model similar to that used in relational database systems. If this were done, the kinds of properties listed above could be written down in the relational calculus much as one expresses consistency constraints in a database setting. Doing so would represent initial steps toward *system definition languages* within which the necessary, desired, or typical characteristics of a computing system could be specified and reasoned about.

How might one use such information, if a system were to make it explicit? One option would be to construct tools for automatically generating management policies, permitting the system to be monitored and controlled in order to enforce the desired behavior through appropriate interventions. Another possibility would be to use such information to preprogram the communication subsystem so that it can allocate the resources needed to assure that a necessary quality of service will be provided and, if that level of service cannot be guaranteed, that an upcall will be issued to warn the system that it is operating in a degraded mode. One could use this information to protect the system against attacks that seek to create an environment violating one or more of the basic assumptions to force the system itself to fail. Information of this sort could be used to simulate and model the system, permitting intelligent comparison of alternative hardware configurations, management policies, and calculation of likely peak loading.

In his work on Horus, Robbert van Renesse was able to characterize the virtual synchrony stack in terms of a fairly small set of properties that the application might or might not require. One could imagine wiring Horus to some form of system definition language so that the user would merely specify the expectations of the system, and the actual construction of an appropriate protocol stack could then be completely automated. A similar approach might be possible with the RMP system which, as we will see, also makes use of a formal specification language and hence could potentially associate specific properties with specific needs of the user.

At the time of this writing, I am not aware of any work on system definition languages, but such a direction seems like a logical path for the CORBA community to pursue as the basic elements of CORBA become more mature and their practical implications better understood. This topic, then, would appear to be a good one for research within the language community or distributed system management community.

25.4 High-Level Languages and Logics

In our review of the Horus system, the reader may have been puzzled by the repeated references to the ML programming language (see Milner et al.), which is one of the options for specifying Horus layers. ML is a LISP-like, high-level language, which has traditionally been interpreted, although compilers for ML are now entering general use, and mechanisms for translation from ML to C exist (indeed, Horus layers coded in ML are currently translated to C and then compiled from that form). Nonetheless, given that C and C++ offer greater control over performance, why would we consider implementing large parts of Horus in ML?

We have seen that the protocols implemented within Horus layers are potentially complex, both in their own terms and also in terms of their interactions. As a consequence, it may be difficult to reason about the expected properties of a Horus stack combining several layers in a nontrivial way. When programming languages such as C or C++ are used to implement a Horus stack, our ability to reason about such layerings is potentially limited by the lack of automated tools for doing so. We are forced to express the behavior of the layers in English or in some form of temporal logic, and may make mistakes in doing so (we cited, for example, some serious problems that occurred when a colleague and I tried to express the GMS service using a temporal logic language, which turned out to be ill-suited to the purpose). An argument can be made that such an approach will lead to unconvincing correctness and behavioral problems.

When a Horus layer is specified in ML, the situation is somewhat brighter in just these respects. ML is a very high level language, and a substantial body of applied mathematics and logic has evolved around the language. In particular, there exist sophisticated proof tools for ML: programming environments that assist the user in proving things about programs written in ML or in deriving ML programs that illustrate the technique used to prove a purely mathematical property. By coding Horus in ML, we can benefit from both of these automated theorem-proving techniques.

If a Horus layer is expressed in ML, we can potentially claim something about that layer (e.g., that its synchronization mechanism is deadlock-free) and establish the precise assumptions that must be made about layers above and below it for these condition to be true. The NuPrl (pronounced "New Pearl") environment is a particularly powerful tool for this sort of application and is designed to accept ML as an input. At Cornell, the Horus project has already been successful in using NuPrl to establish simple properties of Horus layers. The potential of taking this work further and reasoning about composition of layers or complex properties of layers is very exciting. In particular, because NuPrl is automated, it can track very complex properties and automatically complete tedious aspects of such proofs. The degree of confidence one can express in the result is consequently much higher than for a hand-developed proof, and the nature of the questions one can ask is greatly expanded.

At the same time, one can imagine going from the bottom up and developing a formal logic within which a Horus protocol suite would be proved "possible" under a set of assumptions about the environment. With this direction, it would (at least in theory) be practical to extract the Horus layers that implement the proofs directly from the proofs. Such a process is similar to one whereby, given a pair of positive integers, one can find their least common multiple; this embodies an algorithm. NuPrl is designed to extract algorithms for this form of constructive proof and hence could potentially extract Horus layers that are intrinsically trustworthy, because they would be backed by a rigorous mathematical foundation extending down to the first principles upon which the system is based.

While ambitious, we believe that such a goal is also eventually achievable, albeit in limited ways. In a broader sense, this points to the potential for compiling protocols from descriptions of the desired system behavior. For applications requiring, for example, security and trust, one would be much more likely to trust the resulting protocols than protocols that were hand-coded and proved correct after the fact, because such proofs are only as good as the developer's ability to formalize the behavior of the layer.

There are, unfortunately, serious obstacles to both of these viewpoints. Earlier, we cited the difficulty of expressing a problem, such as the group membership problem, in the most widely used temporal logic for distributed systems. As we saw at the time, the core difficulty lies in the fact that at the time an event occurs within a process, one may not yet know if that process will remain a member of the system or be excluded from it as faulty. Indeed, one may not even know if the system as a whole will continue to make progress. Thus, one is forced to make the claim that *if* the system makes progress, either this process will be considered as *faulty*, in which case some set of conditions holds upon the events that took place before it failed, or it will be considered as *correct*, in which case some other set of conditions holds. This is a very clumsy way to express the behavior of a distributed system, but it seems to be the only possible way to do so in the usual temporal logic. Lacking progress on this problem, it is unlikely that NuPrl will be able to fully model the process group mechanisms implemented by Horus and hence unlikely that we can fully verify Horus.

Yet there may be considerable benefit from less ambitious uses of NuPrl—for example, we have begun to make use of security layers that integrate Horus with the Fortezza standard. It may be entirely practical to formalize the resulting authentication and trust properties of the system and in this manner to use NuPrl to convince ourselves that certain styles of distributed service are safe against attack. There appears to be considerable potential for such lightweight uses of NuPrl, even if the larger challenge of using such a system to verify all of Horus remains far in the future.

It should be noted that Horus is not the first to explore the use of ML for improving the clarity of distributed protocols. Earlier work includes that of Biagioni, Biagioni et al., Harper and Lee, Krumvieda (1991). Also, interested readers should look at NASA's work on formalizing the protocols and properties of the RMP (*Reliable Multicast Protocol*) system. As reported in Callahan and Montgomery, Montgomery, Montgomery and Whetten, Wu, the RMP group has achieved considerable success both in expressing properties of their protocol and in formally verifying that these properties actually hold.

CHAPTER 26

Other Distributed and Transactional Systems

CONTENTS

In this chapter we review some of the advanced research efforts in the areas covered in this book. The first section focuses on message-passing and group communication systems, and the second section focuses on transactional systems. The review is not intended to be exhaustive, but we do try to include the major activities that contributed to the technology areas stressed in the book.

26.1 Related Work in Distributed Computing

There have been many distributed systems in which group communication played a role. We now review some of these systems, providing a brief description of the features of each and citing sources of additional information. Our focus is on distributed computing systems and environments with support for some form of process group computing. However, we do not limit ourselves to those systems implementing virtually synchronous process groups or a variation on the model. Our review presents these systems in alphabetical order. If we were to discuss them chronologically, we would start by considering V, then the Isis Toolkit and Delta-4, and then we would turn to the others in a roughly alphabetical ordering. However, it is important to understand that these systems are the output of a vigorous research community and that each of the systems cited included significant research innovations at the time it was developed. It would be simplistic to say that any one of these systems came first and that the remainder are somehow secondary. It would be more accurate to say that each system was innovative in some areas and borrowed ideas from prior systems in other areas.

Readers interested in learning more about this subject may want to start by consulting the articles that appeared in *Communications of the ACM* in a special section of the April 1996 issue (vol. 39, no. 4). David Powell's introduction to this special section is both witty and informative (see Powell [1996]), and there are articles about several of the systems discussed in this book (see Cristian [1996], Dolev and Malkhi, Moser et al. [1996], Schiper and Raynal, van Renesse et al. [1996]).

26.1.1 Amoeba

During the early 1990s, Amoeba (see Mullender et al., van Renesse et al. [1988, 1989]) was one of a few microkernel-based operating systems proposed for distributed computing; others include V (see Cheriton and Zwaenepoel), Mach (see Rashid), Chorus (see Rozier [Fall 1988, December 1988]), and QNX (see Hildebrand). The focus of the project when it was first launched was to develop a distributed system around a nucleus supporting extremely high performance communication, with the remaining system services being implemented using a client/server architecture. In our area of emphasis, process group protocols, Amoeba supports a subsystem developed by Frans Kaashoek that provides group communication using total ordering (see Kaashoek). Message delivery is atomic and totally ordered and implements a form of virtually synchronous addressing. During the early 1990s, Amoeba's sequencer protocols set performance records for throughput and latency, although other systems subsequently bypassed these using a mixture of protocol refinements and new generations of hardware and software.

26.1.2 Chorus

Chorus is an object-oriented operating system for distributed computing (see Rozier et al. [Fall 1988, December 1988]). Developed at INRIA during the 1980s, the technology shifted to a commercial track in the early 1990s and has become one of the major vehicles for commercial UNIX development and for real-time computing products. The system is notable for its modularity and comprehensive use of object-oriented programming techniques. Chorus was one of the first systems to embrace these ideas and is extremely sophisticated in its support for modular application programming and for reconfiguration of the operating system itself.

Chorus implements a process group communication primitive, which is used to assist applications in dealing with services that are replicated for higher availability. When an RPC is issued to such a replicated service, Chorus picks a single member and issues an invocation to it. A feature is also available for sending an unreliable multicast to members of a process group (no ordering or atomicity guarantees are provided).

In its present commercial incarnation, the Chorus operating system is used primarily in real-time settings for applications that occur in telecommunication systems. Running over Chorus is an object request broker technology called Cool-ORB. This system includes a variety of distributed computing services including a replication service capable of being interconnected to a process group technology, such as that used in the Horus system.

26.1.3 Delta-4

Delta-4 was one of the first systematic efforts to address reliability and fault-tolerant concerns (see Powell [1994]). Launched in Europe during the late 1980s, Delta-4 was developed by a multinational team of companies and academic researchers (see Powell [1991], Rodrigues and Verissimo [1989]). The focus of the project was on factory-floor applications, which combine real-time and fault-tolerant requirements. Delta-4 took an approach in which a trusted module was added to each host computer and used to run fault-tolerant protocols. These modules were implemented in software, but they could be included in a specially designed hardware interface to a shared communication bus. The protocols used in the system included process group mechanisms similar to the ones now employed to support virtual synchrony, although Delta-4 did not employ the virtual synchrony computing model.

The project was extremely successful as a research effort and resulted in working prototypes that were indeed fault tolerant and capable of coordinated real-time control in distributed automation settings. Unfortunately, however, this stage was reached as Europe entered a period of economic difficulties, and none of the participating companies was able to pursue the technology base after the research funding for the project ended. Ideas from Delta-4 can now be found in a number of other group-oriented and real-time distributed systems, including Horus.

26.1.4 Harp

The "gossip" protocols of Ladin and Liskov were mentioned in conjunction with our discussion of communication from a nonmember of a process group to that group (see Ladin et al. [1992],

Liskov et al. [1991]). These protocols were originally introduced in a replicated file system project undertaken at MIT in the early 1990s. The key idea of the Harp system was to use a lazy update mechanism as a way of obtaining high performance and tolerance to partitioning failures in a replicated file system. The system was structured as a collection of file servers consisting of multiple processes, each of which maintained a full copy of the file system, and a set of clients, which issued requests to the servers, switching from server to server to balance load or to overcome failures of the network or of a server process. Clients issued read operations, which the system handled locally at whichever server received the request, and update operations, which were performed using a quorum algorithm. Any updates destined for a faulty or unavailable process were spooled for later transmission when the process recovered or communication to it was reestablished. To ensure that when a client issued a series of requests the file servers performed them at consistent (e.g., logically advancing) times, each response from a file server process to a client included a timestamp, which the client could present on subsequent requests. The timestamp was represented as a vector clock and could be used to delay a client's request if it were sent to a server that had not yet seen some updates on which the request might be dependent.

Harp made extensive use of a hardware feature not widely used in modern workstations, despite its low cost and off-the-shelf availability. A so-called nonvolatile or battery-backed RAM (NVRAM) is a small memory, which preserves its contents even if the host computer crashes and later restarts. Finding that the performance of Harp was dominated by the latency associated with forced log writes to the disk, Ladin and Liskov purchased these inexpensive devices for the machines on which Harp ran and modified the Harp software to use the NVRAM area as a persistent data structure, which could hold commit records, locking information, and a small amount of additional commit-related data. Performance of Harp increased sharply, leading these researchers to argue that greater use should be made of NVRAM in reliable systems of all sorts. However, NVRAM is not found on typical workstations or computing systems, and vendors of the major transactional and database products are under great pressure to offer the best possible performance on completely standard platforms, making the use of NVRAM problematic in commercial products. The technology used in Harp would not perform well without NVRAM storage.

26.1.5 The Highly Available System (HAS)

The Highly Available System (HAS) was developed by IBM's Almaden research laboratory under the direction of Cristian and Strong, with involvement by Skeen and Schmuck, in the late 1980s and subsequently contributed technology to a number of IBM products, including the ill-fated Advanced Automation System (AAS) development that IBM undertook for the American Federal Aviation Agency (FAA) in the early 1990s (see Cristian [February 1991], Cristian and Delancy). Unfortunately, relatively little of what was apparently a substantial body of work was published about this system. The most widely known results include the *timed asynchronous communication model*, proposed by Cristian and Schmuck (see Cristian and Schmuck) and used to provide precise semantics for their reliable protocols. Protocols were proposed for synchronizing the clocks in a distributed system (see Cristian [1989]); managing group membership in real-time settings (see Cristian [April 1991]); atomic communication to groups (see Cristian et al. [1985, 1990]), subject to timing bounds; and achieving totally ordered delivery guarantees at the operational members of groups. Details of these protocols were presented in Chapter 20. A shared memory model called *Delta-Common Storage* was proposed as a part of this project and consisted of a tool by which process group members could communicate using a shared memory abstraction, with guarantees that updates would be seen by all operational group members (if by any) within a limited period of time.

26.1.6 The Isis Toolkit

The Isis Toolkit was developed by my colleagues and me between 1985 and 1990. It was the first process group communication system to use the virtual synchrony model (see Birman and Joseph [February 1987, November 1987], Birman and van Renesse [1994]). As its name suggests, Isis is a collection of procedural tools that are linked directly to the application program, providing it with functionality for creating and joining process groups dynamically, multicasting to process groups with various ordering guarantees, replicating data and synchronizing the actions of group members as they access that data, performing operations in a load-balanced or fault-tolerant manner, and so forth (see Birman and van Renesse [1996]). Over time, a number of applications were developed using Isis, and it became widely used through a public software distribution. These developments led to the commercialization of Isis through a company, which today operates as a wholly owned subsidiary of Stratus Computer, Inc. The company continues to extend and sell the Isis Toolkit itself, as well as an object-oriented embedding of Isis called Orbix+Isis (it extends Iona's popular Orbix product with Isis group functionality and fault tolerance [see Iona Ltd. and Isis Distributed Systems, Inc.]), products for database and file system replication, a message bus technology supporting a reliable publish/subscribe interface, and a system management technology for supervising a system and controlling the actions of its components.

Isis introduced the primary partition virtual synchrony model and the *cbcast* primitive. These steps enabled it to support a variety of reliable programming tools, which was unusual for process group systems at the time Isis was developed. Late in the life cycle of the system, it was one of the first (along with Ladin and Liskov's Harp system) to use vector timestamps to enforce causal ordering. In a practical sense, the system represented an advance merely by being a genuinely usable packaging of a reliable computing technology into a form that could be used by a large community.

Successful applications of Isis include components of the New York and Swiss stock exchanges; distributed control in AMD's FAB-25 VLSI fabrication facility; distributed financial databases such as one developed by the World Bank; a number of telecommunication applications involving mobility, distributed switch management, and control; billing and fraud detection; several applications in air traffic control and space data collection; and many others. The major markets to which the technology is currently sold are financial, telecommunication, and factory automation.

26.1.7 Locus

Locus is a distributed operating system developed by Popek's group at UCLA in the mid-1990s (see Walter et al.). Known for such features as transparent process migration and a uniform distributed shared memory abstraction, Locus was extremely influential in the early development of parallel and cluster-style computing systems. Locus was eventually commercialized and is now a product of Locus Computing Corporation. The file system component of Locus was later extended into the Ficus system, which we discussed earlier in conjunction with other stateful file systems.

26.1.8 Sender-Based Logging and Manetho

In writing this book, I was forced to make certain tradeoffs in terms of the coverage of topics. One topic that was not included is that of log-based recovery, whereby applications create checkpoints periodically and log messages sent or received. Recovery is by rollback into a consistent state, after which log replay is used to regain the state as of the instant the failure occurred.

Manetho (see Elnozahy and Zwaenepoel) is perhaps the best known of the log-based recovery systems, although the idea of using logging for fault tolerance is quite a bit older (see Borg et al., Johnson and Zwaenepoel, Koo and Toueg). In Manetho, a library of communication procedures automates the creation of logs, which include all messages sent from application to application. An assumption is made that application programs are deterministic and will reenter the same state if the same sequence of messages is played into them. In the event of a failure, a rollback protocol is triggered, which will roll back one or more programs until the system state is globally consistent—meaning that the set of logs and checkpoints represents a state the system could have entered at some instant in logical time. Manetho then rolls the system forward by redelivery of the logged messages. Because the messages are logged at the sender, the technique is called *sender-based logging*. Experiments with Manetho have confirmed that the overhead of the technique is extremely small. Moreover, working independently, Alvisi has demonstrated that sender-based logging is just one of a very general spectrum of logging methods that can store messages close to the sender, close to the recipient, or even mix these options (see Alvisi and Marzullo).

Although conceptually simple, logging has never played a major role in reliable distributed systems in the field, most likely because of the determinism constraint and the need to use the logging and recovery technique systemwide. This issue, which also makes it difficult to transparently replicate a program to make it fault-tolerant, seems to be one of the fundamental obstacles to software-based reliability technologies. Unfortunately, nondeterminism can creep into a system through a great many interfaces. Use of shared memory or semaphore-style synchronization can cause a system to be nondeterministic, as can any dependency on the order of message reception, the amount of data in a pipe or the time in the execution when the data arrive, the system clock, or the thread scheduling order. This implies that the class of applications for which one can legitimately make a determinism assumption is very small—for example, suppose that the servers used in some system are a mixture of deterministic and nondeterministic programs.

Active replication could be used to replicate the deterministic programs transparently, and the sorts of techniques discussed in previous chapters could be employed in the remainder. However, to use a sender-based logging technique (or any logging technique), the entire group of application programs needs to satisfy this assumption—hence, one would need to recode the nondeterministic servers before any benefit of any kind could be obtained. This obstacle is apparently sufficient to deter most potential users of the technique.

I am aware, however, of some successes with log-based recovery in specific applications that happen to have a very simple structure. One popular approach to factoring very large numbers involves running large numbers of completely independent factoring processes that deal with small ranges of potential factors; such systems are very well suited to a log-based recovery technique because the computations are deterministic and there is little communication between the participating processes. Log-based recovery seems to be more applicable to scientific computing systems or problems such as the factoring problem than to general-purpose distributed computing of the sort seen in corporate environments or the Web.

26.1.9 NavTech

NavTech is a distributed computing environment built using Horus (see Birman and van Renesse [1996], van Renesse et al. [1996]), but with its own protocols and specialized distributed services (see Rodrigues and Verissimo [1995], Rodrigues et al., Verissimo [1993, 1994, 1996], Verissimo and Rodrigues). The group responsible for the system is headed by Verissimo, who was one of the major contributors to Delta-4, and the system reflects many ideas that originated in that earlier effort. NavTech is aimed at wide area applications with real-time constraints, such as banking systems involving a large number of branches and factory-floor applications in which control must be done close to a factory component or device. The issues that occur when real-time and fault-tolerant problems are considered in a single setting thus represent a particular focus of the effort. Future emphasis by the group will be on the integration of graphical user interfaces, security, and distributed fault tolerance within a single setting. Such a mixture of technologies would result in an appropriate technology base for applications such as home banking and distributed game playing—both expected to be popular early uses of the new generation of Internet technologies.

26.1.10 Phoenix

Phoenix is a recent distributed computing effort, which was launched by C. Malloth and A. Schiper of the École Polytechnique de Lausanne jointly with O. Babaoglu and P. Verissimo (see Malloth, Schiper and Raynal). Most work on the project is currently occurring at EPFL. The emphasis of this system is on issues that occur when process group techniques are used to implement wide area transactional systems or database systems. Phoenix has a Horus-like architecture, but uses protocols specialized to the needs of transactional applications and has developed an extension of the virtual synchrony model within which transactional serializability can be elegantly treated.

26.1.11 Psync

Psync is a distributed computing system, which was developed by Peterson at the University of Arizona in the late 1980s and early 1990s (see Mishra et al. [1991], Peterson [1987], Peterson et al. [1989]). The focus of the effort was to identify a suitable set of tools with which to implement protocols such as the ones we have presented in the last few chapters. In effect, Psync set out to solve the same problem as the Express Transfer Protocol, but where XTP focused on point-to-point datagrams and streaming-style protocols, Psync was more oriented toward group communication and protocols with distributed ordering properties. A basic set of primitives was provided for identifying messages and for reasoning about their ordering relationships. Over these primitives, Psync provided implementations of a variety of ordered and atomic multicast protocols.

26.1.12 Rampart

Rampart is a distributed system, which uses virtually synchronous process groups in settings where security is desired even if components fail in arbitrary (Byzantine) ways (see Reiter [1996]). The activity is headed by Reiter at AT&T Bell Laboratories and has resulted in a number of protocols for implementing process groups despite Byzantine failures, as well as a prototype of a security architecture employing these protocols (see Reiter [1993, 1994], Reiter and Birman, Reiter et al. [1992, 1995]). We discuss this system in more detail in Chapter 19. Rampart's protocols are more costly than those we have presented here, but the system would probably not be used to support a complete distributed application. Instead, Rampart's mechanisms could be employed to implement a very secure subsystem, such as a digital cash server or an authentication server in a distributed setting, while other less-costly mechanisms could be employed to implement the applications that make use of these very secure services.

26.1.13 Relacs

The Relacs system is the product of a research effort headed by Ozalp Babaoglu at the University of Bologna (see Babaoglu et al. [1994, 1995]). The activity includes a strong theoretical component, but has also developed an experimental software testbed within which protocols developed by the project can be implemented and validated. The focus of Relacs is on the extension of virtual synchrony to wide area networks in which partial connectivity disrupts communication. Basic results of this effort include a theory that links *reachability* to consistency in distributed protocols, as well as a proposed extension of the view synchrony properties of a virtually synchronous group model to permit safe operation for certain classes of algorithms despite partitioning failures. At the time of this writing, the project was working to identify the most appropriate primitives and design techniques for implementing wide area distributed applications offering strong fault-tolerant and consistency guarantees and formalizing the models and correctness proofs for such primitives.

26.1.14 RMP

The RMP system is a public-domain process group environment implementing virtual synchrony, with a focus on extremely high performance and simplicity. The majority of the development of this system occurred at the University of California, Berkeley, where graduate student Brian Whetten needed such a technology for his work on distributed multimedia applications (see Callahan and Montgomery, Montgomery, Montgomery and Whetten, Whetten). Over time, the project became much broader, as West Virginia University/NASA researchers Jack Callahan and Todd Montgomery became involved. Broadly speaking, RMP is similar to the Horus system, although less extensively layered.

The major focus of the RMP project has been on embedded system applications, which might occur in future space platforms or ground-based computing support for space systems. Early RMP users have been drawn from this community, and the long-term goals of the effort are to develop technologies suitable for use by NASA. As a result, the verification of RMP has become particularly important, since systems of this sort cannot easily be upgraded or serviced while in flight. RMP has pioneered the use of formal verification and software design tools in protocol verification (see Callahan and Montgomery, Wu), and the project is increasingly focused on robustness through formal methods—a notable shift from its early emphasis on setting new performance records.

26.1.15 StormCast

Researchers at the University of Tromsö, within the Arctic circle, launched this effort, which seeks to implement a wide area weather and environmental monitoring system for Norway. StormCast is not a group communication system per se, but rather is one of the most visible and best documented of the major group communication applications (see Asplin and Johansen, Birman and van Renesse [1996], Johansen, Johansen and Hartvigsen, Johansen et al. [May 1995, June 1995, 1996]). Process group technologies are employed within this system for parallelism, fault tolerance, and system management.

The basic architecture of StormCast consists of a set of data archiving sites, located throughout the Far North. At the time of this writing, StormCast had approximately six such sites, with more coming on-line each year. Many of these sites simply gather and log weather data, but some collect radar and satellite imagery and others maintain extensive data sets associated with short- and long-term weather modeling and predictions. StormCast application programs typically draw on this varied data set for purposes such as local weather prediction, tracking environmental problems such as oil spills (or radioactive discharges from Russia), research into weather modeling, and other similar applications.

StormCast is interesting for many reasons. The architecture of the system has received intense scrutiny (see Johansen, Johansen and Hartvigsen) and has evolved over a series of iterations into one in which the application developer is guided to a solution using tools appropriate to the application and following templates that worked successfully for other similar applications. This concept of architecture driving the solution is one that has been lost in many distributed computing environments, which tend to be architecturally flat (presenting the same tools, services, and APIs systemwide even if the applications themselves have very clear architecture, such as a client/server structure, in which different parts of the system need different forms of support). It is interesting to note that early versions of StormCast, which lacked such a strong concept of system architecture, were much more difficult to use than the current one, in which the developer actually has less freedom but much stronger guidance toward solutions.

StormCast has encountered some difficult technical challenges. The very large amounts of data gathered by weather monitoring systems necessarily must be visited on the servers where they reside; it is impractical to move these data to the place where the user requesting a service, such as a local weather forecast, may be working. Thus, StormCast has pioneered in the development of techniques for sending computations to data: the so-called *agent* architecture (see Johansen et al. [1996]) we discussed in Section 10.8 in conjunction with the TACOMA system.

In a typical case, an airport weather prediction for Tromsö might involve checking for incoming storms in the 500-km radius around Tromsö and then visiting one of several other data archives, depending upon the prevailing winds and the locations of incoming weather systems. The severe and unpredictable nature of arctic weather makes these computations equally unpredictable: Data needed for one prediction may be primarily archived in southern Norway, while those needed for some other prediction may be archived in northern Norway or on a system that collects data from trawlers along the coast. Such problems are solved by designing TACOMA agents, which travel to these data, preprocess them to extract needed information, and then return them to the end user for display or further processing. Although such an approach presents challenging software design and management problems, it also seems to be the only viable option for working with such large quantities of data and supporting such a varied and unpredictable community of users and applications.

It should be noted that StormCast maintains an unusually interesting Web page, http://www.cs.uit.no. Readers who have a Web browser will find interactive remote-controlled cameras focused on the ski trails near the university, current environmental monitoring information including data on small oil spills and the responsible vessels, three-dimensional weather predictions intended to aid air traffic controllers in recommending the best approach paths to airports in the region, and other examples of the use of the system. One can also download a version of TACOMA and use it to develop new weather or environmental applications, which can be submitted directly to the StormCast system, load permitting.

26.1.16 Totem

The Totem system is the result of a multiyear project at the University of California, Santa Barbara, focusing on process groups in settings that require extremely high performance and real-time guarantees (see Agarwal, Amir et al., Melliar-Smith and Moser [1989, 1993], Melliar-Smith et al. [1990], Moser et al. [1994, 1996]). The computing model used is the extended virtual synchrony one and was originally developed by this group in collaboration with the Transis project in Israel. Totem has contributed a number of high-performance protocols, including an innovative causal and total ordering algorithm, based on transitive ordering relationships between messages, and a totally ordered protocol with extremely predictable real-time properties. The system differs from a technology such as Horus, since it focuses on a type of distributed system that would result from the interconnection of clusters of workstations using broadcast media within these clusters and some form of bridging technology between them. Most of the protocols are optimized for applications within which communication loads are high and uniformly distributed over the processes in the system or in which messages originate primarily at a single source. The resulting protocols are very efficient in their use of messages, but they sometimes exhibit higher latency than the protocols we presented in earlier chapters of this book. Intended applications include parallel computing on clusters of workstations and industrial control problems.

26.1.17 Transis

The Transis system (see Dolev and Malkhi) is one of the best known and most successful process group-based research projects at the time of this writing. The research group has contributed extensively to the theory of process group systems and virtual synchrony, repeatedly set performance records with its protocols and flow-control algorithms, and developed a remarkable variety of protocols and algorithms in support of such systems (see Amir [1992, 1993], Friedman et al., Keidar and Dolev, Malkhi). Many of the ideas from Transis were eventually ported into the Horus system. Transis was, for example, the first system to show that by exploiting hardware multicast, a reliable group multicast protocol could scale with almost no growth in cost or latency. The primary focus of this effort was initially partitionable environments, and much of what is known about consistent distributed computing in such settings originated either directly or indirectly from this group. The project is also known for its work on transactional applications, which preserve consistency in partitionable settings.

Recently, the project has begun to look at security issues occurring in systems subject to partitioning failures. The effort seeks to provide secure autonomous communication even while subsystems of a distributed system are partitioned away from a central authentication server. The most widely used security architectures do not allow secure operations to be initiated in a partitioned system component and are not able to deal with the revalidation of such a component if it later reconnects to the system and wants to merge its groups into others that remained in the primary component. Mobility is likely to create a need for security of this sort—for example, in financial applications and in military settings, where a team of soldiers may need to operate without direct communication to the central system from time to time.

As noted earlier, another interesting topic under study by the Transis group is that of building systems that combine multiple protocol stacks in which different reliability or quality-of-service properties apply to each stack (see Chockler et al.). In this work, one assumes that a complex distributed system will give rise to a variety of types of reliability requirements: virtual synchrony for its control and coordination logic, isochronous communication for voice and video, and perhaps special encryption requirements for certain sensitive data—each provided through a corresponding protocol stack. However, rather than treating these protocol stacks as completely independent, the Transis system (which should port easily into Horus) deals with the synchronization of streams across multiple stacks. This will greatly simplify the implementation of demanding applications that need to present a unified appearance and yet cannot readily be implemented within a single protocol stack.

26.1.18 The V System

Because of the alphabetic sequence of this chapter, it is ironic that the first system to have used process groups is the last that we review. The V system was the first of the microkernel operating systems intended specifically for distributed environments; it also pioneered the RISC style of operating systems that later swept the research community. V is known primarily for innovations in the virtual memory and message-passing architecture used within the system, which achieved early performance records for its RPC protocol. However, the system also included a process group mechanism, which was used to support distributed services capable of providing a service at multiple locations in a distributed setting (see Cheriton and Zwaenepoel, Deering [1988]).

Although the V system lacked any strong process group computing model or reliability guarantees, its process group tools were considered quite powerful. In particular, this system was the first to support a publish/subscribe paradigm, in which messages to a subject were transmitted to a process group whose name corresponded to that subject. As we saw earlier, such an approach provides a useful separation between the source and destination of messages: The publisher can send to the group without worrying about its current membership, and a subscriber can simply join the group to begin receiving messages published within it.

The V style of process group was not intended for process group computing of the types we have discussed in this book; reliability in the system was purely on a best-effort basis, meaning that the group communication primitives made an effort to track current group membership and to avoid high rates of message loss—without providing real guarantees. When Isis introduced the virtual synchrony model, the purpose was precisely to show that with such a model, a V style of process group could be used to replicate data, balance workload, or provide fault tolerance. None of these problems were believed solvable in the V system itself. V set the early performance standards against which other group communication systems tended to be evaluated, however, and it was not until a second generation of process group computing systems emerged (the commercial version of Isis, the Transis and Totem systems, Horus, and RMP) that these levels of performance were matched and exceeded by systems that also provided reliability and ordering guarantees.

26.2 Systems That Implement Transactions

We end this chapter with a brief review of some of the major research efforts that have explored the use of transactions in distributed settings. As in the case of our review of distributed communication systems, we present these in alphabetical order.

26.2.1 Argus

The Argus system was an early leader among transactional computing systems that considered transactions on abstract objects. Developed by a team led by Liskov at MIT, the Argus system consists of a programming language and an implementation that was used primarily as a research and experimentation vehicle (see Ladin et al. [1990], Liskov and Scheifler, Liskov et al. [1987]). Many credit the idea of achieving distributed reliability through transactions on distributed objects to this project, and it was a prolific source of publications on all aspects of transactional computing, theoretical as well as practical, during its decade or so of peak activity.

The basic Argus data type is the *guardian*: a software module, which defines and implements some form of persistent storage, using transactions to protect against concurrent access and to ensure recoverability and persistence. Similar to a CORBA object, each guardian exports an interface defining the forms of access and operations possible on the object. Through these interfaces, Argus programs (*actors*) invoke operations on the guarded data. Argus treats all such invocations as transactions and also provides explicit transactional constructs in its programming language, including commit and abort mechanisms, a concurrent execution construct, top-level transactions, and mechanisms for exception handling.

The Argus system implements this model in a transparently distributed manner, with full nested transactions and mechanisms to optimize the more costly aspects, such as nested transaction commit. A sophisticated *orphan termination* protocol is used to track down and abort orphaned subtransactions, which can be created when the parent transaction that initiated some action fails and aborts but leaves active child transactions, which may now be at risk of observing system states inconsistent with the conditions under which the child transaction was spawned— for example, a parent transaction might store a record in some object and then spawn a child subtransaction, which will eventually read this record. If the parent aborts and the orphaned child is permitted to continue executing, it may read the object in its prior state, leading to seriously inconsistent or erroneous actions.

Although Argus was never put into widespread practical use, the system was extremely influential. Not all aspects of the system were successful—many commercial transactional systems have rejected distributed and nested transactions as requiring an infrastructure that is relatively more complex, costly, and difficult to use than flat transactions in standard client/server architectures. Other commercial products, however, have adopted parts of this model successfully. The principle of issuing transactions to abstract data types remains debatable. As we have seen, transactional data types can be very difficult to construct, and expert knowledge of the system will often be necessary to achieve high performance. The Argus effort ended in the early 1990s and the MIT group that built the system began work on Thor, a second-generation technology in this area.

26.2.2 Arjuna

Whereas Argus explored the idea of transactions on objects, Arjuna is a system that focuses on the use of object-oriented techniques to customize a transactional system. Developed by Shrivistava at Newcastle, Arjuna is an extensible and reconfigurable transactional system, in which the developer can replace a standard object-oriented framework for transactional access to persistent objects with type-specific locking or data management objects, which exploit semantic knowledge of the application to achieve high performance or special flexibility. The system was one of the first to focus on C++ as a programming language for managing persistent data, an approach that later became widely popular. Recent development of the system has explored the use of replication for increased availability during periods of failure using a protocol called Newtop; the underlying methodology used for this purpose draws on the sorts of process group mechanisms discussed in previous chapters (see Ezhilhelvan, Macedo et al.).

26.2.3 Avalon

Avalon was a transactional system developed at Carnegie Mellon University by Herlihy and Wing during the late 1980s. The system is best known for its theoretical contributions. This project proposed the *linearizability model*, which weakens serializability in object-oriented settings where full nested serializability may excessively restrict concurrency (see Herlihy and Wing). As noted briefly earlier in the chapter, linearizability has considerable appeal as a model potentially capable of integrating virtual synchrony with serializability. A research project, work on Avalon ended in the early 1990s.

26.2.4 Bayou

Bayou is a recent effort at Xerox PARC that uses transactions with weakened semantics in partially connected settings, such as for the management of distributed calendars for mobile users who may need to make appointments and schedule meetings or read electronic mail while in a disconnected or partially connected environment (see Terry et al.). The system provides weak serialization guarantees by allowing the user to schedule meetings even when the full state of the calendar is inaccessible due to a partition. Later, when communication is reestablished, such a transaction is completed with normal serializability semantics.

Bayou makes the observation that transactional consistency may not guarantee that user-specific consistency constraints will be satisfied—for example, if a meeting is scheduled while disconnected from some of the key participants, it may later be discovered that the time conflicts with some other meeting. Bayou provides mechanisms by which the designer can automate both the detection and resolution of these sorts of problems. In this particular example, Bayou will automatically attempt to shift one or the other rather than requiring that a user become directly involved in resolving all such conflicts. The focus of Bayou is very practical: Rather than seeking extreme generality, the technology is designed to solve the specific problems encountered in paperless offices with mobile employees. This domain-specific approach permits Bayou to solve a number of distributed consistency problems that, in the most general sense, are not even tractable. This reconfirms an emerging theme of the book: Theoretically impossible results often need to be reexamined in specific contexts; what cannot be solved in the most general sense or setting may be entirely tractable in a particular application where more is known about the semantics of operations and data.

26.2.5 Camelot and Encina

This system was developed at Carnegie Mellon University in the late 1980s and was designed to provide transactional access to user-developed data structures stored in files (see Spector). The programming model was one in which application programs perform RPCs on servers. Such transactions become nested if these servers are clients of other servers. The ultimate goal is to support transactional semantics for applications that update persistent storage. Camelot introduced a variety of operating system enhancements for maximizing the performance of such applications and was eventually commercialized in the form of the Encina product from Transarc Corporation. Subsequent to this transition, considerable investment in Encina occurred at Transarc and the system is now one of the leaders in the market for OLTP products. Encina provides both nondistributed and distributed transactions, nested transactions if desired, a variety of tools for balancing load and increasing concurrency, prebuilt data structures for common uses, and management tools for system administration. The distributed data mechanisms can also be used to replicate information for high availability.

Industry analysts have commented that although many Encina users select the system in part for its distributed and nested capabilities, in actual practice most applications of Encina make little or no use of these features. If accurate, this observation raises interesting questions about the true characteristics of the distributed transactional market. Unfortunately, however, I am not aware of any systematic study of this question.

Readers interested in Encina should also look at IBM's CICS technology, perhaps the world's most widely used transactional system, and the Tuxedo system, an OLTP product developed originally at AT&T, which became an industry leader in the UNIX OLTP market. Similar to Encina, CICS and Tuxedo provide powerful and complete environments for client/server-style applications requiring transactional guarantees, and Tuxedo includes real-time features required in telecommunication settings.

Appendix: Problems

This book is intended for use by professionals or advanced students, and the material presented is at a level for which simple problems are not entirely appropriate. Accordingly, most of the problems in this appendix are intended as the basis for essay-style responses or for programming projects, which might build upon the technologies we have treated up to now. Some of these projects are best undertaken as group exercises for a group of three or four students; others could be undertaken by individuals.

Professionals may find these problems interesting from a different perspective. Many of them are the sorts of questions that one would want to ask about a proposed distributed solution and hence could be useful as a tool for individuals responsible for the development of a complex system. I am sometimes asked to comment on proposed system designs, and, like many others, have found that it can be difficult to know where to start when the time for questions finally arrives after a two-hour technical presentation. A reasonable suggestion is to begin to pose simple questions aimed at exposing the reliability properties and nonproperties of the proposed system, the assumptions it makes, the dependencies embodied in it, and the cost/benefit tradeoffs reflected in the architecture. Such questions may not lead to a drastically changed system, but they do represent a path toward understanding the mentality of the designer and the philosophical structure of the proposed system. Many of the questions below are of a nature that might be used in such a situation.

1   Write a program to experimentally characterize the packet loss rate, frequency of out-of-order delivery, send-to-receive latency, and byte throughput of the UDP and TCP transport protocols available on your computer system. Evaluate both the local case (source and destination on the same machine) and the remote case (source and destination on different machines).

2   We discussed the concept of a broadcast storm in conjunction with Ethernet technologies. Devise an experiment that will permit you to quantify the conditions under which such a storm might occur on the equipment in your laboratory. Use your findings to arrive at a set of recommendations that should, if followed, minimize the likelihood of a broadcast storm even in applications that make heavy use of broadcast.

3 Devise a method for rapidly detecting the failure of a process on a remote machine and implement it. How rapidly can your solution detect a failure without risk of inaccuracy? Your work should consider one or more of the following cases: a program that runs a protocol you have devised and implemented over UDP, a program that is monitored by a parent program, program on a machine that fails or becomes partitioned from the network. For each case, you may use any system calls or standard communication protocols that are available to you.

4 Suppose that it is your goal to develop a network radio service, which transmits identical data to a large set of listeners, and that you need to pick the best communication transport protocol for this purpose. Evaluate and compare the UDP, TCP, and IP multicast transport protocols on your computer (you may omit IP multicast if this is not available in your testing environment). Your evaluation should look at throughput and latency (focusing on variability of these as a function of throughput presented to the transport). Can you characterize a range of performance within which one protocol is superior to the others in terms of loss rate, achievable throughput, and consistently low latency? Your results will take the form of graphs showing how these attributes scale with increasing numbers of destinations.

5 Develop a simple ping-pong program that bounces a UDP packet back and forth between a source and destination machine. One would expect such a program to give extremely consistent latency measurements when run on idle workstations. In practice, however, your test is likely to reveal considerable variation in latency. Track down the causes of these variations and suggest strategies for developing applications with highly predictable and stable performance properties.

6 One challenge to timing events in a distributed system is that the workstations in that system may be running some form of clock synchronization algorithm, which is adjusting clock values even as your test runs—leading to potentially confusing measurements. From product literature for the computers in your environment or by running a suitable experiment, determine the extent to which this phenomenon occurs in your testing environment. Can you propose ways of measuring performance that are immune to distortions of this nature?

7 Suppose you wish to develop a *topology service* for a local area network, using *only* two kinds of information as input with which to deduce the network topology: IP addresses for machines and measured point-to-point latency (for lightly loaded conditions, measured to a high degree of accuracy). How practical would it be to solve this problem? Ideally, a topology service should be able to produce a map showing how your local area network is interconnected, including bridges, individual Ethernet segments, and so forth.

8 (moderately difficult) If you concluded that you should be able to do a good job on the previous problem, implement such a topology service using your local area network. What practical problems limit the accuracy of your solution? What forms of use could you imagine for your service?

9 In Chapter 5, we saw that stream protocols could fail in inconsistent ways. Develop an application that demonstrates this problem by connecting two programs with multiple TCP streams, running them on multiple platforms and provoking a failure in which some of the streams break and some remain connected. To do this test you may need to briefly disconnect one of the workstations from the network; hence, you should obtain the permission of your network administration staff.

10 Propose a method for passing pointers to servers in an RPC environment, assuming that the source and destination programs are coded in C++ and that pointers are an abstract data type. What costs would a user of your scheme incur? Can you recommend programming styles or new programming constructs to minimize the impact of these costs on the running application? Contrast your solutions with those in Culler and von Eicken's Split C programming environment.

11 (requires sophistication in C++) Suppose that a CORBA implementation of the UNIX compression and decompression utilities is needed, and you have been asked to build it. Your utility needs to operate on arbitrary C++ objects of varied types. The types are not known in advance. Some of these objects will have a *compress_self* and a *decompress_self* interface, but others will not. How could this problem be solved?

12 Can a CORBA application see a difference between CORBA remote invocations implemented directly over UDP and CORBA remote invocations implemented over a TCP-style reliable stream?

13 Suppose one were building a CORBA-based object-oriented system for very long-lived applications. The system needs to remain *continuously operational* for years at a time. Yet it is also expected that it will sometimes be necessary to upgrade software components of the system. Could such a problem be solved in software—that is, can a general-purpose upgrade mechanism be designed as part of an application so that objects can be dynamically upgraded? To make this concrete, you can focus on a system of k objects, $O_1 \ldots O_k$, and consider the case where we want to replace O_i with O_i' while the remaining objects are unchanged. Express your solution by describing a proposed upgrade mechanism and the constraints it imposes on applications that use it.

14 Suppose that a CORBA system is designed to cache information at the clients of a server. The clients would be bound to *client objects*, which would handle the interaction with the remote server. Now, consider the case where the data being cached can be dynamically updated on the server. What options exist for maintaining the coherency of the cached data within the clients? What practical problems might need to be overcome in order to solve such a problem reliably? Does the possibility that the clients, the server, or the communication system might fail complicate your solution?

15 In CORBA we saw that it is possible to trap error conditions, such as server failure. Presumably, one would want to standardize the handling of such conditions. Suppose you are designing a general-purpose mechanism to handle fail-over, whereby a client connected to a server, S, will automatically and transparently rebind itself to server S' in the event that S fails. Under what conditions would this be easy? How would you deal with the possibility that the state of S' might not be identical to that of S? Could one detect such a problem and recover from it transparently?

16 Propose a set of extensions to the C++ IDL used in CORBA for the purpose of specifying reliability properties of a distributed server, such as fault tolerance, real-time guarantees, or security.

17 Discuss options for handling the case where a transactional CORBA application performs operations on a nontransactional CORBA server.

18 (moderately difficult; term project for a group) Build a CORBA-based Web server and browser. What benefits or disadvantages might result from using a replication technology such as Orbix+Isis to replicate the server state and load-share clients among the servers in a process group? Experimentally test your expectations.

19 Each of the following is a potential reliability exposure for CORBA-based applications. Discuss the nature of the problem and the possible remedies. Do you feel that any of these is a "show stopper" for a typical large potential user, such as a bank with worldwide operations or a telecommunication company managing millions of lines of code and application programs?

   ▶ Operator overloading and unexpected consequences of simple operations, such as
     a := b

   ▶ Exception handling when communicating with remote objects

   ▶ The need to use CORBA throughout the distributed environment in order to benefit from the technology in a systemwide manner. Here, the implication might be that large amounts of old or commercially obtained code (some of which may not be well documented or even easily recompiled) may have to be modified to support CORBA IDL-style interface declarations and remotely accessible operations.

20 Suppose that a CORBA rebinding mechanism is to be used to automatically rebind CORBA applications to a working server if the server being used fails. What constraints on the application would make this a safe thing to do without notifying the application when rebinding occurs? Would this form of complete transparency make sense, or are the constraints too severe to use such an approach in practice?

21 A protocol that introduces tolerance to failures will also make the application using it more complex than one making no attempt to tolerate failures. Presumably, this complexity carries with it a cost in decreased application reliability. Discuss the pros and cons of building systems to be robust, in light of the likelihood that doing so will increase the cost of developing the application, the complexity of the resulting system, and the challenge of testing it. Can you suggest a principled way to reach a decision on the appropriateness of hardening a system to provide a desired property?

22 Suppose you are using a conventional client/server application for a banking environment, and the bank requires that there be *absolutely no risk* of authorizing a client to withdraw funds beyond the limit of the account. Considering the possibility that the client systems may sometimes crash and need to be repaired before they restart, what are the practical implications of such a policy? Can you suggest other policies that might be less irritating to the customer while bounding the risk to the bank?

23 Suppose you are developing a medical computing system using a client/server architecture, in which the client systems control the infusion of medication directly into an IV line to the patient. Physicians will sometimes change medication orders by interacting with the server systems. It is *absolutely imperative* that the physician be confident that an order he or she has given will be carried out or that an alarm will be sounded if there is *any uncertainty whatsoever* about the state of the system. Provide an analysis of possible failure modes (client system crashes, server crashes) and the way they should be handled to satisfy this reliability goal. Assume that the software used in the system is correct and that the only failures experienced are due to hardware failures of the machines on which the client and server systems run or communication failures in the network.

24 Consider an air traffic control system in which each flight is under the control of a specific individual at any given point in time. Suppose the system takes the form of a collection of client/server distributed networks—one for each of a number of air traffic control centers. Design a protocol for handing off a flight from one controller to another, considering first the case of a single center and then the case of a multicenter system. Now, analyze the possible failure modes of your protocol under the assumption that client systems, server systems, and the communication network may be subject to failures.

25 (term project) Using the Web, locate the specifications of the Web server protocol (HTTP) over the network. Make a list of the *critical dependencies* of a typical Web browser application—that is, list the technologies and servers that the browser trusts in its normal mode of operation. Now, suppose you were concerned with possible punning attacks, in which a trusted server is replaced with a nontrustworthy server that mimics the behavior of the true one but in fact sets out to compromise the user. What methods could be used to reduce the exposure of your browsers to such attacks?

26 (term project; team of two or more) Copy one of the public-domain Web server sources to your system. In this book we have explored technologies for increasing distributed system reliability using replication, fault tolerance in servers, security tools, and coherent caching. Using protocols of your own, or Cornell's public Horus distribution, extend the Web server to implement one or more of these features. Evaluate the result of your effort by comparing the before and after behavior of the server in the areas that you modified.

27 (term project; team of two or more) Design a wide area service for maintaining directory-style information in very large environments. Such systems implement a mapping from *name* to *value* for potentially large numbers of names. Implement your architecture using existing distributing computing tools. Now evaluate the quality of your solution in

terms of performance, scaling, and reliability attributes. To what degree can your system be trusted in critical settings, and what technology dependencies does it have? Note: The X.500 standard specifies a directory service interface and might be a good basis for your design.

28 Use Horus to implement layers based on two or more of the best known *abcast* ordering protocols. Compare the performance of the resulting implementations as a function of load presented and the number of processes in the group receiving the message.

29 Suppose that a Horus protocol stack implementing Cristian's real-time atomic broadcast protocol will be used side by side with one implementing virtual synchronous process groups with *abcast*, both in the same application. To what degree might inconsistency be visible to the application when group membership changes because of failures of some group members? Can you suggest ways that the two protocol stacks might be linked to limit the time period during which such inconsistencies can occur? (Hard problem: Implement your proposal.)

30 Some authors consider RPC to be an extremely successful protocol, because it is highly transparent, reasonably robust, and can be optimized to run at very high speed—so high that if an application wants stronger guarantees, it makes more sense to layer a protocol over a lower-level RPC facility than to build it into the operating system at potentially high cost. Discuss the pros and cons of this point of view. In the best possible world, how would you design a communication subsystem?

31 Research the *end-to-end argument*. Does the goal of building reliable distributed systems bring aspects of this argument into question? Explain.

32 Review flow-control options for multicast environments in which a small number of data sources send steady streams of data to large numbers of data sinks over hardware that supports a highly (but not perfectly) reliable multicast mechanism. How does the requirement that data be reliably delivered to all data sinks change the problem?

33 A protocol is said to be "acky" if most packets are acknowledged immediately upon reception. Discuss some of the pros and cons of this property. Suppose that a stream protocol could be switched in and out of an acky mode. Under what conditions would it be advisable to operate that protocol with frequent acks?

34 Suppose that a streaming style of multidestination information service, such as the one in problem 32, is to be used in a setting where a small subset of the application programs can be unresponsive for periods of time. A good example of such a setting would be a network in which the client systems run on PCs, because the most popular PC operating systems allow applications to preempt the CPU and inhibit interrupts—a behavior that can delay the system from responding to incoming messages in a timely manner. What options can you propose for ensuring that data delivery will be reliable and ordered *in all cases*, but that small numbers of briefly unresponsive machines will not impact performance for the much larger number of highly responsive machines?

35 Several of the operating system technologies we reviewed gained performance by eliminating copying on the communication path between the communication device and the application that generates or consumes data. Suppose you were building a large-scale distributed system for video playback of short video files on demand—for example, such a system might be used in a large bank to provide brokers and traders with current projections for the markets and trading instruments tracked by the bank. What practical limits can you identify that might make it hard to use "zero copy" playback mechanisms between the file servers on which these video snippets are stored and the end user who will see the result? Assume that the system is intended to work in a very general heterogeneous environment shared with many other applications.

36 Consider the Group Membership Protocol discussed in Section 13.9. Suppose that this protocol was implemented in the address space of an application program and that the application program contained a bug causing it to infrequently but randomly corrupt a few cells of memory. To what degree would this render the assumptions underlying the GMS protocol incorrect? What behaviors might result? Can you suggest practical countermeasures that would overcome such a problem if it were indeed very infrequent?

37 (difficult) Again, consider the Group Membership Protocol discussed in Section 13.9. This protocol has the following property: All participating processes observe *exactly the same sequence* of membership views. The coordinator can add unlimited numbers of processes in each round and can drop any minority of the members each time it updates the system membership view; in both cases, the system is provably immune from partitioning. Would this protocol be simplified by eliminating the property that processes must observe the same view sequence? (Hint: Try to design a protocol that offers this "weaker" behavior.) What about the partition freedom property: Would the protocol be simpler if this were not required?

38 Suppose that the processes in a process group are managing replicated data. Due to a lingering bug, it is known that although the group seems to work well for periods of hours or even days, over very long periods of time the replicated data can become slightly corrupted so that different group members have different values. Discuss the pros and cons of introducing a stabilization mechanism, whereby the members would periodically exchange values and, if an inconsistency develops, arbitrarily switch to the most common value or to the value of an agreed upon leader. What issues might this raise in the application program, and how might they be addressed?

39 Implement a very simple banking application supporting accounts into which money can be deposited and withdrawals can be made. Have your application support a form of *disconnected operation* based on the two-tiered architecture, in which each branch system uses its own set of process groups and maintains information for local accounts. Your application should simulate partitioning failures through a command interface. If branches cache information about remote accounts, what options are there for permitting a client to withdraw funds while the local branch at which the account really resides is unavailable? Consider both the need for safety by the bank and the need for availability, if possible, for the user—for example, it would be silly to refuse a user $250 from an account that had thousands of dollars in it moments earlier when connections were still working! Can you propose a policy that is always safe for the bank and yet also allows remote withdrawals during partition failures?

40 Design a protocol by which a process group implemented using Horus can solve the asynchronous consensus problem. Assume that the environment is one in which Horus can be used, processes only fail by crashing, and the network only fails by losing messages with some low frequency. Your processes should be assumed to start with a variable, $input_i$, which, for each process, p_i, is initially 0 or 1. After deciding, each process should set a variable, $output_i$, to its decision value. The solution should be such that the processes all reach the same decision value, v, and this value is the same as at least one of the inputs.

41 In regard to your solution to problem 40, discuss the sense in which your solution solves the asynchronous consensus problem. Would Horus be guaranteed to make progress under the stated conditions? Do these conditions correspond to the conditions of the asynchronous model used in the FLP and Chandra/Toueg results?

42 Can the virtual synchrony protocols of a system such as Horus be said to guarantee safety and liveness in the general asynchronous model of FLP or the Chandra/Toueg results?

43 Suppose you were responsible for porting the Horus system to a cluster-style processor known to consist of between 16 and 32 identical high-speed computing nodes interconnected by a high-speed ATM-style communication bus, with a reliable mechanism for detecting hardware failures of nodes within a few microseconds after such events occur. Your goal in undertaking this port is to implement a cluster API, providing standard cluster-oriented operating system services to application developers. How would you consider changing Horus itself to adapt it better to this environment? Would the Horus Common Protocol Interface (HCPI) be a suitable cluster API, or would you implement some other layer over Horus; if the latter, what would your API include? Assume that an important goal is that the cluster be highly available, easily serviced and upgraded, and able to support highly available application programs with relative ease.

44 Can the virtual synchrony protocols of a system such as Horus be said to guarantee safety and liveness in a cluster-style computer architecture such as the one described in problem 43?

45 The Horus stability layer operates as follows: Each message is given a unique ID, and is transmitted and delivered using the stack selected by the user. The stability layer expects the processes receiving the message to issue a downcall when they consider the message locally stable. This information is relayed within the group, and each group member can obtain a matrix giving the stabilization status of pending messages originated within the group as needed. Could the stability layer be used in a way that would add the dynamic uniformity guarantee to messages sent in a group?

46 Suppose that a process group is created in which three member processes each implement different algorithms for performing the same computation (so-called "implementation redundancy"). You may assume that these processes interact with the external environment *only using message send and receive primitives*. Design a wrapper that compares the actions of the processes, producing a single output if two of the three or all three processes agree on the action to take for a given input and signaling an exception if all three processes produce different outputs for a given input. Implement your solution using Horus and demonstrate it for a set of fake processes that usually copy their input to their output, but that sometimes make a random change to their output before sending it.

47 A set of processes in a group monitors devices in the external environment, detecting *device service requests* to which they respond in a load-balanced manner. The best way to handle such requests depends upon the frequency with which they occur. Consider the following two extremes: requests that require long computations to handle but that occur relatively infrequently and requests that require very short computations to handle but that occur frequently on the time scale with which communication is done in the system. Assuming that the processes in a process group have identical capabilities (any can respond to any request), how would you solve this problem in the two cases?

48 Design a locking protocol for a virtually synchronous process group. Your protocol should allow a group member to *request* a lock, specifying the name of the object to be locked (the name can be an integer to simplify the problem), and to *release* a lock that it holds. What issues occur if a process holding a lock fails? Recommend a good, general way of dealing with this case and then give a distributed algorithm by which the group members can implement the *request* and *release* interfaces, as well as your solution to the broken lock case.

49 (suggested by Jim Pierce) Suppose we want to implement a system in which n process groups will be superimposed—much like the petals of a flower. Some small set of k processes will belong to all n groups, and each group will have additional members that belong only to it. The problem now occurs of how to handle *join* operations for the processes that belong to the overlapping region and in particular how to deal with state transfers to such a process. Assume that the group states are only updated by "petal" processes, which do not belong to the overlap region. Now, the virtually synchronous state transfer mechanisms we discussed in Section 15.3.2 would operate on a group-by-group basis, but it may be that the states of the processes in the overlap region are a mixture of information arriving from all of the petal processes. For such cases one would want to do a *single* state transfer to the joining process, reflecting the *joint state* of the overlapped groups. Propose a fault-tolerant protocol for joining the overlap region and transferring state to a joining process that will satisfy this objective. (Refer to Figure A.1.)

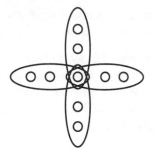

FIGURE A.1. Overlapping process groups for problem 49. In this example there is only a single process in the overlap region; the problem concerns state transfer if we wanted to add another process to this region. Assume that the state of the processes in the overlap region reflects messages sent to it by the outer processes, which belong to the "petals" but not the overlap area. Additionally, assume that this state is not cleanly decomposed group by group and that it is necessary to implement a single state transfer for the entire structure.

50 Discuss the pros and cons of using an *inhibitory* protocol to test for a condition along a consistent cut in a process group. Describe a problem or scenario where such a solution might be appropriate and one where it would not be.

51 Suppose that the processes in a distributed system share a set of resources, which they lock prior to using and then unlock when finished. If these processes belong to a process group, how could deadlock detection be done within that group? Design your deadlock detection algorithm to be completely idle (with no background communication costs) when no deadlocks are suspected; the algorithm should be one that can be launched when a timeout in a waiting process suggests that a deadlock may have occurred. For bookkeeping purposes, you may assume that a process waiting for a resource calls the local procedure *waiting_for(resource)*, a process holds exclusive access to a resource calls the procedure *holding(resource)*, and a process releasing a resource calls the procedure *release(resource)*, where the resources are identified by integers. Each process thus maintains a local database of its resource status. Notice that you are not being asked to implement the actual mutual-exclusion algorithm here: Your goal is to devise a protocol that can interact with the processes in the system as needed to accurately detect deadlocks. Prove that your protocol detects deadlocks if, and only if, they are present.

52 Suppose you wish to monitor a distributed system for an overload condition, defined as follows. The system state is considered normal if no more than one-third of the processes signal that they are overloaded, heavily loaded if more than one-third but less than two-thirds of the processes signal that they are overloaded, and seriously overloaded if two-thirds or more of the processes are overloaded. Assume further that the loading condition does not impact communication performance. If the processes belong to a process group, would it be sufficient to simply send a multicast to all members asking their states and then to compute the state of the system from the vector of replies so obtained? What issues would such an approach raise, and under what conditions would the result be correct?

53 (Joseph and Schmuck) What would be the best way to implement a *predicate addressing* communication primitive for use within virtually synchronous process groups (assume that the group primitives are already implemented and available for you). Such a primitive sends a message to *all the processes in the group for which some acceptance criteria hold* and does so *along a consistent cut.* You may assume that each process contains a predicate, *accept()*, which, at the time it is invoked, returns *true* if the process wishes to accept a copy of the message and *false* if not. (Hint: It is useful to consider two separate cases here—one in which the criteria that determine acceptance change slowly and one in which they change rapidly, relative to the speed of multicasting in the system.)

54 In discussing the concept of wrappers, we developed the example of a *worldwide memory system*, in which shared memory primitives are redefined to permit programs to share access to very large scale distributed memories maintained over an ATM-style network. Suppose you were implementing such a system using Horus over the U-Net system on a wide area ATM, and you knew the expected application to be an in-memory server for Web pages. These pages will in some cases be updated rapidly (at video speeds) and for that purpose your browser will have the ability to memory-map *video image* objects directly to the display of the viewing computer. What special design considerations are implied by this intended application? Recall that the memory architecture we developed had a concept of *prefetching* built into it, much like a traditional virtual memory subsystem would have. How should prefetching be implemented in your mapped memory system?

55 (difficult; team programming project) Implement the architecture you proposed in problem 54, focusing on the case of side-by-side computers with a high-speed link between them.

56 (difficult, research topic) Implement a worldwide memory system such as the one discussed in problem 54 and develop a detailed justification and evaluation of the architecture you used.

57 (Schneider) We discussed two concepts of clock synchronization: *accuracy* and *precision*. Consider the case of aircraft operating under *free-flight* rules, where each pilot makes routing decisions on behalf of his or her plane, using a shared trajectory mapping system. Suppose you faced a fundamental tradeoff between using clocks with high accuracy for such a mapping system or clocks with high precision. Which would you favor and why? Would it make sense to implement two such solutions, side by side?

58 Suppose that a-posteriori clock synchronization using GPS receivers becomes a worldwide standard in the coming decade. The use of temporal information now represents a form of communication channel that can be used in indirect ways—for example, process p, executing in Lisbon, can wait until process q performs a desired operation in New York (or fails) using timer events. Interestingly, such an approach communicates information faster than messages can. What issues do these sorts of hidden information channels raise in regard to the protocols we explored in the book? Could temporal information create hidden causality relationships?

59 Show how tightly synchronized real-time clocks can be made to reflect causality in the manner of Lamport's logical clocks. Would such a clock be preferable in some ways to a purely logical clock? Explain, giving concrete examples to illustrate your points.

60 (difficult) In discussion of the CASD protocols, we saw that if such protocols were used to replicate the state of a distributed system, a mechanism would be needed to overcome inconsistencies occurring when a process is technically considered incorrect according to the definitions of the protocols and therefore does not benefit from the normal guarantees of atomicity and ordering seen by correct processes. In an IBM technical report, Skeen and Cristian once suggested that the CASD protocols could be used in support of an abstraction called Δ-*common storage*; the basic idea is to implement a distributed shared memory, which can be read by any process and updated using the CASD style of broadcast protocol. Such a distributed shared memory would reflect an update within Δ time units after it is initiated, plus or minus a clock skew factor of ε. How might the inconsistency issue of the CASD protocol be visible in a Δ-common storage system? Propose a method for detecting and eliminating such inconsistencies. (Note: This issue was not considered in the technical report.)

61 (Marzullo and Sabel) Suppose you wish to monitor a distributed system to detect situations in which a logical predicate defined over the states of the member processes holds. The predicate may state, for example, that process p_i holds a token and that process p_j is waiting to obtain the token. Under the assumption that the states in question change very slowly in comparison to the communication speeds of the system, design a solution to this problem. You may assume that there is a function, *sample_local_state()*, that can be executed in each process to sample those aspects of its local state referenced in the query, and when the local states have been assembled in one place, a function, *evaluate*, can determine if the predicate holds or not. Now, discuss the modifications needed if the rate of state changes is increased enough so that the state can change in the same order of time as your protocol needs to run. How is your solution affected if you are required to detect *every state in which the predicate holds*, as opposed to just detecting *states in which the predicate happens to hold when the protocol is executed*. Demonstrate that your protocol cannot falsely detect satisfying states.

62 There is increasing interest in building small multiprocessor systems for use in inexpensive communication satellites. Such systems might look similar to a rack containing a small number of conventional workstations or PCs, running software that handles such tasks as maintaining the proper orientation of the satellite by adjusting its position periodically, turning on and off the control circuits that relay incoming messages to outgoing channels, and handling other aspects of satellite function. Now, suppose that it is possible to put highly redundant memory modules on the satellite to protect extremely critical regions of memory, but that it is costly to do so. However, unprotected memory is likely to experience a low level of corruption as a result of the harsh conditions in space, such as cosmic rays and temperature extremes. What sorts of programming considerations would result from using such a model? Propose a software architecture that minimizes the need for redundant memory, but also minimizes the risk that a satellite will be completely lost (e.g., a satellite might be lost if it erroneously fires its positioning rockets and thereby exhausts its supply of fuel). You may assume that the actual rate of corruption of memory is low, but not completely insignificant, and that program instructions are as likely as data to be corrupted. Assume that the extremely reliable memories, however, never experience corruption.

63 Continuing with the topic of problem 62, there is debate concerning the best message-routing architecture for these sorts of satellite systems. In one approach, the satellites maintain a routing network among themselves; a relatively small number of ground stations interact with whatever satellite happens to be over them at a given time, and control and data messages are then forwarded satellite to satellite until they reach the destination. In a second approach, satellites communicate only with ground stations and mobile transmitter/receiver units: Such satellites require a larger number of ground systems, but they do not depend upon a routing transport protocol, which could be a source of unreliability. Considering the conditions cited in problem 62 and your responses, what would be the best design for a satellite-to-satellite routing network? Can you suggest a scientifically sound way to make the design tradeoff between this approach and one using a larger number of potentially costly groundstations?

64 We noted that the theoretical community considers a problem to be impossible in a given environment if, for all proposed solutions to the problem, there exists at least one behavior consistent with the environment that would prevent the proposed solution from terminating or would lead to an incorrect outcome. Later we considered probabilistic protocols, which may be able to guarantee behaviors to very high levels of reliability—higher, in practice, than the reliability of the computers on which the solutions run. Suggest a definition of *impossible* that might reconcile these two perspectives on computing systems.

65 If a message must take d hops to reach its destination and the worst-case delay for a single link is δ, it is common to assume that the worst-case transit time for the network will be $d * \delta$. However, a real link will typically exhibit a distribution of latencies, with the vast majority clustered near some minimum latency, δ_{min}, and only a very small percentage taking as long as δ_{max} to traverse the link. Under the assumption that the links of a routed network provide statistically independent and identical behavior, derive the distribution of expected latencies for a message that must traverse d links of a network. You may assume that the distribution of delays has a convenient form for your analysis.

66 Suppose that a security architecture supports *revocation* of permissions. Thus, XYZ was permitted to access resource ABC, but now has finished the task for which permission was granted and we want to disable future accesses. Would it be safe to use a remote procedure call from the authentication server to the resource manager for resource ABC to accomplish this revocation? Explain.

67 (ethical problem) Suppose that a medical system does something a person would not be able to do, such as continuously monitoring the vital signs of a patient and continuously adjusting some form of medication or treatment in response to the measured values. Now, imagine that we want to attach this device to a distributed system so that physicians and nurses elsewhere in the hospital can remotely monitor the behavior of the medical system and so that they can change the rules that control its actions if necessary (e.g., by changing the dosage of a drug). In this book we have encountered many practical limits to security and reliability. Identify some of the likely limits on the reliability of a technology such as this. What are the ethical issues that need to be balanced in deciding whether or not to build such a system?

68 (ethical problem) An *ethical theory* is a set of governing principles or rules for resolving ethical conflicts such as the one in the previous problem—for example, an ethical theory might stipulate that decisions should be made to favor the maximum benefit for the greatest number of individuals. A theory governing the deployment of technology could stipulate that machines must not replace people if the resulting system is at risk of making erroneous decisions that a person would have avoided. Notice that these particular theories could be in conflict—for example, if a technology that is normally beneficial develops occasional life-threatening complications. Discuss the issues that might occur in developing an ethical theory for the introduction of technologies in life- or safety-critical settings and, if possible, propose such a theory. What tradeoffs are required, and how would you justify them?

Bibliography

Abbott, M., and L. Peterson. "Increasing Network Throughput by Integrating Protocol Layers." *IEEE/ ACM Transactions on Networking* 1:5 (October 1993): 600–610.

Agarwal, D. A. "Totem: A Reliable Ordered Delivery Protocol for Interconnected Local Area Networks." Ph.D. diss., Department of Electrical and Computer Engineering, University of California, Santa Barbara, 1994.

Ahamad, M., J. Burns, P. Hutto, and G. Neiger. "Causal Memory." Technical Report, College of Computing, Georgia Institute of Technology, July 1991.

Alon, N. A. Barak, and U. Manber. "On Disseminating Information Reliably without Broadcasting." *Proceedings of the Seventh International Conference on Distributed Computing Systems* (Berlin, September 1987). New York: IEEE Computer Society Press, 74–81.

Alonso, R., and F. Korth. "Database Issues in Nomadic Computing." *Proceedings of the ACM SIGMOD International Conference on Management of Data* (Washington, DC, May 1993), 388–392.

Alvisi, L., and K. Marzullo. "Message Logging: Pessimistic, Causal, and Optimistic." *Proceedings of the Fifteenth IEEE Conference on Distributed Computing Systems* (Vancouver, 1995), 229–236.

Amir, O., Y. Amir, and D. Dolev. "A Highly Available Application in the Transis Environment." *Proceedings of the Workshop on Hardware and Software Architectures for Fault Tolerance.* Springer-Verlag Lecture Notes in Computer Science, vol. 774, 125–139.

Amir, Y. "Replication Using Group Communication over a Partitioned Network." Ph.D. diss., Hebrew University of Jerusalem, 1995.

Amir, Y., D. Dolev, S. Kramer, and D. Malkhi. "Membership Algorithms in Broadcast Domains." *Proceedings of the Sixth WDAG* (Israel, June 1992). Springer-Verlag Lecture Notes in Computer Science, vol. 647, 292–312.

————. "Transis: A Communication Subsystem for High Availability." *Proceedings of the Twenty-Second Symposium on Fault-Tolerant Computing Systems* (Boston, July 1992). New York: IEEE Computer Society Press, 76–84.

Amir, Y. et al. "The Totem Single-Ring Ordering and Membership Protocol." *ACM Transactions on Computer Systems* 13:4 (November 1995): 311–342.

Anceaume, E., B. Charron-Bost, P. Minet, and S. Toueg. "On the Formal Specification of Group Membership Services." Technical Report 95-1534. Department of Computer Science, Cornell University, August 1995.

Anderson, T., B. Bershad, E. Lazowska, and H. Levy. "Scheduler Activations: Effective Kernel Support for the User-Level Management of Parallelism." *Proceedings of the Thirteenth ACM Symposium on Operating Systems Principles* (Pacific Grove, CA, October 1991), 95–109.

Anderson, T. et al. "Serverless Network File Systems." *Proceedings of the Fifteenth Symposium on Operating Systems Principles* (Copper Mountain Resort, CO, December 1995). New York: ACM Press, 109–126. Also *ACM Transactions on Computing Systems* 13:1 (February 1996).

Andrews, G. R. *Concurrent Programming: Principles and Practice.* Redwood City, CA: Benjamin/Cummings, 1991.

Architecture Projects Management Limited. "The Advanced Networked Systems Architecture: An Application Programmer's Introduction to the Architecture." Technical Report TR-017-00. November 1991.

———. "The Advanced Networked Systems Architecture: A System Designer's Introduction to the Architecture." Technical Report RC-253-00. April 1991.

———. "The Advanced Networked Systems Architecture: An Engineer's Introduction to the Architecture." Technical Report TR-03-02. November 1989.

Armand, F., M. Gien, F. Herrmann, and M. Rozier. "Revolution 89, or Distributing UNIX Brings It Back to Its Original Virtues." Technical Report CS/TR-89-36-1. Chorus Systemes, Paris, France, August 1989.

Asplin, J., and D. Johansen. "Performance Experiments with the StormView Distributed Parallel Volume Renderer." Computer Science Technical Report 95-22. University of Tromsö, June 1995.

Babaoglu, O., A. Bartoli, and G. Dini. "Enriched View Synchrony: A Paradigm for Programming Dependable Applications in Partitionable Asynchronous Distributed Systems." Technical Report. Department of Computer Science, University of Bologna, May 1996.

Babaoglu, O., R. Davoli, L. A. Giachini, and M. B. Baker. "RELACS: A Communications Infrastructure for Constructing Reliable Applications in Large-Scale Distributed Systems." BROADCAST Project Deliverable Report, 1994. Department of Computing Science, University of Newcastle upon Tyne, United Kingdom.

Babaoglu, O., R. Davoli, and A. Montresor. "Failure Detectors, Group Membership, and View-Synchronous Communication in Partitionable Asynchronous Systems." Technical Report UBLCS-95-18. Department of Computer Science, University of Bologna, November 1995.

Babaoglu, O., and K. Marzullo. "Consistent Global States of Distributed Systems: Fundamental Concepts and Mechanisms." In *Distributed Systems* (2d ed.), S. J. Mullender, ed. Reading, MA: Addison-Wesley/ACM Press, 1993.

Bache, T. C. et al. "The Intelligent Monitoring System." *Bulletin of the Seismological Society of America* 80:6 (December 1990): 59–77.

Bailey, N. *The Mathematical Theory of Epidemic Diseases,* 2d ed. London: Charles Griffen and Company, 1975.

Baker, M. G. et al. "Measurements of a Distributed File System." *Proceedings of the Thirteenth ACM Symposium on Operating Systems Principles* (Orcas Island, WA, November 1991), 198–212.

Bal, H. E., M. F. Kaashoek, and A. S. Tanenbaum. "Orca: A Language for Parallel Programming of Distributed Systems." *IEEE Transactions on Software Engineering* (March 1992), 190–205.

Bartlett, J., J. Gray, and B. Horst. "Fault Tolerance in Tandem Computing Systems." *In Evolution of Fault-Tolerant Computing.* Springer-Verlag, 1987, 55–76.

Bartlett, J. F. "A Nonstop Kernel." *Proceedings of the Eighth ACM Symposium on Operating Systems Principles* (Pacific Grove, CA, December 1981). New York: ACM Press, 22–29.

Bellovin, S. M., and M. Merritt. "Limitations of the Kerberos Authentication System." *Computer Communication Review* 20:5 (October 1990): 119–132.

Ben-Or, M. "Fast Asynchronous Byzantine Agreement." *Proceedings of the Fourth ACM Symposium on Principles of Distributed Computing* (Minaki, Canada, August 1985), 149–151.

Berners-Lee, T., C. J-F. Groff, and B. Pollermann. "World Wide Web: The Information Universe." *Electronic Networking Research, Applications and Policy* 2:1 (1992): 52–58.

Berners-Lee, T. et al. *Hypertext Transfer Protocol—HTTP 1.0.* IETF HTTP Working Group Draft 02 (Best Current Practice). August 1995.

———. "The World Wide Web." *Communications of the ACM* 37:8 (August 1994): 76–82.

Bernstein, P. E., V. Hadzilacos, and N. Goodman. *Concurrency Control and Recovery in Database Systems.* Reading, MA: Addison-Wesley, 1987.

Bershad, B., T. Anderson, E. Lazowska, and H. Levy. "Lightweight Remote Procedure Call." *Proceedings of the Eleventh ACM Symposium on Operating Systems Principles* (Litchfield Springs, AZ, December 1989), 102–113. Also *ACM Transactions on Computer Systems* 8:1 (February 1990): 37–55.

Bershad, B. et al. "Extensibility, Safety, and Performance in the SPIN Operating System." *Proceedings of the Fifteenth Symposium on Operating Systems Principles* (Copper Mountain Resort, CO, December 1995), 267–284.

Bhide, A., E. N. Elnozahy, and S. P. Morgan. "A Highly Available Network File Server." *Proceedings of the USENIX Winter Conference* (Austin, December 1991), 199–205.

Biagioni. E. "A Structured TCP in Standard ML." *Proceedings of the 1994 Symposium on Communications Architectures and Protocols* (London, August 1994).

Biagioni, E., R. Harper, and P. Lee. "Standard ML Signatures for a Protocol Stack." Technical Report CS-93-170. Department of Computer Science, Carnegie Mellon University, October 1993.

Birman, K. P. "A Response to Cheriton and Skeen's Criticism of Causal and Totally Ordered Communication." *Operating Systems Review* 28:1 (January 1994): 11–21.

————. "The Process Group Approach to Reliable Distributed Computing." *Communications of the ACM* 36:12 (December 1993).

Birman, K. P., and B. B. Glade. "Consistent Failure Reporting in Reliable Communications Systems." *IEEE Software*, Special Issue on Reliability, April 1995.

Birman, K. P., and T. A. Joseph. "Exploiting Virtual Synchrony in Distributed Systems." *Proceedings of the Eleventh Symposium on Operating Systems Principles* (Austin, November 1987). New York: ACM Press, 123–138.

————. "Reliable Communication in the Presence of Failures." *ACM Transactions on Computer Systems* 5:1 (February 1987): 47–76

Birman, K. P., D. Malkhi, A. Ricciardi, and A. Schiper. "Uniform Action in Asynchronous Distributed Systems." Technical Report TR 94-1447. Department of Computer Science, Cornell University, 1994.

Birman, K. P., A. Schiper, and P. Stephenson. "Lightweight Causal and Atomic Group Communication." *ACM Transactions on Computing Systems* 9:3 (August 1991): 272–314.

Birman, K. P., and R. van Renesse, eds. *Reliable Distributed Computing with the Isis Toolkit*. New York: IEEE Computer Society Press, 1994.

————. "Software for Reliable Networks." *Scientific American* 274:5 (May 1996): 64-69.

Birrell, A. "Secure Communication Using Remote Procedure Calls." *ACM Transactions on Computer Systems* 3:1 (February 1985): 1–14.

Birrell, A., and B. Nelson. "Implementing Remote Procedure Call." *ACM Transactions on Programming Languages and Systems* 2:1 (February 1984): 39–59.

Birrell, A., G. Nelson, S. Owicki, and T. Wobbera. "Network Objects." *Proceedings of the Fourteenth Symposium on Operating Systems Principles* (Asheville, NC, 1993), 217–230.

Black, A., N. Hutchinson, E. Jul, and H. Levy. "Object Structure in the Emerald System." *ACM Conference on Object-Oriented Programming Systems, Languages, and Applications* (Portland, OR, October 1986).

Borg, A., J. Baumbach, and S. Glazer. "A Message System for Supporting Fault Tolerance." *Proceedings of the Ninth Symposium on Operating Systems Principles* (Bretton Woods, NH, October 1983), 90–99.

Borg, A. et al. "Fault Tolerance under UNIX." *ACM Transactions on Computer Systems* 3:1 (February 1985): 1–23.

Borr, A., and C. Wilhelmy. "Highly Available Data Services for UNIX Client/Server Networks: Why Fault-Tolerant Hardware Isn't the Answer." *Hardware and Software Architectures for Fault Tolerance*, M. Banatre and P. Lee, eds. Springer-Verlag Lecture Notes in Computer Science, vol. 774, 385–304.

Brakmo, L., S. O'Malley, and L. Peterson. "TCP Vegas: New Techniques for Congestion Detection and Avoidance." *Proceedings of the ACM SIGCOMM '94* (London, 1994).

Braun, T., and C. Diot. "Protocol Implementation Using Integrated Layer Processing." *Proceedings of SIGCOMM-95* (September 1995).

Bressoud, T. C., and F. B. Schneider. "Hypervisor-based Fault Tolerance." *Proceedings of the Fifteenth Symposium on Operating Systems Principles* (Copper Mountain Resort, CO, December 1995). New York: ACM Press, 1–11. Also *ACM Transactions on Computing Systems* 13:1 (February 1996).

Brockschmidt, K. *Inside OLE-2*. Redmond, WA: Microsoft Press, 1994.

Budhiraja, N. et al. "The Primary-Backup Approach." In *Distributed System,* 2d ed., S. J. Mullender, ed. Reading, MA: Addison-Wesley/ACM Press, 1993.

Burrows, M., M. Abadi, and R. Needham. "A Logic of Authentication." *Proceedings of the Eleventh ACM Symposium on Operating Systems Principles* (Litchfield Springs, AZ, December 1989). New York: ACM Press, 1–13.

Callahan, J., and T. Montgomery. "Approaches to Verification and Validation of a Reliable Multicast Protocol." *Proceedings of the 1996 ACM International Symposium on Software Testing and Analysis,* (San Diego, CA, January 1996), 187–194. Also appears as *ACM Software Engineering Notes,* vol. 21, no. 3, May 1996, 187–194.

Carter, J. "Efficient Distributed Shared Memory Based on Multi-Protocol Release Consistency." Ph.D. diss., Rice University, August 1993.

Chandra, T., V. Hadzilacos, and S. Toueg. "The Weakest Failure Detector for Solving Consensus." *ACM Symposium on Principles of Distributed Computing* (August 1992), 147–158.

Chandra, T., V. Hadzilacos, S. Toueg, and B. Charron-Bost. "On the Impossibility of Group Membership." *Proceedings of the ACM Symposium on Principles of Distributed Computing* (Vancouver, May 1996).

Chandra, T., and S. Toueg. "Time and Message Efficient Reliable Broadcasts." Technical Report TR 90-1094. Department of Computer Science, Cornell University, February 1990.

———. "Unreliable Failure Detectors for Asynchronous Systems." *Journal of the ACM*, in press. Previous version in *ACM Symposium on Principles of Distributed Computing* (Montreal, 1991), 325–340.

Chandy, K. M., and L. Lamport. "Distributed Snapshots: Determining Global States of Distributed Systems." *ACM Transactions on Computer Systems* 3:1 (February 1985): 63–75.

Chang, M., and N. Maxemchuk. "Reliable Broadcast Protocols." *ACM Transactions on Computer Systems* 2:3 (August 1984): 251–273.

Charron-Bost, B. "Concerning the Size of Logical Clocks in Distributed Systems." *Information Processing Letters* 39:1 (July 1991): 11–16.

Chase, J., H. Levy, M. Feeley, and E. Lazowska. "Sharing and Protection in a Single-Address-Space Operating System." *ACM Transactions on Computer Systems* 12:4 (November 1994): 271–307.

Chaum, D. "Untraceable Electronic Mail, Return Addresses, and Digital Pseudonyms." *Communications of the ACM* 24:2 (February 1981): 84-88.

Chen, P. et al. "RAID: High Performance, Reliable, Secondary Storage." *ACM Computing Surveys* 26:2 (June 1994): 45–85.

Cheriton, D., and D. Skeen. "Understanding the Limitations of Causally and Totally Ordered Communication." *Proceedings of the Thirteenth ACM Symposium on Operating Systems Principles* (Asheville, NC, December 1993). New York: ACM Press, 44–57.

Cheriton, D., and W. Zwaenepoel. "Distributed Process Groups in the V Kernel." *ACM Transactions on Computer Systems* 3:2 (May 1985): 77–107.

Chilaragee, R. "Top Five Challenges Facing the Practice of Fault Tolerance." In *Hardware and Software Architectures for Fault Tolerance*, M. Banatre and P. Lee, eds. Springer-Verlag Lecture Notes in Computer Science, vol. 774, 3–12.

Cho, K., and K. P. Birman. "A Group Communication Approach for Mobile Computing." Technical Report TR94-1424. Computer Science Department, Cornell University, May 1994.

Chockler, G. U., N. Huleihel, I. Keidar, and D. Dolev. "Multimedia Multicast Transport Service for Groupware" TINA '96: *The Convergence of Telecommunications and Distributed Computing Technologies* (Heidelberg, September 1996). Berlin: VDE-Verlag.

Clark, D., V. Jacobson, J. Romkey, and H. Salwen. "An Analysis of TCP Processing Overhead." *IEEE Communications* 27:6 (June 1989): 23–29.

Clark, D., and D. L. Tennenhouse. "Architectural Considerations for a New Generation of Protocols." *Proceedings of the 1990 Symposium on Communication Architectures and Protocols* (Philadelpha, September 1990). New York: ACM Press, 200–208.

Clark, D., and M. Tennenhouse. "Architectural Considerations for a New Generation of Protocols." *Proceedings of SIGCOMM-87* (August 1987), 353–359.

Clegg, M., and K. Marzullo. "Clock Synchronization in Hard Real-Time Distributed Systems." Technical Report. Department of Computer Science, University of California, San Diego, March 1996.

Coan, B., B. M. Oki, and E. K. Kolodner. "Limitations on Database Availability When Networks Partition." *Proceedings of the Fifth ACM Symposium on Principles of Distributed Computing* (Calgary, August 1986), 187–194.

Coan, B., and G. Thomas. "Agreeing on a Leader in Real Time." *Proceedings of the Eleventh Real-Time Systems Symposium* (December 1990), 166–172.

Comer., D., and J. Griffioen. "A New Design for Distributed Systems: The Remote Memory Model." *Proceedings of the 1990 Summer USENIX Conference* (June 1990), 127–135.

Comer, D. E. *Internetworking with TCP/IP*. Vol. I: *Principles, Protocols, and Architecture*. Englewood Cliffs, NJ: Prentice Hall, 1991.

Comer, D. E., and D. L. Stevens. *Internetworking with TCP/IP*. Vol. III: *Client/Server Programming and Applications*. Englewood Cliffs, NJ: Prentice Hall, 1993.

————. *Internetworking with TCP/IP*. Vol. II: *Design, Implementation, and Internals*. Englewood Cliffs, NJ: Prentice Hall, 1991.

Cooper, D. A., and K. P. Birman. "The Design and Implementation of a Private Message Service for Mobile Computers." *Wireless Networks* 1:3 (October 1995): 297–309.

Cooper, E. "Replicated Distributed Programs." *Proceedings of the Tenth ACM Symposium on Operating Systems Principles* (Orcas Island, WA, December 1985). New York: ACM Press, 63–78.

Cooper, R. "Experience with Causally and Totally Ordered Group Communication Support—A Cautionary Tale." *Operating Systems Review* 28:1 (January 1994): 28–32.

Coulouris, G., J. Dollimore, and T. Kindberg. *Distributed Systems: Concepts and Design*. Reading, MA: Addison-Wesley, 1994.

Cristian, F. "Synchronous and Asynchronous Group Communication." *Communications of the ACM* 39:4 (April 1996): 88–97.

———. "Reaching Agreement on Processor Group Membership in Synchronous Distributed Systems." *Distributed Computing* 4:4 (April 1991): 175–187.

———. "Understanding Fault-Tolerant Distributed Systems." *Communications of the ACM* 34:2 (February 1991): 57–78.

———. "Probabilistic Clock Synchronization." *Distributed Computing* 3:3 (1989): 146–158.

Cristian, F., H. Aghili, R. Strong, and D. Dolev. "Atomic Broadcast: From Simple Message Diffusion to Byzantine Agreement." *Proceedings of the Fifteenth International Symposium on Fault-Tolerant Computing* (1985). New York: IEEE Computer Society Press, 200–206. Revised as IBM Technical Report RJ5244.

Cristian, F., and R. Delancy. "Fault Tolerance in the Advanced Automation System." IBM Technical Report RJ7424. IBM Research Laboratories, San Jose, CA, April 1990.

Cristian, F., D. Dolev, R. Strong, and H. Aghili. "Atomic Broadcast in a Real-Time Environment." In *Fault-Tolerant Distributed Computing*. Springer-Verlag Lecture Notes in Computer Science, vol. 448, 1990, 51–71.

Cristian, F., and C. Fetzer. "Fault-Tolerant Internal Clock Synchronization." *Proceedings of the Thirteenth Symposium on Reliable Distributed Systems* (October 1994).

Cristian, F., and F. Schmuck. "Agreeing on Process Group Membership in Asynchronous Distributed Systems." Technical Report CSE95-428. Department of Computer Science and Engineering, University of California, San Diego, 1995.

Custer, H. *Inside Windows NT*. Redmond, WA: Microsoft Press, 1993.

Damm, A., J. Reisinger, W. Schwabl, and H. Kopetz. "The Real-Time Operating System of Mars." *ACM Operating Systems Review* 22:3 (July 1989): 141–157.

Davidson, S., H. Garcia-Molina, and D. Skeen. "Consistency in a Partitioned Network: A Survey." *ACM Computing Surveys* 17:3 (September 1985): 341–370.

Deering, S. E. "Host Extensions for IP Multicasting." Technical Report RFC 1112. SRI Network Information Center, August 1989.

———. "Multicast Routing in Internetworks and Extended LANs." *Computer Communications Review* 18:4 (August 1988): 55–64.

Deering, S. E., and D. R. Cheriton. "Multicast Routing in Datagram Internetworks and Extended LANs." *ACM Transactions on Computer Systems* 8:2 (May 1990): 85–110.

Della Ferra, C. A. et al. "The Zephyr Notification Service." *Proceedings of the Winter USENIX Conference* (December 1988).

Demers, A. et al. "Epidemic Algorithms for Replicated Data Management." *Proceedings of the Sixth Symposium on Principles of Distributed Computing*, (Vancouver, August 1987): 1–12. Also *Operating Systems Review* 22:1 (January 1988): 8–32.

Dempsey, B., J. C. Fenton, and A. C. Weaver. "The MultiDriver: A Reliable Multicast Service Using the Xpress Transfer Protocol." *Proceedings of the Fifteenth Conference on Local Computer Networks* (1990). New York: IEEE Computer Society Press, 351–358.

Denning, D. "Digital Signatures with RSA and Other Public-Key Cryptosystems." *Communications of the ACM* 27:4 (April 1984): 388–392.

Denning, D., and D. A. Branstad. "Taxonomy for Key Escrow Encryption Systems." *Communications of the ACM* 39:3 (March 1996): 34–40.

Desmedt, Y. "Society and Group-Oriented Cryptography: A New Concept." *Advances in Cryptology—CRYPTO '87 Proceedings*. Springer-Verlag Lecture Notes in Computer Science, vol. 293, 1988, 120–127.

Desmedt, Y., Y. Frankel, and M. Yung. "Multi-Receiver/Multi-Sender Network Security: Efficient Authenticated Multicast/Feedback." *Proceedings of IEEE INFOCOM* (May 1992).

Diffie, W. "The First Ten Years of Public-Key Cryptography." *Proceedings of the IEEE* 76:5 (May 1988): 560–577.

Diffie, W., and M. E. Hellman. "Privacy and Authentication: An Introduction to Cryptography." *Proceedings of the IEEE* 67:3 (March 1979): 397–427.

Digital Equipment Corporation. "A Technical Description of the DECsafe Available Server Environment (ASE)." *Digital Equipment Corporation Technical Journal* 7:4 (September 1995): 89–100.

Dolev, D., and D. Malkhi. "The Transis Approach to High Availability Cluster Communication." *Communications of the ACM* 39:4 (April 1996): 64–70.

Dolev, D., D. Malkhi, and R. Strong. "A Framework for Partitionable Membership Service." Technical Report TR 95-4. Institute of Computer Science, Hebrew University of Jerusalem, March 1995.

Drushel, P., and L. L. Peterson. "Fbufs: A High-Bandwidth Cross-Domain Transfer Facility." *Proceedings of the Thirteenth ACM Symposium on Operating Systems Principles* (Pacific Grove, CA, December 1993). New York: ACM Press, 189–202.

Elnozahy, E. N., and W. Zwaenepoel. "Manetho: Transparent Rollback-Recovery with Low Overhead, Limited Rollback, and Fast Output Control." *IEEE Transactions on Computers,* Special Issue on Fault-Tolerant Computing, May 1992.

Engler, D. R., M. F. Kaashoek, and J. O'Toole. "Exokernel: An Operating System Architecture for Application-Level Resource Management." *Proceedings of the Fifteenth Symposium on Operating Systems Principles* (Copper Mountain Resort, CO, December 1995). New York: ACM Press, 251–266

Ezhilhevan, P., R. Macedo, and S. Shrivastava. "Newtop: A Fault-Tolerant Group Communication Protocol." *Proceedings of the Fifteenth International Conference on Distributed Systems* (Vancover, May 1995).

Feeley, M. et al. "Implementing Global Memory Management in a Workstation Cluster." *Proceedings of the Fifteenth ACM SIGOPS Symposium on Operating Systems Principles* (Copper Mountain Resort, CO, December 1995), 201–212.

Felton, E., and J. Zahorjan. "Issues in the Implementation of a Remote Memory Paging System." Technical Report 91-03-09. Department of Computer Science and Engineering, University of Washington, March 1991.

Fidge, C. "Timestamps in Message-Passing Systems That Preserve the Partial Ordering." *Proceedings of the Eleventh Australian Computer Science Conference* (1988).

Fisher, M. J., N. A. Lynch, and M. Merritt. "Easy Impossibility Proofs for Distributed Consensus Problems." *Proceedings of the Fourth Annual ACM Symposium on Principles of Distributed Computing* (Minaki, Canada, August 1985). New York: ACM Press.

Fisher, M.J., N. A. Lynch, and M. S. Paterson. "Impossibility of Distributed Computing with One Faulty Process." *Journal of the ACM* 32:2 (April 1985): 374–382.

Floyd, S., V. Jacobson, S. McCanne, C-G. Liu, and L. Zhang. "A Reliable Multicast Framework for Lightweight Sessions and Application-Level Framing." *Proceedings of the '95 Symposium on Communication Architectures and Protocols* (Cambridge, MA August 1995). New York: ACM Press.

Fortezza Application Developers Guide, Version 3.0. Washington, DC: U.S. Government Printing Office, July 1995. Available at Web site www.armadillo.de.us.

Frank, A., L. Wittie, and A. Bernstein. "Multicast Communication on Network Computers." *IEEE Software* (May 1985).

Frankel, Y. "A Practical Protocol for Large Group-Oriented Networks." *Advances in Cryptology— EUROCRYPT '89*. Springer-Verlag Lecture Notes in Computer Science, vol. 434, 56–61.

Frankel, Y., and Y. Desmedt. "Distributed Reliable Threshold Multisignature." Technical Report TR-92-0402. Department of EECS, University of Wisconsin, Milwaukee, June 1992.

Friedman, R., and K. P. Birman. "Using Group Communication Technology to Implement a Reliable and Scalable Distributed IN Coprocessor." *TINA '96: The Convergence of Telecommunications and Distributed Computing Technologies* (Heidelberg, September 1996), 25–42. Berlin: VDE-Verlag. Also Technical Report. Department of Computer Science, Cornell University, March 1996.

Friedman, R., I. Keider, D. Malkhi, K. P. Birman, and D. Dolev. "Deciding in Partitionable Networks." Technical Report 95-1554. Department of Computer Science, Cornell University, October 1995.

Friedman, R., and R. van Renesse. "Strong and Weak Virtual Synchrony in Horus." Technical Report 95-1537. Department of Computer Science, Cornell University, August 1995.

———. "Packing Messages as a Tool for Boosting the Performance of Total Ordering Protocols." Technical Report 95-1527. Department of Computer Science, Cornell University, July 1995. Submitted to *IEEE Transactions on Networking*.

Garcia-Molina, H., and A. Spauster. "Ordered and Reliable Multicast Communication." *ACM Transactions on Computer Systems* 9:3 (August 1991): 242–271.

Geist, G. A. et al. *PVM: A User's Guide and Tutorial for Networked Parallel Computing*. Cambridge, MA: MIT Press, 1994.

Gharachorloo, K. et al. "Memory Consistency and Event Ordering in Scalable Shared-Memory Multiprocessors." *Proceedings of the Seventeenth Annual International Symposium on Computer Architecture* (Seattle, May 1990), 15–26.

Gibbs, B. W. "Software's Chronic Crisis." *Scientific American*, September 1994.

Gifford, D. "Weighted Voting for Replicated Data." *Proceedings of the Seventh ACM Symposium on Operating Systems Principles* (Pacific Grove, CA, December 1979). New York: ACM Press, 150–162.

Glade, B. B. "A Scalable Architecture for Reliable Publish/Subscribe Communication in Distributed Systems." Ph.D. diss., Department of Computer Science, Cornell University, May 1996.

Glade, B. B., K. P. Birman, R. C. Cooper, and R. van Renesse. "Lightweight Process Groups in the Isis System." *Distributed Systems Engineering Journal*, July 1993.

Gleeson, B. "Fault-Tolerant Computer System with Provision for Handling External Events." U.S. Patent 5,363,503, November 1994.

Golding, R., and K. Taylor. "Group Membership in the Epidemic Style." Technical Report UCSC-CRL-92-13. University of California, Santa Cruz, May 1992.

Golding, R. A. "Weak Consistency Group Communication and Membership." Ph.D. diss., Computer and Information Sciences Department, University of California, Santa Cruz, 1992.

———. "Distributed Epidemic Algorithms for Replicated Tuple Spaces." Technical Report HPL-CSP-91-15. June 1991. Concurrent systems project, Hewlett-Packard Laboratories.

Gong, L. "Securely Replicating Authentication Services." *Proceedings of the Ninth International Conference on Distributed Computing Systems* (August 1989), 85–91.

Gopal, A., R. Strong, S. Toueg, and F. Cristian. "Early-Delivery Atomic Broadcast." *Proceedings of the Ninth ACM Symposium on Principles of Distributed Computing* (Toronto, August 1990). New York: ACM Press, 297–309.

Gosling, J., and H. McGilton. "The Java Language Environment: A White Paper." Sun Microsystems, Inc., October 1995a. Available as http://java.sun.com/langEnv/index.html.

———. "The Java Programmer's Guide: A White Paper." Sun Microsystems, Inc. October 1995b. Available as http://java.sun.com/progGuide/index.html.

Govindran, R., and D. P. Anderson. "Scheduling and IPC Mechanisms for Continuous Media." *Proceedings of the Twelfth ACM Symposium on Operating Systems Principles* (Asilomar, CA, October 1991). New York: ACM Press, 68–80.

Gray, J. "High Availability Computer Systems." *IEEE Computer*, September 1991.

———. "A Census of Tandem System Availability between 1985 and 1990." Technical Report 90.1. Tandem Computer Corporation, September 1990.

————. "Notes on Database Operating Systems. Operating Systems: An Advanced Course." Springer-Verlag Lecture Notes in Computer Science, vol. 60, 1978, 393–481.

Gray, J., J. Bartlett, and R. Horst. "Fault Tolerance in Tandem Computer Systems." *The Evolution of Fault-Tolerant Computing*, A. Avizienis, H. Kopetz, and J. C. Laprie, eds. Springer-Verlag, 1987.

Gray, J., and A. Reuter. *Transaction Processing: Concepts and Techniques.* San Mateo, CA: Morgan Kaufmann, 1993.

Guerraoui, R. "Revisiting the Relationship between Nonblocking Atomic Commitment and Consensus." *International Workshop on Distributed Algorithms* (September 1995), 87–100.

Guerraoui, R., and A. Schiper. "Gamma-Accurate Failure Detectors." Technical Report EPFL. Lausanne, Switzerland: Department d'Informatique, 1996.

Gurwitz, R. F., M. Dean, and R. E. Schantz. "Programming Support in the Chronus Distributed Operating System." *Proceedings of the Sixth International Conference on Distributed Computing Systems* (1986). New York: IEEE Computer Society Press, 486–493.

Hagmann, R. "Reimplementing the Cedar File System Using Logging and Group Commit." *Proceedings of the Eleventh ACM Symposium on Operating Systems Principles* (Austin, November 1987). New York: ACM Press, 155–171.

Handel, R., H. Huber, and S. Schroder. *ATM Networks: Concepts, Protocols, Applications.* Reading, MA: Addison-Wesley, 1994.

Harper, R., and P. Lee. "The Fox Project in 1994." Technical Report CS-94-01. Department of Computer Science, Carnegie Mellon University, 1994.

Hartman, J. H., and J. K. Ousterhout. "The Zebra Striped Network File System." *Proceedings of the Thirteenth ACM Symposium on Operating Systems Principles* (Asheville, NC, December 1993). New York: ACM Press, 29–43.

Hayden, M., and K. P. Birman. "Probabilistic Broadcast." Technical Report TR96-1606. Department of Computer Science, Cornell University, September 1996.

————. "Achieving Critical Reliability with Unreliable Components and Unreliable Glue." Technical Report TR95-1493. Department of Computer Science, Cornell University, March 1995. (This paper was subsequently substantially revised; a new version was released in September 1996.)

Heidemann, J., and G. Popek. "Performance of Cache Coherence in Stackable Filing." *Proceedings of the Fifteenth ACM Symposium on Operating Systems Principles* (Copper Mountain Resort, CO, December 1995), 127–42.

————. "File System Development with Stackable Layers." *Communications of the ACM* 12:1 (February 1994): 58–89.

Heinlein, J., K. Garachorloo, S. Dresser, and A. Gupta. "Integration of Message Passing and Shared Memory in the Stanford FLASH Multiprocessor." *Proceedings of the Sixth International Conference om Architectural Support for Programming Languages and Operating Systems* (October 1994), 38–50.

Herlihy, M. "Replication Methods for Abstract Data Types." Ph.D. diss., Massachusetts Institute of Technology, May 1984. Available as Technical Report LCS-84-319.

Herlihy, M., and J. Wing. "Linearizability: A Correctness Condition for Concurrent Objects." *ACM Transactions on Programming Languages and Systems* 12:3 (July 1990): 463–492.

Herlihy, M. P., and J. D. Tygar. "How to Make Replicated Data Secure." *Advances in Cryptography—CRYPTO '87 Proceedings.* Springer-Verlag Lecture Notes in Computer Science, vol. 293, 379–391.

Hildebrand, D. "An Architectural Overview of QNX." *Proceedings of the First USENIX Workshop on Microkernels and Other Kernel Architectures* (Seattle, April 1992), 113–126.

Howard, J. et al. "Scale and Performance in a Distributed File System." *Proceedings of the Eleventh ACM Symposium on Operating Systems Principles* (Austin, November 1987). New York: ACM Press. Also *ACM Transactions on Computing Systems* 5:1 (February 1988).

Hunt, G. D. "Multicast Flow Control on Local Area Networks." Ph.D. diss., Department of Computer Science, Cornell University, February 1995. Also available as Technical Report TR-95-1479.

Internet Engineering Task Force. *Secure Sockets Layer, Version 3.0.* 1995.

Iona Ltd. "Information about Object Transaction Services for Orbix." 1995. Available at info@iona.ie.

Iona Ltd. and Isis Distributed Systems, Inc. "An Introduction to Orbix+Isis." 1995. Available at info@iona.ie.

Jacobson, V. "Compressing TCP/IP Headers for Low-Speed Serial Links." Technical Report RFC 114. Network Working Group, February 1990.

———. "Congestion Avoidance and Control." *Proceedings of the ACM SIGCOMM '88* (Palo Alto, 1988).

Jalote, P. *Fault Tolerance in Distributed Systems.* Englewood Cliffs, NJ: Prentice Hall, 1994.

Johansen, D. "StormCast: Yet Another Exercise in Distributed Computing." In *Distributed Open Systems in Perspective*, D. Johansen and Brazier, eds. New York: IEEE Computer Society Press, 1994.

Johansen, D., and G. Hartvigsen. "Architecture Issues in the StormCast System." Springer-Verlag Lecture Notes in Computer Science, vol. LNCS 938, 1–16.

Johansen, D., R. van Renesse, and F. Schneider. "Supporting Broad Internet Access to TACOMA." Technical Report. February 1996.

———. "An Introduction to the TACOMA Distributed System (Version 1.0)." Computer Science Technical Report 95-23. University of Tromsö, June 1995.

———. "Operating System Support for Mobile Agents." *Proceedings of the Fifth Workshop on Hot Topics in Operating Systems* (Orcas Island, WA, May 1995). New York: IEEE Computer Society Press, 42–45.

Johnson, D. B., and W. Zwaenepoel. "Sender-Based Message Logging." *Proceedings of the Seventeenth Annual International Symposium on Fault-Tolerant Computing* (June 1987). New York: IEEE Computer Society Press, 14–19.

Johnson, K., M. F. Kaashoek, and D. Wallach. "CRL: High-Performance All Software Distributed Shared Memory." *Proceedings of the Fifteenth ACM Symposium on Operating Systems Principles* (Copper Mountain Resort, CO, December 1995), 213–228.

Jones, M. B. "Interposition Agents: Transparent Interposing User Code at the System Interface." *Proceedings of the Fourteenth ACM Symposium on Operating Systems Principles* (Asheville, NC, December 1993). New York: ACM Press, 80–93.

Joseph, T. A. "Low Cost Management of Replicated Data." Ph.D. diss., Cornell University, 1986. Also Technical Report. Department. of Computer Science, Cornell University.

Joseph, T. A., and K. P. Birman. "Low Cost Management of Replicated Data in Fault-Tolerant Distributed Systems." *ACM Transactions on Computer Systems* 4:1 (February 1986): 54–70.

Kaashoek, F. "Group Communication in Distributed Computer Systems." Ph.D. diss., Vrije Universiteit, 1992.

Kaashoek, M. F., and A. S. Tannenbaum. "Group Communication in the Amoeba Distributed Operating System." *Proceedings of the Eleventh International Conference on Distributed Computing Systems.* New York: IEEE Computer Society Press, 222–230.

Kaashoek, M. F. et al. "An Efficient Reliable Broadcast Protocol." *Operating Systems Review* 23:4 (July 1978): 5–19.

Kalantar, M. "Issues in Ordered Multicast Performance: A Simulation Study." Ph.D. diss., Department of Computer Science, Cornell University, August 1995. Also Technical Report TR-95-1531.

Karamcheti, V., and A. A. Chien. "Software Overhead in Messaging Layers: Where Does the Time Go?" *Proceedings of the Sixth ACM Symposium on Principles of Programming Languages and Operating Systems* (San Jose, CA, October 1994). New York: ACM Press.

Kay, J., and J. Pasquale. "The Importance of Nondata Touching Processing Overheads." *Proceedings of SIGCOMM-93* (August 1993), 259–269.

Kay, J. S. "PathIDs: A Mechanism for Reducing Network Software Latency." Ph.D. diss., University of California, San Diego, May 1994.

Keidar, I., and D. Dolev. "Increasing the Resilience of Atomic Commit at No Additional Cost." *Proceedings of the 1995 ACM Symposium on Principles of Database Systems* (May 1995), 245–254.

Keleher, P., A. L. Cox, and W. Zwaenepoel. "Lazy Release Consistency for Software Distributed Shared Memory." *Proceedings of the Ninteenth Annual International Symposium on Computer Architecture* (May 1992), 13–21.

Khalidi, Y. A. et al. "Solaris MC: A Multicomputer OS." Technical Report 95-48. Sun Microsystems Laboratories, November 1995.

Kistler, J. J., and M. Satyanarayanan. "Disconnected Operation in the Coda File System." *Proceedings of the Twelfth ACM Symposium on Operating Systems Principles* (Asilomar, CA, October 1991). New York: ACM Press, 213–225. Also *ACM Transactions on Computing Systems* 10:1 (February 1992): 3–25.

Koo, R., and S. Toueg. "Checkpointing and Rollback Recovery for Distributed Systems." *IEEE Transactions on Software Engineering* SE-13:1 (January 1990): 23–31.

Kopetz, H. "Sparse Time versus Dense Time in Distributed Systems." *Proceedings of the Twelfth International Conference on Distributed Computing Systems* (Yokohama, June 1992). New York: IEEE Computer Society Press.

Kopetz, H., and W. Ochsenreiter. "Clock Synchronization in Distributed Real-Time Systems." *IEEE Transactions on Computers* C36:8 (August 1987): 933–940.

Kopetz, H., and P. Verissimo. "Real-Time Dependability Concepts." In *Distributed Systems,* 2d ed., S. J. Mullender, ed. Reading, MA: Addison-Wesley/ACM Press, 1993, 411–446.

Kronenberg, N., H. Levy, and W. Strecker. "VAXClusters: A Closely-Coupled Distributed System." *Proceedings of the Tenth ACM Symposium on Operating Systems Principles* (Orcas Island, WA, December 1985). Also *ACM Transactions on Computer Systems* 4:2 (May 1986): 130–146.

Krumvieda, C. "Expressing Fault-Tolerant and Consistency Preserving Programs in Distributed ML." *Proceedings of the ACM SIGPLAN Workshop on ML and its Applications* (June 1992), 157–162.

———. "DML: Packaging High-Level Distributed Abstractions in SML." *Proceedings of the Third International Workshop on Standard ML* (Pittsburgh, September 1991). New York: IEEE Computer Society Press.

Ladin, R., B. Liskov, L. Shrira, and S. Ghemawat. "Providing Availability Using Lazy Replication." *ACM Transactions on Computer Systems* 10:4 (November 1992): 360–391.

———. "Lazy Replication: Exploiting the Semantics of Distributed Services." *Proceedings of the Tenth ACM Symposium on Principles of Distributed Computing* (Quebec, August 1990). New York: ACM Press, 43–58.

Laih, C. S., and L. Harn. "Generalized Threshold Cryptosystems." *Proceedings of ASIACRYPT '91* (1991).

Lamport, L. "Using Time Instead of Timeout for Fault-Tolerant Distributed Systems." *ACM Transactions on Programming Languages and Systems* 6:2 (April 1984): 254–280.

———. "Time, Clocks, and the Ordering of Events in a Distributed System." *Communications of the ACM* 21:7 (July 1978): 558–565.

———. "The Implementation of Reliable Distributed Multiprocess Systems." *Computing Networks* 2 (March 1978): 95–114.

Lamport, L., and P. M. Melliar-Smith. "Synchronizing Clocks in the Presence of Faults." *Journal of the ACM* 32:1 (January 1985): 52–78.

Lampson, B. "Hints for Computer System Design." *Proceedings of the Ninth Symposium on Operating Systems Principles* (Bretton Woods, NH, October 1993), 33–48.

———. "Designing a Global Name Service." Paper presented at the 1985 ACM PODC. Also *Proceedings of the Sixth ACM Symposium on Principles of Distributed Computing* (Calgary, 1986), 1–10.

Lampson, B., M. Abadi, M. Burrows, and E. Wobber. "Authentication in Distributed Systems: Theory and Practice." *ACM Transactions on Computer Systems* 10:4 (November 1992): 265–434.

Leffler, S. J. et al. *4.3BSD UNIX Operating System.* Reading, MA: Addison-Wesley, 1989.

Lenoski, D. et al. "The Stanford DASH Multiprocessor." *Computer* 25:3 (March 1992): 63–79.

Leroy, X. *The Caml Light System, Release 0.7*. France: INRIA July 1993.

Li, K., and P. Hudak. "Memory Coherence in a Shared Virtual Memory System." *ACM Transactions on Computer Systems* 7:4 (November 1989): 321–359.

Liskov, B. "Practical Uses of Synchronized Clocks in Distributed Systems." *Distributed Computing* 6:4 (November 1993): 211–219.

Liskov, B., D. Curtis, P. Johnson, and R. Scheifler. "Implementation of Argus." *Proceedings of the Eleventh ACM Symposium on Operating Systems Principles* (Austin, November 1987). New York: ACM Press, 111–122.

Liskov, B., and R. Ladin. "Highly Available Distributed Services and Fault-Tolerant Garbage Collection." *Proceedings of the Fifth ACM Symposium on Principles of Distributed Computing* (Calgary, August 1986). New York: ACM Press, 29–39.

Liskov, B., and R. Scheifler. "Guardians and Actions: Linguist Support for Robust, Distributed Programs." *ACM Transactions on Programming Languages and Systems* 5:3 (July 1983): 381–404.

Liskov, B. et al. "Replication in the Harp File System." *Proceedings of the Twelfth ACM Symposium on Operating Systems Principles* (Asilomar, CA, October 1991). New York: ACM Press, 226–238.

Lynch, N. *Distributed Algorithms*. San Mateo, CA: Morgan Kaufmann, 1996.

Lyu, M. R., ed. *Software Fault Tolerance*. New York: John Wiley & Sons, 1995.

Macedo, R. A., P. Ezhilchlvan, and S. Shrivastava. "Newtop: A Total Order Multicast Protocol Using Causal Blocks." BROADCAST Project Technical Reports, vol. I. Department of Computer Science, University of Newcastle upon Tyne, October 1993.

Maffeis, S. "Adding Group Communication and Fault Tolerance to CORBA." *Proceedings of the 1995 USENIX Conference on Object-Oriented Technologies*. (Monterey, CA, June 1995).

Makpangou, M., and K. P. Birman. "Designing Application Software in Wide Area Network Settings." Technical Report 90-1165. Department of Computer Science, Cornell University, 1990.

Malkhi, D. "Multicast Communication for High Availability." Ph.D. diss., Hebrew University of Jerusalem, 1994.

Malkhi, D., K. P. Birman, A. Ricciardi, and A. Schiper. "Uniform Actions in Asynchronous Distributed Systems." Technical Report TR 94-1447. Department of Computer Science, Cornell University, September 1994.

Malloth, C. "Conception and Implementation of a Toolkit for Building Fault-Tolerant Distributed Applications in Large-Scale Networks." Ph.D. diss., Swiss Federal Institute of Technology, Lausanne (EPFL), 1996.

Malloth, C. P., P. Felher, A. Schiper, and U. Wilhelm. "Phoenix: A Toolkit for Building Fault-Tolerant Distributed Applications in Large-Scale Networks." *Proceedings of the Workshop on Parallel and Distributed Platforms in Industrial Products* (held during the Seventh IEEE Symposium on Parallel and Distributed Processing) (San Antonio, TX, October 1995). New York: IEEE Computer Society Press.

Marzullo, K. "Tolerating Failures of Continuous Valued Sensors." *ACM Transactions on Computer Systems* 8:4 (November 1990): 284–304.

———. "Maintaining the Time in a Distributed System." Ph.D. diss., Department of Electrical Engineering, Stanford University, June 1984.

Marzullo, K., R. Cooper, M. Wood, and K. P. Birman. "Tools for Distributed Application Management." *IEEE Computer*, August 1991.

Marzullo, K., and M. Wood. "Tools for Constructing Distributed Reactive Systems." Technical Report TR91-1193. Department. of Computer Science, Cornell University, February 1991.

Marzullo, K., M. Wood, K. P. Birman, and R. Cooper. "Tools for Monitoring and Controlling Distributed Applications." *Spring 1991 Conference Proceedings* (Bologna, Italy, May 1991). EurOpen, 185–196. Revised and extended as *IEEE Computer* 24:8 (August 1991): 42–51.

Mattern, F. "Time and Global States in Distributed Systems." *Proceedings of the International Workshop on Parallel and Distributed Algorithms*. Amsterdam: North-Holland, 1989.

Meldal, S., S. Sankar, and J. Vera. "Exploiting Locality in Maintaining Potential Causality." *Proceedings of the Tenth Symposium on Principles of Distributed Computing* (Montreal, August 1991), 231–239.

Melliar-Smith, P. M., and L.E. Moser. "Trans: A Reliable Broadcast Protocol." *IEEE Transactions on Communications* 140:6 (December 1993): 481–493.

———. "Fault-Tolerant Distributed Systems Based on Broadcast Communication." *Proceedings of the Ninth International Conference on Distributed Computing Systems* (June 1989), 129–133.

Melliar-Smith, P. M., L. E. Moser, and V. Agrawala. "Membership Algorithms for Asynchronous Distributed Systems." *Proceedings of the IEEE Eleventh ICDCS* (May 1991), 480–488.

———. "Broadcast Protocols for Distributed Systems." *IEEE Transactions on Parallel and Distributed Systems* 1:1 (January 1990): 17–25.

Milner, R., M. Tofte, and R. Harper. *The Definition of Standard ML*. Cambridge, MA: MIT Press, 1990.

Mishra, S., L. L. Peterson, and R. D. Schlichting. "Experience with Modularity in Consul." *Software—Practice and Experience* 23:10 (October 1993): 1050–1075.

———. "A Membership Protocol Based on Partial Order." *Proceedings of the IEEE International Working Conference on Dependable Computing for Critical Applications* (February 1991), 137–145.

Mogul, J., R. Rashid, and M. Accetta. "The Packet Filter: An Efficient Mechanism for User-Level Network Code." *Proceedings of the Eleventh ACM Symposium on Operating Systems Principles* (Austin, November 1987). New York: ACM Press, 39–51.

Montgomery, T. "Design, Implementation, and Verification of the Reliable Multicast Protocol." Master's thesis, Department of Electrical and Computer Engineering, West Virginia University, December 1994.

Montgomery, T., and B. Whetten. "The Reliable Multicast Protocol Application Programming Interface." Technical Report NASA-IVV-94-007. NASA/WVU Software Research Laboratory, August 1994.

Moser, L. E., Y. Amir, P. M. Melliar-Smith, and D. A. Agarwal. "Extended Virtual Synchrony." *Proceedings of the Fourteenth International Conference on Distributed Computing Systems* (June 1994). New York: IEEE Computer Society Press, 56–65. Also Technical Report TR-93-22. Department of ECE, University of California, Santa Barbara, December 1993.

Moser, L. E., P. M. Melliar-Smith, D. A. Argarwal, R. K. Budhia, and C. A. Lingley-Papadopoulos. "Totem: A Fault-Tolerant Multicast Group Communication System." *Communications of the ACM* 39:4 (April 1996): 54–63.

Moser, L. E., P. M. Melliar-Smith, and U. Agarwal. "Processor Membership in Asynchronous Distributed Systems." *IEEE Transactions on Parallel and Distributed Systems* 5:5 (May 1994): 459–473.

Moss, J. E. "Nested Transactions and Reliable Distributed Computing." *Proceedings of the Second Symposium on Reliability in Distributed Software and Database Systems* (1982), 33–39.

Mullender, S. J. et al. "Amoeba—A Distributed Operating System for the 1990s." *IEEE Computer* 23:5 (May 1990): 44–53.

Mummert, L. B., M. R. Ebling, and M. Satyanarayanan. "Exploiting Weak Connectivity for Mobile File Access." *Proceedings of the Fifteenth Symposium on Operating Systems Principles* (Copper Mountain Resort, CO, December 1995). New York: ACM Press, 143–155. Also *ACM Transactions on Computing Systems* 13:1 (February 1996).

National Bureau of Standards. *Data Encryption Standard*. Federal Information Processing Standards Publication 46. Washington, DC: U.S. Government Printing Office, 1977.

Needham, R. M., and M.D. Schroeder. "Using Encryption for Authentication in Large Networks of Computers." *Communications of the ACM* 21:12 (December 1988): 993–999.

Neiger, G. "A New Look at Membership Services." *Proceedings of the Fifteenth ACM Symposium on Principles of Distributed Computing* (Vancouver, 1996). In press.

Nelson, M., B. Welsh, and J. Ousterhout. "Caching in the Sprite Network File System." *Proceedings of the Eleventh ACM Symposium on Operating Systems Principles* (Austin, November 1987). New York: ACM Press. Also *ACM Transactions on Computing Systems* 6:1 (February 1988).

Object Management Group and X/Open. "Common Object Request Broker: Architecture and Specification." Reference OMG 91.12.1, 1991.

Oki, B., M. Pfluegl, A. Siegel, and D. Skeen. "The Information Bus—An Architecture for Extensible Distributed Systems." *Proceedings of the Thirteenth ACM Symposium on Operating Systems Principles* (Asheville, NC, December 1993). New York: ACM Press, 58–68.

Open Software Foundation. *Introduction to OSF DCE*. Englewood Cliffs, NJ: Prentice Hall, 1994.

Ousterhout, J. *TCL and the TK Toolkit*. Reading, MA: Addison-Wesley, 1994.

———. "Why Aren't Operating Systems Getting Faster as Fast as Hardware?" *USENIX Summer Conference Proceedings* (Anaheim, CA, 1990), 247–256.

Ousterhout, J. et al. "The Sprite Network Operating System." *Computer* 21:2 (February 1988): 23–36.

———. "A Trace-Driven Analysis of the UNIX 4.2 BSD File System." *Proceedings of the Tenth ACM Symposium on Operating Systems Principles* (Orcas Island, WA, December 1985). New York: ACM Press, 15–24.

Partridge, C., and S. Pink. "A Faster UDP." *IEEE/ACM Transactions on Networking* 1:4 (August 1993): 429–440.

Patterson, D., G. Gibson, and R. Katz. "A Case for Redundant Arrays of Inexpensive Disks (RAID)." *Proceedings of the 1988 ACM Conference on Management of Data (SIGMOD)* (Chicago, June 1988), 109–116.

Peterson, I. *Fatal Defect: Chasing Killer Computer Bugs.* New York: Time Books/Random House, 1995.

Peterson, L. "Preserving Context Information in an IPC Abstraction." *Proceedings of the Sixth Symposium on Reliability in Distributed Software and Database Systems* (March 1987). New York: IEEE Computer Society Press, 22–31.

Peterson, L., N. C. Buchholz, and R. D. Schlicting. "Preserving and Using Context Information in Interprocess Communication." *ACM Transactions on Computing Systems* 7:3 (August 1989): 217–246.

Peterson, L., N. Hutchinson, S. O'Malley, and M. Abbott. "RPC in the *x*-Kernel: Evaluating New Design Techniques." *Proceedings of the Twelfth Symposium on Operating Systems Principles* (Litchfield Park, AZ, November 1989). New York: ACM Press, 91–101.

Pfister, G. F. *In Search of Clusters.* Englewood Cliffs, NJ: Prentice Hall, 1995.

Pittel, B. "On Spreading of a Rumor." *SIAM Journal of Applied* Mathematics 47:1 (1987): 213–223.

Powell, D. "Introduction to Special Section on Group Communication." *Communications of the ACM* 39:4 (April 1996): 50–53.

———. "Lessons Learned from Delta-4." *IEEE Micro* 14:4 (February 1994): 36–47.

———, ed. *Delta-4: A Generic Architecture for Dependable Distributed Computing.* Springer-Verlag ESPRIT Research Reports, vol. I, Project 818/2252, 1991.

Pradhan, D. *Fault-Tolerant Computer System Design.* Englewood Cliffs, NJ: Prentice Hall, 1996.

Pradhan, D., and D. Avresky, eds. *Fault-Tolerant Parallel and Distributed Systems.* New York: IEEE Computer Society Press, 1995.

Pu, D. "Relaxing the Limitations of Serializable Transactions in Distributed Systems." *Operating Systems Review* 27:2 (April 1993): 66–71. (Special issue on the Workshop on Operating Systems Principles at Le Mont St. Michel, France.)

Rabin, M. "Randomized Byzantine Generals." *Proceedings of the Twenty-Fourth Annual Symposium on Foundations of Computer Science* (1983). New York: IEEE Computer Society Press, 403–409.

Rangan, P. V., and H. M. Vin. "Designing File Systems for Digital Video and Audio." *Proceedings of the Twelfth ACM Symposium on Operating Systems Principles* (Asilomar, CA, October 1991). New York: ACM Press, 81–94.

Rashid, R. F. "Threads of a New System." *UNIX Review* 4 (August 1986): 37–49.

Reed, D. P., and R. K. Kanodia. "Synchronization with Eventcounts and Sequencers." *Communications of the ACM* 22:2 (February 1979): 115–123.

Reiher, P. et al. "Resolving File Conflicts in the Ficus File System." *Proceedings of the Summer USENIX Conference* (June 1994), 183–195.

Reiter, M. K. "Distributing Trust with the Rampart Toolkit." *Communications of the ACM* 39:4 (April 1996): 71-75.

———. "Secure Agreement Protocols: Reliable and Atomic Group Multicast in Rampart." *Proceedings of the Second ACM Conference on Computer and Communications Security* (Oakland, November 1994), 68–80.

———. "A Secure Group Membership Protocol." *Proceedings of the 1994 Symposium on Research in Security and Privacy* (Oakland, May 1994). New York: IEEE Computer Society Press, 89–99.

———. "A Security Architecture for Fault-Tolerant Systems." Ph.D. diss., Cornell University, August 1993. Also Technical Report. Department of Computer Science, Cornell University.

Reiter, M. K., and K. P. Birman. "How to Securely Replicate Services." *ACM Transactions on Programming Languages and Systems* 16:3 (May 1994): 986–1009.

Reiter, M. K., K. P. Birman, and L. Gong. "Integrating Security in a Group-Oriented Distributed System." *Proceedings of the IEEE Symposium on Research in Security and Privacy* (Oakland, May 1992). New York: IEEE Computer Society Press, 18–32.

Reiter, M. K., K. P. Birman, and R. van Renesse. "A Security Architecture for Fault-Tolerant Systems." *ACM Transactions on Computing Systems*, May 1995.

Ricciardi, A. "The Impossibility of (Repeated) Reliable Broadcast." Technical Report TR-PDS-1996-003. Department of Electrical and Computer Engineering, University of Texas, Austin, April 1996.

Ricciardi, A., and K. P. Birman. "Using Process Groups to Implement Failure Detection in Asynchronous Environments." *Proceedings of the Eleventh ACM Symposium on Principles of Distributed Computing* (Quebec, August 1991). New York: ACM Press, 341–351.

Ricciardi, A.M. "The Group Membership Problem in Asynchronous Systems." Ph.D. diss., Cornell University, January 1993.

Riecken, D. "Intelligent Agents." *Communications of the ACM* 37:7 (July 1994): 19–21.

Ritchie, D. M. "A Stream Input-Output System." *Bell Laboratories Technical Journal, AT&T* 63:8 (1984): 1897–1910.

Rivest, R. L., A. Shamir, and L. Adleman. "A Method for Obtaining Digital Signatures and Public Key Cryptosystems." *Communications of the ACM* 22:4 (December 1978): 120–126.

Rodrigues, L., and P. Verissimo. "Causal Separators for Large-Scale Multicast Communication." *Proceedings of the Fifteenth International Conference on Distributed Computing Systems* (May 1995), 83–91.

———. "xAMP: A MultiPrimitive Group Communications Service." *Proceedings of the Eleventh Symposium on Reliable Distributed Systems* (Houston, October 1989). New York: IEEE Computer Society Press.

Rodrigues, L., P. Verissimo, and J. Rufino. "A Low-Level Processor Group Membership Protocol for LANs." *Proceedings of the Thirteenth International Conference on Distributed Computing Systems* (May 1993), 541–550.

Rosenblum, M., and J. K. Ousterhout. "The Design and Implementation of a Log-Structured File System." *Proceedings of the Twelfth ACM Symposium on Operating Systems Principles* (Asilomar, CA October 1991). New York: ACM Press, 1–15. Also *ACM Transactions on Computing Systems* 10:1 (February 1992): 26–52.

Rowe, L. A., and B. C. Smith. "A Continuous Media Player." *Proceedings of the Third International Workshop on Network and Operating Systems Support for Digital Audio and Video* (San Diego, CA, November 1992).

Rozier, M. et al. "Chorus Distributed Operating System." *Computing Systems Journal* 1:4 (December 1988): 305–370.

———. "The Chorus Distributed System." *Computer Systems*, Fall 1988: 299–328.

Sabel, L., and K. Marzullo. "Simulating Fail-Stop in Asynchronous Distributed Systems." *Proceedings of the Thirteenth Symposium on Reliable Distributed Systems* (Dana Point, CA, October 1994). New York: IEEE Computer Society Press, 138–147.

Saltzer, J. H., D. P. Reed, and D. D. Clark. "End-to-End Arguments in System Design." *ACM Transactions on Computer Systems* 39:4 (April 1990).

Satyanarayanan, M. et al. "Integrating Security in a Large Distributed System." *ACM Transactions on Computer Systems* 7:3 (August 1989): 247–280.

———. "The ITC Distributed File System: Principles and Design." *Proceedings of the Tenth ACM Symposium on Operating Systems Principles* (Orcas Island, WA, December 1985). New York: ACM Press, 35–50.

Schantz, R. E., R. H. Thomas, and G. Bono. "The Architecture of the Chronus Distributed Operating System." *Proceedings of the Sixth International Conference on Distributed Computing Systems* (New York, June 1986). New York: IEEE Computer Society Press, 250–259.

Schiller, J. I. "Secure Distributed Computing." *Scientific American* (November 1994): 72–76.

Schiper, A., J. Eggli, and A. Sandoz. "A New Algorithm to Implement Causal Ordering." *Proceedings of the Third International Workshop on Distributed Algorithms* (1989). Springer-Verlag Lecture Notes in Computer Science, vol. 392, 219–232.

Schiper, A., and M. Raynal. "From Group Communication to Transactions in Distributed Systems." *Communications of the ACM* 39:4 (April 1996): 84–87.

Schiper, A., and A. Sandoz. "Uniform Reliable Multicast in a Virtually Synchronous Environment." *Proceedings of the Thirteenth International Conference on Distributed Computing Systems* (May 1993). New York: IEEE Computer Society Press, 561–568.

Schlicting, R. D., and F. B. Schneider. "Fail-Stop Processors: An Approach to Designing Fault-Tolerant Computing Systems." *ACM Transactions on Computer Systems* 1:3 (August 1983): 222–238.

Schmuck, F. "The Use of Efficient Broadcast Primitives in Asynchronous Distributed Systems." Ph.D. diss., Cornell University, August 1988. Also Technical Report. Department of Computer Science, Cornell University.

Schmuck, F., and J. Wyllie. "Experience with Transactions in QuickSilver." *Proceedings of the Twelfth ACM Symposium on Operating Systems Principles* (Asilomar, CA, October 1991). New York: ACM Press, 239–252.

Schneider, F. B. *On Concurrent Programming*. New York: Springer-Verlag, in press.

―――. "Implementing Fault-Tolerant Services Using the StateMachine Approach." *ACM Computing Surveys* 22:4 (December 1990): 299–319.

―――. "The StateMachine Approach: A Tutorial." *Proceedings of the Workshop on Fault-Tolerant Distributed Computing* (Asilomar, CA, 1988). Springer-Verlag Lecture Notes on Computer Science, vol. 448, 18–41.

―――. "Byzantine Generals in Action: Implementing Fail-Stop Processors." *ACM Transactions on Computer Systems* 2:2 (May 1984): 145–154.

―――. "Synchronization in Distributed Programs." *ACM Transactions on Programming Languages and Systems* 4:2 (April 1982): 179–195.

Schneider, F. B., D. Gries, and R. D. Schlicting. "Fault-Tolerant Broadcasts." *Science of Computer Programming* 3:2 (March 1984): 1–15.

Schwarz, R., and F. Mattern. "Detecting Causal Relationships in Distributed Computations." Technical Report 215-91. Department of Computer Science, University of Kaiserslautern, 1991.

Seltzer, M. "Transaction Support in a Log-Structured File System." *Proceedings of the Ninth International Conference on Data Engineering* (April 1993).

Shroeder, M., and M. Burrows. "Performance of Firefly RPC." *Proceedings of the Eleventh ACM Symposium on Operating Systems Principles* (Litchfield Springs, AZ, December 1989), 83–90. Also *ACM Transactions on Computing Systems* 8:1 (February 1990): 1–17.

Siegal, A. "Performance in Flexible Distributed File Systems." Ph.D. diss., Cornell University, February 1992. Also Technical Report TR-92-1266. Department of Computer Science, Cornell University.

Siegel, A., K. P. Birman, and K. Marzullo. "Deceit: A Flexible Distributed File System." Technical Report 89-1042. Department of Computer Science, Cornell University, 1989.

Simons, B., J. N. Welch, and N. Lynch. "An Overview of Clock Synchronization." In *Fault-Tolerant Distributed Computing,* (B. Simons and A. Spector, eds), Springer-Verlag Lecture Notes in Computer Science, vol. 448, 1990, 84–96.

Skeen, D. "Determining the Last Process to Fail." *ACM Transactions on Computer Systems* 3:1 (February 1985): 15–30.

―――. "Crash Recovery in a Distributed Database System." Ph.D. diss., Department of EECS, University of California, Berkeley, June 1982.

―――. "A Quorum-Based Commit Protocol." *Proceedings of the Berkeley Workshop on Distributed Data Management and Computer Networks* (Berkeley, CA, February 1982), 69–80.

Spasojevic, M., and M. Satyanarayanan. "An EmpIrical Study of a Wide Area Distributed File System." *ACM Transactions on Computer Systems* 14:2 (May 1996).

Spector, A. "Distributed Transactions for Reliable Systems." *Proceedings of the Tenth ACM Symposium on Operating Systems Principles* (Orcas Island, WA, December 1985), 12–146.

Srikanth, T. K., and S. Toueg. "Optimal Clock Synchronization." *Journal of the ACM* 34:3 (July 1987): 626–645.

Srinivasan, V., and J. Mogul. "Spritely NFS: Experiments with Cache Consistency Protocols." *Proceedings of the Eleventh ACM Symposium on Operating Systems Principles* (Litchfield Springs, AZ; December 1989), 45–57.

Steiner, J. G., B. C. Neuman, and J. I. Schiller. "Kerberos: An Authentication Service for Open Network Systems." *Proceedings of the 1988 USENIX Winter Conference* (Dallas, February 1988), 191–202.

Stephenson, P. "Fast Causal Multicast." Ph.D. diss., Cornell University, February 1991. Also Technical Report. Department of Computer Science, Cornell University.

Strayer, W. T., R. J. Dempsey, and A. C. Weaver. *XTP: The Xpress Transfer Protocol*. Reading, MA: Addison-Wesley, 1992.

Strayer, W. T., G. Simon, and R. E. Cline, Jr. "An Object-Oriented Implementation of the Xpress Transfer Protocol." XTP Forum Research Affiliate Annual Report, 1994, 53–66.

Tanenbaum, A. *Computer Networks,* 2d ed. Englewood Cliffs, NJ: Prentice Hall, 1988.

Tanenbaum, A., and R. van Renesse. "A Critique of the Remote Procedure Call Paradigm." *Proceedings of the EUTECO '88 Conference* (Vienna, April 1988), 775–783.

Telecommunications Information Network Architecture Conference, Proceedings of (Heidelberg, September 3–5, 1996). Berlin: VDE-Verlag.

Tennenhouse, D. "Layered Multiplexing Considered Harmful." In *Protocols for High Speed Networks*. Elsevier, 1990.

Terry, D. B. et al. "Managing Update Conflicts in a Weakly Connected Replicated Storage System." *Proceedings of the Fifteenth Symposium on Operating Systems Principles* (Copper Mountain Resort, CO, December 1995). New York: ACM Press, 172–183.

Thekkath, C. A., and H. M. Levy. "Limits to Low-Latency Communication on High-Speed Networks." *ACM Transactions on Computer Systems* 11:2 (May 1993): 179–203.

Thekkath, C. A., T. Nguyen, E. Moy, and E. Lazowska. "Implementing Network Protocols at User Level." *IEEE Transactions on Networking* 1:5 (October 1993): 554–564.

Thomas, T. "A Majority Consensus Approach to Concurrency Control for Multiple Copy Databases." *ACM Transactions on Database Systems* 4:2 (June 1979): 180–209.

Torrellas, J., and J. Hennessey. "Estimating the Performance Advantages of Relaxing Consistency in a Shared-Memory Multiprocessor." Technical Report CSL-TN-90-265. Stanford University Computer Systems Laboratory, February 1990.

Turek, J., and D. Shasha. "The Many Faces of Consensus in Distributed Systems." *IEEE Computer* 25:6 (1992): 8–17.

van Renesse, R. "Why Bother with CATOCS?" *Operating Systems Review* 28:1 (January 1994): 22–27.

———. "Causal Controversy at Le Mont St.-Michel." *Operating Systems Review* 27:2 (April 1993): 44–53.

van Renesse, R., K. P. Birman, R. Cooper, B. Glade, and P. Stephenson. "Reliable Multicast between Microkernels." *Proceedings of the USENIX Workshop on Microkernels and Other Kernel Architectures* (Seattle, April 1992).

van Renesse, R., K. P. Birman, R. Friedman, M. Hayden, and D. Karr. "A Framework for Protocol Composition in Horus." *Proceedings of the Fourteenth Symposium on the Principles of Distributed Computing* (Ottawa, August 1995). New York: ACM Press, 80–89.

van Renesse, R., K. P. Birman, and S. Maffeis. "Horus: A Flexible Group Communication System." *Communications of the ACM* 39:4 (April 1996): 76–83.

van Renesse, R., H. van Staveren, and A. Tanenbaum. "The Performance of the Amoeba Distributed Operating System." *Software—Practice and Experience* 19:3 (March 1989): 223–234.

———. "Performance of the World's Fastest Operating System." *Operating Systems Review* 22:4 (October 1988): 25–34.

Verissimo, P. "Causal Delivery in Real-Time Systems: A Generic Model." *Real-Time Systems Journal* 10:1 (January 1996).

———. "Ordering and Timeliness Requirements of Dependable Real-Time Programs." *Journal of Real-Time Systems* 7:2 (September 1994): 105–128.

———. "Real-Time Communication." In *Distributed Systems,* 2d ed., 1993, S. J. Mullender, ed. Reading, MA: Addison-Wesley/ACM Press, 1993, 447–490.

Verissimo, P., and L. Rodrigues. "A-Posteriori Agreement for Fault-Tolerant Clock Synchronization on Broadcast Networks." *Proceedings of the Twenty-Second International Symposium on Fault-Tolerant Computing* (Boston, July 1992).

Vogels, W. "The Private Investigator." Technical Report. Department of Computer Science, Cornell University, April 1996.

von Eicken, T., A. Basu, V. Buch, and W. Vogels. "U-Net: A User-Level Network Interface for Parallel and Distributed Computing." *Proceedings of the Fifteenth Symposium on Operating Systems Principles* (Copper Mountain Resort, CO, December 1995). New York: ACM Press, 40–53.

von Eicken, T., D. E. Culler, S. C. Goldstein, and K. E. Schauser. "Active Messages: A Mechanism for Integrated Communication and Computation." *Proceedings of the Nineteenth International Symposium on Computer Architecture* (May 1992), 256–266.

Voydock, V. L., and S. T. Kent. "Security Mechanisms in High-Level Network Protocols." *ACM Computing Surveys* 15:2 (June 1983): 135–171.

Wahbe, R., S. Lucco, T. Anderson, and S. Graham. "Efficient Software-Based Fault Isolation." *Proceedings of the Thirteenth ACM Symposium on Operating Systems Principles* (Asheville, NC, December 1993). New York: ACM Press, 203–216.

Walter, B. et al. "The Locus Distributed Operating System." *Proceedings of the Ninth ACM Symposium on Operating Systems Principles* (Bretton Woods, NH, October 1993), 49–70.

Whetten, B. "A Reliable Multicast Protocol." In *Theory and Practice in Distributed Systems*, K. Birman, F. Mattern, and A. Schiper, eds. Springer-Verlag Lecture Notes on Computer Science, vol. 938, July 1995.

Wilkes, J. et al. "The HP AutoRAID Hierarchical Storage System." *Proceedings of the Fifteenth Symposium on Operating Systems Principles* (Copper Mountain Resort, CO, December 1995). New York: ACM Press, 96–108. Also *ACM Transactions on Computing Systems* 13:1 (February 1996).

Wong, T. Private communication, May 1995.

Wood, M. D. "Replicated RPC Using Amoeba Closed-Group Communication." *Proceedings of the Twelfth International Conference on Distributed Computing Systems* (Pittsburgh, 1993).

————. "Fault-Tolerant Management of Distributed Applications Using a Reactive System Architecture." Ph.D. diss., Cornell University, December 1991. Also Technical Report TR 91-1252. Department of Computer Science, Cornell University.

Wu, Y. "Verification-Based Analysis of RMP." Technical Report NASA-IVV-95-003. NASA/WVU Software Research Laboratory, December 1995.

XTP Forum. *Xpress Transfer Protocol Specification*. XTP Rev. 4.0, 95-20, March 1995.

Index

A

DNS 69. *See also* Domain Name Service
Domain Name Service 40–43
DSM 386
dynamic membership model 225, 249–253, 267–269, 336–345
dynamic uniformity 251–268, 272, 282, 293, 299, 317–334, 465, 550
   performance implications 340–341

E

e-mail 194–195
Electra 401–403
electronic mail. *See* e-mail
embedded systems 495
Encina 139, 460, 467, 542
encryption used in virtual private networks 419
end-to-end argument 48–49, 346
ENS. *See* event notification service
enterprise Web servers 391–393
equity trading system based on Horus 412
error correction 84–85
Ethernet 25–27
event dispatch 72–74
event notification service 106–108, 198–201.
   See also CORBA
exactly once semantics 76
exponential convergence of gossip protocols 474
extended virtual synchrony 318–322
external data representation. *See* XDR

F

F/C field (ATM header) 33–34
fail-over in CORBA 354
fail-stop failures 5
fail-stop model 225–226, 265
failure detectors 210
failures 5–6, 206–213
   detecting 210
false sharing 386
fault tolerance 4, 88–90, 163, 359, 387–388, 497
   primary backup 89

fault-tolerant real-time control 434
fault-tolerant tools **363**
fbcast 283, 297–299
fbufs 146–148
FDDI. *See* Fiber Distributed Data Interface
Fiber Distributed Data Interface 27–28
Ficus 126
file handle 118
file servers 57
   stateful 122–129
file transfer protocols 194
Firefly RPC costs 143
firewall protection (with wrappers) 360
firewalls 201–202, 353, 417–419
flow control 85–86, 413
flush 309
Fortezza 427–429
fourth-generation languages (4GLs) 364–371
fragmentation 12

G

gap-freedom guarantee 317
gateway 202
GIF 174
Global Positioning System. *See* GPS receivers
globally total order 307
GMS. *See* group membership service
gossip protocols 473–485
GPS receivers 70, 433–437, 553–554
group address 397
group communication and Java applets 391–393
group communication in Web applications 366–367
group membership protocol 80–81, 255, 345
group membership service 253–266, 280–281, 442
   extensions to allow partition and merge 266
   primary partition properties 264–266
   summary of properties 280–281
group object 397
groupware video protocols 267
guaranteed execution tools **363**
guardians 540
GUI builders 184–185, 190

H

halting failures 5, 225
hardware cryptographic protection 419
hardware fault tolerance 512
Harp 121, 468, 530–531
HAS 531–532
header (of a message) 11
Heisenbugs 207–208, 316
high availability 4
Horus system
    basic performance 403–404
    protocol accelerator 405–409
    real-time protocols 445
    replication in the Web 394–397
    robust groupware application 399–401
    scalability 410
    story behind name 391
    virtually synchronous process groups 397
hostile environments 211–213
HotJava 179–184
how computers fail 206–213
HTML 168–169
HTTP 170–173
    commands 171–173
HyperText Markup Language 168–169
HyperText Transport Protocol 170–173

I

IDL 98. *See also* CORBA, remote procedure call
IIOP 106
impossibility results for the asynchronous
    model 248, 341
IN coprocessor fault tolerance 448
inconsistent failure detection in available copies
    470–471
information warfare 211–213, 499–503
inode 118
instrumentation of a distributed system 489–490
integration of process groups with database
    proper 471
intentional threats 211
interface definition language. *See* IDL

Internet

Internet
    domain name service 40–43
    protocol 39
        IP 44
        IP multicast 46–55
Internet Inter-ORB Protocol. *See* IIOP
Internet Packet Multicast Protocol. *See* IP
    Multicast
Interobject Broker Protocol 109
IOB. *See* Interobject Broker Protocol
IP address 40
IP multicast 46–55
    and scalability of reliable group process
        412–413
IP over ATM (reliability issues raised) 35
IP protocol 44
ISDN 28–31
Isis toolkit 532
    story behind name 391
iterated multicast 295

J

Java 179–184
    applets structured as object groups 366–367
    groupware opportunities 366–367
    integrated with group communication tools
        391–393
JPEG 174

K

Kerberos 71–72, 123, 422–430
key escrow 424, 428

L

layered protocol architectures (pros and cons) 10
LCS. *See* life-cycle service
LFS 128–129
life-cycle service 108
lightweight process groups 310
lightweight remote procedure call 143–146,
lightweight tasks. *See* threads

W

Web proxy 178–179, 394–397
Web server
   replication and load-balancing 366–367
   wrapped for fault tolerance 364–371
Web, the xvii, 41–42, 57–59, 115, 167,
   177, 194, 364–371
   Active/X 188–189
   agent programming languages 186
   agent-based browsers 179–184
   architectural structures and reliability tools
     391–393
   banking on the Web 189–190
   basic authentication protocol 175
   browser technologies 167–192
   commerce servers 189–190
   commercial use of 162, 177–178
   consistency issues 192
   database servers 189–190
   exchange servers 189–190
   fault tolerance and load-balancing 366–367
   firewalls 201–202
   groupware tools and solutions 366–367
   HotJava browser 179–184
   HTML 168–169
   HTTP 170–173
   Java 179–184
   Java applets structured as object groups
     366–367
   military use 163
   other agent languages 185–187
   plug-in technologies 188–189
   proxy 167–192
   reliability 163–164

Web, the *(continued)*
   replicated data 226–227
   replication and reliability 394–397
   search engines and Web crawlers 187–188
   secure sockets layer 175–176
   security 163–164, 188–189, 197
   security and privacy issues 177–178
   security with digital signatures 366–367
   transactional uses 189–190
   URL. *See* Uniform Resource Locators
   Web Proxies 178–179
white pages 70
whole file transfer compared with
   prefetching 120
wide area group communication for the Web
   366–367
World Wide Web 41–42, 57–58.
   See also Web, the
worm 211
wrappers 352–364
write-ahead log 453–454
write-through policy 118–119

X

x-Kernel 19, 146–147, 394–395, 398, 406
X.500 69–70, 428–429, 547
XDR 66
Xpress Transfer Protocol. *See* XTP
XTP 52–54

Y

yellow pages 69